AF879267

Springer Series in the Data Sciences

Springer Series in the Data Sciences focuses primarily on monographs and graduate level textbooks. The target audience includes students and researchers working in and across the fields of mathematics, theoretical computer science, and statistics. Data Analysis and Interpretation is a broad field encompassing some of the fastest-growing subjects in interdisciplinary statistics, mathematics and computer science. It encompasses a process of inspecting, cleaning, transforming, and modeling data with the goal of discovering useful information, suggesting conclusions, and supporting decision making. Data analysis has multiple facets and approaches, including diverse techniques under a variety of names, in different business, science, and social science domains. Springer Series in the Data Sciences addresses the needs of a broad spectrum of scientists and students who are utilizing quantitative methods in their daily research. The series is broad but structured, including topics within all core areas of the data sciences. The breadth of the series reflects the variation of scholarly projects currently underway in the field of machine learning.

Partha Pratim Ray

Foundations of Federated Learning

Springer

Partha Pratim Ray
Computer Applications
Sikkim University
Gangtok, Sikkim, India

ISSN 2365-5674 ISSN 2365-5682 (electronic)
Springer Series in the Data Sciences
ISBN 978-3-032-34083-2 ISBN 978-3-032-34084-9 (eBook)
https://doi.org/10.1007/978-3-032-34084-9

Mathematics Subject Classification: 68T07, 68T09, 68T05, 68W15, 90C30, 94A17

This Springer imprint is published by the registered company Springer Nature Switzerland AG
The registered company address is: Gewerbestrasse 11, 6330 Cham, Switzerland

sad eva somya idam agra āsīd,
ekam evādvitīyam.

"O gentle one, in the beginning this universe was Existence alone—
One only, without a second."

— Chāndogya Upaniṣad 6.2.1

yasmin idaṃ sarvaṃ protam otaṃ ca,
tam evaikaṃ jānatha ātmānam,
anyā vāco vimuñcatha.

"Know that One alone as the Self,
in whom all this universe is woven and interwoven;
abandon all other discourse."

— Bṛhadāraṇyaka Upaniṣad 4.4.22

sarvaṃ khalv idaṃ brahma,
taj jalān iti śānta upāsīta.

"Indeed, all this is Brahman.
From Him it arises, in Him it is sustained, and into Him it dissolves;
therefore, one should meditate in serenity."

— Chāndogya Upaniṣad 3.14.1

yaḥ prāṇo nimīlayati,
yaḥ prāṇeṣu pratiṣṭhitaḥ,
taṃ devā upāsate prāṇam.

"That which moves and sustains all life,
which abides in every vital force—
that Prāṇa the gods worship."

— *Atharva Veda 11.4.1*

ātmaivedam agra āsīd,
puruṣa-vidhaḥ.

"In the beginning, this universe was the Self alone,
in the form of the Cosmic Person."

— *Bṛhadāraṇyaka Upaniṣad 1.4.1*

Upaniṣadic Invocation

pūrṇam adaḥ pūrṇam idaṃ pūrṇāt pūrṇam udacyate,
pūrṇasya pūrṇam ādāya pūrṇam evāvaśiṣyate.

"That is whole; this is whole.
From the whole, the whole emerges.
Even when the whole is taken from the whole,
the whole alone remains."

— *Īśā Upaniṣad, Invocation (Śānti Mantra)*

Upaniṣadic Invocation

saha nāv avatu,
saha nau bhunaktu,
saha vīryaṃ karavāvahai,
tejasvināv adhītam astu mā vidviṣāvahai.

"May He protect us both (teacher and student).
May He nourish us both.
May we work together with great energy.
May our study be luminous; may we not oppose one another."

— *Taittirīya Upaniṣad, 2.1.0*

Dedicated to

Maa & Baba

Preface

"Before intelligence could be distributed, it had to be trusted."

The rapid expansion of data-driven artificial intelligence has been accompanied by equally significant concerns related to privacy, data ownership, regulatory compliance, accountability, and trust. While modern machine learning systems have achieved remarkable success by leveraging large, centralized datasets, such assumptions increasingly conflict with the realities of contemporary data ecosystems. In practice, data is fragmented across organizations, devices, legal jurisdictions, and administrative boundaries, often governed by strict privacy regulations and institutional constraints. This fundamental tension between *data utility* and *data sovereignty* has given rise to *federated learning*, a paradigm that enables collaborative model training without requiring centralized data collection.

Federated learning represents a profound conceptual shift in how learning systems are designed and deployed. Rather than moving data to computation, federated learning moves computation to data. Models are trained across decentralized clients—such as mobile devices, edge nodes, hospitals, enterprises, or governmental institutions—while sensitive data remains local and under the control of its owner. Only model updates or carefully protected information are exchanged. This shift fundamentally alters the assumptions underlying traditional learning systems and raises new questions that are simultaneously mathematical, algorithmic, systems-oriented, and ethical. How can learning converge reliably under heterogeneous and non-IID data distributions? How should communication protocols be structured under bandwidth, latency, and participation constraints? What guarantees of privacy, robustness, fairness, and accountability can be provided? And how can such large-scale collaborative systems be governed responsibly across multiple stakeholders?

This book—*Foundations of Federated Learning*—is devoted to a systematic and rigorous exploration of these questions. It presents federated learning not merely as an engineering workaround for privacy preservation, but as a principled computational framework that integrates distributed optimization, statistical learning, cryptographic protection, and collaborative system design. The text emphasizes clarity of formulation, rigor of analysis, and pedagogical coherence, with the objective of building strong conceptual foundations alongside practical insight into real-world deployment.

Federated learning is not a single algorithm, nor is it a narrowly defined technique. It is an ecosystem of interrelated ideas spanning centralized and decentralized architectures, synchronous and asynchronous training protocols, optimization under statistical heterogeneity, privacy-preserving mechanisms, robustness against adversarial behavior, and domain-specific adaptations. This book therefore treats federated learning as a unified discipline—one that requires reasoning across theory, systems, security, and governance, rather than isolated algorithmic solutions.

Why Federated Learning Matters

Federated learning has emerged as a critical enabling technology for privacy-aware, regulation-compliant, and trustworthy artificial intelligence. In domains such as healthcare, finance, telecommunications, defense, and the Internet of Things, centralized data aggregation is often infeasible, legally restricted, or ethically undesirable. Federated learning offers a principled alternative, allowing knowledge to be shared and models to improve collectively while sensitive data remains local.

Beyond privacy preservation, federated learning addresses several fundamental challenges inherent to modern AI systems, including:

- learning under statistical heterogeneity and non-IID data distributions,
- communication-efficient optimization at large scale,
- robustness against unreliable, partial, or adversarial participants,
- personalization across diverse clients, devices, and operational environments,
- governance, incentive alignment, and accountability in collaborative learning ecosystems.

A rigorous understanding of federated learning equips readers to design intelligent systems that are not only accurate and efficient, but also transparent, resilient, and socially responsible.

How to Use This Book

This book is written for students, researchers, and practitioners seeking a deep, structured, and principled understanding of federated learning. It emphasizes conceptual clarity, formal modeling, and algorithmic reasoning, supported by mathematical formulations, illustrative examples, system-level discussions, and critical analysis of assumptions and limitations.

Each chapter is designed to be largely self-contained while contributing to a coherent progression from foundational concepts to advanced methods and applications. The material is suitable for advanced undergraduate and postgraduate courses in artificial intelligence, machine learning, and data science, as well as for researchers and practitioners entering the field of federated learning or seeking to strengthen their theoretical grounding.

Relationship to Distributed and Modern Machine Learning

Federated learning is closely related to distributed machine learning, yet differs fundamentally in its assumptions, objectives, and design constraints. While traditional distributed learning focuses primarily on computational scalability and throughput, federated learning prioritizes data locality, privacy preservation, robustness to heterogeneity, and cross-organizational collaboration.

Modern artificial intelligence increasingly integrates federated learning with deep learning, reinforcement learning, and generative models, particularly in settings where data cannot be freely shared. Understanding federated learning is therefore essential for developing next-generation AI systems that operate across decentralized, regulated, and multi-stakeholder environments, where technical, organizational, and ethical considerations are deeply intertwined.

Scope and Structure of the Book

The book progresses systematically from foundational principles to advanced techniques and real-world applications. It begins with the motivation, definitions, and mathematical formulation of federated learning, followed by detailed discussions of system architectures and training protocols. Subsequent chapters address data heterogeneity, optimization algorithms, privacy and security mechanisms, governance and incentive structures, and practical deployment across diverse application domains.

Throughout the book, emphasis is placed on principled analysis, algorithmic transparency, and the interplay between theory and practice. The goal is not only to explain how federated learning systems operate, but also to equip readers with the intellectual tools required to reason about their design, limitations, and responsible use.

Brief Overview of the Chapters

- **Chapter 1: Introduction to Federated Learning** Introduces the motivation for decentralized and collaborative learning and explains why traditional centralized machine learning assumptions break down in modern data ecosystems. The chapter discusses privacy, regulatory, communication, and robustness limitations of centralized learning and motivates federated learning through real-world application scenarios.
- **Chapter 2: Fundamentals and Definitions of Federated Learning** Formalizes federated learning by defining its core entities, including clients, servers, and communication rounds. The chapter establishes data locality principles, local and global objectives, and provides an intuitive consensus-based perspective on federated optimization.

- **Chapter 3: Mathematical Formulation of Federated Learning** Develops the mathematical foundations of federated learning, including global optimization objectives, client-wise loss functions, aggregation mechanisms, and consensus constraints. The relationship between local and global minima under non-IID data is carefully analyzed.
- **Chapter 4: Federated Learning System Architectures** Examines centralized federated learning architectures with emphasis on the central server's role as an orchestrator. Client selection, communication bottlenecks, fault tolerance, and scalability considerations are systematically discussed.
- **Chapter 5: Decentralized and Peer-to-Peer Federated Learning** Explores alternatives to centralized orchestration, including orchestrator-less learning, gossip-based protocols, consensus-driven aggregation, and blockchain-enabled federated systems. The impact of network topology on learning dynamics and convergence is analyzed.
- **Chapter 6: Federated Learning Training Process** Provides a detailed walk-through of the federated training lifecycle, from model initialization and client participation to local training, aggregation, synchronization, and termination criteria. Both synchronous and asynchronous training paradigms are examined.
- **Chapter 7: Data Heterogeneity and Non-IID Challenges** Analyzes the nature of statistical heterogeneity in federated learning, including non-IID data distributions, covariate shift, prior probability shift, concept drift, and data imbalance. Their impact on convergence behavior and model stability is studied.
- **Chapter 8: Techniques for Handling Non-IID Data** Presents principled strategies for mitigating heterogeneity, including adaptive aggregation, normalization techniques, client clustering, regularization-based methods, and robust optimization. Theoretical and practical trade-offs are highlighted.
- **Chapter 9: Foundations of Federated Optimization Algorithms** Introduces the core optimization algorithms underlying federated learning, with a focus on federated stochastic gradient descent. Communication–computation trade-offs and convergence limitations are analyzed in depth.
- **Chapter 10: Federated Averaging** Examines the Federated Averaging (FedAvg) algorithm in detail, contrasting weight averaging and gradient averaging approaches. Communication efficiency, adaptive optimizers, and performance under heterogeneous data are discussed.
- **Chapter 11: Advanced Federated Optimization Methods** Covers advanced optimization techniques such as FedProx, dynamic regularization, one-gradient methods, and inverse-distance aggregation. Convergence guarantees, robustness properties, and practical limitations are analyzed.
- **Chapter 12: Personalized Federated Learning** Explores personalization strategies that balance global generalization with client-specific adaptation. Topics include personalized objectives, sub-federated averaging, model pruning, and the personalization–generalization trade-off.

- **Chapter 13: Heterogeneous Federated Learning** Addresses heterogeneity in computation, communication capabilities, and model architectures across clients. Resource-aware training, dynamic model scaling, and partial participation frameworks are presented.
- **Chapter 14: Hybrid Federated Learning** Develops hybrid federated learning approaches that combine sample-wise and feature-wise data partitioning. Primal–dual optimization frameworks, convergence behavior, and comparisons with classical federated averaging are discussed.
- **Chapter 15: Privacy-Preserving Techniques in Federated Learning** Presents privacy-enhancing mechanisms including differential privacy, secure aggregation, secure multi-party computation, and homomorphic encryption. Their guarantees, limitations, and deployment considerations are examined.
- **Chapter 16: Security Threats and Robust Federated Learning** Analyzes security vulnerabilities such as model poisoning, backdoor attacks, inference attacks, and Byzantine failures. Robust aggregation methods, defense strategies, and trust assumptions are systematically discussed.
- **Chapter 17: Governance, Ethics, and Incentive Mechanisms** Discusses governance models for federated learning, including model ownership, incentive design, regulatory compliance, and ethical considerations in collaborative artificial intelligence systems.
- **Chapter 18: Applications, Emerging Trends, and Tools** Surveys real-world applications of federated learning across domains such as healthcare, finance, IoT, robotics, and education. The chapter concludes with emerging research trends, open challenges, and an overview of practical federated learning frameworks and tools.

Intended Audience

This book is intended for undergraduate and postgraduate students of computer science, artificial intelligence, data science, and related disciplines. It is also suitable for researchers, educators, and practitioners who seek a rigorous, structured, and coherent understanding of federated learning and its role in modern artificial intelligence.

Closing Reflections

Federated learning reflects a broader transformation in artificial intelligence—from centralized computation toward collaborative, privacy-aware, and trustworthy systems. As artificial intelligence increasingly intersects with society, governance, and human values, such principled foundations become indispensable.

If this book enables readers to reason more carefully about decentralized learning, to design federated systems more responsibly, and to engage critically with the technical and ethical challenges of collaborative AI, then it has achieved its purpose.

Any errors or omissions remain the responsibility of the author. Suggestions, corrections, and constructive feedback are warmly welcomed and will be gratefully acknowledged in future editions.

Partha Pratim Ray

Acknowledgements The writing of this book, Foundations of Federated Learning, has been guided by the intellectual evolution of a field that seeks to reconcile large-scale machine learning with privacy, decentralization, and trust. Federated learning represents a convergence of ideas from optimization, distributed systems, cryptography, and ethical AI, and this book is shaped by the collective scholarship that has emerged at this intersection. I wish to acknowledge the many individuals, communities, and traditions of inquiry that have influenced this work, both directly and indirectly.

I am deeply grateful to teachers, mentors, colleagues, and academic peers whose discussions, lectures, and research have shaped my understanding of distributed and privacy-aware learning systems. Their emphasis on mathematical rigor, system-level thinking, and principled design reinforced the importance of presenting federated learning not merely as an engineering workaround for data silos, but as a coherent theoretical and algorithmic framework. Their insights into optimization under heterogeneity, communication constraints, and privacy-preserving computation have left a lasting imprint on this book.

I extend my sincere appreciation to the broader research community working in federated learning, privacy-preserving machine learning, and distributed optimization. The foundational contributions of early pioneers—through seminal papers, benchmark studies, and open frameworks—have established the conceptual scaffolding upon which this text is built. This book draws from decades of cumulative progress in stochastic optimization, secure aggregation, non-IID learning, and collaborative modeling, refined through rigorous debate and empirical validation.

I am thankful for the academic environments and institutions that foster interdisciplinary inquiry across machine learning, systems, and data governance. Engagement with students, researchers, and fellow educators consistently reaffirmed the relevance of federated learning in real-world settings, particularly in healthcare, finance, IoT, and cross-organizational collaboration. Their questions and perspectives helped shape the pedagogical structure of this book, ensuring that theory, systems, and applications are presented in balanced harmony.

I gratefully acknowledge the authors and contributors of open-source software, experimental platforms, and open educational resources related to federated learning. Frameworks, libraries, and shared benchmarks have played a crucial role in transforming federated learning from a conceptual paradigm into a deployable technology. This spirit of openness and collaboration mirrors the very philosophy of federated learning itself—progress through shared models without shared data.

My heartfelt thanks go to friends and colleagues whose encouragement, thoughtful conversations, and quiet support sustained me throughout the writing process. Their presence provided intellectual balance and personal motivation during the extended and often demanding journey of completing this manuscript.

Above all, I express my deepest gratitude to my family. Their patience, understanding, and unwavering support made it possible to devote sustained attention to this work. Their faith and encouragement formed the emotional foundation upon which this book was written.

Finally, with humility and reverence, I offer my gratitude to Bhagawan Sri Vishnu for granting clarity of thought, inner discipline, and perseverance. It is through divine grace that reason finds direction and effort finds completion.

Partha Pratim Ray

Competing Interests The author has no competing interests to declare that are relevant to the content of this manuscript.

About This Book

Foundations of Federated Learning presents a systematic, rigorous, and pedagogically structured introduction to federated learning as a modern paradigm for privacy-aware, decentralized, and trustworthy artificial intelligence. Written by Partha Pratim Ray, the book develops federated learning not merely as a technical workaround for data privacy, but as a principled computational framework that integrates machine learning, distributed optimization, communication-efficient training, privacy preservation, security, governance, and real-world deployment.

The book begins with the motivation for federated learning in contemporary data ecosystems, where data is often distributed across devices, institutions, organizations, and jurisdictions. It explains why traditional centralized machine learning assumptions become inadequate when data cannot be freely collected, transferred, or pooled. From this foundation, the text introduces the core concepts of federated learning, including clients, servers, communication rounds, local objectives, global optimization goals, and model aggregation.

A major strength of the book is its balanced treatment of theory, algorithms, and systems. The early chapters establish the mathematical and architectural foundations of federated learning, while later chapters examine decentralized and peer-to-peer architectures, federated training processes, data heterogeneity, non-IID learning challenges, and methods for improving robustness under practical constraints. Particular attention is given to federated optimization, including federated stochastic gradient descent, Federated Averaging, advanced federated optimization methods, personalization, heterogeneous federated learning, and hybrid federated learning.

The book also addresses the trust dimension of federated learning in depth. Dedicated chapters examine privacy-preserving techniques such as differential privacy, secure aggregation, secure multi-party computation, and homomorphic encryption. Security threats, including model poisoning, backdoor attacks, inference attacks, and Byzantine failures, are discussed alongside robust aggregation and defense strategies. Governance, ethics, incentive mechanisms, model ownership, regulatory compliance, and accountability are treated as integral components of responsible federated learning rather than peripheral concerns.

The final part of the book connects federated learning with practical applications and emerging trends. It discusses use cases in healthcare, finance, the Internet of Things, smart environments, robotics, education, telecommunications, and related domains. The book also introduces important federated learning tools and frameworks, supporting readers who wish to move from conceptual understanding to implementation-oriented exploration.

This book is intended for advanced undergraduate students, postgraduate students, researchers, educators, and practitioners in artificial intelligence, machine learning, data science, computer science, distributed systems, privacy-preserving computation, and intelligent systems. Thus, by combining conceptual clarity, mathematical formulation, algorithmic explanation, system-level analysis, and ethical reflection, *Foundations of Federated Learning* aims to equip readers with the intellectual foundation required to understand, design, analyze, and responsibly deploy federated learning systems.

Contents

About the Author

Partha Pratim Ray is an Assistant Professor (Stage 2, Academic Level 11) in the Department of Computer Applications, Sikkim University, Gangtok, Sikkim, India. He has more than fourteen years of experience in higher education, including teaching, research, curriculum development, postgraduate student mentoring, academic administration, and institutional service. His research broadly focuses on the Internet of Things, Edge Artificial Intelligence, Pervasive Generative AI, and resource-constrained machine learning systems, with particular emphasis on reproducible deployment in real-world environments.

He is currently at the pre-submission stage of his Ph.D. in Computer Applications at Sikkim University. His doctoral research is titled "Enabling Large Language Model on Resource-Constrained Edge: A Multi-Faceted Approach," and investigates the deployment of localized large language models on low-resource edge computing platforms. He holds an M.Tech in Electronics and Communication Engineering with specialization in Embedded Systems and a B.Tech in Computer Science and Engineering. He also qualified the Graduate Aptitude Test in Engineering (GATE) in Computer Science and Engineering.

Mr. Ray's major research interests include Pervasive Generative AI, Internet of Things, Edge Computing, Edge AI, resource-constrained intelligent systems, Pervasive Biomedical Informatics, and intelligent computing for societal applications. He is credited with conceptualizing and advancing the research paradigms of the Internet of AI Things (IoAT) and Pervasive Generative AI. Along with his colleague, he has also contributed to the formulation of the Pervasive Biomedical Informatics (PBI) paradigm. These ideas share a common focus on intelligent, context-aware, resource-efficient, and deployable computing systems for real-world environments.

He has an extensive scholarly record comprising 131 SCI-indexed journal publications, 2 SSCI-indexed journal publications, 34 Scopus-indexed journal publications, 10 books, several book chapters, and numerous international conference papers. His research impact is reflected in 14,655 Google Scholar citations, with an h-index of 43 and an i10-index of 90. He has received several national and international recognitions, including inclusion among the World's Top 2% Scientists by Stanford University from 2020 to 2025.

His other recognitions include Fellow status in the Institution of Engineers in Scotland (FIES), Thinking About Thinking, and the Institution of Electronics and Telecommunication Engineers (FIETE), as well as Senior Membership of IEEE

(SMIEEE) and INAE (M-INAE). He has also received the IEI Young Engineers Award, the Sikkim University Publication Award, and multiple best paper and young scientist awards.

Mr. Ray actively contributes to the academic and professional community through editorial, reviewing, and professional service. He serves as Editor of the *IETE Journal of Education*, Associate Editor of *Frontiers in Communications and Networks* and *Frontiers in Bioinformatics*, and Editorial Board Member of multiple EAI Endorsed Transactions. He is associated with several professional bodies, including IEEE, INAE, IETE, IES, IET, IEI, ACM, CSI, ISOC, AAM, IMS, and others.

Beyond his technical research, he is committed to interdisciplinary scholarship connecting artificial intelligence, IoT, edge intelligence, healthcare, biomedical informatics, intelligent systems, and the societal applications of emerging technologies. He is also a strong advocate of Indian Knowledge Systems and has informally established the Indian Knowledge Forum to promote and disseminate India's indigenous knowledge traditions in contemporary academic and public contexts.

Part I

Foundations of Federated Learning

Introduction to Federated Learning 1

Federated learning (FL) represents a fundamental paradigm shift in the design, training, and deployment of modern machine learning systems. Classical machine learning methodologies are largely built upon the assumption that data can be freely collected from multiple sources, aggregated into a centralized repository, and processed within a single computational infrastructure. Under this paradigm, learning is performed by optimizing a global model over a unified dataset, with little consideration for where or how the data was generated.

In contemporary data ecosystems, however, this assumption is increasingly violated. Data is no longer produced in a small number of controlled environments but is generated continuously across geographically distributed, organizationally isolated, and resource-constrained settings. Examples include personal data generated on mobile and wearable devices, sensitive medical records maintained by hospitals, financial transaction data stored within banking institutions, and large volumes of sensor data produced by Internet-of-Things (IoT) platforms. In such settings, unrestricted data aggregation is often infeasible, undesirable, or explicitly prohibited.

Federated learning challenges the conventional centralized training model by fundamentally reversing the traditional flow of computation and data. Rather than transferring raw data from its point of origin to a central server for training, FL enables learning to occur directly at the data source. A collection of autonomous participants—commonly referred to as *clients*—independently train a shared model using their local, private datasets. Each client computes local updates, such as gradients or parameter changes, and transmits only these intermediate representations to a coordinating entity, which aggregates them to form an improved global model. Through this process, collaborative intelligence is achieved without explicit exchange of raw data, thereby reducing privacy risks, communication overhead, and regulatory exposure.

P. P. Ray, *Foundations of Federated Learning*,
Springer Series in the Data Sciences,
https://doi.org/10.1007/978-3-032-34084-9_1

Beyond its algorithmic formulation, federated learning reflects a broader evolution in the philosophy of distributed intelligence. It aligns naturally with emerging principles of data sovereignty, user autonomy, and privacy-by-design, recognizing that control over data is a first-class concern rather than a secondary constraint. In federated systems, predictive performance is no longer the sole objective; instead, learning must be balanced with considerations of trust, accountability, and ethical data stewardship. Consequently, federated learning should be viewed not merely as an extension of distributed computing techniques, but as a principled response to the societal, legal, and institutional forces that increasingly shape the deployment of intelligent systems.

This chapter introduces the foundational concepts, motivations, and defining characteristics of federated learning. It situates FL within the broader historical trajectory of distributed and collaborative learning paradigms, illustrating how shifts in data generation patterns, computational infrastructure, and regulatory frameworks have motivated decentralized approaches to learning. The chapter systematically examines the limitations of centralized machine learning, particularly in environments dominated by privacy, scalability, and governance constraints. Together, these discussions establish the conceptual groundwork for the formal models, system architectures, and optimization techniques that are developed in the chapters that follow.

1.1 Background and Motivation for Decentralized Learning

The unprecedented growth in data generation over the past decade has profoundly reshaped the foundational assumptions under which machine learning systems are designed and deployed. Advances in sensing technologies, ubiquitous connectivity, and embedded computation have enabled continuous data production from a diverse range of sources, including smartphones, wearable health devices, smart home appliances, industrial control systems, autonomous vehicles, and large-scale Internet-of-Things (IoT) infrastructures. Unlike earlier data-centric paradigms, this data is generated not in centralized data centers but predominantly *at the periphery of the network*, in close proximity to users, physical environments, and organizational boundaries.

This shift in data generation fundamentally alters the nature of the learning problem. In contemporary machine learning ecosystems, data is rarely created or stored at a single, unified location. Instead, it is inherently *distributed across a large number of devices and institutions*, each operating under distinct computational, administrative, and regulatory constraints. Moreover, the data produced by these sources is typically *heterogeneous* in both structure and statistical properties, reflecting differences in user behavior, sensor modalities, operating conditions, and data collection processes.

Equally important is the increasingly *privacy-sensitive* nature of modern data. Information associated with personal behavior, medical history, financial transactions, and proprietary industrial processes carries significant ethical, legal, and commercial implications. The decentralized, heterogeneous, and sensitive characteristics of such data directly challenge the classical assumptions of centralized machine learning, which presuppose unrestricted data access, seamless data aggregation, and centralized governance. These emerging realities motivate the exploration of learning paradigms that can operate effectively under decentralized data generation and ownership, laying the conceptual groundwork for decentralized and federated learning frameworks.

1.1.1 Centralized Learning Assumption

Classical machine learning frameworks are fundamentally grounded in the *centralized learning assumption*, which presumes that all training data generated across different sources can be collected, stored, and processed within a single computational repository. Under this assumption, learning is performed by optimizing a model using a unified dataset that is fully accessible to a central authority. Formally, the centralized learning paradigm is built on the assumption that all data generated across different sources can be collected into a single, unified dataset. This dataset is represented as

$$\mathcal{D} = \{(x_i, y_i)\}_{i=1}^{N},$$

where each ordered pair (x_i, y_i) denotes an individual training example. Here, $x_i \in \mathcal{X}$ represents an input instance drawn from the input space $\mathcal{X}$—such as an image, a text sequence, or a sensor measurement—while $y_i \in \mathcal{Y}$ denotes the corresponding target output or label from the label space $\mathcal{Y}$. The index i simply enumerates the samples, and N denotes the total number of training examples aggregated from all data-producing entities.

This formulation implicitly treats the dataset as *homogeneous* and *centrally governed*. That is, once data enters $\mathcal{D}$, distinctions regarding its origin, ownership, or collection context are ignored. All samples are assumed to be equally accessible, equally reliable, and equally governed by a single authority. Such an abstraction simplifies analysis and algorithm design, but it conceals the complexity of real-world data generation, where samples may originate from diverse devices, users, institutions, or environments with differing constraints and characteristics.

The goal of centralized learning is to determine a set of model parameters θ that yields good predictive performance over the entire dataset. This objective is typically formalized through *empirical risk minimization*, which seeks to minimize the average loss incurred by the model across all training samples:

$$\min_{\theta} \frac{1}{N} \sum_{i=1}^{N} \ell\big(f_\theta(x_i), y_i\big).$$

In this expression, $f_\theta : \mathcal{X} \to \mathcal{Y}$ denotes a predictive model—such as a linear classifier, a neural network, or a probabilistic estimator—parameterized by $\boldsymbol{\theta}$. The loss function $\ell(\cdot)$ quantifies the discrepancy between the model—prediction $f_\theta(x_i)$ and the true label y_i, assigning a numerical penalty that reflects prediction error.

The averaging over all N samples reflects a key assumption of centralized learning: each data point contributes equally to the global learning objective. This implicitly presumes that samples are drawn from a common distribution and that no individual data source should exert greater influence than another. While this assumption is mathematically convenient and often reasonable in controlled settings, it becomes increasingly restrictive in large-scale, decentralized environments where data distributions, sampling rates, and data quality may vary substantially across sources.

Interpretation. The centralized empirical risk minimization framework rests on several strong and often implicit assumptions. First, it assumes that all data samples are readily accessible at a single central location, implying unrestricted data collection and aggregation. Second, it presumes that transferring data from distributed sources to the central server is inexpensive, reliable, and unconstrained by bandwidth, latency, or energy limitations. Third, it assumes that a single entity possesses complete authority over data storage, processing, access control, and governance. While these assumptions were reasonable in early machine learning applications operating on modest datasets and controlled infrastructure, they increasingly fail to hold in modern, large-scale data ecosystems.

As data volume grows, transmitting raw data from millions of distributed devices—such as mobile phones, sensors, and edge systems—to a central server imposes prohibitive bandwidth and storage costs. As data velocity increases, centralized aggregation introduces latency that limits the responsiveness and adaptability of learning systems, particularly in time-sensitive applications. Most critically, as data sensitivity increases, privacy regulations, ethical considerations, and institutional policies restrict or explicitly prohibit the movement of raw data across organizational, jurisdictional, or geographical boundaries.

Motivation for Decentralized Learning

Decentralized learning arises from the necessity to perform machine learning *at the location where data is generated*. Rather than transferring raw data to a central repository, learning systems must adapt to privacy regulations, communication constraints, data ownership boundaries, and the practical infeasibility of large-scale data aggregation.

Figure 1.1 illustrates the classical centralized learning paradigm. Data generated at multiple distributed sources $\{\mathcal{D}_1, \mathcal{D}_2, \ldots, \mathcal{D}_K\}$ is first transferred in raw form to a single central repository, where it is merged into a unified dataset $\mathcal{D}$. A central server

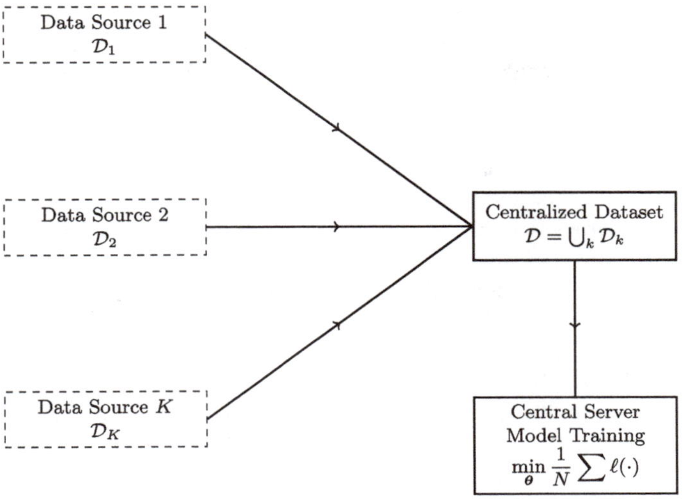

Fig. 1.1 Centralized learning architecture illustrating raw data aggregation from multiple sources into a single central repository, followed by model training at a central server

then performs model training by optimizing an empirical risk objective over the aggregated data. This architecture assumes unrestricted data movement, centralized governance, and homogeneous access to all training samples, thereby placing data availability and control entirely under a single computational authority.

1.1.2 From a Single Dataset to Multiple Local Datasets

In decentralized learning environments, the foundational assumption of a single, globally accessible training dataset no longer holds. Unlike centralized learning—where data from all sources is aggregated into a unified repository—the data required for learning is inherently distributed across multiple autonomous entities. These entities may correspond to individual users, devices, organizations, or institutions, each operating under its own administrative, legal, and operational constraints.

Formally, the global dataset $\mathscr{D}$ assumed in centralized learning is replaced by a collection of local datasets distributed across K distinct entities, commonly referred to as *clients*:

$$\mathscr{D}_1, \mathscr{D}_2, \ldots, \mathscr{D}_K.$$

Here, $\mathscr{D}_k$ denotes the dataset stored locally at client k. Depending on the application, a client may correspond to an individual mobile device, a hospital information system, a financial institution, or an industrial monitoring site. Crucially, these datasets are not simply logical partitions of a larger dataset maintained for convenience. Rather, they are physically separated, administratively independent, and tightly coupled to the entities that generate, own, and govern the data.

This distinction has profound implications. Because data remains local to each client, it cannot be freely pooled, rebalanced, or preprocessed at a global level. The learning system must therefore operate directly on these distributed datasets without assuming centralized access or uniform control.

A defining characteristic of decentralized data is that local datasets are typically *unequal in size* and *non-identically distributed*. In practice, the number of samples stored at different clients, denoted by $|\mathscr{D}_k|$, may vary by orders of magnitude. Some clients may generate data continuously, while others may contribute only sporadically. Moreover, the underlying data-generating distribution at client k, denoted by $P_k(x,y)$, may differ substantially from those of other clients. This directly violates the independent and identically distributed (IID) assumption that underpins much of classical machine learning theory.

Statistical heterogeneity arises naturally in real-world systems. For example, in a mobile text prediction task, one user may generate thousands of messages per day, using a rich and personalized vocabulary, while another may generate only a few short messages. Similarly, in healthcare applications, hospitals located in different regions may serve patient populations with distinct demographic, genetic, and clinical profiles. As a result, the corresponding local datasets $\mathscr{D}_k$ differ not only in size but also in feature distributions, label frequencies, and noise characteristics.

This heterogeneity has important consequences for learning. When data is non-IID, model updates computed at individual clients tend to reflect client-specific patterns that may not generalize uniformly across the broader population. Naïvely combining such updates can lead to slow convergence, unstable training dynamics, or biased global models that overrepresent certain clients or data regimes. Consequently, learning in decentralized environments requires careful coordination mechanisms that balance local specialization with global model performance, ensuring that the resulting model benefits from diverse data sources without being dominated by any single client.

Illustrative Example. Consider a predictive keyboard model collaboratively trained across millions of smartphones. Each device continuously collects typing data that reflects the unique vocabulary, writing style, and language preferences of its user. Centralizing this raw text data would not only incur prohibitive communication costs but would also raise serious privacy concerns. Decentralized learning allows each device to locally improve the shared model using its private data, while contributing only abstract learning signals—such as parameter updates or gradients—to a central aggregator. In this manner, the global model benefits from diverse user behavior without exposing individual text inputs.

Figure 1.2 illustrates the fundamental transition from centralized data aggregation to decentralized data ownership. In classical learning settings, all training samples are assumed to be merged into a single global dataset $\mathscr{D}$, over which model training is performed. In decentralized environments, this assumption no longer holds.

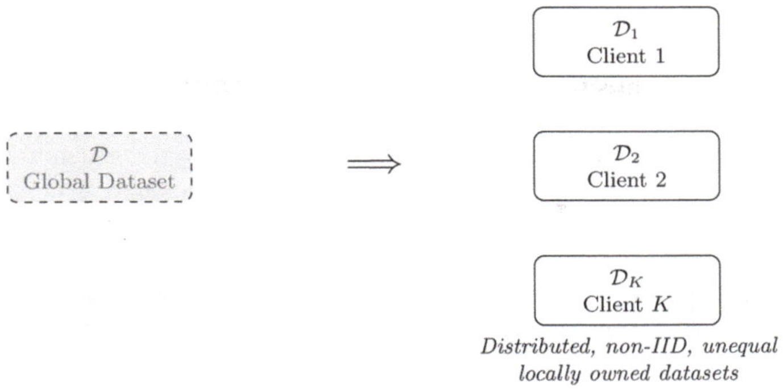

Fig. 1.2 Transition from a single centralized dataset to multiple local datasets in decentralized learning environments

Instead, data is partitioned across multiple clients, each maintaining its own local dataset $\mathscr{D}_k$. These datasets are physically and administratively bound to the entities that generate the data—such as individual users, institutions, or edge devices—and cannot be freely pooled. As indicated in the figure, local datasets may differ significantly in size and statistical properties, reflecting real-world heterogeneity in user behavior, demographics, and operating conditions.

This distributed data configuration breaks the classical IID assumption and introduces non-identically distributed (non-IID) data across clients. Consequently, learning algorithms must aggregate knowledge from heterogeneous sources while preserving local autonomy. This visual abstraction captures the core data-centric challenge that motivates federated and decentralized learning paradigms.

> **Key Aspect: Decentralized and Non-IID Data**
> In decentralized learning, the single global dataset assumed by centralized machine learning is replaced by multiple local datasets distributed across clients. These datasets are typically unequal in size and follow distinct, non-identically distributed data distributions. As a result, learning must account for statistical heterogeneity, client-specific data ownership, and the absence of direct access to raw data, fundamentally altering both model optimization and system design.

1.1.3 Shift in Learning Paradigm

The motivation for decentralized learning extends far beyond the goal of computational scalability. While early distributed learning systems were primarily designed to accelerate training by parallelizing computation, contemporary decentralized learning paradigms emerge from fundamentally different pressures. These include the growing importance of data privacy laws, ethical constraints on data usage, institutional governance policies, and principles of data sovereignty that grant individuals and organizations explicit control over their data.

In such environments, the design of machine learning systems can no longer be guided solely by predictive accuracy or computational efficiency. Instead, learning algorithms must be compatible with legal frameworks that restrict data movement, ethical norms that prioritize user trust, and organizational boundaries that limit centralized data ownership. Decentralized learning responds to these constraints by enabling participants to retain full authority over their local data while still contributing to the improvement of a shared model. In this sense, decentralization represents a shift from *data-centric aggregation* to *knowledge-centric collaboration*.

From a system design perspective, decentralized learning frameworks are characterized by the strategic exploitation of local computational resources. Edge processors, mobile CPUs, embedded accelerators, and on-device neural processing units increasingly possess sufficient computational capability to perform meaningful local training. Rather than treating these resources as passive data generators, decentralized learning elevates them to active participants in the learning process.

Global knowledge formation in such systems is achieved through *model-level coordination* rather than data exchange. Each participant performs local optimization using its private dataset and communicates only derived information—such as parameter updates, gradients, or sufficient statistics—to a coordinating mechanism. These abstract representations capture useful learning signals while substantially reducing the exposure of sensitive information and the cost of communication. As a result, decentralized learning systems decouple learning progress from direct access to raw data.

This shift in learning paradigm has profound implications for how machine learning systems are conceptualized, implemented, and governed. Learning becomes an inherently collaborative process in which autonomy and coordination coexist. Participants are no longer passive data providers but active agents that influence the global model while enforcing local constraints.

This design philosophy provides the conceptual foundation upon which federated learning is constructed. Federated learning can therefore be understood as a principled and structured realization of decentralized learning, explicitly designed to balance model performance, privacy preservation, and system scalability. By formalizing coordination mechanisms, communication protocols, and aggregation strategies, federated learning transforms the broader idea of decentralized learning into a practical and deployable framework.

Figure 1.3 illustrates the federated learning paradigm using a hub-and-spoke architecture. A central federated server coordinates learning by distributing the current global model to a set of heterogeneous clients. Each client performs local training using its private dataset and returns only model updates to the server. The server aggregates these updates to refine the global model, which is then redistributed in subsequent rounds. At no point is raw data transmitted, thereby preserving privacy, ownership, and data locality.

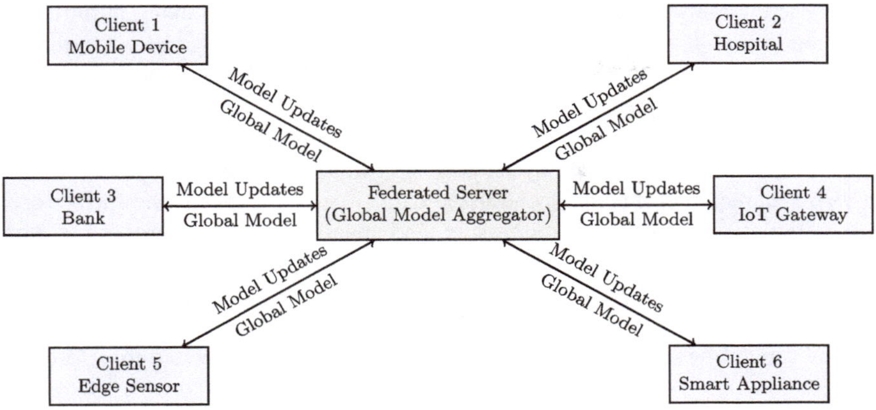

Fig. 1.3 Federated learning scenario illustrated in a hub-and-spoke (boom) style. Each client performs local training on private data and shares only model updates with the federated server, which aggregates them to produce a global model. Raw data never leaves the client devices

1.2 Limitations of Centralized Machine Learning

Centralized machine learning is based on the assumption that all training data can be freely collected, aggregated, and processed within a single computational infrastructure. Under this paradigm, learning is performed by optimizing a global model over a unified dataset stored at a central server. While this approach offers conceptual simplicity and algorithmic convenience, it becomes increasingly inadequate in modern data-driven environments.

As data generation shifts toward distributed, heterogeneous, and privacy-sensitive settings, the limitations of centralized machine learning become both practical and structural. These limitations arise not merely from scale, but from fundamental mismatches between centralized assumptions and real-world constraints.

1.2.1 Privacy and Security Risks

Centralized machine learning architectures rely on the aggregation of large volumes of raw data into a single repository. While this design simplifies model training and data management, it simultaneously creates a concentration of sensitive information that inevitably becomes a high-value target for security breaches. Data such as personal identifiers, medical histories, financial transactions, behavioral logs, and proprietary industrial records are often stored together, magnifying the potential impact of any compromise.

From a system security perspective, centralized storage amplifies risk by introducing a single point at which confidentiality, integrity, and availability can be violated. A single successful cyberattack—whether through external intrusion, malware infection, or exploitation of software vulnerabilities—can lead to the exposure of data belonging to millions of users or multiple organizations. The scale of such breaches is not merely additive but systemic, as the compromise of a central repository undermines trust in the entire learning infrastructure.

Privacy risks in centralized systems are not limited to malicious attacks. Even in well-secured environments, centralized data collection expands the surface area for internal misuse, accidental disclosure, and unauthorized secondary use of data. Access privileges, logging failures, or human error can result in data being viewed, retained, or repurposed beyond its original and intended scope. Long-term storage of centralized datasets further increases the risk of future misuse, particularly as analytical capabilities evolve and data can be re-identified or correlated in unforeseen ways.

These concerns are especially pronounced in domains where data sensitivity is intrinsic rather than incidental. In healthcare, centralized aggregation of patient records raises critical concerns related to confidentiality, consent, and clinical ethics. In finance, centralized transaction data can expose individuals and institutions to fraud, identity theft, and market manipulation. Similarly, in personalized consumer services, centralized behavioral data can enable intrusive profiling and erosion of user autonomy.

Collectively, these privacy and security risks reveal a fundamental tension between centralized data aggregation and the principles of responsible data stewardship. As machine learning systems increasingly operate on sensitive and personal data, architectures that require extensive raw data centralization become progressively harder to justify, motivating the exploration of learning paradigms that reduce data exposure while preserving analytical capability.

Consider a centralized machine learning system used to train a predictive model for disease risk across multiple hospitals. Patient records from different institutions are transferred to a single central server for aggregation and analysis. While this approach may simplify model training, a security breach or misconfiguration at the central repository could expose sensitive medical histories of patients across all participating hospitals. Even without external attacks, centralized access by analysts or administrators may allow patient data to be accessed beyond its original clinical purpose, raising serious ethical and legal concerns. Such scenarios illustrate how centralized aggregation transforms localized privacy risks into system-wide vulnerabilities.

1.2.2 Regulatory and Legal Constraints

Modern data ecosystems operate within an increasingly complex landscape of legal and regulatory frameworks that govern how data may be collected, stored, transferred, and processed. These frameworks are designed to protect individual privacy, ensure data accountability, and preserve institutional and national sovereignty over sensitive information. As a result, data is no longer treated as a freely transferable resource, but as an asset subject to explicit legal, ethical, and jurisdictional controls.

Privacy and data protection regulations impose strict limitations on the sharing of data across organizational, institutional, and geographical boundaries. Such regulations often specify where data may be stored, who may access it, how long it may be retained, and for what purposes it may be used. Importantly, these constraints apply not only to personal data but also to sensitive corporate, industrial, and governmental datasets.

Centralized machine learning architectures frequently require the movement of raw data from its point of origin to a central training facility. In practice, this often entails cross-border or cross-institutional data transfer, which may directly conflict with regulatory requirements. For example, medical institutions are commonly prohibited from sharing patient records outside their administrative or national boundaries, even for research collaboration. Similarly, financial organizations may be legally restricted from exporting transaction data to third-party analytics platforms.

In healthcare, hospitals operating in different regions may each possess valuable clinical data, yet privacy regulations can prevent these data from being centralized into a single repository for model training. In finance, banks may be unable to pool customer transaction histories across institutions due to confidentiality and anti-misuse laws. In the public sector, government agencies may be barred from sharing citizen data with external entities or cloud providers located in other jurisdictions.

These regulatory constraints imply that centralized learning frameworks face challenges that extend well beyond computational efficiency. Even when centralized aggregation is technically feasible, it may be legally impermissible. Compliance requirements cannot be addressed through algorithmic optimization alone, as they are rooted in governance, accountability, and legal responsibility rather than model performance.

Consequently, learning paradigms that rely on unrestricted data aggregation become increasingly difficult to deploy in regulated environments. This mismatch between centralized learning assumptions and regulatory realities motivates the development of decentralized approaches that enable collaborative model training without requiring the transfer of raw data.

1.2.3 Communication and Infrastructure Overhead

Centralized machine learning architectures require the continuous transmission of raw data from geographically distributed sources to a central server. As data generation becomes increasingly decentralized, this requirement introduces substantial communication overhead. The volume, dimensionality, and frequency of modern data streams—such as high-resolution sensor readings, images, audio, and behavioral logs—cause bandwidth requirements to grow rapidly, often exceeding the capacity of available network infrastructure. As a consequence, centralized aggregation can lead to increased latency, network congestion, and reduced system responsiveness.

These challenges are particularly pronounced in environments involving mobile devices, edge sensors, and remote or resource-constrained installations. In such settings, network connectivity may be intermittent, costly, or energy-limited. Frequent transmission of raw data not only consumes bandwidth but also drains device batteries and increases operational costs. As a result, centralized data upload strategies become inefficient and, in many cases, practically infeasible.

Beyond communication costs, centralized learning architectures impose significant demands on storage and computational infrastructure. A central server must be provisioned to ingest, store, and process continuous data streams originating from potentially millions of sources. Scaling storage capacity, memory, and compute resources to meet these demands introduces substantial financial and operational burdens. Moreover, peak data arrival times can create bottlenecks that degrade system performance and reliability.

Consider a smart city deployment in which thousands of cameras and environmental sensors continuously generate high-dimensional data. Uploading all raw video frames and sensor readings to a central server for machine learning analysis would require sustained high-bandwidth connectivity and large-scale storage infrastructure. In practice, network limitations, latency constraints, and energy costs make such centralized data transfer impractical. These limitations highlight how communication and infrastructure overhead can become a primary bottleneck in centralized learning systems, motivating approaches that reduce data movement while preserving learning capability.

1.2.4 Single Point of Failure and Limited Robustness

Centralized machine learning architectures typically depend on a single central server, or a tightly coupled cluster of servers, to coordinate data storage, model training, and update dissemination. While this design simplifies control and management, it introduces a fundamental vulnerability: the entire learning pipeline becomes critically dependent on the availability and integrity of the central infrastructure. Any failure or compromise at this point can disrupt the functioning of the complete system.

Outages caused by hardware malfunctions, software bugs, network disruptions, or targeted cyberattacks can halt model training, delay the distribution of model updates, or render learning-based services temporarily or permanently unavailable. Because all coordination is centralized, recovery from such failures is often slow and resource-intensive. In mission-critical applications—including healthcare monitoring, financial transaction processing, and industrial automation—these interruptions can have severe operational, economic, and safety-related consequences.

Beyond availability concerns, centralized architectures exhibit limited robustness in dynamic or adversarial environments. The learning process is tightly coupled to the reliability of central infrastructure, making it difficult to adapt to fluctuating network conditions, partial system failures, or adversarial interference. Even when individual data sources remain operational, the inability of the central node to coordinate learning effectively can stall or degrade overall system performance.

Consider an industrial monitoring system in which sensor data from multiple factories is sent to a central server to train predictive maintenance models. If the central server experiences a network outage or hardware failure, model updates cannot be generated or distributed, even though local sensors continue to collect data. As a result, the system becomes blind to emerging faults during the outage period, potentially leading to undetected equipment failures. This example illustrates how centralized coordination transforms local failures into system-wide disruptions, highlighting the limited robustness of centralized learning architectures.

1.2.5 Single Point of Failure and Limited Robustness

Centralized machine learning architectures typically rely on a single central server, or a tightly coupled cluster of servers, to coordinate core functions such as data storage, model training, parameter aggregation, and update dissemination. Although this design simplifies system control, scheduling, and model management, it introduces a fundamental structural weakness: the entire learning pipeline becomes critically dependent on the availability, correctness, and security of the central infrastructure. Any failure, degradation, or compromise at this point can propagate through the system and disrupt learning at a global scale.

Outages may arise from a variety of sources, including hardware malfunctions, software bugs, configuration errors, network disruptions, or targeted cyberattacks. When such failures occur, centralized learning systems may be unable to ingest new data, perform model updates, or distribute trained models to downstream applications. Because coordination logic and state are concentrated at the center, recovery often requires manual intervention, infrastructure repair, or system-wide rollback, making fault recovery slow and resource-intensive.

The consequences of such disruptions are particularly severe in mission-critical applications. In healthcare monitoring systems, delayed or unavailable model updates can impair early warning mechanisms for patient deterioration. In financial systems, interruptions in learning-based fraud detection or risk assessment can expose institutions to financial loss. In industrial automation, failures in predictive models may lead to undetected equipment degradation, safety hazards, or costly downtime.

Beyond availability concerns, centralized machine learning architectures exhibit limited robustness in dynamic or adversarial environments. The learning process is tightly coupled to the reliability of the central node, reducing the system—ability to tolerate partial failures, fluctuating network conditions, or adversarial interference. Even if individual data sources or sensors remain fully operational, the inability of the central server to coordinate training effectively can stall learning progress or degrade overall model quality.

Consider an industrial monitoring system in which sensor data from multiple factories is transmitted to a central server for training predictive maintenance models. If the central server experiences a network outage or hardware failure, model updates cannot be generated or disseminated, even though local sensors continue to collect data. During this outage period, emerging equipment faults may go undetected, increasing the risk of unexpected failures and production losses. Here, a localized infrastructure issue escalates into a system-wide learning failure.

As another example, consider a centralized traffic prediction system used to manage congestion across a large metropolitan area. If the central learning server becomes unavailable due to a software crash or cyberattack, real-time model updates cannot be performed, despite traffic sensors and cameras continuing to stream data. The system—inability to adapt to rapidly changing traffic conditions can result in inefficient signal control, increased congestion, and reduced emergency response effectiveness. This scenario highlights how centralized coordination can undermine robustness in time-sensitive, data-intensive environments.

1.3 Concept of Collaborative and Federated Learning

Collaborative learning refers to a broad class of machine learning paradigms in which multiple participants jointly contribute to the training of a model by sharing *knowledge* rather than raw data. The central idea is that useful learning signals can be extracted locally and combined in a coordinated manner, allowing multiple data owners to benefit from collective intelligence while preserving local data autonomy.

Unlike traditional centralized learning, collaborative learning does not assume that data from different sources can be freely aggregated into a single repository. Instead, each participant retains control over its local dataset and contributes to the learning process through abstract representations, such as model updates or statistical summaries. This paradigm is particularly well suited to environments where data is distributed across devices, organizations, or jurisdictions, and where data sharing is constrained by privacy, ownership, or regulatory considerations.

Federated learning represents a structured, systematic, and privacy-aware realization of collaborative learning. It formalizes how local training, communication, and aggregation should be performed so that a shared global model can be learned without exposing raw data. In federated learning, each participant—commonly referred to as a *client*—maintains its own local dataset and performs local model optimization using that data. Rather than transmitting data to a central server, each client computes local updates and shares only these updates with a coordinating entity, often called the *server* or *aggregator*.

From a procedural perspective, federated learning typically proceeds in iterative rounds. In each round, a subset of clients receives the current global model, trains it locally for a limited number of steps, and returns the resulting parameter updates or gradients. The coordinating entity then aggregates these contributions to produce an improved global model, which is redistributed to clients in the next round. Throughout this process, raw data never leaves the client—local environment.

Federated Learning: Core Idea
Federated learning enables multiple autonomous clients to collaboratively train a shared global model by exchanging model updates instead of raw data, thereby preserving data privacy, ownership, and locality by design.

The distinction between collaborative learning in general and federated learning in particular lies in the level of formalization and system support. While collaborative learning captures the high-level notion of knowledge sharing without data exchange, federated learning specifies concrete mechanisms for client participation, update aggregation, communication efficiency, and privacy preservation. As such, federated learning can be viewed as an operational framework that transforms the abstract idea of collaboration into a deployable learning system.

Consider a network of hospitals that wish to train a diagnostic model using patient data. Due to privacy regulations and ethical constraints, patient records cannot be transferred to a central location. Through federated learning, each hospital trains the model locally using its own clinical data and shares only model updates. The aggregated model benefits from diverse patient populations without any hospital revealing sensitive records.

As another example, modern smartphone keyboards continuously adapt to user typing behavior. Each device collects highly personal text input that must remain private. Federated learning allows each phone to locally improve a shared language model while contributing only abstract updates, enabling global model improvement without exposing individual messages.

Why Collaborative and Federated Learning Matter
By decoupling learning from raw data exchange, collaborative and federated learning enable model training across organizational, geographical, and device boundaries while respecting privacy, data ownership, and regulatory constraints.

This collaborative structure fundamentally expands the scope of machine learning. Models can now be trained across institutions that do not trust each other with data, across devices that cannot afford continuous data uploads, and across jurisdictions where data movement is legally restricted. Federated learning thus serves as a cornerstone paradigm for scalable, privacy-preserving intelligence in modern distributed data ecosystems.

Terminological Note: Collaborative Learning vs. Federated Learning
In some introductory sources, the term *federated learning* is occasionally used interchangeably with *collaborative learning*, reflecting their shared emphasis on joint model training without centralized data aggregation. At this level of abstraction, both paradigms acknowledge decentralized, often heterogeneous data held by multiple participants.

In this chapter, a more precise distinction is adopted for analytical clarity. *Collaborative learning* is used as an umbrella term encompassing any approach in which multiple entities contribute knowledge rather than raw data. *Federated learning* is treated as a specific, formalized subclass of collaborative learning, distinguished by explicit client–server orchestration, structured communication rounds, support for partial participation, and privacy-by-design guarantees.

This refined usage does not contradict broader definitions but instead provides the conceptual resolution required for rigorous system design, optimization analysis, and real-world deployment.

The distinction between collaborative learning in general and federated learning in particular lies in the level of formalization and system support.

Table 1.1 provides a comparative overview of centralized learning, collaborative learning, and federated learning, highlighting their fundamental differences across data handling, privacy, communication, and system design dimensions. The table illustrates how centralized learning is inherently data-centric, relying on unrestricted access to raw data stored in a central repository, which leads to high communication costs, weak privacy guarantees, and limited robustness in large-scale or regulated environments.

Collaborative learning represents an intermediate paradigm in which participants aim to improve a shared model by exchanging knowledge rather than raw data. While this approach alleviates some privacy and regulatory concerns, its effectiveness depends heavily on the chosen collaboration protocol and often lacks standardized mechanisms for coordination, robustness, and scalability.

Federated learning, in contrast, emerges as a principled and structured learning paradigm explicitly designed for decentralized and privacy-sensitive settings. As shown in Table 1.1, federated learning enforces data locality by design, prohibits raw data sharing, and supports heterogeneous and non-identically distributed data across clients. By coordinating local model updates through a formal client–server orchestration mechanism, federated learning achieves scalable collaborative intelligence while maintaining strong privacy preservation and regulatory compliance.

Table 1.1 Comparison of Centralized Learning, Collaborative Learning, and Federated Learning

Aspect	Centralized Learning	Collaborative Learning	Federated Learning
Data location	All data is collected and stored in a central repository	Data remains distributed across participants	Data remains strictly local to each client
Raw data sharing	Required	Typically avoided	Explicitly prohibited
Learning objective	Optimize a global model using centralized data	Joint model improvement through shared knowledge	Train a global model via coordinated local updates
Privacy preservation	Weak; relies on post-hoc security measures	Moderate; depends on collaboration protocol	Strong; privacy enforced by design
Regulatory compliance	Difficult in regulated environments	Easier than centralized learning	Well suited for regulated and sensitive domains
Communication cost	High due to raw data transfer	Moderate; depends on shared information	Reduced; only model updates are exchanged
Statistical assumptions	Typically IID data	May involve heterogeneous data	Explicitly supports non-IID data
System robustness	Vulnerable to single point of failure	Partially decentralized	High resilience through decentralized participation
Coordination mechanism	Central server controls all processes	Flexible or loosely coordinated	Formal client–server orchestration
Scalability	Limited by central infrastructure	Scales with participant cooperation	Designed for large-scale, cross-device settings
Typical applications	Traditional analytics, small-scale ML	Cross-institutional research	Mobile devices, healthcare, finance, IoT
Design philosophy	Data-centric aggregation	Knowledge-centric collaboration	Privacy-first, decentralized intelligence
Energy consumption	High due to continuous data transmission and centralized processing	Moderate; depends on participant computation and communication patterns	Optimized for edge and mobile devices with localized computation
Fault tolerance	Low; failure of central server disrupts the entire system	Moderate; partial failures may be tolerated	High; learning can proceed despite partial client failures
Data ownership and control	Central authority assumes control over aggregated data	Data ownership remains distributed but loosely governed	Full data ownership retained by individual clients
Trust assumptions	Requires high trust in central data custodian	Requires mutual trust among collaborators	Minimizes trust requirements through data locality

(Continued)

Table 1.1 (continued)

Aspect	Centralized Learning	Collaborative Learning	Federated Learning
Adaptability to dynamic environments	Limited adaptability due to centralized coordination	Moderate adaptability depending on collaboration protocol	High adaptability to changing client availability and network conditions
Suitability for edge environments	Poor; relies on stable connectivity and powerful servers	Limited; depends on coordination overhead	Excellent; explicitly designed for edge and on-device learning
Learning latency	High due to centralized data transfer and batch processing	Moderate; depends on coordination and communication frequency	Low to moderate; local training reduces end-to-end update latency

1.4 Data Decentralization and Client-Based Learning

A defining characteristic of federated learning is *data decentralization*, wherein training data remains distributed across a large number of autonomous participants, referred to as *clients*. Unlike centralized learning, no single entity has direct access to the entire dataset. Instead, each client maintains its own local dataset, which may differ substantially from others in terms of size, statistical distribution, data quality, and temporal characteristics.

This decentralized data landscape fundamentally alters the nature of the learning problem. Rather than optimizing a model over a single, homogeneous dataset, federated learning must coordinate learning across a population of clients that are diverse both in the data they generate and in the resources they possess. Client-based learning thus introduces new challenges and opportunities that do not arise in traditional centralized settings.

1.4.1 Statistical Heterogeneity in Client Data

One of the most significant and distinctive challenges introduced by data decentralization in federated learning is *statistical heterogeneity*. Unlike classical centralized learning, where training data is typically assumed to be drawn independently and identically from a single underlying distribution, federated learning operates on collections of local datasets whose statistical properties may differ substantially across clients.

Formally, let $\mathscr{D}_k$ denote the local dataset associated with client k. In many federated learning scenarios, the data-generating distribution $P_k(x, y)$ underlying $\mathscr{D}_k$

differs from the distributions $P_j(x, y)$ of other clients $j \neq k$. This *non-identically distributed* (non-IID) nature of client data is not an anomaly but a natural consequence of decentralized data generation.

Statistical heterogeneity arises organically in real-world applications. For instance, in a mobile keyboard prediction task, each user–device collects text data that reflects individual language preferences, vocabulary usage, writing habits, and contextual behavior. Some users may predominantly type in one language, while others switch frequently between multiple languages. Similarly, in healthcare applications, hospitals located in different geographical regions may serve patient populations with distinct demographic profiles, disease prevalence, and treatment protocols. As a result, the local datasets $\mathscr{D}_k$ can vary widely in terms of feature distributions, label frequencies, and noise characteristics.

This form of heterogeneity fundamentally complicates the learning process. Local models trained on client-specific data may converge toward solutions that are optimal for their respective distributions but suboptimal or even conflicting from a global perspective. When such locally optimized updates are aggregated without careful coordination, the resulting global model may exhibit slow convergence, oscillatory behavior, or systematic bias toward overrepresented or dominant client distributions. In extreme cases, naïve aggregation can lead to model divergence or degraded performance on underrepresented data groups.

Consequently, addressing non-IID data is a central concern in the design of federated learning algorithms. Federated optimization methods must balance the need to respect local data characteristics with the objective of learning a model that generalizes well across the entire population of clients. This challenge has motivated the development of specialized aggregation strategies, adaptive weighting schemes, and regularization techniques tailored to statistically heterogeneous environments.

Consider a federated learning system deployed across multiple hospitals to train a disease diagnosis model. A hospital specializing in geriatric care may generate data dominated by elderly patients, while another hospital may primarily treat younger individuals with different clinical conditions. Local models trained at these sites will naturally emphasize different diagnostic patterns and risk factors. Aggregating such updates without accounting for these differences can bias the global model toward certain patient groups, underscoring the need for federated learning techniques that explicitly handle statistical heterogeneity.

1.4.2 System Heterogeneity in Federated Learning

Beyond statistical heterogeneity in data, federated learning systems must also contend with pronounced *system heterogeneity*. Unlike centralized environments where computation occurs on uniform and well-provisioned servers, federated learning operates across a highly diverse set of client devices. These clients may differ substantially in terms of computational capabilities, memory capacity, energy availability, operating systems, and network connectivity.

Such heterogeneity arises naturally in real-world deployments. For example, modern smartphones may be equipped with multi-core CPUs, dedicated neural processing units, and stable network access, enabling them to perform multiple local training iterations efficiently. In contrast, low-power IoT sensors or embedded devices may possess only limited processing capability, constrained memory, and intermittent connectivity, allowing for minimal local computation and infrequent communication.

System heterogeneity directly influences how much computation a client can perform, how often it can participate in training, and how reliably it can communicate updates. If not carefully accounted for, these differences can lead to uneven training contributions, increased straggler effects, and inefficient use of system resources. Federated learning algorithms must therefore be designed to accommodate a wide range of client capabilities without penalizing weaker devices or overburdening stronger ones.

In a smart IoT deployment, industrial gateways with powerful processors may be capable of training complex local models, while battery-powered sensors may only contribute lightweight updates. A federated learning framework must adapt to both, ensuring that system heterogeneity does not stall or bias the overall learning process.

1.4.3 Partial and Intermittent Client Participation

In addition to heterogeneity in hardware and resources, federated learning systems must operate under conditions of *partial and intermittent client participation*. At any given training round, only a subset of clients may be available, responsive, or willing to participate in the learning process. Client availability can fluctuate due to network conditions, energy constraints, user behavior, or device scheduling policies.

For instance, mobile devices may participate in federated training only when connected to a stable Wi-Fi network and sufficient battery power. Other devices may drop out unexpectedly due to connectivity loss, background application constraints, or user activity. As a result, federated learning systems cannot rely on consistent or synchronous participation from all clients.

This intermittent participation introduces additional complexity into the learning process. The set of participating clients may change from one training round to the next, leading to variability in the aggregated updates and potentially uneven representation of client data. Federated learning algorithms must therefore be robust to client dropout and designed to make meaningful progress even when participation is sparse or unpredictable.

In a federated learning system deployed across millions of smartphones, some devices may participate only occasionally, while others may remain offline for extended periods. Despite this variability, the system must continue to refine the global model using whichever subset of client updates is available at each round, without assuming universal or continuous participation.

These challenges require federated learning frameworks to carefully balance efficiency, robustness, and fairness, ensuring that learning progress does not depend on the persistent availability of any particular subset of clients.

Client-Centric Learning Paradigm

In federated learning, clients are not passive data providers but active participants with diverse capabilities and availability. They perform local training, enforce privacy constraints, and collectively shape the evolution of the global model under decentralized, heterogeneous, and dynamic system conditions.

Despite the challenges introduced by system heterogeneity and partial client participation, client-based learning enables scalable and adaptive intelligence across diverse environments. By leveraging local computation and coordinating knowledge at the model level, federated learning makes it possible to learn from vast, distributed datasets that would otherwise remain isolated.

Table 1.2 summarizes the principal patterns of client participation observed in federated learning systems and highlights their implications for learning dynamics and system design. Full participation, while conceptually simple, is largely confined to controlled experimental settings and does not reflect the realities of large-scale

Table 1.2 Comparison of Client Participation Patterns in Federated Learning

Aspect	Full Participation	Partial Participation	Intermittent Participation
Client availability	All clients participate in every round	Only a subset of clients participate per round	Client participation varies unpredictably over time
Typical environment	Controlled or simulated settings	Large-scale real-world deployments	Mobile, edge, and energy-constrained systems
Participation predictability	High and deterministic	Moderate; often sampled or scheduled	Low; driven by external conditions
Communication reliability	Stable and continuous	Generally reliable for active clients	Intermittent and often unreliable
Impact on convergence	Fast and stable convergence	Slightly slower but manageable	Potentially unstable without robust aggregation
System robustness requirement	Low	Moderate	High
Representative example	Laboratory experiments	Enterprise or institutional FL systems	Smartphone and IoT-based FL deployments

deployments. In contrast, partial participation–where only a subset of clients contributes updates in each round–represents the dominant operational mode in practical federated learning systems.

Intermittent participation, as shown in Table 1.2, poses the greatest challenge. Client availability in such settings is highly unpredictable and driven by external factors such as network connectivity, energy constraints, and user behavior. This variability introduces fluctuations in the aggregated updates and can lead to uneven representation of client data across training rounds. Consequently, federated learning algorithms must be explicitly designed to tolerate client dropout, adapt to changing participation patterns, and maintain learning progress despite sparse or unreliable client engagement.

The comparison underscores that partial and intermittent participation are not exceptional cases but defining characteristics of real-world federated learning. Robust aggregation strategies, adaptive client selection, and fairness mechanisms are therefore essential to ensure that learning performance does not depend on persistent or universal client availability.

Privacy, data protection, and control over information have emerged as central concerns in modern data-driven systems. As machine learning models increasingly rely on data derived from personal behavior, sensitive records, and proprietary processes, the traditional practice of aggregating raw data into centralized repositories has come under growing scrutiny. High-profile data breaches, misuse of personal information, and the introduction of stringent data protection regulations have exposed the limitations of centralized data-centric learning architectures and have prompted a fundamental rethinking of how learning systems should be designed.

Privacy preservation is therefore one of the most fundamental motivations behind the development of federated learning. Unlike traditional centralized machine learning, which requires raw data to be transferred, stored, and processed at a central location, federated learning deliberately restricts the movement of data from the very beginning of the learning process. By ensuring that raw data never leaves the client device or institution, federated learning substantially reduces the risk of data leakage, unauthorized access, and secondary use of information beyond its original purpose.

This approach represents a decisive shift in philosophy. Conventional privacy mechanisms are often reactive in nature, relying on post-hoc safeguards such as encryption, anonymization, or access controls applied after data has already been collected and centralized. Federated learning, in contrast, adopts a *privacy-by-design* perspective, in which privacy constraints are embedded directly into the system architecture. Data locality, limited information exchange, and client-controlled participation are treated not as optional features, but as foundational design principles.

Beyond technical considerations, this privacy-centric architecture aligns naturally with broader societal, legal, and ethical expectations. Principles such as data minimization, user consent, and data ownership are increasingly codified in regulatory frameworks and organizational policies. Federated learning provides a concrete mechanism for operationalizing these principles in large-scale machine learning systems, enabling collaborative intelligence without requiring unrestricted access to raw data.

1.4.4 Local Data Retention and Privacy Preservation

At the core of federated learning is the principle of *local data retention*. Each client performs training locally using its private dataset, and only derived information—such as model updates or gradients—is shared with a coordinating entity. Raw data, including personal records, sensor readings, or proprietary logs, remains confined to the local environment under the control of the data owner.

This architectural choice substantially reduces exposure risks. Even if the central coordinating server is compromised, attackers gain access only to aggregated or intermediate model information rather than sensitive raw data. As a result, the impact of potential security breaches is significantly mitigated compared to centralized data aggregation approaches.

In a healthcare federated learning system, patient records remain stored within individual hospitals. Each hospital trains a local model on its own clinical data and shares only model updates. At no point are individual patient records transmitted to an external server, preserving confidentiality and aligning with medical ethics and privacy regulations.

1.4.5 Principle of Data Minimization

Federated learning is inherently aligned with the principle of *data minimization*, a foundational concept in modern data governance and privacy regulation. Data minimization advocates that learning systems should collect, process, and share only the information that is strictly necessary to accomplish a specified task, and nothing more. This principle stands in contrast to traditional centralized learning approaches, which often rely on large-scale data aggregation under the assumption that more data invariably leads to better models.

In federated learning, data minimization is achieved through architectural design rather than post-processing constraints. Instead of transferring raw data to a central repository, each client performs learning locally and communicates only abstracted learning signals, such as model parameters, gradients, or update vectors. These representations encode information relevant to model improvement while deliberately excluding identifiable or context-rich raw data.

Data Minimization as an Architectural Property

Federated learning minimizes data exposure by design. Only model-relevant information is exchanged, while raw data remains local and inaccessible to external entities.

From a learning perspective, data minimization encourages efficiency and generalization. By focusing on parameters that contribute directly to the optimization objective, federated learning avoids unnecessary data transfer and reduces the risk of overfitting to idiosyncratic or irrelevant details. The learning process is thus guided by task-relevant signals rather than exhaustive data accumulation.

From a governance and compliance perspective, data minimization reduces legal and operational risks. Limiting data exchange decreases the scope of regulatory obligations, simplifies data lifecycle management, and lowers long-term storage and retention requirements. In environments governed by strict data protection laws, this reduction in data handling complexity can be decisive for practical deployment.

Consider a mobile keyboard application that adapts to user typing behavior. Rather than uploading raw text messages or keystroke logs to a central server, each device locally updates a language model based on its private input data. Only parameter updates reflecting learned patterns are transmitted. This approach ensures that sensitive personal text never leaves the device, while the global model continues to improve through collective learning.

Data minimization therefore serves both technical and ethical objectives. It enables scalable and efficient learning while respecting individual privacy, organizational data boundaries, and regulatory constraints. In federated learning, minimizing data exposure is not a limitation but a defining strength that enables responsible and sustainable collaborative intelligence.

1.4.6 Access Rights and Data Ownership

A defining strength of federated learning lies in its explicit support for *granular access control* and *data ownership*. Unlike centralized machine learning architectures, where data is often transferred to and governed by a single authority, federated learning preserves the autonomy of individual data owners throughout the learning process. Data owners–whether individual users, institutions, or organizations– retain full authority over their local datasets and determine how their data is accessed, processed, and utilized.

In federated settings, participation in learning is voluntary and configurable. Each client decides when and how its data contributes to model training, which models it participates in, and under what operational conditions updates are shared. This level of control allows participants to enforce internal policies, comply with organizational governance frameworks, and adapt participation based on evolving legal, ethical, or operational requirements.

Data Ownership in Federated Learning
Federated learning preserves data ownership by design. Control over data access, usage, and participation remains with the data owner rather than being transferred to a central authority.

This decentralized governance model stands in sharp contrast to centralized learning architectures. In centralized systems, once data is ingested into a shared repository, ownership boundaries often become blurred. Access policies may be dictated by the central operator, and data may be repurposed, retained, or analyzed beyond its original scope. Federated learning avoids this erosion of control by ensuring that data never leaves its original administrative domain.

Furthermore, federated learning enables transparent auditing and accountability. Because data remains local, organizations can track exactly how and when local training occurs, verify compliance with internal standards, and withdraw from collaborative training if conditions change. This ability to disengage without data loss or exposure is particularly important in long-term collaborations and multi-stakeholder settings.

Consider a cross-institutional collaboration among financial institutions seeking to improve fraud detection models. Each bank retains exclusive control over its transaction records, which are subject to strict confidentiality and regulatory requirements. Through federated learning, each institution trains a local model using its own data and shares only abstract model updates. Participation does not require transferring ownership of sensitive records, allowing institutions to collaborate without violating confidentiality agreements, competitive boundaries, or legal obligations.

Access rights and data ownership therefore form a cornerstone of the federated learning paradigm. By enabling collaboration without surrendering control, federated learning reconciles collective intelligence with institutional sovereignty, trust, and accountability in distributed data ecosystems.

1.4.7 Privacy-by-Design as an Architectural Principle

The integration of local data retention, data minimization, and participant-controlled access rights reflects a broader *privacy-by-design* philosophy that underpins federated learning. Privacy-by-design emphasizes that privacy should not be treated as an auxiliary feature added after system deployment, but as a foundational requirement that shapes system architecture from the outset. Federated learning embodies this philosophy by constraining data access and data movement at the architectural level.

In conventional machine learning systems, privacy is often addressed through post-hoc mechanisms such as data anonymization, access control layers, or encryption applied after data has already been centralized. While these measures can reduce risk, they do not eliminate the fundamental exposure created by aggregating raw data in a shared repository. Federated learning adopts a fundamentally different approach: it prevents raw data exposure by design, ensuring that sensitive data never leaves its original administrative or physical boundary.

By performing learning locally and exchanging only abstract model updates, federated learning reduces both the probability and the impact of privacy breaches. Even in the presence of system failures or adversarial attacks, the absence of centralized raw data storage significantly limits the scope of potential harm. Privacy, in this context, is achieved not through trust in a central authority, but through architectural constraints that minimize what can be accessed in the first place.

Privacy-by-Design in Federated Learning
Federated learning enforces privacy as a core architectural constraint. By keeping raw data local, minimizing information exchange, and preserving data ownership, privacy protection is achieved by construction rather than through auxiliary or post-hoc safeguards.

This architectural emphasis enables federated learning systems to scale across domains where trust, compliance, and accountability are critical. In healthcare, patient confidentiality must be preserved while enabling collective learning across institutions. In finance, transaction data must remain confidential even as models adapt to emerging fraud patterns. In personalized consumer services, user behavior data must be protected to maintain trust and regulatory compliance. Federated learning provides a principled framework for meeting these requirements without sacrificing analytical capability.

Thus, privacy preservation, data minimization, and access rights form a coherent foundation for responsible and sustainable collaborative intelligence. Federated learning demonstrates that high-quality machine learning models can be trained without compromising privacy or relinquishing control over sensitive data. By reconciling technological advancement with ethical, legal, and societal constraints, privacy-by-design positions federated learning as a cornerstone paradigm for the next generation of intelligent systems.

1.5 Federated Learning versus Distributed Learning

Federated learning is frequently conflated with traditional distributed learning because both paradigms involve multiple computational entities participating in a shared learning task. Despite this superficial similarity, the two approaches differ fundamentally in their underlying assumptions, design objectives, and operational constraints. Understanding this distinction is essential for correctly positioning federated learning within the broader landscape of large-scale machine learning systems.

Traditional distributed learning primarily emerged as a response to *computational scalability*. As datasets and models grew too large to be processed on a single machine, distributed learning frameworks were developed to partition computation across multiple nodes. In such settings, data is typically assumed to be centrally owned, logically shared, or easily transferable among participating nodes. The primary goal is to accelerate training by parallelizing computation while maintaining tight synchronization and centralized control.

Federated learning, in contrast, arises from a different set of constraints. Rather than assuming shared data and stable infrastructure, federated learning explicitly models environments in which data is owned by independent entities, cannot be freely transferred, and is subject to privacy, legal, and organizational restrictions. The objective is not merely to scale computation, but to enable collaborative learning *without violating data isolation and ownership.*

Key Distinction
Distributed learning focuses on *computational scalability* under centralized control, whereas federated learning prioritizes *data privacy, ownership, and decentralization* as first-class design constraints.

These differing priorities lead to substantial differences in system architecture, communication patterns, optimization strategies, and robustness requirements. While distributed learning assumes reliable connectivity and persistent node availability, federated learning must tolerate partial participation, intermittent connectivity, and heterogeneous client capabilities.

Figure 1.4 illustrates a classical distributed learning architecture with centralized coordination. Each worker node maintains a local data shard and exchanges information bidirectionally with a central parameter server. Workers transmit gradients or parameter updates derived from local training, while the server aggregates these contributions and broadcasts updated model parameters. The bidirectional arrows emphasize the tight synchronization between workers and the server, while the ellipsis indicates scalability to a large number of participating nodes.

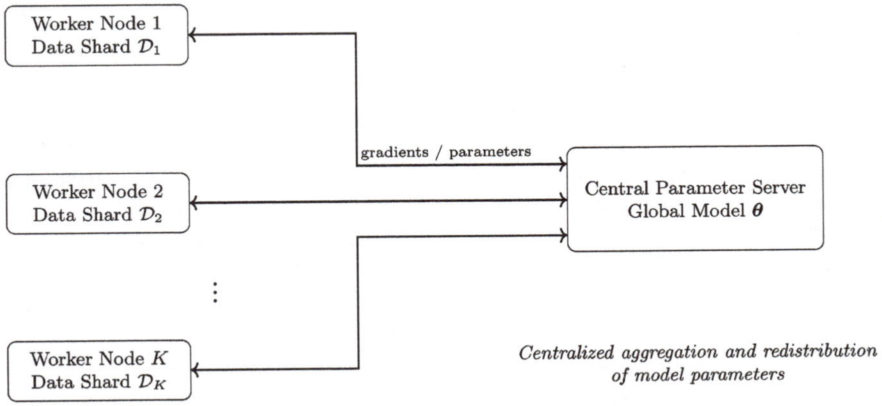

Fig. 1.4 Distributed learning architecture with centralized coordination, showing bidirectional exchange of gradients and model parameters between worker nodes and the parameter server

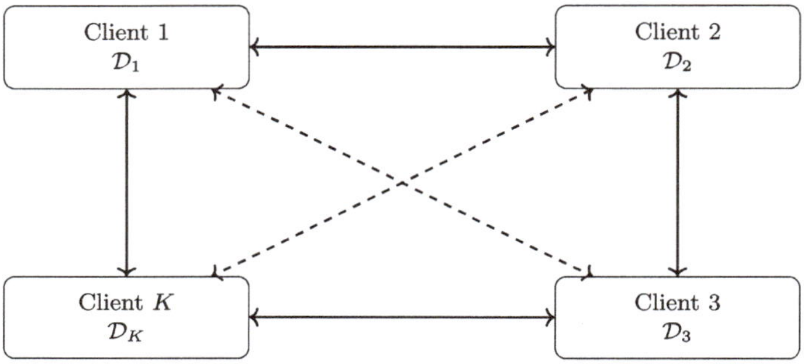

Fig. 1.5 Decentralized learning architecture with peer-to-peer coordination. Clients exchange model updates directly without relying on a central parameter server

Figure 1.5 illustrates a fully decentralized learning architecture in which multiple clients collaboratively train a model without the presence of a central coordinating server. Each client k maintains its own local dataset $\mathscr{D}_k$, and all learning-related communication occurs directly between peers.

The solid bidirectional links represent primary peer-to-peer exchanges of model updates, such as gradients or parameter vectors, between neighboring clients. These connections may follow a logical topology such as a ring, mesh, or graph, enabling structured but decentralized coordination. The dashed bidirectional links indicate additional or opportunistic communication paths that facilitate faster information diffusion, consensus formation, or robustness against node or link failures.

In this architecture, model updates propagate gradually across the network through repeated local interactions rather than being aggregated instantaneously at a central location. As a result, the global model emerges implicitly from the collective dynamics of the system rather than from explicit centralized averaging. This design eliminates single points of failure, preserves full data locality, and aligns naturally with environments where centralized control is infeasible or undesirable.

However, the absence of a central coordinator also introduces challenges related to convergence guarantees, synchronization, and communication overhead. Ensuring consistency across clients often requires additional consensus, gossip, or averaging protocols, particularly in the presence of heterogeneous data distributions and unreliable network connectivity.

Table 1.3 summarizes the key conceptual and operational differences between distributed learning and federated learning across multiple dimensions relevant to real-world deployment. Distributed learning and federated learning differ not merely in implementation details, but in their foundational assumptions.

Distributed learning presumes that data can be centralized or freely exchanged and that participating nodes operate under a single administrative domain.

These assumptions enable efficient synchronization and fast convergence but limit applicability in privacy-sensitive or regulated environments.

Table 1.3 Comparison of Distributed Learning and Federated Learning

Aspect	Distributed Learning	Federated Learning
Primary objective	Accelerate training through parallel computation	Enable collaborative learning under privacy and ownership constraints
Data ownership	Centrally owned or freely shared	Decentralized and locally owned
Raw data movement	Common and often required	Explicitly avoided or prohibited
System control	Centralized orchestration	Decentralized participation with coordination
Client availability	Assumed stable and reliable	Partial, intermittent, and unpredictable
Statistical assumptions	Often IID or globally shuffled data	Explicitly non-IID across clients
Robustness requirement	Moderate; controlled infrastructure	High; must tolerate dropout and heterogeneity
Typical deployment	Data centers, HPC clusters	Mobile devices, healthcare, finance, IoT
Design philosophy	Compute-centric optimization	Privacy-first, client-centric intelligence

Federated learning, by contrast, is explicitly designed for scenarios where such assumptions do not hold. Data remains local, participation is voluntary and intermittent, and learning must proceed despite statistical and system heterogeneity.

As a result, federated learning algorithms emphasize robustness, fairness, and privacy preservation alongside model accuracy.

In a distributed learning setup for image classification, a large dataset may be partitioned across multiple GPUs within a data center, with frequent synchronization of gradients.

In a federated learning setup for the same task deployed on mobile devices, each device trains locally on private images and contributes only abstract updates, without ever sharing raw data. Although both systems train a shared model, the design constraints and optimization challenges differ fundamentally.

This comparison clarifies that federated learning is not a special case of distributed learning, but a distinct paradigm tailored to modern data ecosystems where privacy, ownership, and decentralization are central concerns.

Table 1.4 contrasts two fundamentally different realizations of distributed learning: systems with centralized coordination and fully decentralized, peer-to-peer approaches. The table highlights how centralized coordination simplifies aggregation, synchronization, and convergence analysis by relying on a parameter server, but at the cost of scalability limitations and a single point of failure. In contrast, decentralized distributed learning eliminates reliance on a central entity, improving robustness and autonomy, while introducing greater algorithmic and communication

Table 1.4 Comparison of distributed learning with centralized coordination and decentralized learning

Aspect	Distributed Learning (Centralized Coordination)	Distributed Learning (Decentralized Coordination)
Coordination entity	Central parameter server coordinates aggregation and updates	No central entity; coordination emerges through peer-to-peer interactions
Model aggregation	Explicit global aggregation (e.g., parameter server, AllReduce)	Implicit aggregation via consensus, gossip, or local averaging
Communication pattern	Star topology centered at the server	Mesh, ring, or graph-based peer-to-peer topology
Single point of failure	Present; failure of the central server disrupts training	Absent; system remains operational under partial node failures
Synchronization	Often tightly synchronized across workers	Typically asynchronous or weakly synchronized
Scalability	Limited by server capacity and network bottlenecks	Highly scalable with increasing number of participants
System complexity	Simpler convergence analysis and control	More complex coordination and convergence guarantees
Typical use cases	Data-center training, HPC clusters, cloud ML platforms	Edge networks, multi-institutional collaboration, ad hoc systems

complexity. These trade-offs make centralized approaches more suitable for controlled data-center environments, whereas decentralized learning is better aligned with large-scale, heterogeneous, and infrastructure-constrained settings such as edge networks and cross-organizational collaborations.

Distributed learning with centralized coordination prioritizes efficient and well-controlled optimization by leveraging a global parameter server, making it suitable for homogeneous, high-bandwidth environments. In contrast, decentralized distributed learning trades centralized control for robustness, autonomy, and scalability, enabling collaborative model training across loosely connected and independently governed participants. This distinction is fundamental when transitioning from cloud-centric machine learning to edge-native and privacy-sensitive learning systems.

1.6 Federated Learning in Practice: The Google Mobile Scenario

Conventional machine learning pipelines are built on the assumption that training data can be centrally collected, stored, and processed within large-scale data centers. In such systems, powerful servers execute highly iterative optimization

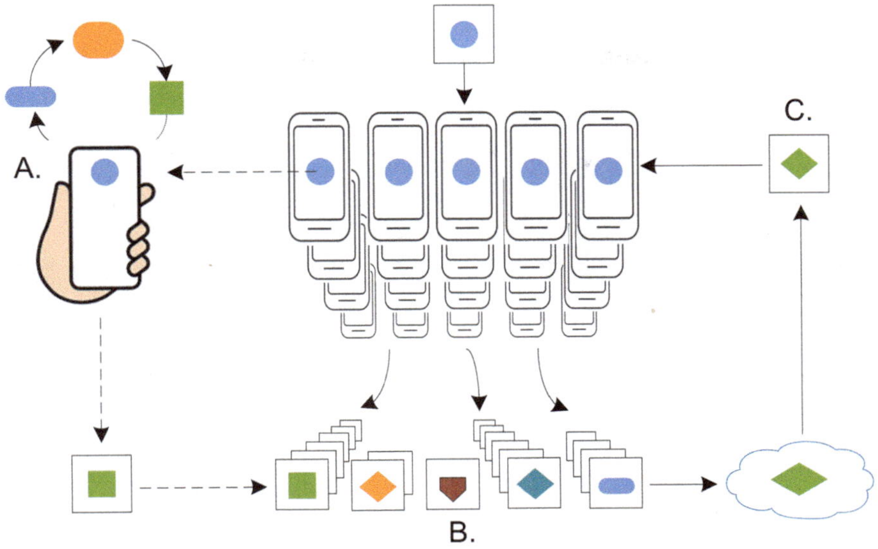

Fig. 1.6 System-level view of federated learning across mobile devices, illustrating local personalization (A), aggregation of updates (B), and consensus global model updates (C)

algorithms—most commonly stochastic gradient descent (SGD)—over massive, centrally aggregated datasets. This paradigm has been remarkably successful for tasks where data is readily shareable and connectivity is stable.[1]

However, this centralized assumption becomes increasingly problematic when training data originates from *personal, sensitive, and widely distributed sources*, such as mobile phones. User-generated data on smartphones reflects private behavior, preferences, and contextual information, making large-scale data centralization undesirable from both privacy and regulatory perspectives. Federated learning directly addresses this challenge by fundamentally decoupling *model training* from *data centralization*, thereby allowing learning to occur where the data is generated.

Figures 1.6[2] and 1.7[3] together illustrate the operational principles of federated learning as deployed in large-scale mobile ecosystems, most notably within Google's on-device learning infrastructure. Viewed jointly, these figures reveal how federated learning reconciles the competing demands of scalability, privacy, communication efficiency, and personalization.

[1] https://research.google/blog/federated-learning-collaborative-machine-learning-without-centralized-training-data/

[2] https://research.google/blog/federated-learning-collaborative-machine-learning-without-centralized-training-data/

[3] https://research.google/blog/federated-learning-collaborative-machine-learning-without-centralized-training-data/

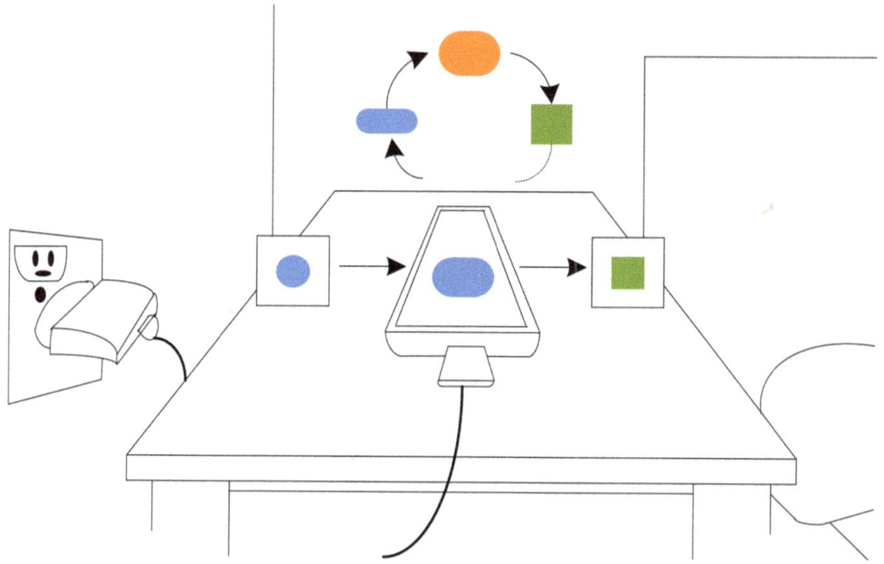

Fig. 1.7 Device-level view of federated learning, showing continuous on-device training and feedback driven by local user interactions

End-to-end Federated Learning Workflow. Figure 1.6 presents a system-level perspective of federated learning operating across millions of heterogeneous mobile devices. Each participating device, referred to as a *client*, maintains a private local dataset derived from user interactions, such as typing behavior, query selections, or contextual feedback. Crucially, this raw data is never uploaded to the cloud.

Instead of transmitting data, each client periodically downloads the current global model from the coordinating server and performs local optimization using its private dataset. This local training step leverages the computational capabilities of modern mobile processors while preserving strict data locality.

The figure highlights three essential stages of the federated learning cycle:

- **Local personalization (A):** Each device adapts the shared model to its unique usage patterns. This step allows the model to capture individual-specific linguistic habits, preferences, or contextual signals. Importantly, the improved local model can be used immediately on the device, enabling real-time personalization without waiting for global model updates.
- **Aggregation of updates (B):** After local training, devices compute compact summaries of their learning progress, such as parameter updates or gradient statistics. Only these abstract updates are transmitted to the coordinating server, ensuring that no raw user data or sensitive information leaves the device.
- **Consensus model update (C):** The server aggregates updates received from a large population of devices—typically using weighted averaging—to produce an

improved global model. This aggregation step extracts population-level learning signals while suppressing individual-specific noise. The updated model is then redistributed to clients in the next training round.

Key Insight from Fig. 1.6
Federated learning converts millions of isolated, privacy-sensitive local training processes into a unified global learning system by exchanging only model updates, thereby enabling large-scale collaboration without centralizing raw data.

On-device Learning and User Context. While Fig. 1.6 emphasizes global coordination, Fig. 1.7 provides a complementary device-centric perspective. It illustrates how federated learning is embedded seamlessly into the everyday operation of a user's mobile phone, operating largely in the background without explicit user intervention.

In this setting, all user interaction data–such as keystrokes, context signals, or feedback on suggestions–remains strictly local. Local training is executed opportunistically, typically only when the device satisfies favorable conditions such as being idle, connected to a power source, and operating on an unmetered network. This adaptive scheduling ensures that federated learning imposes minimal impact on device performance, battery life, or user experience.

The cyclic arrows in Fig. 1.7 emphasize that learning is *continuous and iterative*. Local interaction data improves the model, the refined model enhances future predictions, and this positive feedback loop repeats over time, leading to progressively better personalization.

On-Device Intelligence
Federated learning extends intelligence to the edge by enabling both inference and training directly on user devices, allowing models to adapt continuously to individual behavior while maintaining strong privacy guarantees.

Communication Efficiency and Federated Averaging. A key challenge implicitly revealed by both figures is the scale and heterogeneity of participating devices. Unlike traditional distributed learning environments—where servers communicate over reliable, high-bandwidth networks—mobile devices are characterized by intermittent connectivity, limited upload capacity, and unpredictable availability.

To address these constraints, Google introduced the *Federated Averaging* (FedAvg) algorithm. Rather than transmitting gradients after every local update, each device performs multiple local optimization steps before sending a single, higher-quality update. This strategy significantly reduces communication frequency and allows federated learning to operate efficiently even under severe bandwidth limitations. Empirically, federated averaging can reduce communication costs by one to two orders of magnitude compared to naïvely federated SGD.

Beyond data locality, federated learning strengthens privacy through secure aggregation protocols. The coordinating server is able to observe only the aggregated model update and only when contributions from a sufficiently large number of clients are available. Individual updates are cryptographically protected and never stored or inspected in isolation. As a result, no single client's contribution can be reconstructed, providing strong protection against inference attacks and data leakage.

A prominent real-world deployment of federated learning is Google's use of the paradigm in Gboard on Android.[4] When users interact with keyboard suggestions, contextual information about acceptance or rejection of suggestions is processed locally to refine query prediction models. Local improvements benefit the user immediately, while aggregated updates improve the global model over successive federated training rounds.

Taken together, Figs. 1.6 and 1.7 illustrate a fundamental shift in the design of machine learning systems. Federated learning enables intelligent models to scale across millions of devices while preserving privacy, reducing end-to-end latency, and lowering energy consumption.

This paradigm challenges long-standing assumptions about centralized data access and requires practitioners to rethink model development, evaluation, and optimization under strict communication, privacy, and availability constraints. Consequently, federated learning should be viewed not merely as an optimization technique, but as a new architectural foundation for large-scale, privacy-preserving machine intelligence.

1.7 Key Application Domains of Federated Learning

Federated learning has gained prominence across a diverse range of application domains in which data sensitivity, decentralization, and governance constraints are intrinsic rather than incidental. In these settings, traditional centralized machine learning approaches are often impractical, legally restricted, or ethically problematic. Federated learning enables collaborative intelligence while respecting domain-specific constraints on data access and ownership.

The most prominent application domains are discussed below.

- **Healthcare and Medical Informatics:** Healthcare data is among the most sensitive categories of information, encompassing patient records, diagnostic images,

[4] https://blog.google/products/search/gboard-now-on-android/

genetic data, and clinical histories. Federated learning enables hospitals, clinics, and research institutions to collaboratively train predictive and diagnostic models without transferring patient data outside institutional boundaries. Typical applications include disease risk prediction, medical image analysis, personalized treatment recommendation, and early warning systems for patient deterioration. By keeping patient records local, federated learning supports compliance with medical ethics and data protection regulations while enabling population-scale learning.

- **Mobile and Edge Devices:** Modern smartphones, wearables, and edge devices continuously generate personal and contextual data, such as text input, location traces, sensor readings, and usage patterns. Federated learning has been successfully applied to tasks such as keyboard prediction, speech recognition, personalized recommendation, and adaptive user interfaces. Each device improves a shared model using local data and contributes only abstract updates, enabling personalization without exposing private user behavior. This domain represents one of the largest real-world deployments of federated learning at global scale.
- **Financial Services and Banking:** Financial institutions operate under strict confidentiality and regulatory requirements that limit data sharing across organizations. Federated learning enables collaborative fraud detection, credit risk assessment, and anti-money laundering systems across banks and payment providers without exposing transaction-level data. By aggregating learning signals rather than raw financial records, institutions can identify emerging fraud patterns and systemic risks while preserving competitive boundaries and customer confidentiality.
- **Internet of Things (IoT) and Smart Systems:** IoT environments consist of large numbers of geographically distributed sensors and devices that generate continuous streams of data. Centralizing this data is often infeasible due to bandwidth limitations, latency constraints, and energy consumption. Federated learning enables smart cities, industrial monitoring systems, and environmental sensing platforms to learn from distributed data sources while minimizing communication overhead. Applications include predictive maintenance, anomaly detection, traffic optimization, and energy management.
- **Industrial and Manufacturing Systems:** In industrial settings, operational data is often proprietary and strategically sensitive. Federated learning allows multiple factories or production sites to jointly improve models for quality control, fault detection, and process optimization without revealing internal operational details. This enables cross-site learning while preserving intellectual property and organizational autonomy.
- **Smart Grids and Energy Systems:** Energy consumption data collected from smart meters is highly sensitive and subject to regulatory oversight. Federated learning enables utilities to forecast demand, detect faults, and optimize energy distribution by learning from decentralized meter data without centralizing individual household consumption records. This approach improves grid efficiency while protecting consumer privacy.

- **Personalized Education and Recommendation Systems:** Educational platforms and content providers can use federated learning to personalize learning materials, assessments, and recommendations based on user interaction data. By keeping learner data local to devices or institutions, federated learning supports personalization while respecting student privacy and institutional data governance policies.

Why Federated Learning Matters Across Domains
Federated learning enables scalable, privacy-preserving intelligence in domains where centralized data aggregation is impractical, legally restricted, or ethically unacceptable. By decoupling learning from raw data exchange, it expands the reach of machine learning into sensitive, regulated, and decentralized environments.

Across these domains, federated learning demonstrates that high-quality models can be trained without compromising privacy, ownership, or institutional autonomy. This chapter establishes the conceptual foundation upon which the mathematical formulations, system architectures, and optimization strategies of federated learning will be developed in subsequent chapters, linking real-world needs with principled algorithmic design.

Exercises

This exercise set reinforces the foundational concepts, motivations, system challenges, and application domains of federated learning, emphasizing decentralized data, privacy-by-design, and collaborative model training.

A. Short Answer Questions

Answer each question briefly and precisely.

1. Define federated learning and explain how it differs from centralized machine learning.
2. What is meant by *data decentralization* in federated learning?
3. Explain the term *non-identically distributed (non-IID)* data in the context of federated learning.
4. Why is raw data sharing prohibited in federated learning?
5. What role does the central server (aggregator) play in federated learning?
6. State two reasons why centralized machine learning poses privacy risks.
7. What is meant by *partial client participation*?
8. Define the principle of data minimization.

9. What is meant by privacy-by-design?
10. List any two application domains where federated learning is preferred over centralized learning.

B. Long Answer Questions

Answer the following questions with clear explanations, equations, and diagrams wherever appropriate.

1. Explain the limitations of centralized machine learning in modern data ecosystems, focusing on privacy, communication overhead, and robustness.
2. Describe the federated learning training workflow, including client selection, local training, and model aggregation.
3. Discuss statistical heterogeneity in federated learning. Why does non-IID data complicate convergence?
4. Compare centralized learning, collaborative learning, and federated learning with respect to data ownership and governance.
5. Explain system heterogeneity in federated learning and its implications for algorithm design.
6. Discuss partial and intermittent client participation with real-world examples.
7. Explain how federated learning satisfies regulatory and legal constraints in sensitive domains.
8. Contrast distributed learning and federated learning in terms of assumptions, objectives, and deployment environments.
9. Explain the concept of client-based learning and its role in scalable intelligence.
10. Discuss why federated learning is considered a privacy-preserving paradigm by design rather than by enforcement.

C. True/False Questions

State whether each statement is **True** or **False**.

1. In federated learning, raw data is transmitted to a central server for aggregation.
2. Federated learning assumes that client data is IID by default.
3. Centralized machine learning introduces a single point of failure.
4. Federated learning supports intermittent client participation.
5. Data minimization reduces regulatory and privacy risks.
6. Distributed learning and federated learning are identical paradigms.
7. In federated learning, clients act as active learners rather than passive data providers.
8. Communication cost in federated learning is typically lower than centralized learning.

9. Federated learning is unsuitable for mobile and edge devices.
10. Privacy-by-design implies embedding privacy constraints into system architecture.

Answers:

1. False
2. False
3. True
4. True
5. True
6. False
7. True
8. True
9. False
10. True

D. Fill in the Blanks

Fill in the blanks with the most appropriate term or expression.

1. Federated learning keeps training data __________ to the client.
2. Data distributions across clients in federated learning are often __________.
3. The central coordinating entity in federated learning is called the __________.
4. Sharing raw data is __________ in federated learning.
5. The principle of __________ advocates collecting only necessary data.
6. Federated learning reduces __________ overhead by exchanging model updates.
7. Client dropout leads to __________ participation.
8. Privacy-by-design treats privacy as an __________ constraint.
9. Learning based on knowledge exchange rather than data exchange is called __________ learning.
10. Edge and mobile environments strongly benefit from __________ learning paradigms.

Answers:

1. local
2. non-IID
3. server / aggregator
4. prohibited
5. data minimization
6. communication
7. intermittent

8. architectural
9. collaborative
10. federated

E. Multiple Choice Questions (MCQs)

Choose the **most appropriate** answer.

1. The primary motivation behind federated learning is:
   (a) Faster computation
   (b) Centralized data storage
   (c) Privacy-preserving collaboration
   (d) Model compression
2. Which entity performs local training in federated learning?
   (a) Central server
   (b) Cloud provider
   (c) Client device
   (d) Data broker
3. Non-IID data primarily affects:
   (a) Storage cost
   (b) Model convergence
   (c) Encryption strength
   (d) Network routing
4. Which learning paradigm assumes stable connectivity?
   (a) Federated learning
   (b) Distributed learning
   (c) Edge learning
   (d) Personalized learning
5. In federated learning, what is exchanged?
   (a) Raw data
   (b) Sensor logs
   (c) Model updates
   (d) User identities
6. Privacy-by-design means:
   (a) Encrypting stored data
   (b) Centralizing access control
   (c) Architectural privacy enforcement
   (d) Removing identifiers
7. Which domain is least suitable for centralized learning?
   (a) Image datasets
   (b) Public benchmarks
   (c) Healthcare records
   (d) Academic datasets

8. Federated learning is best described as:
   (a) Compute-centric
   (b) Data-centric
   (c) Privacy-first
   (d) Server-centric
9. Client participation in federated learning is:
   (a) Mandatory
   (b) Continuous
   (c) Optional and dynamic
   (d) Fixed
10. The main coordination mechanism in federated learning is:
    (a) Data pooling
    (b) Gradient descent
    (c) Model aggregation
    (d) Feature extraction

Answers:

1. (c)
2. (c)
3. (b)
4. (b)
5. (c)
6. (c)
7. (c)
8. (c)
9. (c)
10. (c)

F. Application and Project-Based Questions

These problems emphasize modeling, reasoning, and implementation.

1. Design a federated learning system for disease prediction across multiple hospitals. Identify clients, server, and data flow.
2. Propose a federated learning architecture for smartphone keyboard personalization.
3. Simulate the impact of non-IID data on federated model convergence using synthetic datasets.
4. Compare centralized and federated learning for IoT sensor data in smart cities.
5. Design a client selection strategy for energy-constrained devices in federated learning.
6. Analyze communication cost trade-offs between centralized and federated learning.

7. Implement a basic federated averaging algorithm and evaluate its performance.
8. Propose fairness metrics for client contribution in federated learning.
9. Study regulatory constraints (e.g., healthcare or finance) and explain how federated learning addresses them.
10. Critically evaluate federated learning as a sustainable AI paradigm for the next decade.

Fundamentals and Definitions of Federated Learning

2

Federated learning emerges from a growing recognition that many modern machine learning problems cannot be adequately addressed within the traditional paradigm of centralized data collection and processing. As discussed in Chapter 1, contemporary data is increasingly generated across large numbers of geographically distributed, administratively independent, and resource-constrained entities. In such environments, the assumptions that underpin classical machine learning-namely, unrestricted data aggregation, uniform governance, and global accessibility-no longer hold. Federated learning responds to this shift by redefining how learning systems are structured when data must remain local to its source.

While the previous chapter focused on the motivation for federated learning and the practical limitations of centralized approaches, the goal of the present chapter is to place federated learning on firm conceptual and mathematical footing. This chapter does not yet concern itself with specific algorithms, training protocols, or system implementations. Instead, it addresses a more fundamental question: *what does it mean to learn collaboratively when data is decentralized and cannot be shared?* Answering this question requires careful definitions, explicit modeling assumptions, and a clear separation between local computation and global coordination.

To this end, the chapter introduces the core abstractions that define a federated learning system. These include the notions of clients and servers, locally stored datasets, shared model parameters, and iterative communication rounds. By formalizing these entities and their interactions, federated learning can be expressed as a well-defined optimization problem whose objective is constructed implicitly from multiple local loss functions rather than a single global dataset. This formulation highlights a key departure from centralized learning: the global learning objective is never evaluated directly, but is instead approximated through coordinated local computations.

P. P. Ray, *Foundations of Federated Learning*,
Springer Series in the Data Sciences,
https://doi.org/10.1007/978-3-032-34084-9_2

A central theme throughout the chapter is the principle of data locality. Because raw data remains confined to individual clients, coordination among participants must occur through indirect means such as parameter updates or model deltas. This constraint fundamentally alters the learning dynamics, introducing new sources of uncertainty arising from statistical heterogeneity, unequal data volumes, and partial client participation. As a result, many intuitions drawn from classical empirical risk minimization must be re-examined in the federated setting.

The chapter further develops an intuitive understanding of federated learning as a form of consensus-driven optimization. Rather than enforcing strict synchronization across all participants, federated learning seeks gradual alignment among independently evolving model instances. Convergence, therefore, should be interpreted not as exact agreement at every iteration, but as the emergence of a shared model that performs well across a diverse population of clients. This perspective provides a conceptual bridge between optimization theory, distributed systems, and real-world federated deployments.

Finally, the chapter positions federated learning within the broader landscape of parallel and distributed learning paradigms. By explicitly contrasting their assumptions, objectives, and operational constraints, it becomes clear that federated learning is not merely a variant of distributed optimization designed for scalability. Rather, it constitutes a distinct learning paradigm shaped by concerns of privacy, data ownership, regulatory compliance, and system heterogeneity. The definitions and principles introduced here therefore form the theoretical backbone for the algorithmic methods, convergence analyses, and application case studies developed in the subsequent chapters.

2.1 Formal Definition of Federated Learning

Federated learning provides a formal framework for collaborative model training in environments where data is inherently decentralized and cannot be freely shared. In contrast to traditional machine learning, which assumes that all training data can be aggregated into a single repository, federated learning explicitly recognizes that data is often generated, stored, and governed by multiple autonomous entities. These entities may represent individual users, edge devices, organizations, or institutions, each operating under distinct privacy, regulatory, and operational constraints.

Unlike classical distributed optimization-where datasets are typically partitioned solely to improve computational scalability-federated learning is fundamentally motivated by *data locality* rather than computational efficiency. The primary objective is not merely to distribute computation, but to enable learning while respecting data ownership, minimizing data exposure, and adhering to legal or ethical restrictions. As a result, federated learning must be defined in a manner that decouples learning from centralized data access and replaces direct data aggregation with indirect coordination mechanisms.

System Entities. A federated learning system consists of a collection of K autonomous participants, commonly referred to as *clients*, indexed by $k \in \{1, 2, \dots, K\}$. Each client k maintains a local dataset

$$\mathscr{D}_k = \{(x_{k,i}, y_{k,i})\}_{i=1}^{n_k},$$

where $n_k = |\mathscr{D}_k|$ denotes the number of training samples available at that client. These datasets are stored locally and are not shared with other clients or with any coordinating entity.

As established in Chapter 1, the local datasets $\{\mathscr{D}_k\}$ are generally *heterogeneous* in both size and statistical characteristics. In particular, the data-generating distribution at client k, denoted by $P_k(x, y)$, may differ significantly across clients. Consequently, federated learning does not assume that samples are independent and identically distributed across the system. This departure from the IID assumption has far-reaching implications for optimization, convergence behavior, and model generalization.

Model Parameterization. Despite the decentralization of data, federated learning assumes a *shared model hypothesis space*. All clients collaboratively train a common predictive model

$$f_{\theta} : \mathscr{X} \to \mathscr{Y},$$

parameterized by a global parameter vector $\theta \in \mathbb{R}^d$. The model architecture-such as a linear predictor, neural network, or probabilistic estimator-is identical across all clients, ensuring that locally computed updates are compatible and can be meaningfully combined.

While the parameter vector θ is conceptually global, each client maintains and updates its own local copy during training. These local model instances evolve according to client-specific data and computation, reflecting the underlying heterogeneity of the learning environment. The challenge of federated learning lies in reconciling these divergent local updates into a single global model that performs well across the entire population of clients.

Federated Optimization Objective. The goal of federated learning is to learn a parameter vector θ that minimizes a global objective function defined implicitly over the union of all local datasets. Formally, this objective is expressed as

$$\min_{\theta} F(\theta) \triangleq \sum_{k=1}^{K} \frac{n_k}{N} F_k(\theta), \quad \text{where } N = \sum_{k=1}^{K} n_k.$$

Here, the weight $\frac{n_k}{N}$ reflects the relative contribution of client k based on its data volume, ensuring that clients with larger datasets exert proportionally greater influence on the global objective.

The local objective function at client k is given by

$$F_k(\theta) = \frac{1}{n_k} \sum_{(x,y)\in\mathscr{D}_k} \ell(f_\theta(x), y),$$

where $\ell(\cdot)$ denotes a suitable loss function measuring the discrepancy between model predictions and true labels. This formulation mirrors empirical risk minimization in centralized learning, but with the crucial distinction that each local risk is evaluated only on client-specific data.

Interpretation. The global objective $F(\theta)$ can be interpreted as the empirical risk over the conceptual union of all client datasets, even though such a union is never explicitly constructed. Federated learning therefore replaces direct access to a global dataset with an implicit aggregation of local empirical risks. This abstraction allows learning to proceed while respecting strict data locality constraints.

Key Distinction
In federated learning, the global objective $F(\theta)$ is never evaluated on a centralized dataset. Instead, it emerges from the weighted combination of locally computed objectives, each defined over private, non-IID data that remains under the control of individual clients.

Conceptual Implications. This formulation highlights the fundamental tension at the heart of federated learning. While the objective resembles classical empirical risk minimization in form, the constraints under which it is optimized are radically different. Learning must proceed under limited visibility, partial participation, and heterogeneous data distributions. These characteristics distinguish federated learning as a unique and non-trivial extension of traditional machine learning, motivating the specialized algorithms and convergence analyses developed in later chapters.

Worked Example: Federated Learning with Three Clients
We now extend the previous illustration to a slightly richer toy scenario involving three clients. This example demonstrates how the federated objective naturally scales with the number of participating clients while preserving data locality and proportional contribution.

Toy Learning Scenario. Consider a federated learning system with $K = 3$ clients. Each client holds a small local dataset consisting of scalar input–output pairs. As before, we consider a simple linear regression model

$$f_\theta(x) = \theta x,$$

where $\theta \in \mathbb{R}$ is a scalar model parameter shared across all clients.

Local Datasets (Tabular Representation). The datasets stored at the three clients are shown in Tables 2.1–2.3.

The corresponding dataset sizes are

$$n_1 = 2, \qquad n_2 = 3, \qquad n_3 = 3, \qquad N = n_1 + n_2 + n_3 = 8.$$

Loss Function. We again use the squared error loss

$$\ell\big(f_\theta(x), y\big) = (\theta x - y)^2.$$

Step 1: Local Empirical Risk at Client 1. Using the definition

$$F_k(\theta) = \frac{1}{n_k} \sum_{(x,y)\in\mathscr{D}_k} (\theta x - y)^2,$$

the local objective at Client 1 is

$$\begin{aligned} F_1(\theta) &= \frac{1}{2}\Big[(\theta \cdot 1 - 2)^2 + (\theta \cdot 2 - 4)^2\Big] \\ &= \frac{1}{2}\Big[(\theta - 2)^2 + (2\theta - 4)^2\Big]. \end{aligned}$$

Table 2.1 Local dataset at Client 1

x	y
1	2
2	4

Table 2.2 Local dataset at Client 2

x	y
1	1
2	2
3	3

Table 2.3 Local dataset at Client 3

x	y
2	3
3	4
4	5

Step 2: Local Empirical Risk at Client 2. For Client 2, we obtain

$$F_2(\theta) = \frac{1}{3}\left[(\theta - 1)^2 + (2\theta - 2)^2 + (3\theta - 3)^2\right].$$

Step 3: Local Empirical Risk at Client 3. Similarly, the local objective at Client 3 is

$$F_3(\theta) = \frac{1}{3}\left[(2\theta - 3)^2 + (3\theta - 4)^2 + (4\theta - 5)^2\right].$$

Step 4: Global Federated Objective. Referring to the main federated learning formulation, the global objective is defined as

$$F(\theta) = \sum_{k=1}^{3} \frac{n_k}{N} F_k(\theta).$$

Substituting $n_1 = 2$, $n_2 = 3$, $n_3 = 3$, and $N = 8$, we obtain

$$F(\theta) = \frac{2}{8}F_1(\theta) + \frac{3}{8}F_2(\theta) + \frac{3}{8}F_3(\theta).$$

Step 5: Explicit Expansion of the Global Objective. Substituting the expressions for $F_1(\theta)$, $F_2(\theta)$, and $F_3(\theta)$ yields

$$\begin{aligned} F(\theta) = {} & \frac{1}{8}\left[(\theta - 2)^2 + (2\theta - 4)^2\right] + \frac{1}{8}\left[(\theta - 1)^2 + (2\theta - 2)^2 + (3\theta - 3)^2\right] \\ & + \frac{1}{8}\left[(2\theta - 3)^2 + (3\theta - 4)^2 + (4\theta - 5)^2\right]. \end{aligned}$$

Remark. The cancellation of the local normalization factors with the client weights yields a uniform contribution of $\frac{1}{N}$ per data sample. This confirms that federated learning recovers the centralized empirical risk objective in form, while preserving strict data locality in computation.

Interpretation. The resulting global objective corresponds to the empirical risk over all eight data points distributed across the three clients. Importantly, this objective is never evaluated on a centralized dataset. Instead, it is constructed implicitly through weighted aggregation of local empirical risks computed independently at each client.

Conceptual Insight
As the number of clients increases, the federated learning objective retains the same mathematical structure as centralized empirical risk minimization. The key difference lies not in the form of the objective, but in how it is computed-through decentralized, privacy-preserving local evaluations rather than centralized data access.

2.2 Clients, Servers, and Communication Rounds

Federated learning systems are structured around a clear logical separation between entities that *own data* and entities that *coordinate learning*. This separation is architectural rather than hierarchical and is fundamental to preserving data locality. Unlike centralized machine learning pipelines-where data flows toward a central processor-federated learning reverses this flow by moving computation toward data while limiting information exchange to carefully controlled updates.

At the core of this architecture lie two conceptual components: *clients*, which perform local computation on private data, and an optional coordinating entity known as the *server*, which facilitates collaboration without accessing raw data. Learning progresses through a sequence of *communication rounds*, during which models are repeatedly refined via decentralized computation and centralized aggregation.

2.2.1 Clients

A client is any autonomous computational entity that participates in the federated learning process. Formally, a client is characterized by three essential capabilities:

1. Local storage of a private dataset $\mathscr{D}_k$,
2. Local computation of model updates using $\mathscr{D}_k$,
3. Communication of selected information derived from local computation.

Clients may correspond to personal mobile devices, edge computing nodes, medical institutions, or enterprise databases. Importantly, clients operate under heterogeneous conditions, differing in computational power, network connectivity, data volume, and data distribution. These differences play a significant role in shaping the learning dynamics of federated systems.

2.2.2 Server

The server acts as a coordination mechanism rather than a data repository. Its primary role is to maintain a global model state θ and orchestrate interaction among clients. Specifically, the server is responsible for:

- Storing the current global model parameters,
- Selecting a subset of clients for participation in each round,
- Aggregating locally computed updates into a new global model.

Crucially, the server never receives or stores raw data samples. All coordination occurs through model parameters or parameter updates, ensuring that sensitive data remains confined to client devices.

2.2.3 Communication Rounds

Federated learning unfolds over a sequence of discrete communication rounds indexed by $t = 0, 1, 2, \ldots$. At the beginning of round t, the server holds a global model $\theta^{(t)}$. A typical communication round consists of the following steps:

1. **Broadcast:** The server sends $\theta^{(t)}$ to a selected subset of clients.
2. **Local computation:** Each participating client performs local optimization using its private dataset and produces an update.
3. **Aggregation:** The server aggregates the received updates to obtain the next global model $\theta^{(t+1)}$.

These rounds continue until a convergence criterion or stopping condition is satisfied.

Communication-Centric View
Federated learning is fundamentally constrained by communication rather than computation. Because client devices may be bandwidth-limited or intermittently connected, federated algorithms are designed to minimize the frequency, size, and sensitivity of exchanged information.

Worked Example: Clients, Server, and Three Communication Round
We now present a complete toy example that explicitly illustrates the roles of clients, the server, and communication rounds using concrete datasets and numerical calculations.

Toy Learning Setup. Consider a federated learning system with $K = 3$ clients collaboratively training a simple linear regression model

$$f_\theta(x) = \theta x,$$

where $\theta \in \mathbb{R}$ is a scalar model parameter. We use the squared error loss

$$\ell\big(f_\theta(x), y\big) = (\theta x - y)^2.$$

Local Datasets. Each client holds a private dataset, summarized in Tables 2.4, 2.5, and 2.6.

The dataset sizes are

$$n_1 = 2, \quad n_2 = 3, \quad n_3 = 3, \quad N = 8.$$

Initialization at the Server. At communication round $t = 0$, the server initializes the global model as

$$\theta^{(0)} = 0,$$

and broadcasts this value to all three clients.

Local Computation at Clients. Each client computes the gradient of its local empirical risk function

$$F_k(\theta) = \frac{1}{n_k} \sum_{(x,y)\in\mathcal{D}_k} (\theta x - y)^2$$

Table 2.4 Local dataset at Client 1

x	y
1	2
2	4

Table 2.5 Local dataset at Client 2

x	y
1	1
2	2
3	3

Table 2.6 Local dataset at Client 3

x	y
2	3
3	4
4	5

with respect to the model parameter θ, evaluated at the initial global model $\theta^{(0)} = 0$.

Taking the derivative with respect to θ, we obtain

$$\nabla F_k(\theta) = \frac{1}{n_k} \sum_{(x,y)\in\mathscr{D}_k} 2(\theta x - y)\, x.$$

Communication Round 1 ($t = 0$).

Client 1. The local dataset at Client 1 is

$$\mathscr{D}_1 = \{(1,2),(2,4)\}, \quad n_1 = 2.$$

Substituting into the gradient expression,

$$\begin{aligned}\nabla F_1(\theta) &= \frac{1}{2}\Big[2(\theta\cdot 1 - 2)\cdot 1 + 2(\theta\cdot 2 - 4)\cdot 2\Big]\\ &= \frac{1}{2}\Big[2(\theta - 2) + 4(2\theta - 4)\Big].\end{aligned}$$

Evaluating at $\theta = 0$,

$$\nabla F_1(0) = \frac{1}{2}\big[-4 - 16\big] = -10.$$

Client 2. The local dataset at Client 2 is

$$\mathscr{D}_2 = \{(1,1),(2,2),(3,3)\}, \quad n_2 = 3.$$

The gradient of the local objective is

$$\begin{aligned}\nabla F_2(\theta) &= \frac{1}{3}\Big[2(\theta\cdot 1 - 1)\cdot 1 + 2(\theta\cdot 2 - 2)\cdot 2 + 2(\theta\cdot 3 - 3)\cdot 3\Big]\\ &= \frac{1}{3}\Big[2(\theta - 1) + 4(2\theta - 2) + 6(3\theta - 3)\Big].\end{aligned}$$

Evaluating at $\theta = 0$,

$$\nabla F_2(0) = \frac{1}{3}\big[-2 - 8 - 18\big] = -\frac{28}{3}.$$

Client 3. The local dataset at Client 3 is

$$\mathscr{D}_3 = \{(2,3),(3,4),(4,5)\}, \quad n_3 = 3.$$

The gradient of the local objective is

$$\nabla F_3(\theta) = \frac{1}{3}\Big[2(\theta\cdot 2-3)\cdot 2 + 2(\theta\cdot 3-4)\cdot 3 + 2(\theta\cdot 4-5)\cdot 4\Big]$$
$$= \frac{1}{3}\Big[4(2\theta-3) + 6(3\theta-4) + 8(4\theta-5)\Big].$$

Evaluating at $\theta = 0$,

$$\nabla F_3(0) = \frac{1}{3}\big[-12-24-40\big] = -\frac{76}{3}.$$

Aggregation at the Server. Referring to the federated learning objective

$$F(\theta) = \sum_{k=1}^{3}\frac{n_k}{N}F_k(\theta),$$

the gradient of the global objective satisfies

$$\nabla F(\theta) = \sum_{k=1}^{3}\frac{n_k}{N}\nabla F_k(\theta).$$

With $n_1 = 2$, $n_2 = 3$, $n_3 = 3$, and $N = 8$, the aggregated gradient at $\theta^{(0)} = 0$ is

$$\nabla F(0) = \frac{2}{8}(-10) + \frac{3}{8}\left(-\frac{28}{3}\right) + \frac{3}{8}\left(-\frac{76}{3}\right)$$
$$= -\frac{20}{8} - \frac{28}{8} - \frac{76}{8} = -\frac{124}{8} = -15.5.$$

Global Model Update. Using a learning rate $\eta = 0.1$, the server updates the global model parameter according to

$$\theta^{(1)} = \theta^{(0)} - \eta\nabla F(0) = 0 - 0.1(-15.5) = 1.55.$$

Interpretation. This worked example illustrates a complete communication round in a federated learning system. Each client performs local computation using private data, while the server aggregates only numerical updates. Despite the absence of centralized data access, the resulting update is equivalent to performing a gradient step on the global empirical risk over all data samples.

Key Insight
Communication rounds enable decentralized learning to be transformed into coordinated global optimization. The server serves as an aggregation point for information, not data, preserving privacy while enabling collective intelligence.

Communication Round 2 ($t = 1$). At the beginning of the second communication round, the server holds the updated global model

$$\theta^{(1)} = 1.55,$$

which is broadcast to all three clients.

Local Gradient Evaluation. Each client evaluates its previously derived gradient expression at $\theta^{(1)} = 1.55$.

Client 1.

$$\nabla F_1(1.55) = \frac{1}{2}\Big[2(1.55 - 2) + 4(2 \cdot 1.55 - 4)\Big] = -2.25.$$

Client 2.

$$\nabla F_2(1.55) = \frac{1}{3}\Big[2(1.55 - 1) + 4(2 \cdot 1.55 - 2) + 6(3 \cdot 1.55 - 3)\Big] = \frac{77}{15} \approx 5.13.$$

Client 3.

$$\nabla F_3(1.55) = \frac{1}{3}\Big[4(2 \cdot 1.55 - 3) + 6(3 \cdot 1.55 - 4) + 8(4 \cdot 1.55 - 5)\Big] = \frac{139}{30} \approx 4.63.$$

Aggregation at the Server. Using the same weighted aggregation rule,

$$\begin{aligned}\nabla F(1.55) &= \frac{2}{8}(-2.25) + \frac{3}{8}(5.13) + \frac{3}{8}(4.63) \\ &= 3.10.\end{aligned}$$

Global Model Update. With learning rate $\eta = 0.1$,

$$\theta^{(2)} = \theta^{(1)} - 0.1\,\nabla F(1.55) = 1.55 - 0.31 = 1.24.$$

Round 2 Insight
The reduced magnitude of the aggregated gradient indicates that the global model is moving closer to a stationary point of the federated objective.

Communication Round 3 ($t = 2$). At the start of the third communication round, the global model is

$$\theta^{(2)} = 1.24.$$

Local Gradient Evaluation.

Client 1.

$$\nabla F_1(1.24) = \frac{1}{2}\Big[2(1.24-2) + 4(2 \cdot 1.24 - 4)\Big] = -3.80.$$

Client 2.

$$\nabla F_2(1.24) = \frac{1}{3}\Big[2(1.24-1) + 4(2 \cdot 1.24 - 2) + 6(3 \cdot 1.24 - 3)\Big] = 2.24.$$

Client 3.

$$\nabla F_3(1.24) = \frac{1}{3}\Big[4(2 \cdot 1.24 - 3) + 6(3 \cdot 1.24 - 4) + 8(4 \cdot 1.24 - 5)\Big] = -1.36.$$

Aggregation at the Server.

$$\begin{aligned}\nabla F(1.24) &= \frac{2}{8}(-3.80) + \frac{3}{8}(2.24) + \frac{3}{8}(-1.36)\\ &= -0.62.\end{aligned}$$

Global Model Update.

$$\theta^{(3)} = \theta^{(2)} - 0.1\,\nabla F(1.24) = 1.24 + 0.062 = 1.302.$$

Table 2.7 Evolution of model parameter across communication rounds

Round t	$\theta^{(t)}$	$\nabla F(\theta^{(t)})$	$\theta^{(t+1)}$
0	0.000	–15.50	1.550
1	1.550	+3.10	1.240
2	1.240	–0.62	1.302

Round 3 Insight
Oscillation around a narrow parameter range reflects stabilization of the global model under repeated client-server coordination.

Summary of Three Communication Rounds (Table 2.7)

Overall Observation
The sequence of communication rounds demonstrates gradual convergence of the global model through decentralized updates. This behavior mirrors centralized gradient descent while strictly preserving data locality.

2.3 Data Locality and Model Parameter Exchange

A defining and non-negotiable characteristic of federated learning is the principle of *strict data locality*. Unlike centralized or conventional distributed learning paradigms, federated learning explicitly prohibits the transfer of raw training data outside the entity on which it is generated or stored. This constraint is not merely an implementation choice, but a foundational design requirement driven by privacy concerns, regulatory restrictions, ownership rights, and practical limitations on data movement.

In a federated system, learning must therefore proceed under the assumption that data remains permanently confined to client devices. All coordination among participants is achieved through the exchange of model-related information rather than direct access to data. This constraint fundamentally reshapes the learning pipeline, replacing data aggregation with carefully controlled information exchange.

2.3.1 Local Computation

Each client performs all learning-related computations exclusively on its local dataset $\mathcal{D}_k$. Given a shared global model parameter vector $\theta^{(t)}$ received at communication round t, the client applies local optimization procedures-such as gradient descent or its variants-using only locally available data. The outcome of this computation is not a trained model in isolation, but an *intermediate representation* of learning progress.

Typical locally computed quantities include gradients of the local objective, locally updated model parameters $\theta_k^{(t)}$, or compressed summaries thereof. These quantities capture how the global model should be adjusted based on client-specific data, while avoiding the disclosure of individual training samples.

2.3.2 Parameter Exchange

Instead of sharing raw data, clients communicate information derived from local optimization. A commonly exchanged quantity is the *model update* or *parameter delta*, defined as

$$\Delta\theta_k^{(t)} = \theta_k^{(t)} - \theta^{(t)},$$

where $\theta^{(t)}$ denotes the global model at the start of round t, and $\theta_k^{(t)}$ represents the locally updated model after client-side computation.

These parameter deltas encode the direction and magnitude of improvement suggested by client k's data. By aggregating such updates across clients, the server can construct a new global model without ever observing the underlying data. This mechanism allows federated learning to emulate the effect of centralized optimization while maintaining strict data isolation.

Key Concept: Indirect Coordination
Model parameters and updates act as a proxy for data in federated learning. They enable collaboration through *indirect coordination*, where clients influence the global model without revealing their private datasets.

Privacy Implications and Limitations. While parameter exchange significantly reduces the risk of direct data exposure, it does not in itself guarantee formal privacy. Model updates may still leak information about local datasets through statistical inference or adversarial analysis. Consequently, federated learning should be viewed as a *privacy- enabling architecture* rather than a complete privacy solution.

In practice, federated systems often incorporate additional mechanisms such as update clipping, noise injection, secure aggregation, or cryptographic protocols to further mitigate privacy risks. These enhancements operate on top of the basic data locality and parameter exchange framework described here and are discussed in detail in later chapters.

Data Locality Principle
In federated learning, *data moves computation to itself*, not the other way around. This inversion of the classical data pipeline underpins privacy preservation, regulatory compliance, and scalability across decentralized environments.

Broader Implications. The combination of strict data locality and parameter-based communication distinguishes federated learning from all earlier machine learning paradigms. It introduces new challenges in optimization, convergence, robustness, and system design, but simultaneously enables learning in settings where centralized approaches are infeasible or unacceptable. Understanding this trade-off is essential for appreciating both the power and the limitations of federated learning systems.

Worked Example: Data Locality and Parameter Exchange
This example extends the data locality discussion to a federated learning system with four clients, highlighting how parameter exchange scales naturally with the number of participants while preserving strict data isolation.

Toy Learning Setup. Consider a federated learning system with $K = 4$ clients collaboratively training a linear regression model

$$f_\theta(x) = \theta x,$$

where $\theta \in \mathbb{R}$. We employ the squared error loss

$$\ell\big(f_\theta(x), y\big) = (\theta x - y)^2.$$

Local Datasets (Tabular Form). Each client stores its data locally, as summarized in Tables 2.8–2.11.

Table 2.8 Local dataset at Client 1

x	y
1	1
2	2

Table 2.9 Local dataset at Client 2

x	y
1	2
3	6

Table 2.10 Local dataset at Client 3

x	y
2	3
4	6
6	9

Table 2.11 Local dataset at Client 4

x	y
1	0.5
2	1.0
3	1.5

The dataset sizes are

$$n_1 = 2, \quad n_2 = 2, \quad n_3 = 3, \quad n_4 = 3, \qquad N = 10.$$

Global Model Broadcast. At communication round t, the server broadcasts the global model

$$\theta^{(t)} = 1.$$

No raw data is transmitted-only the shared parameter value.

Step 1: Local Gradient Computation. Each client evaluates the gradient of its local empirical risk

$$F_k(\theta) = \frac{1}{n_k} \sum_{(x,y)\in\mathcal{D}_k} (\theta x - y)^2$$

at $\theta^{(t)} = 1$, using the general gradient expression

$$\nabla F_k(\theta) = \frac{1}{n_k} \sum_{(x,y)\in\mathcal{D}_k} 2(\theta x - y)x.$$

Client 1.

$$\nabla F_1(1) = \frac{1}{2}\Big[2(1 \cdot 1 - 1) \cdot 1 + 2(1 \cdot 2 - 2) \cdot 2\Big] = 0.$$

Client 2.

$$\nabla F_2(1) = \frac{1}{2}\Big[2(1 \cdot 1 - 2) \cdot 1 + 2(1 \cdot 3 - 6) \cdot 3\Big] = -14.$$

Client 3.

$$\nabla F_3(1) = \frac{1}{3}\Big[2(2-3)\cdot 2 + 2(4-6)\cdot 4 + 2(6-9)\cdot 6\Big] = -\frac{56}{3}.$$

Client 4.

$$\nabla F_4(1) = \frac{1}{3}\Big[2(1-0.5)\cdot 1 + 2(2-1)\cdot 2 + 2(3-1.5)\cdot 3\Big] = \frac{14}{3}.$$

Step 2: Local Model Updates. Using a local learning rate $\eta = 0.1$, each client updates its local model:

$$\theta_k^{(t)} = \theta^{(t)} - \eta \nabla F_k(\theta^{(t)}).$$

Thus,

$$\begin{aligned}
\theta_1^{(t)} &= 1.00,\\
\theta_2^{(t)} &= 1 + 1.40 = 2.40,\\
\theta_3^{(t)} &= 1 + 1.87 \approx 2.87,\\
\theta_4^{(t)} &= 1 - 0.47 \approx 0.53.
\end{aligned}$$

Step 3: Parameter Delta Exchange. Each client transmits only its parameter delta

$$\Delta\theta_k^{(t)} = \theta_k^{(t)} - \theta^{(t)}.$$

Hence,

$$\Delta\theta_1^{(t)} = 0, \quad \Delta\theta_2^{(t)} = 1.40, \quad \Delta\theta_3^{(t)} \approx 1.87, \quad \Delta\theta_4^{(t)} \approx -0.47.$$

Step 4: Server Aggregation. The server aggregates parameter deltas using data-proportional weights:

$$\Delta\theta^{(t)} = \sum_{k=1}^{4} \frac{n_k}{N} \Delta\theta_k^{(t)}.$$

Substituting values,

$$\begin{aligned}
\Delta\theta^{(t)} &= \frac{2}{10}(0) + \frac{2}{10}(1.40) + \frac{3}{10}(1.87) + \frac{3}{10}(-0.47)\\
&\approx 0.42.
\end{aligned}$$

Step 5: Global Model Update. The server updates the global model as

$$\theta^{(t+1)} = \theta^{(t)} + \Delta\theta^{(t)} = 1.42.$$

Interpretation. This example demonstrates that even with four heterogeneous clients, learning proceeds without any raw data exchange. Each client's influence on the global model is mediated solely through numerical parameter updates, preserving data locality while enabling collaborative optimization.

Key Insight
As the number of clients increases, federated learning maintains the same optimization structure while scaling coordination through weighted parameter exchange rather than centralized data aggregation.

2.4 Local Objectives and Global Optimization Goal

In a federated learning system, each participating client performs learning by optimizing a *local objective function* $F_k(\theta)$ defined over its private dataset $\mathscr{D}_k$. These local objectives reflect client-specific data distributions, sampling frequencies, and noise characteristics. However, federated learning is not merely a collection of independent learning tasks executed in parallel. Rather, it is a coordinated optimization process whose ultimate goal is to produce a *single shared model* that generalizes well across the entire population of clients.

This distinction is fundamental. While optimization is carried out locally, the notion of success is defined globally. The federated system therefore operates under a dual perspective: each client seeks to improve performance on its own data, while the server orchestrates these local improvements toward a unified global objective.

2.4.1 Local Objective Functions

For each client k, the local objective $F_k(\theta)$ measures the empirical risk of the shared model on the locally available dataset. Because datasets differ across clients, these objectives are generally heterogeneous. Even when clients use identical model architectures and loss functions, the resulting optimization landscapes may vary significantly due to differences in feature distributions, label frequencies, and data quality. As a result, the direction that most rapidly decreases F_k for one client may be suboptimal-or even detrimental-for another.

2.4.2 Global Optimization Objective

To coordinate learning across heterogeneous clients, federated learning defines a global objective function that aggregates local objectives in a data-weighted manner:

$$F(\theta) = \sum_{k=1}^{K} \frac{n_k}{N} F_k(\theta),$$

where n_k denotes the number of samples at client k and $N = \sum_{k=1}^{K} n_k$ is the total number of samples across all clients. This formulation ensures that clients contribute to the global objective in proportion to the amount of data they hold, thereby preventing small datasets from exerting disproportionate influence.

While data-proportional weighting partially mitigates imbalance across clients, it does not eliminate all sources of bias. In particular, when client datasets are non-identically distributed, the aggregated objective may still favor dominant data regimes or frequent client behaviors. Consequently, the global objective should be interpreted as a population-level compromise rather than a perfect representation of any individual client's learning goal.

Tension Between Local and Global Objectives. A central challenge in federated learning arises from the inherent tension between local optimization and global generalization. The parameter values that minimize a local objective F_k may differ substantially across clients. In extreme cases, the local minimizers may lie in entirely different regions of the parameter space. As a result, updates computed to reduce local loss can conflict with one another when aggregated, slowing convergence or inducing oscillatory behavior in the global model.

This phenomenon, often referred to as *client drift*, is especially pronounced in non-IID settings and when clients perform multiple local optimization steps before communication. Understanding and controlling this tension is critical for the design of stable and efficient federated learning algorithms.

Optimization Challenge in Federated Learning
Federated learning must continuously balance two competing objectives:

- **Local fidelity**, which encourages models to adapt closely to client-specific data distributions,
- **Global generalization**, which requires the shared model to perform well across a diverse and decentralized population.

Reconciling these objectives lies at the heart of federated optimization and motivates many of the algorithmic innovations discussed in later chapters.

Table 2.12 Comparison between Local Objectives and the Global Objective in Federated Learning

Aspect	**Local Objective** $F_k(\theta)$	**Global Objective** $F(\theta)$
Definition	Empirical risk computed using only the private dataset $\mathscr{D}_k$ of client k	Weighted aggregation of all local objectives across clients
Data Scope	Restricted to a single client's data	Represents the entire population of participating clients
Optimization Goal	Minimize error on client-specific data distribution	Minimize population-level empirical risk
Sensitivity to Non-IID Data	Highly sensitive to local data bias and skew	Partially mitigates bias through data-proportional weighting
Minimizer Location	Local minimizers may differ significantly across clients	Global minimizer reflects a compromise across heterogeneous clients
Update Direction	Optimized for local performance improvement	Optimized for collective generalization performance
Role in Learning Process	Drives client-side updates and local adaptation	Guides server-side aggregation and global coordination

Implications for Algorithm Design. The interplay between local objectives and the global optimization goal has far-reaching consequences. It influences how many local updates clients should perform, how client contributions should be weighted, and how communication rounds should be scheduled. Many federated learning algorithms can be viewed as different strategies for managing this trade-off-seeking to exploit the efficiency of local learning while preserving the stability and convergence properties of global optimization.

Table 2.12 highlights the conceptual and operational differences between local objectives optimized at individual clients and the global objective coordinated by the server. While local objectives capture client-specific learning dynamics, the global objective serves as a unifying optimization target that balances heterogeneity and promotes generalization across the federated system.

2.5 Consensus Learning and Convergence Intuition

Federated learning can be fruitfully interpreted through the lens of *consensus optimization*, a perspective that provides deep insight into both its strengths and its challenges. Rather than viewing learning as a single optimizer operating on a centralized dataset, federated learning should be understood as a distributed collection of autonomous learners that repeatedly negotiate, adjust, and align their local model estimates through structured and restricted information exchange. The goal is not to enforce strict synchronization or identical parameter values at every step, but to guide the collective learning dynamics toward a common and stable solution over time.

This consensus-based interpretation is especially important in federated settings, where data heterogeneity, system variability, communication constraints, and partial client participation are the norm rather than the exception. In such environments, perfect agreement at each iteration is neither feasible nor desirable. Instead, consensus emerges gradually, shaped by repeated rounds of local computation followed by global aggregation, with alignment improving as learning progresses.

2.5.1 Consensus Perspective

At any communication round t, each participating client maintains a local view of the model parameters, either implicitly through its local update trajectory or explicitly through a locally updated model $\theta_k^{(t)}$. These local models encode client-specific information, reflecting the underlying data distribution, optimization dynamics, and noise characteristics present at each client. Because clients operate independently and observe different data, their local models may diverge substantially from one another at intermediate stages of training.

The server aggregation step can be interpreted as a *consensus operator* that maps this collection of heterogeneous local models onto a single global estimate $\theta^{(t+1)}$. Rather than selecting one client's model or enforcing strict equality, the server combines local contributions-typically through weighted averaging-to produce a parameter update that reflects the collective information available across clients. This operation acts as a stabilizing force, pulling the system toward a shared representation while preserving the diversity of local learning signals.

From this viewpoint, federated learning progresses through repeated applications of local optimization followed by consensus enforcement. Local steps allow each client to adapt the model to its private data, capturing fine-grained and personalized patterns. Aggregation then reconciles these adaptations, preventing the global model from drifting too far toward any single client's objective. Over successive communication rounds, this push-and-pull dynamic encourages gradual alignment among client models, even when they originate from strongly heterogeneous data sources.

2.5.2 Convergence Intuition

Under idealized conditions-such as convex and smooth loss functions, bounded gradients, sufficiently small learning rates, and frequent client participation-the federated optimization process exhibits convergence behavior analogous to centralized empirical risk minimization. In these settings, the aggregated client updates provide an unbiased or low-variance approximation to the gradient of the global objective. As a result, the global model can be expected to converge toward a stationary point of the federated loss, mirroring classical gradient descent behavior.

However, federated learning rarely operates under such ideal conditions in practice. Client datasets are typically non-identically distributed, local updates may be computed over multiple steps before aggregation, communication may be delayed or lossy, and only a subset of clients may participate in any given round. These factors introduce both bias and variance into the optimization process. Consequently, convergence must be interpreted in a weaker and more realistic sense: not as monotonic decrease of a global loss at every communication round, but as the gradual stabilization of the global model within a neighborhood of an optimal or near-optimal solution.

2.5.3 Impact of Heterogeneity

Statistical heterogeneity across clients plays a central role in shaping convergence behavior. When local objectives differ substantially, the gradients computed by clients may point in conflicting directions. Aggregating such updates can reduce the net progress made in any single round, slowing convergence or producing oscillatory dynamics in the global model. This effect is particularly pronounced when local learning rates are large or when clients perform many local updates before synchronization.

Consensus learning in federated systems therefore requires carefully designed aggregation strategies, learning rate schedules, and sometimes regularization or correction mechanisms that temper the influence of extreme or highly specialized client updates. Managing heterogeneity is one of the defining challenges of federated optimization and a key focus of algorithmic innovation.

2.5.4 Role of Participation and Stochasticity

Unlike classical distributed optimization, federated learning must operate under partial, intermittent, and often unpredictable client participation. At any given round, only a subset of clients may be available, willing, or selected to contribute updates. As a result, the optimization process is inherently stochastic at the system level, even when local optimization is deterministic.

This stochastic participation prevents convergence guarantees from being stated in a strict, iteration-by-iteration sense. Instead, convergence must be understood statistically, typically in expectation over randomly sampled client subsets and stochastic local updates. While this introduces additional variability into the learning process, it also enables scalability. Over many communication rounds, repeated exposure to diverse subsets of clients allows the global model to accumulate information from across the broader population, driving long-term alignment despite short-term fluctuations.

Key Insight
Convergence in federated learning is not defined by exact agreement at every communication round. Rather, it emerges as the gradual alignment of model trajectories across heterogeneous clients through repeated cycles of local optimization and global consensus enforcement.

2.5.5 Implications for Algorithm Design

Viewing federated learning as a consensus-driven optimization process provides valuable intuition for algorithm design. It clarifies why overly aggressive local updates can destabilize global learning, why adaptive aggregation and client weighting schemes are often necessary under heterogeneity, and why convergence guarantees are typically formulated in expectation rather than deterministically. These insights motivate many practical refinements of federated algorithms-such as controlled local update frequencies, adaptive learning rates, and robustness-enhancing aggregation rules-which are examined in detail in subsequent chapters.

2.6 Comparison with Parallel and Distributed Optimization

At first glance, federated learning appears closely related to classical parallel and distributed optimization techniques, as all these paradigms involve multiple computational entities collaborating to solve a learning problem. However, this resemblance is largely superficial. The assumptions that underpin federated learning differ fundamentally from those of traditional distributed optimization, leading to substantially different algorithmic designs, system architectures, and theoretical challenges.

Understanding these differences is essential for appreciating why methods developed for conventional distributed systems cannot be directly applied to federated learning without significant adaptation.

2.6.1 Parallel and Distributed Optimization

Classical distributed optimization emerged primarily as a response to computational and memory constraints. Large datasets are partitioned across multiple workers to accelerate training and to enable scaling beyond the capacity of a single machine. These workers typically operate under a unified administrative domain and follow centrally enforced protocols.

Standard assumptions in this setting include:

- **Centralized control over data partitions**, where data shards are created intentionally and can be reassigned or reshuffled as needed;
- **Reliable and synchronous communication**, often supported by high-bandwidth, low-latency networks;
- **Homogeneous or predictable computational resources**, allowing load balancing and coordinated execution.

Under these assumptions, the primary objective of distributed optimization is to improve efficiency-reducing training time while preserving the convergence guarantees of centralized learning.

2.6.2 Federated Learning

Federated learning operates in a markedly different regime. Data partitions are not created for computational convenience but arise naturally from data ownership and locality constraints. Clients are autonomous, heterogeneous, and often outside the direct control of the coordinating server.

Key assumptions in federated learning include:

- **Strict data ownership and isolation**, where raw data cannot be accessed or redistributed;
- **Intermittent, partial, and unreliable participation**, with clients joining and leaving unpredictably;
- **Strong statistical and system heterogeneity**, including non-IID data, variable compute power, and diverse network conditions.

As a result, federated learning prioritizes robustness, privacy, and autonomy over raw computational efficiency. Communication becomes a scarce resource, and convergence must be interpreted statistically rather than deterministically.

2.6.3 Conceptual Distinction

The fundamental conceptual difference between the two paradigms lies in their primary motivation. Distributed optimization is driven by the need to scale computation and speed up training. Federated learning, by contrast, is driven by the need to learn under constraints imposed by data sovereignty, privacy regulation, and organizational autonomy.

This distinction reshapes the role of coordination. In distributed optimization, coordination enforces efficiency. In federated learning, coordination enables collaboration under limited trust and incomplete observability.

Table 2.13 Comparison between Parallel/Distributed Optimization and Federated Learning

Aspect	Parallel/Distributed Optimization	Federated Learning
Primary Motivation	Scalability and computational efficiency	Privacy, data ownership, and decentralized collaboration
Data Control	Centrally managed and freely partitioned	Privately owned and strictly local to clients
Data Distribution	Typically IID or intentionally balanced	Naturally non-IID and highly heterogeneous
System Control	Centrally coordinated workers	Autonomous and loosely coordinated clients
Client Participation	Stable and predictable	Intermittent and partially observable
Communication Assumptions	Reliable, high-bandwidth, low-latency	Limited, costly, and often unreliable
Convergence Interpretation	Deterministic and iteration-based	Statistical and expectation-based
Failure Model	Rare and often handled centrally	Common and intrinsic to the system

Paradigm Shift
Federated learning is not merely distributed learning executed at large scale; it represents a redefinition of machine learning under data sovereignty, privacy, and decentralization constraints.

Table 2.13 summarizes the key differences between parallel and distributed optimization and federated learning. While both paradigms distribute computation across multiple entities, they are shaped by fundamentally different constraints and objectives. Federated learning must operate in environments characterized by limited trust, heterogeneity, and partial observability, requiring algorithms that emphasize robustness and privacy preservation rather than strict synchronization and speed. These distinctions explain why federated learning constitutes a separate research paradigm rather than a straightforward extension of classical distributed optimization.

Exercises

This chapter established the conceptual, mathematical, and system-level foundations of federated learning, including data locality, client–server coordination, global optimization objectives and convergence intuition. The following exercises are designed to reinforce definitions, deepen theoretical understanding, and encourage applied reasoning.

A. Short Answer Questions

Answer each question briefly and precisely.

1. Define federated learning and state its primary motivation.
2. What is meant by *data locality* in federated learning?
3. Distinguish between a client and a server in a federated system.
4. What is a communication round?
5. Define the local objective function $F_k(\theta)$.
6. Why is the IID assumption typically violated in federated learning?
7. What role does data-proportional weighting play in the global objective?
8. Explain the concept of client drift.
9. What is meant by consensus learning in federated systems?
10. Why is convergence in federated learning often analyzed in expectation?

B. Long Answer Questions

Answer the following questions with clear explanations, equations, and diagrams wherever appropriate.

1. Formally define a federated learning system and explain its core components.
2. Derive the global federated objective function from local empirical risks.
3. Explain the impact of non-IID data on federated optimization and convergence.
4. Describe the client–server interaction model across communication rounds.
5. Discuss federated learning as a form of consensus optimization.
6. Compare centralized empirical risk minimization and federated learning.
7. Explain why communication is a primary bottleneck in federated learning.
8. Analyze the tension between local objectives and the global optimization goal.
9. Discuss how partial client participation affects convergence behavior.
10. Explain why federated learning is considered a paradigm shift rather than a simple extension of distributed optimization.

C. True/False Questions

State whether each statement is **True** or **False**.

1. Federated learning requires centralized storage of all training data.
2. Clients in federated learning may hold datasets of unequal size.
3. The global objective in federated learning is evaluated on a single dataset.
4. Federated learning assumes homogeneous data distributions across clients.
5. The server in federated learning never accesses raw client data.
6. Communication rounds involve both local computation and global aggregation.
7. Client drift arises primarily due to statistical heterogeneity.

8. Federated convergence guarantees are typically deterministic.
9. Parameter updates act as a proxy for data in federated learning.
10. Federated learning prioritizes privacy over computational efficiency.

Answers:

1. False
2. True
3. False
4. False
5. True
6. True
7. True
8. False
9. True
10. True

D. Fill in the Blanks

Fill in the blanks with the most appropriate term or expression.

1. In federated learning, raw data remains on the __________.
2. The coordinating entity in a federated system is called the __________.
3. The global objective is a weighted sum of __________ objectives.
4. Data-proportional weights are given by $\frac{\underline{\hspace{3cm}}}{N}$.
5. Communication rounds are indexed by $t =$ __________.
6. Client datasets are typically __________ distributed.
7. Parameter deltas encode __________ progress.
8. Federated learning replaces data aggregation with __________ exchange.
9. Convergence is interpreted as gradual __________ of models.
10. Federated learning emphasizes __________ and data sovereignty.

Answers:

1. client
2. server
3. local
4. n_k
5. $0, 1, 2, \ldots$
6. non-IID
7. learning
8. parameter
9. alignment
10. privacy

E. Multiple Choice Questions (MCQs)

Choose the **most appropriate** answer.

1. The primary motivation for federated learning is:
   (a) faster computation
   (b) reduced model size
   (c) data privacy and locality
   (d) simpler optimization
2. In federated learning, the global objective is:
   (a) optimized independently at each client
   (b) evaluated on centralized data
   (c) an unweighted average of local objectives
   (d) a data-weighted aggregation of local objectives
3. Which entity aggregates model updates?
   (a) client
   (b) data owner
   (c) server
   (d) optimizer
4. Client drift is mainly caused by:
   (a) small learning rates
   (b) homogeneous data
   (c) non-IID local datasets
   (d) centralized control
5. Which quantity is typically exchanged instead of raw data?
   (a) Training samples
   (b) Feature vectors
   (c) Model parameters or updates
   (d) Labels
6. Federated learning convergence is usually:
   (a) Exact at every iteration
   (b) Deterministic
   (c) Independent of participation
   (d) Statistical and expectation-based
7. Which assumption is weakest in federated learning?
   (a) Data locality
   (b) Partial participation
   (c) Reliable synchronization
   (d) Model sharing
8. Federated learning differs from distributed optimization mainly due to:
   (a) Higher computation cost
   (b) Stronger privacy constraints
   (c) Smaller datasets
   (d) Simpler objectives

9. The global model is updated after:
   (a) Each local step
   (b) Each data sample
   (c) Each communication round
   (d) Each epoch
10. Consensus learning aims to:
   (a) Enforce identical local models instantly
   (b) Eliminate local objectives
   (c) Gradually align local model trajectories
   (d) Avoid aggregation

Answers:

1. (c)
2. (d)
3. (c)
4. (c)
5. (c)
6. (d)
7. (c)
8. (b)
9. (c)
10. (c)

F. Application and Project-Based Questions

These problems emphasize modeling, reasoning, and implementation.

1. Design a federated learning setup for mobile keyboard prediction.
2. Simulate federated averaging for a simple linear regression problem.
3. Analyze the effect of non-IID data using synthetic client distributions.
4. Implement a basic client–server federated learning loop in Python.
5. Study the impact of partial client participation on convergence.
6. Compare centralized and federated training on the same dataset split.
7. Design a communication-efficient federated learning strategy.
8. Investigate client drift under multiple local update steps.
9. Propose a modification to improve robustness under heterogeneity.
10. Critically evaluate federated learning for a healthcare application.

Mathematical Formulation of Federated Learning

3

The preceding chapters introduced federated learning from conceptual, architectural, and systems-level perspectives. We examined how federated learning departs fundamentally from centralized and classical distributed learning by enforcing strict data locality, autonomy of participants, and communication-centric coordination. While these high-level principles motivate the paradigm, a precise understanding of federated learning requires a rigorous mathematical formalization.

This chapter provides that formal foundation. Its purpose is to translate the qualitative notions of collaboration, local computation, and global coordination into a unified optimization framework. In doing so, federated learning is revealed not merely as a heuristic protocol for training models across devices, but as a structured and constrained optimization problem operating under decentralization, heterogeneity, and partial observability.

From a mathematical standpoint, federated learning occupies a unique position between centralized empirical risk minimization and fully decentralized consensus optimization. On one hand, the learning objective remains global: a single model is trained to perform well across the collective data distribution. On the other hand, this objective must be achieved without direct access to a unified dataset, relying instead on locally computed quantities that are periodically aggregated. The tension between these two requirements gives rise to novel optimization dynamics that are absent in classical learning settings.

This chapter systematically develops the mathematical structure underlying federated learning. We begin by formalizing the global federated optimization objective as a weighted composition of local empirical risks. We then examine the role of client-specific loss functions, local model updates, and aggregation mechanisms that implicitly enforce coordination among autonomous learners. Subsequently, we interpret federated learning through the lens of consensus constraints and analyze the relationship between local and global optima. Finally, we expose the fundamental optimization challenges induced by non-identically distributed (non-IID)

P. P. Ray, *Foundations of Federated Learning*,
Springer Series in the Data Sciences,
https://doi.org/10.1007/978-3-032-34084-9_3

data, which lie at the core of many theoretical and practical difficulties in federated learning.

Throughout this chapter, we emphasize continuity with classical optimization theory while carefully highlighting where federated learning deviates from its assumptions. All notation introduced here is consistent with the formulations established in earlier chapters, and the mathematical abstractions are chosen to support both theoretical analysis and algorithmic design. The results of this chapter serve as the analytical backbone for the federated optimization algorithms, convergence analyses, and robustness techniques developed in subsequent chapters.

3.1 Global Federated Optimization Objective

At the core of federated learning lies a *global optimization objective* that binds together a collection of learning problems defined locally at multiple, autonomous clients. This objective plays a role analogous to empirical risk minimization in centralized machine learning, but it is constructed under fundamentally different constraints. In particular, federated learning abandons the assumption that training data can be pooled into a single repository and instead seeks to achieve global learning through coordination of decentralized computations.

In a classical centralized setting, a learning algorithm directly minimizes a single empirical risk defined over all available data. Federated learning, by contrast, must reconcile multiple empirical risks, each computed on a distinct local dataset, into a unified optimization goal. The global objective therefore emerges as an *implicit aggregation* of locally defined loss functions rather than an explicitly computable quantity.

Formally, consider a federated system consisting of K clients. Each client $k \in \{1, 2, \dots, K\}$ holds a private dataset $\mathscr{D}_k$ containing n_k training samples. The total number of samples across all participating clients is

$$N = \sum_{k=1}^{K} n_k.$$

This quantity serves as a normalization constant that links local datasets to the overall scale of the learning problem.

Each client defines its own local empirical risk function $F_k(\theta)$, reflecting both the size and the statistical properties of its dataset. The federated learning problem is then formulated as the minimization of a weighted sum of these local objectives:

$$\min_{\theta \in \mathbb{R}^d} F(\theta) \triangleq \sum_{k=1}^{K} \frac{n_k}{N} F_k(\theta).$$

This expression defines the *global federated objective*. The weighting factor $\frac{n_k}{N}$ assigns influence to each client in proportion to the amount of data it contributes.

Clients with larger datasets therefore exert greater impact on the global optimization trajectory, while clients with fewer samples have a correspondingly smaller effect. This choice of weighting mirrors centralized empirical risk minimization, where each data point contributes equally to the objective, and ensures consistency between centralized and federated formulations when data is evenly distributed.

It is crucial to emphasize that the global objective $F(\theta)$ is never evaluated directly. No entity in the system has access to all datasets $\{\mathcal{D}_k\}_{k=1}^K$, and therefore no entity can compute $F(\theta)$ in its entirety. Instead, $F(\theta)$ exists as a *conceptual construct* that guides the design and analysis of federated algorithms. All practical learning dynamics arise from local evaluations of $F_k(\theta)$ and their subsequent aggregation.

Federated Objective Principle
The global objective in federated learning is a *virtual* objective: it defines what the system aims to optimize, even though no participant can compute or observe it in full.

This formulation highlights a defining characteristic of federated learning: global optimization is achieved without global visibility. The learning system must rely on partial, locally computed information to approximate descent directions for a loss function that is never explicitly instantiated. As a result, the global objective provides a unifying mathematical lens for analysis, while simultaneously exposing the intrinsic difficulty of optimization under data locality, heterogeneity, and limited coordination.

By preserving strict data locality and allowing each client to retain control over its own dataset, this global objective formulation enables collaborative model training across decentralized and autonomous entities. At the same time, it introduces new challenges absent in centralized learning, setting the stage for the aggregation mechanisms, consensus constraints, and convergence issues explored in the remainder of this chapter.

Worked Example: Constructing the Global Federated Objective with Absolute Loss
We now revisit the construction of the global federated optimization objective using a different loss function. Instead of the squared loss, we employ the *absolute error loss*, which is widely used in robust regression and median-based estimation. This example demonstrates that the federated formulation remains valid independently of the specific choice of loss function.

Federated Setting. Consider a federated system with $K = 3$ clients. As before, each client k holds a private dataset $\mathcal{D}_k$. We use a simple linear prediction model

$$f_\theta(x) = \theta x,$$

where $\boldsymbol{\theta} = \theta \in \mathbb{R}$. The absolute loss function is defined as

$$\ell\big(f_{\theta}(x), y\big) = |\theta x - y|.$$

Local Datasets. The datasets distributed across the three clients are identical to those shown in Table 6.3. The dataset sizes are therefore (Table 3.1)

$$n_1 = 2, \quad n_2 = 3, \quad n_3 = 1, \qquad N = 6.$$

Local Empirical Risk Functions. Using the federated definition of local empirical risk,

$$F_k(\theta) = \frac{1}{n_k} \sum_{(x,y)\in\mathscr{D}_k} |\theta x - y|,$$

we obtain the client-wise objectives.

Client 1:

$$F_1(\theta) = \frac{1}{2}\big(|\theta \cdot 1 - 2| + |\theta \cdot 2 - 4|\big).$$

Client 2:

$$F_2(\theta) = \frac{1}{3}\big(|\theta \cdot 1 - 1| + |\theta \cdot 2 - 2| + |\theta \cdot 3 - 3|\big).$$

Client 3:

$$F_3(\theta) = |\theta \cdot 2 - 5|.$$

Table 3.1 Local datasets distributed across three federated clients

Client	Sample Index	x	y
Client 1 ($\mathscr{D}_1$)	1	1	2
	2	2	4
Client 2 ($\mathscr{D}_2$)	1	1	1
	2	2	2
	3	3	3
Client 3 ($\mathscr{D}_3$)	1	2	5

Each client independently minimizes a piecewise-linear, convex objective defined solely over its local dataset.

Global Federated Objective. The global federated optimization objective is constructed as a weighted sum of the local risks:

$$F(\theta) = \sum_{k=1}^{3} \frac{n_k}{N} F_k(\theta) = \frac{2}{6}F_1(\theta) + \frac{3}{6}F_2(\theta) + \frac{1}{6}F_3(\theta).$$

Substituting the explicit expressions yields

$$F(\theta) = \frac{1}{6}\Big(|\theta - 2| + |2\theta - 4| + |\theta - 1| + |2\theta - 2| \\ + |3\theta - 3| + |2\theta - 5|\Big).$$

As in the squared-loss case, this expression is algebraically identical to the absolute-error empirical risk that would be obtained by pooling all six samples centrally. However, it has been constructed without transferring or revealing any raw data.

Loss-Agnostic Federated Formulation
The federated optimization framework is independent of the specific loss function used; smooth and non-smooth objectives can both be expressed as weighted compositions of local empirical risks.

Interpretation. This example highlights several important points:

- Federated learning supports a broad class of loss functions, including non-smooth and robust losses.
- The global objective remains a virtual construct assembled from local computations.
- Statistical heterogeneity across clients directly influences the shape and minimizers of the global objective.

The use of absolute loss also foreshadows optimization challenges that arise in federated settings when gradients are undefined or unstable, motivating the advanced methods discussed in later chapters.

3.2 Local Loss Functions and Empirical Risk Minimization

In federated learning, optimization is fundamentally decomposed across clients. Rather than minimizing a single empirical risk defined over a globally accessible dataset, each client independently performs *empirical risk minimization* (ERM) using only its locally available data. This local optimization step constitutes the primary computational workload in federated learning and serves as the sole mechanism through which raw data influences the training process.

Formally, let client k possess a local dataset

$$\mathscr{D}_k = \{(x_{k,1}, y_{k,1}), \ldots, (x_{k,n_k}, y_{k,n_k})\},$$

where $n_k = |\mathscr{D}_k|$ denotes the number of samples at client k. Given a shared model f_θ parameterized by θ and a task-specific loss function $\ell(\cdot)$, the local empirical risk at client k is defined as

$$F_k(\theta) = \frac{1}{n_k} \sum_{i=1}^{n_k} \ell\big(f_\theta(x_{k,i}), y_{k,i}\big).$$

This definition mirrors the classical empirical risk formulation used in centralized learning. From the perspective of an individual client, the learning problem is therefore indistinguishable from standard supervised learning: a model is trained by minimizing an average loss over observed data samples. No assumptions about other clients, global coordination, or data heterogeneity are required at this stage.

However, the apparent simplicity of local ERM is deceptive. Because each client optimizes its objective in isolation, the resulting updates reflect client-specific data distributions, sampling biases, and noise characteristics. Crucially, local ERM does not directly optimize the global federated objective introduced in the previous section. Instead, it provides only a partial and potentially biased approximation of the descent direction for the global loss.

Local ERM vs Global ERM
In federated learning, each client minimizes a *local* empirical risk defined over private data, while the system as a whole seeks to minimize a *global* objective that no participant can compute or observe in full.

The discrepancy between locally optimal solutions and globally optimal solutions is a defining feature of federated learning. When data distributions differ significantly across clients, local ERM updates may conflict with one another, leading to slow convergence, instability, or biased global models. Understanding this tension is essential for analyzing federated optimization algorithms and motivates many of the techniques introduced in later chapters.

Worked Example: Local Empirical Risk Minimization with Larger Client Datasets

To further illustrate the mechanics and implications of local empirical risk minimization in federated learning, we now consider a richer example involving larger local datasets at each client. This expanded setting more closely reflects practical federated environments, where clients may hold multiple observations drawn from distinct local distributions.

Federated Setup. Assume a federated system with $K = 3$ clients. Each client k holds a private dataset $\mathscr{D}_k$. We again consider a simple linear prediction model

$$f_\theta(x) = \theta x, \qquad \theta \in \mathbb{R},$$

and employ the absolute error loss

$$\ell\big(f_\theta(x), y\big) = |\theta x - y|.$$

Local Datasets. The local datasets stored at the three clients are shown in Table 3.2. The dataset sizes are therefore

$$n_1 = 4, \quad n_2 = 5, \quad n_3 = 3.$$

Table 3.2 Larger local datasets distributed across federated clients

Client	Sample	x	y
Client 1 ($\mathscr{D}_1$)	1	1	2
	2	2	4
	3	3	6
	4	4	8
Client 2 ($\mathscr{D}_2$)	1	1	1
	2	2	2
	3	3	3
	4	4	4
	5	5	5
Client 3 ($\mathscr{D}_3$)	1	1	3
	2	2	5
	3	3	7

Local Empirical Risk Functions.

Client 1. The local empirical risk at Client 1 is

$$\begin{aligned} F_1(\theta) &= \frac{1}{4}\sum_{i=1}^{4} |\theta x_{1,i} - y_{1,i}| \\ &= \frac{1}{4}\big(|\theta - 2| + |2\theta - 4| + |3\theta - 6| + |4\theta - 8|\big). \end{aligned}$$

This objective is minimized when θ is close to 2, reflecting the consistent linear relationship present in Client 1's data.

Client 2. The local empirical risk at Client 2 is

$$\begin{aligned} F_2(\theta) &= \frac{1}{5}\sum_{i=1}^{5} |\theta x_{2,i} - y_{2,i}| \\ &= \frac{1}{5}\big(|\theta - 1| + |2\theta - 2| + |3\theta - 3| + |4\theta - 4| + |5\theta - 5|\big). \end{aligned}$$

This client's objective favors values of θ near 1, consistent with its locally observed data distribution.

Client 3. The local empirical risk at Client 3 is

$$\begin{aligned} F_3(\theta) &= \frac{1}{3}\sum_{i=1}^{3} |\theta x_{3,i} - y_{3,i}| \\ &= \frac{1}{3}\big(|\theta - 3| + |2\theta - 5| + |3\theta - 7|\big). \end{aligned}$$

This objective pulls θ toward values near 2.5, illustrating a client whose data distribution differs markedly from the others.

Local Minimizers and Interpretation. Each client independently minimizes its local empirical risk $F_k(\theta)$. Due to the piecewise-linear nature of the absolute loss, the minimizers correspond to median-consistent slopes implied by the local datasets:

$$\theta_1^\star \approx 2, \quad \theta_2^\star \approx 1, \quad \theta_3^\star \approx 2.5.$$

These locally optimal solutions are clearly misaligned. Consequently, local empirical risk minimization produces update directions that reflect client-specific data characteristics rather than a unified global objective.

Insight from Larger Local Datasets
As local dataset sizes grow, client-specific empirical risks become more stable, but their optima may diverge further under non-IID data, intensifying the tension between local optimization and global generalization.

This expanded example demonstrates that increasing local dataset size does not eliminate the fundamental challenges of federated learning. Instead, it sharpens the distinction between locally optimal behavior and globally optimal model performance, reinforcing the need for careful aggregation and coordination mechanisms.

3.3 Client-Wise Model Parameters and Aggregation

Although all clients in a federated learning system employ an identical model architecture and share a common parameter space, their learning trajectories typically diverge soon after training begins. This divergence is a natural and unavoidable consequence of the decentralized setting. Each client performs optimization using its own private dataset, which may differ significantly from others in terms of size, feature distribution, label frequency, noise level, and temporal dynamics. Consequently, even when clients start from the same initial model, locally optimized parameters tend to evolve in different directions.

Let $\theta^{(t)}$ denote the global model parameters at the beginning of communication round t. This global model represents the system's current best approximation to the solution of the federated optimization problem. At the start of the round, the server broadcasts $\theta^{(t)}$ to a subset of clients selected for participation. Each participating client k initializes its local model as

$$\theta_k^{(t,0)} = \theta^{(t)}.$$

This shared initialization ensures that all clients begin local training from a common reference point, even though their subsequent optimization paths may diverge.

Each client then performs τ steps of local optimization using its own empirical risk function $F_k(\theta)$. In practice, this local training phase is often implemented using stochastic gradient descent or one of its variants, operating on mini-batches sampled from $\mathscr{D}_k$. Because the optimization is entirely local, no assumptions are made about the availability, distribution, or quality of data at other clients. After completing the prescribed number of local updates, client k obtains a locally optimized parameter vector

$$\theta_k^{(t,\tau)}.$$

Rather than transmitting raw data, gradients at every step, or detailed optimization traces, clients communicate only a compact summary of their local learning progress. This summary is typically expressed as a parameter difference, or *model update*, defined by

$$\Delta\theta_k^{(t)} = \theta_k^{(t,\tau)} - \theta^{(t)}.$$

This representation highlights how far, and in which direction, the client's locally trained model has moved away from the global initialization during the current communication round. By communicating updates instead of full models or raw data, federated learning significantly reduces communication overhead while preserving data locality.

The server aggregates the received client updates to construct the next global model. Let $\mathcal{S}^{(t)} \subseteq \{1, \ldots, K\}$ denote the set of clients that successfully participated in round t. The aggregation rule is given by

$$\theta^{(t+1)} = \theta^{(t)} + \sum_{k\in\mathcal{S}^{(t)}} \frac{n_k}{\sum_{j\in\mathcal{S}^{(t)}} n_j} \Delta\theta_k^{(t)}.$$

This update rule computes a weighted average of the client-proposed parameter changes and applies it to the current global model.

The weighting factor $\frac{n_k}{\sum_{j\in\mathcal{S}^{(t)}} n_j}$ plays a critical role in the aggregation process. It ensures that clients with larger local datasets exert proportionally greater influence on the global update, thereby approximating the effect of centralized empirical risk minimization where each data sample contributes equally. At the same time, this weighting scheme prevents clients with very small datasets from disproportionately steering the global model, which could otherwise lead to instability or bias.

Aggregation as Implicit Coordination
In federated learning, aggregation replaces explicit synchronization: independent local optimizers are coordinated solely through the weighted averaging of their parameter updates, without direct interaction or data sharing among clients.

Importantly, the aggregation mechanism must operate reliably under a range of practical constraints. In real-world federated systems, client participation is often partial and unpredictable, communication may be delayed or unreliable, and data heterogeneity may be severe. The global model therefore evolves as a compromise among multiple, sometimes conflicting, local objectives rather than as the result of a single, centrally controlled optimization process. Understanding this aggregation step is essential, as it constitutes the primary mechanism through which federated learning reconciles autonomy, privacy, and collective model improvement.

Worked Example: Client-Wise Parameters and Aggregation with Five Clients
We now present a detailed numerical example illustrating how client-wise model parameters are aggregated to form a new global model.

Federated Setup. Consider a federated system with $K = 5$ clients. We assume a scalar linear model

$$f_\theta(x) = \theta x, \quad \theta \in \mathbb{R}.$$

Suppose the global model at round t is initialized as

$$\theta^{(t)} = \theta^{(t)} = 1.5.$$

Each client performs local training and obtains an updated parameter value $\theta_k^{(t,\tau)}$.

Client Data and Local Updates. Assume the dataset sizes and locally trained parameters shown in Table 3.3.

The total number of samples across participating clients is

$$\sum_{k=1}^{5} n_k = 10 + 20 + 5 + 15 + 10 = 60.$$

Weighted Aggregation. Using the federated aggregation rule, the global update is computed as

$$\theta^{(t+1)} = \theta^{(t)} + \sum_{k=1}^{5} \frac{n_k}{60} \Delta\theta_k^{(t)}.$$

Substituting numerical values yields

$$\theta^{(t+1)} = 1.5 + \left(\frac{10}{60} \cdot 0.5 + \frac{20}{60} \cdot (-0.3) + \frac{5}{60} \cdot 1.0 + \frac{15}{60} \cdot 0.3 + \frac{10}{60} \cdot (-0.5)\right).$$

Table 3.3 Client-wise dataset sizes and local model parameters

Client k	n_k	$\theta_k^{(t,\tau)}$	$\Delta\theta_k^{(t)}$
1	10	2.0	+0.5
2	20	1.2	−0.3
3	5	2.5	+1.0
4	15	1.8	+0.3
5	10	1.0	−0.5

Evaluating each term,

$$\begin{aligned}\theta^{(t+1)} &= 1.5 + (0.0833 - 0.1000 + 0.0833 + 0.0750 - 0.0833)\\ &= 1.5 + 0.0583\\ &\approx 1.558.\end{aligned}$$

Interpretation. Although some clients push the model upward and others downward, the aggregated update reflects a data-weighted compromise. Client 3 proposes the largest positive change, but its influence is limited by its small dataset. Conversely, Client 2 exerts a strong downward pull due to its larger data volume. The resulting global model balances these competing effects.

Key Insight from Aggregation
Federated aggregation does not average client models uniformly; it weights their influence according to data volume, ensuring that global updates approximate centralized empirical risk minimization under data locality constraints.

This example demonstrates how client-wise parameter updates are combined to form a new global model and illustrates the central role of aggregation in reconciling heterogeneous local learning dynamics.

3.4 Consensus Constraints and Convergence Conditions

Federated learning can be rigorously interpreted as a *constrained optimization problem* in which multiple autonomous clients jointly seek to learn a shared model while operating on distributed, private datasets. Unlike centralized optimization—where a single parameter vector is updated using all available data—federated learning introduces *client-wise model parameters* that must be coordinated indirectly through communication and aggregation.

To formalize this idea, consider introducing a local copy of the model parameters at each client, denoted by θ_k. The objective of federated learning is not to allow these parameters to evolve independently, but to ensure that they ultimately agree on a common solution. This requirement can be expressed through explicit consensus constraints:

$$\theta_k = \theta, \quad \forall k \in \{1, 2, \dots, K\},$$

where θ represents the global model shared across clients.

Under this interpretation, the federated learning problem can be written as a constrained optimization problem:

$$\min_{\{\theta_k\}} \sum_{k=1}^{K} \frac{n_k}{N} F_k(\theta_k) \quad \text{subject to } \theta_k = \theta.$$

This formulation highlights two key aspects of federated learning. First, optimization is performed locally, as each client minimizes its own empirical risk function $F_k(\theta_k)$. Second, coordination is achieved by enforcing agreement among the local parameter vectors. In contrast to classical distributed optimization, these constraints are not enforced through continuous synchronization or shared memory. Instead, they are satisfied only approximately through periodic model aggregation.

In practical federated systems, exact enforcement of the consensus constraints at every iteration is neither feasible nor desirable. Communication is costly, client participation is intermittent, and local optimization may proceed for multiple steps between synchronization events. As a result, local parameters θ_k are allowed to temporarily deviate from one another, and consensus is achieved gradually over successive communication rounds.

Consensus via Iterative Alignment
Federated learning achieves consensus not by enforcing strict equality among local models at all times, but by repeatedly reducing disagreement through periodic aggregation and alignment.

Convergence analysis of federated learning algorithms therefore differs fundamentally from that of centralized methods. Rather than tracking the behavior of a single optimization trajectory, one must reason about the evolution of a collection of coupled local trajectories. Typical convergence guarantees rely on assumptions such as convexity of the loss functions, bounded gradient variance, sufficient client participation over time, and controlled levels of statistical heterogeneity. When these conditions hold, the iterative alignment induced by aggregation can drive the system toward a stationary point of the global objective.

Consensus constraints and convergence conditions play complementary but fundamentally different roles in federated learning (Table 3.4). Consensus constraints describe the structural requirement that local client models should ultimately agree with a shared global model, while convergence conditions specify the analytical assumptions under which this agreement leads to stable and meaningful optimization. In practice, consensus is enforced approximately through aggregation, whereas convergence conditions are invoked to reason about long-term behavior. Understanding the distinction between these two concepts is essential for interpreting both federated learning algorithms and their theoretical guarantees.

Table 3.4 Comparison between Consensus Constraints and Convergence Conditions in Federated Learning

Aspect	Consensus Constraints	Convergence Conditions
Conceptual role	Enforce agreement among client-wise model parameters by requiring $\theta_k = \theta$	Ensure that the iterative learning process approaches a stationary point or optimum of the global objective
Nature	Structural and deterministic in formulation; expressed as equality constraints in the optimization problem	Analytical and probabilistic; expressed as assumptions on functions, data, and system behavior
Mathematical formulation	Explicit constraints in the constrained optimization view of federated learning	Inequalities or boundedness assumptions used in convergence proofs
Enforcement mechanism	Approximately enforced through periodic aggregation and model synchronization	Not enforced algorithmically; assumed to hold to guarantee theoretical results
Time scale	Operates at the level of each communication round	Operates across many rounds, governing long-term learning dynamics
Dependence on data distribution	Independent of data distribution; applies regardless of IID or non-IID data	Strongly dependent on data heterogeneity, gradient variance, and sampling bias
Typical examples	Equality of local and global parameters, consensus penalties, alignment constraints	Convexity, smoothness, bounded gradients, sufficient client participation
Failure implication	Leads to persistent model disagreement across clients	Leads to divergence, oscillation, or failure to reach optimality

Worked Example: Consensus Formation through Iterative Aggregation

We now present a fully worked numerical example to illustrate, in a step-by-step manner, how consensus constraints are approximately enforced in federated learning through iterative aggregation. The objective of this example is not merely to compute the next global model, but to explicitly demonstrate how the abstract consensus formulation introduced earlier translates into concrete numerical operations.

Federated Setup. Consider a federated learning system consisting of $K = 4$ clients. We assume a scalar linear prediction model

$$f_\theta(x) = \theta x, \qquad \theta \in \mathbb{R},$$

which allows the evolution of model parameters to be visualized easily along a single dimension.

Each client k holds a private dataset of size n_k and defines a local empirical risk function $F_k(\theta)$. The dataset sizes across clients are:

$$n_1 = 20, \quad n_2 = 15, \quad n_3 = 25, \quad n_4 = 40, \qquad N = \sum_{k=1}^{4} n_k = 100.$$

These values imply that Client 4 has the largest influence on the global objective, followed by Clients 3, 1, and 2.

Suppose the global model at the start of communication round t is

$$\theta^{(t)} = 1.5.$$

Step 1: Local Initialization. At the beginning of round t, the server broadcasts the global model $\theta^{(t)}$ to all participating clients. Each client initializes its local parameter vector identically:

$$\theta_k^{(t,0)} = \theta^{(t)} = 1.5, \quad k = 1, 2, 3, 4.$$

At this point, the consensus constraints $\theta_k = \theta$ are satisfied exactly. All local models coincide with the global model.

Step 2: Local Optimization and Divergence. Each client now performs τ steps of local optimization using its own dataset and empirical risk function $F_k(\theta)$. Because the datasets differ in both size and distribution, the local optimization trajectories diverge.

After completing local training, the clients obtain the following locally optimized parameters:

$$\theta_1^{(t,\tau)} = 1.9, \quad \theta_2^{(t,\tau)} = 1.2, \quad \theta_3^{(t,\tau)} = 2.1, \quad \theta_4^{(t,\tau)} = 1.4.$$

At this stage, the consensus constraints

$$\theta_k = \theta$$

are violated. Each local parameter reflects the statistics of its corresponding dataset rather than a globally shared solution.

Step 3: Quantifying Disagreement via Local Deviations. To communicate local learning progress, each client computes the difference between its locally optimized parameter and the global model from which it was initialized. This quantity is defined as

$$\Delta\theta_k^{(t)} = \theta_k^{(t,\tau)} - \theta^{(t)}.$$

Substituting numerical values, we obtain:

$$\Delta\theta_1^{(t)} = 1.9 - 1.5 = 0.4,$$
$$\Delta\theta_2^{(t)} = 1.2 - 1.5 = -0.3,$$
$$\Delta\theta_3^{(t)} = 2.1 - 1.5 = 0.6,$$
$$\Delta\theta_4^{(t)} = 1.4 - 1.5 = -0.1.$$

These deviations provide a precise numerical measure of client-wise disagreement relative to the global model. Positive values indicate upward pressure on the global parameter, whereas negative values indicate downward pressure.

Step 4: Weighted Aggregation and Approximate Consensus. According to the federated aggregation rule, the global model for the next round is computed as

$$\theta^{(t+1)} = \theta^{(t)} + \sum_{k=1}^{4} \frac{n_k}{N} \Delta\theta_k^{(t)}.$$

We first compute the aggregation weights:

$$\frac{n_1}{N} = 0.20, \quad \frac{n_2}{N} = 0.15, \quad \frac{n_3}{N} = 0.25, \quad \frac{n_4}{N} = 0.40.$$

Multiplying each deviation by its corresponding weight yields the weighted contributions:

$$0.20 \cdot 0.4 = 0.08,$$
$$0.15 \cdot (-0.3) = -0.045,$$
$$0.25 \cdot 0.6 = 0.15,$$
$$0.40 \cdot (-0.1) = -0.04.$$

Summing these contributions gives the net update:

$$0.08 - 0.045 + 0.15 - 0.04 = 0.145.$$

Finally, the updated global model is

$$\theta^{(t+1)} = 1.5 + 0.145 = 1.645.$$

Step 5: Interpretation and Reduction of Disagreement. The new global model $\theta^{(t+1)} = 1.645$ lies strictly between the locally optimized parameters proposed by the clients. Clients with larger datasets exert stronger influence on the aggregated update, while clients with smaller datasets have proportionally reduced impact.

Although exact consensus is not achieved in a single communication round, the aggregation step produces a global parameter that is closer to the majority of client solutions. In the next round, all participating clients will reinitialize their local models as

$$\theta_k^{(t+1,0)} = \theta^{(t+1)},$$

thereby reducing the magnitude of subsequent deviations.

Key Insight from the Example
Consensus in federated learning is not enforced instantaneously. Instead, disagreement is allowed to emerge locally but is progressively attenuated through repeated, data-weighted aggregation across communication rounds.

This example demonstrates concretely how the abstract consensus constraints in the federated optimization formulation are realized in practice and why convergence must be understood as a gradual, iterative, and statistical process rather than immediate agreement across clients.

3.5 Relationship Between Local and Global Minima

A defining theoretical and practical challenge in federated learning stems from the fact that optimization is carried out locally, while performance is evaluated globally. Although all clients collaborate toward a shared objective, the minimizers of their local empirical risk functions generally do not coincide with the minimizer of the global federated objective. This fundamental misalignment distinguishes federated learning from centralized empirical risk minimization and lies at the heart of many observed convergence and stability issues.

Formally, recall that each client k defines a local empirical risk function $F_k(\theta)$, while the global federated objective is defined as

$$F(\theta) = \sum_{k=1}^{K} \frac{n_k}{N} F_k(\theta), \quad N = \sum_{k=1}^{K} n_k.$$

Let the minimizers of the local and global objectives be given by

$$\theta_k^\star = \arg\min_\theta F_k(\theta), \qquad \theta^\star = \arg\min_\theta F(\theta).$$

In centralized learning with IID data, all local datasets can be viewed as samples from a common distribution, and local minimizers tend to cluster around the global minimizer. In federated learning, however, data is often statistically heterogeneous.

Different clients may observe distinct feature distributions, label frequencies, noise levels, or task contexts. Consequently, the shapes of the local loss landscapes $F_k(\theta)$ may differ substantially across clients.

Under such non-IID conditions, the local minimizers $\theta_k^\star$ can be widely dispersed in parameter space. Each local optimizer therefore identifies a solution that is optimal for its own data distribution but potentially suboptimal—or even detrimental—for the population-level objective. When these locally optimal updates are aggregated, they may exert competing forces on the global model, slowing convergence or inducing oscillatory behavior.

Local–Global Mismatch
In federated learning, local optimality does not imply global optimality. The discrepancy between $\theta_k^\star$ and $\theta^\star$ is a defining feature of decentralized optimization under data heterogeneity.

Understanding the relationship between local and global minima is therefore essential for designing robust federated optimization algorithms. Many advanced methods—such as proximal regularization, drift correction, and personalized models—can be interpreted as mechanisms for controlling or compensating for this mismatch.

Figure 3.1 illustrates the geometric relationship between local and global minima in federated learning. Each dashed curve represents a client-specific empirical risk function $F_k(\theta)$ with its own minimizer $\theta_k^\star$. Due to statistical heterogeneity, these minima are spatially separated. The solid curve corresponds to the global federated objective $F(\theta)$, whose minimizer $\theta^\star$ lies between the local optima, reflecting a weighted compromise across clients. This visualization highlights why locally optimal updates may conflict and why federated optimization must carefully balance local fidelity and global generalization.

Table 3.5 highlights the fundamental distinction between local and global optimality in federated learning. Each client optimizes its own empirical risk and therefore converges toward a local minimum that reflects its private data distribution. In contrast, the global minimum corresponds to the minimizer of the weighted federated objective and represents a population-level compromise across clients.

Under non-IID data, the dispersion of local minima in parameter space can be substantial, causing locally optimal updates to conflict with global optimization. This mismatch explains why federated learning cannot simply aggregate locally optimal models and motivates the need for coordination mechanisms, regularization, and adaptive aggregation strategies to guide the system toward the global minimum.

Worked Example: Local Minima versus Global Minimum
We now present a fully worked numerical example to explicitly demonstrate how local minimizers differ from the global minimizer and how the federated objective reconciles these competing solutions.

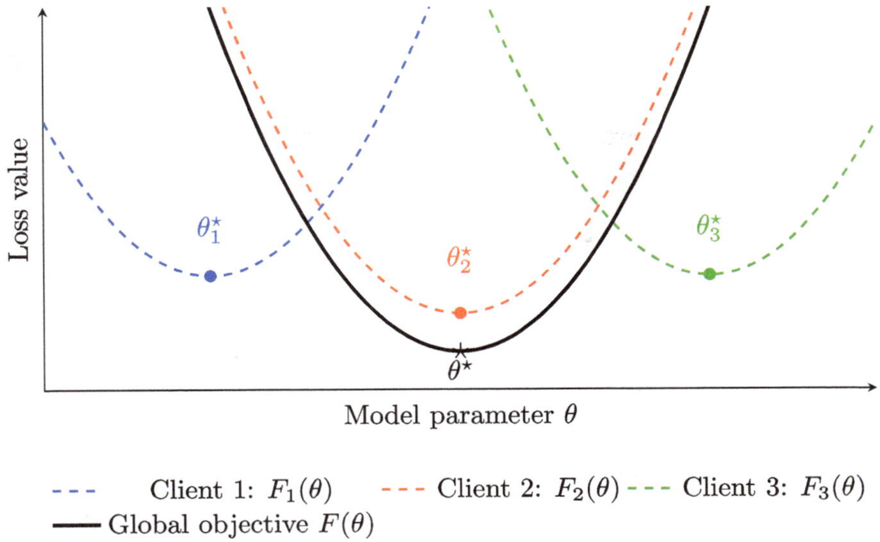

Fig. 3.1 Illustration of the relationship between local minima and the global minimum in federated learning. Each client's empirical risk function has a distinct minimizer $\theta_k^\star$, while the global objective attains its minimum at $\theta^\star$, representing a weighted compromise across clients

Table 3.5 Comparison between Local Minima and Global Minimum in Federated Learning

Aspect	Local Minimum ($\theta_k^\star$)	Global Minimum ($\theta^\star$)
Definition	Minimizer of the client-specific empirical risk $F_k(\theta)$	Minimizer of the federated objective $F(\theta) = \sum_{k=1}^{K} \frac{n_k}{N} F_k(\theta)$
Scope of optimization	Single client's local dataset $\mathcal{D}_k$	Aggregated contribution of all client datasets
Dependence on data distribution	Strongly dependent on client-specific (possibly non-IID) data	Reflects a weighted combination of heterogeneous client distributions
Uniqueness across clients	Generally different for each client	Unique with respect to the global objective (under standard assumptions)
Role in learning dynamics	Drives local update directions during on-device training	Represents the target solution that federated learning seeks to approximate
Effect under non-IID data	May be widely dispersed in parameter space	Lies between local minima as a compromise solution
Optimization implication	Locally optimal but not globally optimal	Globally optimal but not optimal for any single client

Problem Setup. Consider a federated system with $K = 3$ clients. We assume a scalar linear model

$$f_\theta(x) = \theta, \quad \theta \in \mathbb{R},$$

and use a simple squared-error loss

$$\ell(f_\theta(x), y) = (\theta - y)^2,$$

which allows closed-form solutions for all minimizers.

Each client k observes labels drawn from a different distribution. The local datasets are summarized as follows:

Client k	Dataset $\mathscr{D}_k$ (labels only)	n_k
1	{1, 2, 2}	3
2	{4, 5}	2
3	{7, 8, 9}	3

The total number of samples is

$$N = 3 + 2 + 3 = 8.$$

Step 1: Local Empirical Risk Functions. For each client, the local empirical risk is

$$F_k(\theta) = \frac{1}{n_k} \sum_{i=1}^{n_k} (\theta - y_{k,i})^2.$$

Step 2: Compute Local Minimizers. Each local objective is minimized when θ equals the mean of the client's labels.

- Client 1:

$$\theta_1^\star = \frac{1 + 2 + 2}{3} = \frac{5}{3} \approx 1.67.$$

- Client 2:

$$\theta_2^\star = \frac{4 + 5}{2} = 4.5.$$

- Client 3:

$$\theta_3^\star = \frac{7 + 8 + 9}{3} = 8.$$

These values represent the locally optimal solutions preferred by each client in isolation.

Step 3: Global Federated Objective. The global federated objective is

$$F(\theta) = \sum_{k=1}^{3} \frac{n_k}{N} F_k(\theta).$$

For squared loss, the global minimizer is the weighted average of all labels across clients:

$$\theta^\star = \frac{1}{N} \sum_{k=1}^{3} \sum_{i=1}^{n_k} y_{k,i}.$$

Substituting numerical values:

$$\theta^\star = \frac{1+2+2+4+5+7+8+9}{8} = \frac{38}{8} = 4.75.$$

Step 4: Comparing Local and Global Minima. We observe that

$$\theta_1^\star \approx 1.67, \quad \theta_2^\star = 4.5, \quad \theta_3^\star = 8, \quad \theta^\star = 4.75.$$

The global minimizer does not coincide with any local minimizer. Instead, it lies between them, balancing the influence of all clients according to their dataset sizes.

Step 5: Interpretation. Each local minimizer is optimal only for its corresponding client. If the global model were to move too close to $\theta_1^\star$, it would perform poorly on Clients 2 and 3. Conversely, moving toward $\theta_3^\star$ would disadvantage Clients 1 and 2. The global minimizer $\theta^\star$ represents a compromise solution that minimizes the total weighted empirical risk across all clients.

Insight from the Example
The global minimizer in federated learning is generally not equal to any local minimizer. Instead, it emerges as a weighted compromise shaped by data volume and distribution across clients.

This example makes explicit why federated optimization cannot simply follow locally optimal directions and why naïve aggregation of local solutions may lead to instability. Subsequent chapters build upon this insight to introduce algorithmic mechanisms that control local drift and improve global convergence under heterogeneous data.

3.6 Optimization Challenges Under Non-IID Data

One of the most profound departures of federated learning from classical centralized optimization arises from the presence of *non-identically distributed (non-IID) data* across clients. In centralized learning, stochastic gradients computed on mini-batches are typically assumed to be unbiased estimators of the global gradient. This assumption underpins much of the convergence theory for gradient-based optimization methods. In federated learning, however, this assumption is routinely violated.

Formally, recall that the global federated objective is defined as

$$F(\theta) = \sum_{k=1}^{K} \frac{n_k}{N} F_k(\theta), \quad N = \sum_{k=1}^{K} n_k.$$

The gradient of the global objective can therefore be written as

$$\nabla F(\theta) = \sum_{k=1}^{K} \frac{n_k}{N} \nabla F_k(\theta).$$

In an idealized IID setting, local gradients $\nabla F_k(\theta)$ would fluctuate randomly around the global gradient $\nabla F(\theta)$, and their aggregation would yield a reliable descent direction. Under non-IID data, however, local gradients may be systematically biased. Each client's gradient reflects the statistical properties of its own data distribution, which may differ substantially from those of other clients. As a result, local gradients are no longer interchangeable samples of a common underlying gradient field.

This gradient bias has several important consequences. First, the variance between local gradients can become large, even when each client computes its gradient accurately. Second, locally optimal descent directions may conflict, causing the aggregated update to be suboptimal or unstable. Third, increasing the number of local training steps can exacerbate these effects by allowing local models to drift further toward client-specific optima before aggregation.

When data heterogeneity is severe, these phenomena can dominate the optimization process, leading to a range of pathological behaviors, including:

- **Slow or oscillatory convergence:** Under non-IID data, local gradient directions may be poorly aligned across clients. As a result, successive aggregation steps may alternately favor different subsets of clients, causing the global model to oscillate or converge very slowly. This behavior is especially pronounced when local training is performed for multiple steps between communication rounds.
- **Client drift during local training:** Because each client optimizes its own empirical risk, local models may drift toward client-specific optima that differ substantially from the global solution. Excessive local updates amplify this drift, increasing disagreement at aggregation time and reducing the effectiveness of parameter averaging as a consensus mechanism.

- **Degradation of global generalization:** The aggregated model may become biased toward dominant clients with larger datasets or more frequent participation. This can degrade performance on minority clients or rare data regimes, leading to poor population-level generalization despite strong local performance.
- **Gradient bias and update inconsistency:** Local gradients may act as biased estimators of the global gradient, causing aggregated updates to deviate systematically from the true descent direction. This bias undermines classical convergence guarantees and complicates the analysis of learning dynamics.
- **Sensitivity to client participation patterns:** Partial, intermittent, or skewed client participation can interact with non-IID data to further destabilize optimization. Clients that participate infrequently may be underrepresented in the global model, even if their data distributions are important.
- **Trade-off between communication efficiency and stability:** Increasing the number of local optimization steps reduces communication overhead but exacerbates drift and instability under heterogeneous data. Finding an appropriate balance between local computation and global aggregation is therefore a central design challenge.

These challenges highlight a fundamental tension in federated learning. Allowing extensive local computation improves communication efficiency, but it also amplifies the divergence between local and global objectives under non-IID data. Conversely, frequent aggregation reduces drift but increases communication cost. Balancing this trade-off is central to the design of effective federated optimization algorithms.

Non-IID Data as the Core Difficulty
Most theoretical limitations and practical instabilities in federated learning arise from statistical heterogeneity across clients, which breaks the unbiased gradient assumptions underlying classical optimization theory.

Importantly, non-IID data is not an anomaly but a defining feature of real-world federated systems. Differences in user behavior, device usage patterns, geographic location, institutional practices, and temporal dynamics all contribute to heterogeneity. As a result, federated learning algorithms must be designed to operate robustly under persistent and structured distributional mismatch rather than assuming it away.

Subsequent chapters build directly upon this observation. Techniques such as proximal regularization, adaptive and robust aggregation, client clustering, and personalized modeling can all be interpreted as mechanisms for mitigating the adverse effects of non-IID data while preserving the core principles of federated learning.

Exercises

This chapter developed a rigorous mathematical foundation for federated learning, formalizing global objectives, local empirical risks, aggregation mechanisms, consensus constraints, and the impact of non-IID data on optimization and convergence.

A. Short Answer Questions

Answer each question briefly and precisely.

1. What is meant by the *global federated optimization objective*?
2. Why is the global objective in federated learning described as a *virtual* objective?
3. Define the local empirical risk function $F_k(\theta)$.
4. What role does the weighting factor $\frac{n_k}{N}$ play in the global objective?
5. What is meant by *client drift* in federated learning?
6. Explain the purpose of model aggregation in federated learning.
7. What are consensus constraints in the context of federated optimization?
8. Why do local minimizers generally differ from the global minimizer under non-IID data?
9. How does partial client participation affect convergence?
10. What is meant by statistical heterogeneity across clients?

B. Long Answer Questions

Answer the following questions with clear explanations, equations, and diagrams wherever appropriate.

1. Derive the global federated optimization objective starting from client-wise empirical risk minimization.
2. Explain the relationship between local ERM and global ERM in federated learning, highlighting the sources of mismatch.
3. Discuss the constrained optimization view of federated learning using consensus constraints.
4. With the help of a worked example, explain how weighted aggregation approximates centralized empirical risk minimization.
5. Analyze the relationship between local minima and the global minimum under non-IID data.
6. Explain why increasing the number of local update steps may degrade convergence under data heterogeneity.

7. Discuss the key convergence assumptions commonly used in federated learning theory.
8. Explain the role of aggregation frequency in balancing communication cost and optimization stability.

C. True/False Questions

State whether each statement is **True** or **False**.

1. In federated learning, the global objective can be computed exactly at the server.
2. Local empirical risk minimization is identical to global empirical risk minimization.
3. Consensus constraints require that all local models remain exactly equal at every iteration.
4. Under IID data, local minimizers tend to cluster around the global minimizer.
5. Non-IID data can introduce bias in local gradients.
6. Clients with larger datasets have greater influence during aggregation.
7. Increasing local computation always improves convergence speed.
8. Federated learning can be viewed as a form of constrained optimization.
9. Client drift becomes more severe when local training is extended.
10. The global minimizer is always optimal for every individual client.

Answers:

1. False
2. False
3. False
4. True
5. True
6. True
7. False
8. True
9. True
10. False

D. Fill in the Blanks

Fill in the blanks with the most appropriate term or expression.

1. The total number of samples across all clients is denoted by ____________.
2. The local empirical risk at client k is represented by ____________.
3. In federated learning, raw data always remains ____________.
4. Model aggregation is typically performed using a ____________ average.

5. Statistical differences across clients are referred to as ____________ data.
6. Client-wise parameter updates are denoted by ____________.
7. The equality $\theta_k = \theta$ represents a ____________ constraint.
8. The phenomenon where local models move toward client-specific optima is called ____________.
9. Under non-IID data, local gradients may be ____________ estimators of the global gradient.
10. The global minimizer represents a weighted ____________ across clients.

Answers:

1. N
2. $F_k(\theta)$
3. local
4. weighted
5. non-IID
6. $\Delta\theta_k$
7. consensus
8. client drift
9. biased
10. compromise

E. Multiple Choice Questions (MCQs)

Choose the **most appropriate** answer.

1. The global federated objective is defined as:
   (a) Sum of unweighted local losses
   (b) Maximum of all local losses
   (c) Weighted sum of local empirical risks
   (d) Product of local objectives
2. Which quantity is never directly computed in federated learning?
   (a) Local gradient
   (b) Local empirical risk
   (c) Global objective
   (d) Model update
3. Client drift primarily arises due to:
   (a) Model initialization
   (b) Excessive aggregation
   (c) Local optimization on heterogeneous data
   (d) Secure communication
4. Consensus in federated learning is achieved through:
   (a) Exact synchronization at every step
   (b) Data sharing across clients

   (c) Periodic aggregation
   (d) Centralized retraining
5. Under non-IID data, local gradients are:
   (a) Always unbiased
   (b) Identical across clients
   (c) Independent of data distribution
   (d) Potentially biased
6. Increasing local steps typically:
   (a) Reduces client drift
   (b) Improves stability under non-IID data
   (c) Increases communication cost
   (d) Amplifies divergence across clients
7. The weighting factor in aggregation depends on:
   (a) Learning rate
   (b) Dataset size
   (c) Model architecture
   (d) Communication delay
8. The global minimizer $\theta^\star$ is:
   (a) Equal to all local minimizers
   (b) Optimal for one client only
   (c) A compromise across clients
   (d) Independent of data distribution
9. Which assumption is commonly required for convergence analysis?
   (a) Infinite bandwidth
   (b) Exact consensus at each step
   (c) Bounded gradient variance
   (d) Identical client datasets
10. Federated learning lies conceptually between:
   (a) Supervised and unsupervised learning
   (b) Online and batch learning
   (c) Centralized ERM and decentralized consensus optimization
   (d) Convex and non-convex optimization

Answers:

1. (c)
2. (c)
3. (c)
4. (c)
5. (d)
6. (d)
7. (b)
8. (c)
9. (c)
10. (c)

F. Application and Project-Based Questions

These problems emphasize modeling, reasoning, and implementation.

1. Simulate federated learning with two clients and IID data and compare convergence with centralized training.
2. Design a toy federated dataset with non-IID label distributions and analyze client drift numerically.
3. Implement weighted aggregation and study the effect of unequal dataset sizes.
4. Compare convergence behavior for different numbers of local update steps τ.
5. Visualize local and global loss landscapes for heterogeneous clients.
6. Implement a simple federated averaging algorithm from scratch.
7. Study the impact of partial client participation on convergence.
8. Compare absolute loss and squared loss in a federated setting.
9. Propose a regularization strategy to reduce local–global mismatch.
10. Design an experiment to evaluate generalization under non-IID data.

Part II

Federated Learning Architectures and Training Protocols

Federated Learning System Architectures 4

The mathematical formulation of federated learning developed in the preceding chapters deliberately abstracts away many system-level details in order to foreground the essential optimization structure of the paradigm. By expressing federated learning in terms of local empirical risk functions, weighted aggregation, and approximate consensus constraints, the analysis isolates the fundamental learning dynamics that govern convergence under data decentralization and statistical heterogeneity. Such abstraction is indispensable for theoretical clarity, allowing the behavior of federated algorithms to be reasoned about independently of any particular implementation.

In practice, however, federated learning is not realized as a purely mathematical construct. It unfolds within complex, large-scale computational systems composed of autonomous and heterogeneous clients, unreliable networks, and coordination mechanisms operating under strict privacy and resource constraints. Concrete *system architectures* determine how and when clients participate, how models and updates are communicated, how failures and delays are absorbed, and how learning remains stable as the number of participants grows. These architectural decisions are inseparable from learning behavior: they influence the effective optimization trajectory, the degree of model drift, and the statistical representativeness of aggregated updates.

This chapter therefore serves as a critical bridge between federated learning as an optimization problem and federated learning as an operational system. We examine the architectural foundations that enable collaborative learning to proceed without centralized data access, translating abstract notions such as local objectives, aggregation operators, and consensus alignment into concrete mechanisms involving communication protocols, orchestration strategies, and fault-tolerant system design. Rather than treating architecture as an implementation detail layered on top of theory, we emphasize that architectural structure actively shapes the mathematical properties of federated optimization.

P. P. Ray, *Foundations of Federated Learning*,
Springer Series in the Data Sciences,
https://doi.org/10.1007/978-3-032-34084-9_4

Throughout this chapter, we highlight the bidirectional interaction between system architecture and learning dynamics. Decisions regarding client selection, communication topology and synchronization frequency directly affect convergence speed, robustness under non-IID data, and scalability to massive client populations. Conversely, insights from optimization theory constrain architectural design, informing how much local computation can be performed, how updates should be aggregated, and how partial participation can be handled without compromising global learning objectives. Understanding this interplay is essential for designing federated learning systems that are not only theoretically well-founded, but also robust, efficient, and deployable in real-world environments.

4.1 Centralized Federated Learning Architecture

The most widely adopted and practically deployed form of federated learning is the *centralized federated learning architecture*, often described as a centralized orchestration model. Despite the presence of a central coordinating entity, this architecture differs fundamentally from classical centralized machine learning. Its defining feature is that *coordination is centralized, while data remains fully decentralized.*

In this architecture, a single logical entity—commonly referred to as the *central server*—is responsible for coordinating the learning process across a population of autonomous clients. The server does not store, inspect, or process any raw training data. Instead, it acts as an intermediary that facilitates communication, synchronization, and aggregation among clients that otherwise operate independently and in isolation.

Formally, consider a federated learning system consisting of K clients indexed by $k \in \{1, 2, \dots, K\}$ and a single server. The server maintains the global model parameters $\theta^{(t)}$ at the beginning of communication round t. Each client k holds a private local dataset $\mathcal{D}_k$, which remains permanently stored on the client device or institutional infrastructure.

Clients and the server are connected through a star-shaped communication topology. All information exchange occurs exclusively between individual clients and the server; direct communication among clients is neither required nor assumed. This topology simplifies coordination and allows the system to scale to large numbers of participants, albeit at the cost of increased reliance on the server as a communication hub.

The training process unfolds over a sequence of discrete communication rounds. At each round t, the following steps are performed:

1. The server broadcasts the current global model $\theta^{(t)}$ to a selected subset of available clients. This broadcast establishes a common reference point from which all participating clients begin local optimization.
2. Each selected client performs local training using its private dataset $\mathcal{D}_k$, typically by minimizing its local empirical risk function $F_k(\theta)$ through one or more steps

of a gradient-based optimization algorithm. No assumptions are made about the size, distribution, or quality of data held by different clients.
3. Upon completion of local training, each client computes a compact representation of its learning progress, commonly expressed as a parameter update $\Delta\theta_k^{(t)}$, and transmits this update to the server. Raw data samples and intermediate training artifacts are never shared.
4. The server aggregates the received client updates using a predefined aggregation rule to construct the next global model $\theta^{(t+1)}$, which serves as the starting point for the subsequent communication round.

This centralized federated architecture preserves strict data locality while enabling coordinated global learning across heterogeneous and autonomous clients. Although the server occupies a central position in the communication graph, it does not possess any privileged access to training data and cannot reconstruct client datasets under the standard federated learning model. As a result, centralized federated learning should not be conflated with centralized data processing or data aggregation.

Architectural Principle
Centralized federated learning centralizes *coordination*, not *data*. The server functions as an orchestration and aggregation mechanism, while all raw data remains permanently local to the clients.

From a broader perspective, the centralized federated architecture represents a pragmatic compromise between full decentralization and centralized control. It provides a clear coordination mechanism, simplifies algorithm design, and enables theoretical analysis, while simultaneously respecting data sovereignty and privacy constraints. For these reasons, it serves as the architectural foundation for many foundational federated learning algorithms and practical deployments, and forms the baseline against which decentralized and peer-to-peer federated architectures are evaluated in subsequent chapters.

Figure 4.1 illustrates the centralized federated learning architecture and the associated flow of information between the coordinating server and a population of distributed clients. The diagram provides a symbolic representation of how global model parameters are disseminated and how locally computed updates are returned for aggregation, while preserving strict data locality.

At the top of the architecture, the *central server* maintains the current global model parameters $\theta^{(t)}$ at communication round t. The server functions as a coordination hub and does not possess or access any client-side datasets. Its role is limited to broadcasting model parameters and aggregating client updates.

The lower portion of the figure depicts a set of representative clients, each associated with its own private local dataset $\mathcal{D}_k$, for $k \in \{1, 2, \dots, N\}$. These datasets remain

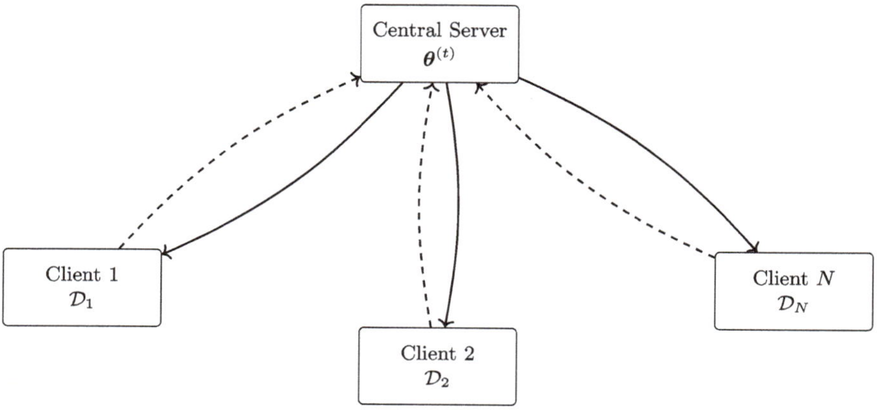

Fig. 4.1 Centralized Federated Learning Architecture with Symbolic Communication Channels

permanently local and are never transmitted outside the client boundary. Clients are depicted as independent entities, emphasizing the absence of direct peer-to-peer communication in the centralized architecture.

Solid arrows originating from the server and pointing toward the clients represent the *broadcast operation*. During each communication round, the server transmits the current global model $\theta^{(t)}$ to the selected clients. This broadcast establishes a common initialization point for local training and ensures that all participating clients begin optimization from the same model state.

Dashed arrows pointing from the clients back to the server represent the *update transmission phase*. After performing local optimization on their respective datasets $\mathscr{D}_k$, clients compute model updates $\Delta\theta_k^{(t)}$ that summarize their learning progress. Only these updates—rather than raw data or intermediate gradients—are communicated to the server.

The bidirectional communication pattern highlighted in the figure captures the iterative nature of federated learning. Each round consists of a downward flow of global parameters followed by an upward flow of locally derived updates. The server aggregates the received updates to produce the next global model $\theta^{(t+1)}$, thereby partially realigning local models and advancing the system toward consensus.

4.2 Role of the Central Server as Orchestrator

In centralized federated learning architectures, the central server plays a fundamentally different role from that of a traditional optimizer in centralized machine learning. Rather than directly computing gradients or updating model parameters using access to training data, the server functions as an *orchestrator* that coordinates a collection of autonomous and heterogeneous learners. Its primary responsibility is to regulate the flow of information and synchronize learning progress across clients, while strictly respecting data locality, privacy constraints, and system-level limitations.

The orchestration role of the server encompasses a range of coordinated tasks that collectively enable global learning to emerge from independent local computations. Concretely, the server is responsible for:

- Maintaining and disseminating the current global model parameters $\theta^{(t)}$ that serve as a common reference point for all participating clients,
- Selecting an appropriate subset of clients for participation in each communication round, taking into account availability, resource constraints, and system scalability,
- Scheduling communication rounds and managing temporal aspects of the training process, including synchronization windows and update deadlines,
- Aggregating client-provided updates using a predefined rule that approximates progress toward the global learning objective,
- Managing model versioning, convergence monitoring, and termination conditions that determine when training should stop.

From a mathematical standpoint, the server's aggregation operation serves as an approximation to a descent step for the *virtual global objective*

$$F(\theta) = \sum_{k=1}^{K} \frac{n_k}{N} F_k(\theta),$$

even though this objective is never evaluated explicitly by any single entity in the system. The server relies on locally computed information—often delayed, partial, and statistically biased due to non-IID data—to infer a direction of improvement for the shared model.

Crucially, the server does not compute gradients of $F(\theta)$ directly, nor does it enforce strict synchronization among clients. Local optimization proceeds independently at each client, potentially for multiple steps between communication rounds. The server merely combines the resulting updates using weighted aggregation, thereby implicitly coordinating the trajectories of multiple local optimizers without dictating their internal behavior.

This separation of responsibilities is central to the federated learning paradigm. By relinquishing direct control over optimization and instead focusing on orchestration, the server enables scalable, privacy-preserving, and robust collaborative learning across diverse and unreliable participants. At the same time, this indirect coordination introduces new challenges, as the server must operate under uncertainty, partial participation, and statistical heterogeneity.

Orchestration vs Optimization
In federated learning, the server governs *when* communication occurs and *how* client updates are combined, while the *process of optimization itself* remains fully local and client-driven.

From this perspective, the central server can be viewed as a mediator between optimization theory and system realities. It translates abstract global objectives into coordinated local activities, ensuring that decentralized learning dynamics remain aligned over time without compromising data autonomy. Understanding the server's orchestration role is therefore essential for analyzing convergence behavior, system robustness, and scalability in federated learning.

Table 4.1 clarifies the fundamental distinction between *orchestration* and *optimization* in federated learning by separating system-level coordination from algorithmic learning dynamics. Although both concepts contribute to the overall training process, they operate at different levels of abstraction and are governed by fundamentally different constraints.

Table 4.1 Difference between Orchestration and Optimization in Federated Learning

Aspect	Orchestration (Server-Centric)	Optimization (Client-Centric)
Primary role	Coordinates the learning process across multiple autonomous clients	Minimizes a loss function using locally available data
Control authority	Governs communication, scheduling, and aggregation	Governs parameter updates and gradient computation
Access to data	No access to raw training data or local samples	Full access to private local dataset $\mathcal{D}_k$
Mathematical object handled	Virtual global objective $F(\theta) = \sum_{k=1}^{K} \frac{n_k}{N} F_k(\theta)$	Local empirical risk $F_k(\theta)$
Gradient computation	Does not compute gradients explicitly	Computes gradients or subgradients of $F_k(\theta)$
Synchronization behavior	Manages partial, intermittent, and asynchronous participation	Performs local updates independently of other clients
Time scale of operation	Operates at the level of communication rounds	Operates at the level of local optimization steps
Responsiveness to heterogeneity	Aggregates statistically heterogeneous updates	Directly reflects client-specific data distributions
Failure tolerance	Designed to tolerate dropouts, delays, and partial updates	Sensitive only to local computation and data quality
Typical mechanisms	Client selection, aggregation rules, scheduling policies	Gradient descent, stochastic optimization, regularization
Conceptual viewpoint	System-level coordination under decentralization	Algorithmic minimization of a loss function

Orchestration is primarily a server-centric, system-level function. As shown in the table, the server does not minimize a loss function or compute gradients directly. Instead, it coordinates when communication occurs, which clients participate, and how locally computed updates are aggregated. The server's interaction with the learning objective is indirect: it operates on a *virtual global objective* $F(\Theta)$ that is never explicitly evaluated, relying instead on partial, delayed, and statistically heterogeneous updates received from clients. This makes orchestration inherently tolerant to failures, delays, and intermittent participation.

In contrast, optimization is a client-centric, algorithmic process. Each client performs empirical risk minimization on its private dataset $\mathscr{D}_k$ by directly minimizing the local objective $F_k(\Theta)$. Clients compute gradients or subgradients, apply local update rules, and control the fine-grained dynamics of parameter adjustment. These operations are sensitive to data distribution, noise, and model architecture, but largely independent of the global system state.

The table further highlights that orchestration and optimization operate on different time scales. Orchestration unfolds over communication rounds and is constrained by network conditions and system scalability, whereas optimization proceeds over local iteration steps governed by algorithmic design. This separation of concerns is a defining characteristic of federated learning and explains why classical centralized optimization concepts cannot be directly applied at the system level.

Taken together, Table 4.1 reinforces a key conceptual insight of federated learning: global learning emerges from the interaction of locally optimized models that are coordinated, but not controlled, by a central orchestrator. Understanding this distinction is essential for analyzing convergence, designing robust aggregation strategies, and reasoning about scalability and fault tolerance in federated systems.

4.3 Client Selection and Scheduling Mechanisms

In large-scale federated learning systems, the population of potential participants may consist of thousands or even millions of clients. Under such conditions, it is neither feasible nor desirable to involve all clients in every communication round. Limitations in communication bandwidth, device availability, energy constraints, and system responsiveness necessitate selective participation. As a result, *client selection and scheduling* become central architectural components that directly influence both system efficiency and learning dynamics.

Formally, let $\mathscr{S}^{(t)} \subseteq \{1, \dots, K\}$ denote the set of clients selected to participate in communication round t. Typically, the number of participating clients satisfies $|\mathscr{S}^{(t)}| \ll K$, reflecting the fact that only a small fraction of the overall client population contributes updates at any given round. The composition of $\mathscr{S}^{(t)}$ may vary from round to round, depending on system conditions and scheduling policies.

A wide range of client selection strategies has been proposed in the literature, each reflecting different trade-offs between fairness, efficiency, and convergence:

- **Uniform random sampling**, in which clients are selected independently with equal probability, providing simplicity and unbiased participation over time;
- **Availability-based selection**, where only clients that are online, idle, or connected through reliable communication links are considered eligible;
- **Resource-aware or priority-based scheduling**, which favors clients with sufficient computational power, energy reserves, or favorable network conditions;
- **Data-aware or performance-driven selection**, in which clients are chosen based on properties of their local datasets, past contribution quality, or impact on global model performance.

From a mathematical perspective, partial participation introduces additional stochasticity into the aggregation process. The global model update at round t is computed as

$$\theta^{(t+1)} = \theta^{(t)} + \sum_{k \in \mathcal{S}^{(t)}} \frac{n_k}{\sum_{j \in \mathcal{S}^{(t)}} n_j} \Delta\theta_k^{(t)},$$

where the randomness arises both from the composition of $\mathcal{S}^{(t)}$ and from the statistical heterogeneity of client datasets. As a consequence, the aggregated update represents only a stochastic approximation of the descent direction associated with the global federated objective.

While partial participation is essential for scalability and practical deployment, it also complicates theoretical analysis. Because not all clients contribute updates at every round, the global optimization trajectory becomes a random process rather than a deterministic one. Convergence guarantees must therefore be expressed in probabilistic terms, such as convergence in expectation or convergence with high probability, rather than exact per-round improvement.

Statistical Participation

Federated learning relies on repeated, randomized participation of clients over time. Convergence emerges statistically from the aggregate effect of many partial updates, rather than from full synchronization at each communication round.

Viewed collectively, client selection and scheduling mechanisms act as the interface between system-level constraints and optimization theory. They determine which local objectives influence the global model at each round and, over time, shape the effective optimization landscape explored by the federated learning process. Careful design of these mechanisms is therefore essential for achieving both scalable deployment and reliable convergence under heterogeneous and decentralized conditions.

Worked Example: Client Selection and Scheduled Aggregation

We now present a detailed numerical example to illustrate how client selection and scheduling affect the aggregation process in federated learning. The example explicitly demonstrates how partial participation leads to a stochastic approximation of the global update.

System Setup. Consider a federated learning system with $K = 5$ clients. Each client k holds a private dataset $\mathscr{D}_k$ containing n_k samples. At communication round t, the server selects a subset of clients $\mathscr{S}^{(t)}$ for participation.

Table 4.2 summarizes the client datasets and their locally computed parameter updates after local training.

Step 1: Client Selection. Assume that due to communication and availability constraints, the server selects only a subset of clients for round t. Let

$$\mathscr{S}^{(t)} = \{2, 4, 5\}.$$

Only these clients will contribute updates during this round.

Step 2: Compute Total Participating Data Size. The total number of data samples among the selected clients is

$$\sum_{k \in \mathscr{S}^{(t)}} n_k = n_2 + n_4 + n_5 = 200 + 250 + 300 = 750.$$

Step 3: Compute Aggregation Weights. Each participating client is assigned a weight proportional to its dataset size:

$$w_k = \frac{n_k}{\sum_{j \in \mathscr{S}^{(t)}} n_j}.$$

Thus,

$$w_2 = \frac{200}{750} \approx 0.267, \quad w_4 = \frac{250}{750} \approx 0.333, \quad w_5 = \frac{300}{750} = 0.400.$$

Table 4.2 Client dataset sizes and local updates at round t

Client k	**Dataset size n_k**	**Local update $\Delta\theta_k^{(t)}$**
1	100	[0.20, –0.10]
2	200	[0.05, –0.05]
3	150	[0.10, –0.20]
4	250	[–0.05, –0.10]
5	300	[0.00, –0.15]

Step 4: Weighted Aggregation of Updates. The global model update is computed as

$$\Delta\theta^{(t)} = \sum_{k\in\mathcal{S}^{(t)}} w_k \Delta\theta_k^{(t)}.$$

Substituting the values:

$$\begin{aligned}\Delta\theta^{(t)} &= 0.267[\,0.05, -0.05\,] + 0.333[-0.05, -0.10\,] + 0.400[\,0.00, -0.15\,]\\ &= [\,0.0134, -0.0134\,] + [\,-0.0167, -0.0333\,] + [\,0.0000, -0.0600\,].\end{aligned}$$

Summing component-wise:

$$\Delta\theta^{(t)} = [\,-0.0033, -0.1067\,].$$

Step 5: Global Model Update. Let the current global model be

$$\theta^{(t)} = [\,1.20, -0.80\,].$$

The updated global model becomes

$$\theta^{(t+1)} = \theta^{(t)} + \Delta\theta^{(t)} = [\,1.1967, -0.9067\,].$$

Interpretation. This example illustrates several key properties of client selection and scheduling in federated learning:

- Only a subset of clients influences the global update at each round, introducing randomness into the optimization trajectory.
- Clients with larger datasets exert greater influence through data-weighted aggregation.
- The resulting update is a stochastic approximation of the descent direction that would have been obtained using full participation.

Over multiple communication rounds, repeated randomized selection ensures that all clients contribute statistically, allowing the global model to converge despite partial participation.

4.4 Communication Topology and Bottlenecks

In federated learning systems, *communication rather than computation* typically constitutes the dominant cost and the primary scalability bottleneck. While modern client devices often possess sufficient computational capability to perform local optimization, the repeated exchange of model information across distributed and heterogeneous environments imposes significant overhead. Unlike centralized machine learning—where data is co-located and parameter updates occur within a single, tightly coupled computational infrastructure—federated learning unfolds over wide-area networks that connect geographically dispersed, intermittently available, and resource-constrained clients. Consequently, the efficiency, reliability, and scalability of federated learning are shaped far more by communication constraints than by raw computational limits.

The communication topology defines how information flows between participants and plays a central role in determining system behavior. In the *centralized federated learning architecture*, communication follows a *star topology*, with the central server acting as the sole coordination hub. All participating clients communicate directly with the server, while direct client-to-client communication is absent. This design greatly simplifies orchestration, aggregation, access control, and security enforcement, since all coordination logic is localized at a single point. At the same time, the star topology concentrates communication load at the server, making it a potential performance bottleneck as the number of participating clients increases.

The implications of this topology are both architectural and algorithmic. Architecturally, the server must support concurrent communication with many clients, manage heterogeneous network conditions, and enforce protocol-level guarantees. Algorithmically, the learning process must tolerate delays, packet loss, and partial participation that naturally arise in such an environment. As a result, communication topology cannot be treated as a neutral implementation detail; it directly influences learning dynamics and convergence behavior.

4.4.1 Core Communication Characteristics

Several communication characteristics are intrinsic to federated learning deployments and collectively distinguish them from classical distributed learning systems:

- **Uplink transmission of client updates:** At the end of each communication round, participating clients transmit locally computed model updates $\Delta\theta_k^{(t)}$ to the server. These uplink transmissions are often the most constrained component of communication, as clients may operate over limited bandwidth, metered connections, or energy-sensitive wireless links. For mobile and edge devices, uplink communication may dominate energy consumption and directly affect device usability.
- **Downlink broadcast of global models:** The server broadcasts the global model $\theta^{(t)}$ to selected clients at the beginning of each communication round. Although downlink bandwidth is typically more abundant than uplink, broadcasting large models to many clients simultaneously can still incur substantial cost, particularly when model sizes are large or when participation rates are high.

- **Limited and unreliable communication links:** Federated learning systems operate over networks with widely varying quality. Packet loss, latency, jitter, and temporary disconnections are common, especially in wireless and mobile environments. These effects introduce asynchrony, uncertainty, and randomness into the learning process, making strict synchronization impractical.
- **Heterogeneous communication capabilities:** Clients differ significantly in network bandwidth, latency, and availability. Some clients may respond quickly with high-quality updates, while others may be slow or frequently unavailable. This heterogeneity complicates scheduling decisions and often leads to the emergence of stragglers or missing updates.

4.4.2 Types of Communication Bottlenecks

Communication bottlenecks in federated learning arise at multiple layers of the system stack and reflect the interaction between network infrastructure, client heterogeneity, and model complexity. These bottlenecks are not isolated phenomena; rather, they jointly influence system throughput, learning stability, and convergence speed. They can be broadly categorized as follows:

- **Server-side bottlenecks:** In a centralized federated architecture, the server acts as the sole communication hub and must simultaneously manage uplink and downlink traffic for a potentially large number of clients. Limitations in server network bandwidth, message processing capacity, or protocol-level overhead (such as encryption, authentication, and connection management) can restrict the number of clients that can be served in each communication round. As the client population grows, these constraints can become the dominant factor limiting scalability, even when individual clients are capable of fast local computation.
- **Client-side bottlenecks:** Federated learning clients operate under widely varying conditions, ranging from high-performance servers to mobile or edge devices with limited connectivity and energy resources. Slow, unstable, or energy-constrained connections can delay or prevent the transmission of updates, causing some clients to become stragglers or to drop out entirely. Such variability increases round latency, reduces effective participation, and introduces additional variance into the aggregated model updates, thereby affecting learning dynamics.
- **Network-level bottlenecks:** Federated learning systems typically rely on wide-area networks whose performance fluctuates across time, geography, and usage patterns. Latency, jitter, packet loss, and temporary outages are common, particularly when clients are geographically dispersed or connected via wireless links. These network-level effects exacerbate asynchrony, complicate scheduling decisions, and make strict synchronization impractical in large-scale deployments.
- **Model-size bottlenecks:** Modern machine learning models often consist of millions or even billions of parameters, resulting in large communication payloads. Repeatedly transmitting full model parameters or updates can quickly become infeasible under bandwidth constraints, especially in uplink communication.

Model size therefore directly amplifies communication bottlenecks and motivates the use of compressed, sparsified, or low-precision representations to reduce transmission cost.

From a systems perspective, the central server functions as a communication hub whose capacity directly constrains scalability. As the number of clients grows, the server's ability to broadcast models and collect updates becomes a limiting factor. This observation motivates architectural strategies that reduce per-round communication load, limit the number of participating clients, or decentralize coordination in more advanced architectures.

From an optimization perspective, communication constraints fundamentally reshape the learning process itself. To reduce communication overhead, federated learning algorithms commonly employ a combination of architectural and algorithmic strategies that trade exact synchronization for efficiency:

- **Infrequent synchronization:** Instead of communicating after every local update, clients are allowed to perform multiple local optimization steps before synchronizing with the server. For example, in the FedAvg algorithm, each selected client may run several epochs of stochastic gradient descent on its local dataset before transmitting an update. This approach amortizes the cost of communication over more local computation, significantly reducing the number of communication rounds. However, excessive local training can cause local models to drift farther from the global trajectory, particularly under non-IID data distributions.
- **Compressed, quantized, or sparsified updates:** To reduce message size, clients may transmit approximate representations of their updates rather than full-precision parameters. For instance, updates can be quantized to low-bit representations, sparsified by transmitting only the largest-magnitude components, or compressed using sketching techniques. As an example, a client may transmit only the top-k elements of $\Delta\theta_k^{(t)}$, discarding small updates that contribute little to immediate progress. While such techniques substantially reduce bandwidth usage, they introduce additional noise or bias into the optimization process, which must be carefully controlled to preserve convergence.
- **Adaptive or partial participation:** At each communication round, the server may select only a fraction of available clients based on availability, resource constraints, or scheduling policies. For example, in a mobile device setting, only clients that are plugged in, idle, and connected via Wi-Fi may be eligible to participate. By limiting the number of active clients per round, the system reduces aggregate communication load while still achieving global progress through repeated, randomized participation over time. This strategy introduces stochasticity into the learning process but is essential for scalability in large federated deployments.

While these strategies substantially improve scalability and practical deployability, they also modify the mathematical properties of federated optimization.

Infrequent synchronization increases *local model drift*, as client models diverge from the global trajectory. Compression and sparsification introduce biased or noisy updates. Partial participation reduces the instantaneous representativeness of aggregated gradients. Together, these effects amplify the challenges posed by non-IID data distributions and complicate convergence analysis.

Communication-Centric Design
Federated learning algorithms are constrained more strongly by communication limitations than by computational capacity. Architectural and algorithmic choices are therefore driven primarily by the need to reduce communication cost while preserving learning stability and convergence.

In summary, communication topology and bottlenecks form a central axis along which federated learning systems must be designed and analyzed. The star topology of centralized federated learning offers simplicity and control, but introduces scalability limits that must be carefully managed. Understanding how communication constraints interact with optimization dynamics is essential for developing federated learning algorithms that are not only scalable, but also mathematically well-behaved under real-world deployment conditions.

4.5 Failure Handling and Fault Tolerance

Federated learning systems are inherently exposed to frequent, unpredictable, and heterogeneous failures. Unlike traditional distributed computing environments—such as data centers or high-performance clusters—where nodes are centrally managed, relatively stable, and connected through reliable networks, federated learning operates over large populations of autonomous clients. These clients may include mobile devices, edge sensors, personal computers, or institutional servers, each subject to distinct hardware limitations, network conditions, energy constraints, and usage patterns. As a result, failures are not exceptional corner cases but a defining feature of the federated learning operational environment.

In centralized federated learning architectures, fault tolerance is therefore treated as a first-class design requirement rather than an auxiliary concern. Clients may disconnect unexpectedly, suspend computation, exhaust battery resources, or experience intermittent connectivity at any point during training. Moreover, participation is often opportunistic: clients join only when local conditions permit, and their availability cannot be enforced by the system. The learning process must therefore continue to make progress despite highly irregular, asynchronous, and partially observable participation, without relying on strict synchronization or guaranteed responses from any individual client.

4.5.1 Common Failure Scenarios

Several distinct classes of failures commonly arise in federated learning systems, each affecting learning dynamics in different ways. These failures stem from the autonomy of clients, heterogeneity of resources, and the unreliability of wide-area communication networks:

- **Client unavailability during a round:** Clients may be offline, engaged in foreground user tasks, or subject to energy-saving or thermal policies at the time they are selected for participation. For example, a mobile device may temporarily suspend background computation when the battery level drops or when a user initiates an interactive application. In such cases, the client produces no update at all, effectively reducing the active participant set for that round and altering the statistical composition of the aggregation.
- **Straggler clients:** Some clients may complete local training significantly more slowly than others due to limited computational resources, heavy system load, or poor network connectivity. For instance, an older device or a client connected through a high-latency cellular network may take much longer to finish local updates. Waiting for such stragglers can dramatically increase round latency, motivating the server to proceed with aggregation using only the faster responses.
- **Partial transmission failures:** Model updates may be corrupted, truncated, or lost during uplink communication. This can occur due to packet loss, intermittent connectivity, or bandwidth limitations, particularly in wireless or mobile environments. As an example, a client may successfully compute a local update but fail to transmit the full parameter vector before a connection drop, resulting in an unusable or incomplete update.
- **Transient network outages:** Short-lived disruptions in connectivity may prevent clients from receiving the global model or returning updates, even if they are otherwise capable of participating. Such outages may be caused by network handovers, signal interference, or temporary infrastructure failures. These events introduce temporal gaps in client involvement and contribute to asynchronous participation patterns.
- **Delayed or stale updates:** Updates computed on outdated global models may arrive after one or more aggregation steps have already occurred. For example, a client that experiences a long delay may return an update based on $\theta^{(t)}$ after the server has already advanced to $\theta^{(t+1)}$ or beyond. Such stale updates raise questions about their validity and usefulness, and may need to be discarded or down-weighted to maintain learning stability.

To accommodate these conditions, federated learning systems typically employ a *best-effort aggregation strategy*. Rather than requiring all selected clients to respond, the server defines a fixed or adaptive time window for each communication round and aggregates only the updates received within this window. Clients whose updates arrive late are either discarded or deferred to subsequent rounds, depending on system policy. This approach prioritizes throughput and responsiveness over per-round completeness.

Mathematically, this behavior implies that the effective participating client set $\mathscr{S}^{(t)}$ is both random and time-varying. Even if a larger set of clients is initially selected, only a subset ultimately contributes updates to the aggregation step. The global model update therefore takes the form

$$\theta^{(t+1)} = \theta^{(t)} + \sum_{k \in \mathscr{S}^{(t)}} \frac{n_k}{\sum_{j \in \mathscr{S}^{(t)}} n_j} \Delta\theta_k^{(t)},$$

where $\mathscr{S}^{(t)}$ reflects both selection and successful participation. This introduces additional stochasticity beyond that induced by data sampling and non-IID distributions.

4.5.2 Fault Tolerance through Statistical Design

Unlike classical fault-tolerant systems, which aim to guarantee exact execution through replication, rollback, or consensus protocols, federated learning adopts a fundamentally statistical notion of resilience. Fault tolerance is achieved not by preventing failures, but by designing learning dynamics that remain stable and convergent under repeated, independent disruptions.

Convergence analysis in federated learning therefore relies on assumptions of *long-term participation coverage*. Rather than requiring every client to contribute in every round, it is assumed that each client participates sufficiently often over the course of training, possibly according to some probability distribution. Under these conditions, the aggregate effect of many partial, noisy, and incomplete updates approximates the behavior of full participation in expectation.

This perspective aligns naturally with stochastic optimization theory. Client dropouts, delayed updates, and partial participation are modeled as sources of noise in the optimization process. Provided this noise is unbiased or bounded over time, the learning process can still converge to a stationary point of the global objective. Fault tolerance is thus embedded directly into the mathematical structure of federated optimization rather than treated as an external system requirement.

Resilience through Redundancy
Federated learning systems achieve fault tolerance by embracing redundancy, stochastic participation, and approximate execution. Robust convergence emerges from repeated, imperfect interactions rather than strict per-round reliability.

Viewed through this lens, failures are not merely tolerated but implicitly modeled as part of the learning process. The ability of federated learning systems to progress under unreliable conditions is a direct consequence of their statistical foundations, which prioritize long-term consistency over short-term exactness. This design philosophy explains both the practical robustness of large-scale federated deployments and the form of convergence guarantees established in federated optimization theory.

4.6 Scalability Considerations

Scalability is a defining and non-negotiable requirement of federated learning systems. Many real-world deployments are expected to operate across populations consisting of hundreds of thousands or even millions of clients, such as mobile devices, edge sensors, or institutional data silos. Unlike traditional distributed learning environments, federated learning must scale under severe constraints: limited and unreliable communication, heterogeneous hardware capabilities, intermittent availability, and strict privacy requirements. Consequently, scalability in federated learning is not simply a matter of increasing computational capacity, but of sustaining stable learning dynamics under extreme decentralization.

Architectural scalability in federated learning emerges from a delicate interaction between system-level design choices and the mathematical properties of federated optimization. From the system perspective, the architecture must support large and evolving client populations without overwhelming communication infrastructure or centralized coordination mechanisms. From the optimization perspective, learning algorithms must remain robust to partial participation, delayed updates, and statistical heterogeneity. Effective scalability therefore requires a careful co-design of orchestration mechanisms, communication strategies, and optimization procedures.

Several interrelated dimensions jointly determine the scalability of a federated learning system:

- **Number of clients K:** As the total client population grows, the system must avoid linear increases in communication and coordination overhead. Scalable designs ensure that per-round communication cost depends on the number of selected clients rather than on K itself.
- **Frequency of communication rounds:** Frequent synchronization accelerates information sharing and may improve convergence speed, but it places heavy demands on network resources. Infrequent communication improves scalability by reducing overhead, at the cost of increased local model drift and slower alignment across clients.
- **Size of transmitted model updates:** Modern models often contain millions of parameters, making naive transmission impractical at scale. This motivates techniques such as update compression, sparsification, quantization, and selective transmission, each of which introduces trade-offs between communication efficiency and optimization accuracy.
- **Heterogeneity of client resources:** Clients differ widely in computational power, memory capacity, energy availability, and network bandwidth. Scalable systems must accommodate this heterogeneity through adaptive scheduling, flexible participation rules, and tolerance to stragglers and dropouts.

From a mathematical standpoint, scalable federated learning relies on the fundamental observation that the global optimization objective

$$F(\theta) = \sum_{k=1}^{K} \frac{n_k}{N} F_k(\theta)$$

can be approximated effectively using updates from only a small, randomly selected subset of clients at each communication round. Under suitable assumptions on client participation and data coverage, the stochastic aggregation of partial updates yields convergence in expectation. This statistical property allows the learning process to scale without requiring full synchronization or universal participation.

From a systems standpoint, scalability depends on efficient orchestration, load balancing, and communication management. The central server must coordinate large numbers of clients without becoming a bottleneck, dynamically adapt to changing availability, and limit communication overhead through selective participation and asynchronous execution. Rather than enforcing rigid schedules, scalable federated systems favor flexible, best-effort coordination that prioritizes throughput and robustness over strict consistency.

Importantly, scalability in federated learning is achieved by embracing approximation. The system deliberately sacrifices per-round completeness and exact coordination in favor of long-term statistical consistency. Learning progress is measured not by instantaneous global alignment, but by gradual improvement over many rounds of partial, noisy, and heterogeneous updates. This perspective represents a departure from classical distributed optimization, where synchronization and determinism are often central design goals.

Scalability Insight
Federated learning achieves scalability by relaxing coordination requirements while preserving statistical representativeness. Global learning emerges from many incomplete and approximate interactions rather than from tightly synchronized execution.

Viewed holistically, scalability in federated learning is an emergent property rather than a single architectural feature. It arises from the combined effect of partial participation, stochastic aggregation, communication-efficient design, and tolerance to heterogeneity. The centralized federated architecture discussed in this chapter provides a practical and theoretically tractable foundation for large-scale deployment under real-world constraints.

Exercises

This chapter examined the system architectures that enable federated learning in practice, highlighting the interaction between optimization theory and real-world constraints such as communication, scalability, and fault tolerance.

A. Short Answer Questions

Answer each question briefly and precisely.

1. What is meant by centralized coordination in federated learning?
2. Why is centralized federated learning not equivalent to centralized data processing?
3. Define the role of the central server in federated learning.
4. What is a star communication topology?
5. Why is partial client participation necessary in large-scale federated systems?
6. What is meant by statistical participation?
7. How does client heterogeneity affect federated optimization?
8. What is a straggler client?
9. Why is communication considered the dominant cost in federated learning?
10. What does scalability mean in the context of federated learning architectures?

B. Long Answer Questions

Answer the following questions with clear explanations, equations, and diagrams wherever appropriate.

1. Explain the centralized federated learning architecture with a neatly labeled diagram and discuss its advantages and limitations.
2. Describe the orchestration role of the central server and explain how it differs from direct optimization in centralized learning.
3. Derive the global aggregation rule for federated learning under partial participation and explain its stochastic nature.
4. Discuss different client selection and scheduling strategies and analyze their impact on convergence behavior.
5. Explain communication bottlenecks in federated learning and discuss algorithmic strategies used to mitigate them.
6. Analyze how non-IID data and infrequent communication jointly affect local model drift.
7. Explain failure handling in federated learning and describe how statistical redundancy enables fault tolerance.
8. Compare scalability challenges in centralized distributed learning and federated learning.

C. True/False Questions

State whether each statement is **True** or **False**.

1. In federated learning, the central server stores all client data.

2. Clients communicate directly with each other in centralized federated learning.
3. Partial client participation introduces stochasticity into the optimization process.
4. Communication cost in federated learning is typically higher than computation cost.
5. The server computes gradients of the global objective function.
6. Fault tolerance in federated learning relies on exact execution.
7. Straggler clients can slow down communication rounds.
8. Federated learning requires all clients to participate in every round for convergence.
9. Update compression may introduce additional optimization noise.
10. Scalability in federated learning is achieved by increasing synchronization frequency.

Answers:

1. False
2. False
3. True
4. True
5. False
6. False
7. True
8. False
9. True
10. False

D. Fill in the Blanks

Fill in the blanks with the most appropriate term or expression.

1. In centralized federated learning, coordination is centralized while data remains ____________.
2. The set of clients selected in round t is denoted by ____________.
3. The global objective function in federated learning is a ____________ sum of local objectives.
4. A communication pattern where all clients communicate only with the server is called a ____________ topology.
5. Clients that respond very slowly are referred to as ____________ clients.
6. Proceeding without waiting for all clients is known as ____________ aggregation.
7. Transmitting reduced-size updates is called update ____________.
8. Scalability relies on selecting only a ____________ fraction of clients per round.

9. Fault tolerance in federated learning is achieved through statistical ____________.
10. Local model divergence caused by infrequent synchronization is called model ____________.

Answers:

1. local
2. $\mathcal{S}^{(t)}$
3. weighted
4. star
5. straggler
6. best-effort
7. compression
8. small
9. redundancy
10. drift

E. Multiple Choice Questions (MCQs)

Choose the **most appropriate** answer.

1. The primary role of the server in federated learning is to:
   (a) store client datasets
   (b) compute gradients
   (c) coordinate communication and aggregation
   (d) train local models
2. Which topology is used in centralized federated learning?
   (a) Ring
   (b) Mesh
   (c) Star
   (d) Peer-to-peer
3. Partial participation mainly improves:
   (a) data centralization
   (b) scalability
   (c) gradient accuracy
   (d) determinism
4. Which factor most strongly constrains federated learning systems?
   (a) CPU speed
   (b) Memory size
   (c) Communication bandwidth
   (d) Model depth

5. Which of the following introduces local model drift?
   (a) Frequent synchronization
   (b) IID data
   (c) Multiple local updates
   (d) Centralized training
6. A straggler client primarily affects:
   (a) data privacy
   (b) convergence guarantees
   (c) round latency
   (d) model size
7. Best-effort aggregation helps to:
   (a) enforce synchronization
   (b) tolerate delays
   (c) eliminate noise
   (d) centralize data
8. Scalability in federated learning relies on:
   (a) full participation
   (b) deterministic execution
   (c) statistical approximation
   (d) centralized optimization
9. Update compression primarily reduces:
   (a) computation cost
   (b) communication cost
   (c) memory usage
   (d) data bias
10. Federated learning convergence is typically analyzed:
   (a) Deterministically
   (b) Per iteration
   (c) In expectation
   (d) Without randomness

Answers:

1. (c)
2. (c)
3. (b)
4. (c)
5. (c)
6. (c)
7. (b)
8. (c)
9. (b)
10. (c)

F. Application and Project-Based Questions

These problems emphasize modeling, reasoning, and implementation.

1. Design a centralized federated learning system for a mobile keyboard prediction task and justify the architectural choices.
2. Simulate client dropouts in a federated learning setup and study their impact on convergence.
3. Implement a client selection strategy and compare uniform random selection with resource-aware scheduling.
4. Analyze communication overhead for different model sizes and synchronization frequencies.
5. Study the effect of non-IID data distributions on local model drift.
6. Implement update compression and evaluate the trade-off between communication efficiency and accuracy.
7. Design a fault-tolerant aggregation mechanism using time-window based updates.
8. Compare centralized and decentralized federated architectures for scalability.
9. Propose a hybrid orchestration strategy that adapts to client availability.
10. Conduct an empirical study on how partial participation affects statistical representativeness over long training horizons.

5 Decentralized and Peer-to-Peer Federated Learning

The federated learning architectures examined in the previous chapter are primarily grounded in the assumption of a *centralized coordination model*, in which a designated server orchestrates the training process by collecting, aggregating, and redistributing model updates. Such architectures have proven highly effective in practice and form the backbone of many large-scale industrial deployments. By centralizing coordination while preserving data locality, they offer a pragmatic compromise between efficiency, scalability, and privacy.

Despite these advantages, centralized federated learning is not a universally applicable solution. In many real-world environments, reliance on a single coordinating entity introduces a range of limitations that extend beyond mere system performance. From a technical perspective, the server represents a potential communication bottleneck and a single point of failure. From an organizational standpoint, centralized control may be incompatible with collaborative learning scenarios involving multiple institutions, competing stakeholders, or decentralized governance structures. From a trust and security perspective, the existence of a central orchestrator raises concerns regarding neutrality, robustness against compromise, and long-term system resilience.

These limitations become especially pronounced in settings where participants are geographically dispersed, operate under heterogeneous administrative domains, or lack a trusted coordinating authority. Examples include inter-organizational collaborations, ad hoc edge networks, peer-operated Internet of Things (IoT) systems, and cross-border data-sharing environments subject to differing regulatory constraints. In such scenarios, the assumption that all participants are willing—or able—to delegate coordination responsibilities to a single server is often untenable.

Motivated by these considerations, this chapter explores *decentralized and peer-to-peer federated learning* paradigms, in which collaborative model training proceeds without a central orchestrator. Instead of relying on a globally distinguished server, coordination is distributed across clients themselves. Learning emerges through structured local interactions, direct exchange of model information among

P. P. Ray, *Foundations of Federated Learning*,
Springer Series in the Data Sciences,
https://doi.org/10.1007/978-3-032-34084-9_5

peers, and algorithmic mechanisms that promote agreement across a network of autonomous learners.

In decentralized federated learning, aggregation is no longer a single global operation performed at a fixed location. Rather, it is realized implicitly through repeated local averaging, gossip-based exchanges, consensus dynamics, or ledger-mediated coordination. Each client acts simultaneously as a learner and a coordinator, contributing to global model evolution while retaining full control over its data and participation. As a result, the notion of a *global model* becomes an emergent property of the system rather than an explicitly maintained artifact.

These approaches extend the federated learning paradigm toward fully distributed intelligence operating under minimal coordination and trust assumptions. They blur the boundary between federated learning and classical distributed optimization while preserving the defining principles of data locality, autonomy, and privacy. At the same time, they introduce new challenges related to convergence, communication efficiency, network topology, and robustness that require careful theoretical and algorithmic treatment.

This chapter provides a systematic introduction to decentralized and peer-to-peer federated learning. We examine the motivations for removing centralized orchestration, formalize orchestrator-less learning models, analyze gossip and consensus-based aggregation mechanisms, and explore blockchain-enabled coordination as a trust infrastructure. Finally, we investigate how network topology fundamentally shapes learning dynamics and performance. Together, these perspectives reveal how federated learning can function as a truly decentralized optimization paradigm in environments where central coordination is impractical, undesirable, or impossible.

5.1 Motivation for Decentralization

The motivation for decentralizing federated learning arises from a convergence of *systemic constraints*, *organizational realities*, and *conceptual principles* that limit the applicability of centralized coordination. While centralized federated learning provides an effective abstraction for many deployments, it embeds assumptions about trust, infrastructure, and scalability that do not hold universally. Decentralized federated learning emerges as a response to these limitations, seeking to preserve collaborative optimization while relaxing reliance on a single coordinating entity.

5.1.1 Single Point of Failure

Centralized federated learning architectures concentrate critical orchestration responsibilities—such as global model maintenance, client selection, update aggregation, and synchronization—within a single coordinating server. While this design greatly simplifies system management and algorithmic control, it also introduces a fundamental structural vulnerability: the server becomes a *single point of failure* for the entire learning process.

Any disruption affecting the server can stall or terminate training, even if all participating clients remain fully operational. Such disruptions may arise from hardware faults, transient or prolonged network outages, denial-of-service attacks, software bugs, or erroneous configuration changes. Because all coordination flows through the server, failures at this node propagate system-wide, preventing further model updates and effectively decoupling clients from the collaborative learning process.

Beyond catastrophic failure, centralized coordination is also susceptible to performance degradation under scale. As the number of clients increases, the server must simultaneously manage a growing volume of inbound model updates and outbound broadcasts of global parameters. This dual communication burden can saturate network links, exhaust computational resources, and introduce significant latency. Even when the server remains operational, overload conditions may force the system to reduce client participation, delay aggregation, or discard updates, thereby slowing convergence and degrading model quality.

Decentralized federated learning architectures mitigate these risks by eliminating reliance on any single coordinating entity. Orchestration responsibilities are distributed across clients, and coordination is achieved through local interactions, peer-to-peer communication, or collective consensus mechanisms. As a result, no individual node is indispensable for progress: learning can continue despite failures, delays, or departures of individual participants. This structural redundancy confers increased resilience, adaptability, and fault tolerance, enabling federated learning systems to operate robustly under dynamic, large-scale, and adversarial conditions.

5.1.2 Trust and Governance

In many federated learning scenarios, participating entities operate under distinct administrative, legal, and organizational domains. Typical examples include collaborations among hospitals belonging to different healthcare systems, financial institutions subject to independent regulatory regimes, academic research laboratories with separate governance structures, or competing service providers seeking to jointly train models without exposing proprietary data. In such environments, the assumption that all participants are willing to delegate coordination responsibilities to a single, centrally controlled server is often unrealistic or unacceptable.

Although a centralized server in federated learning is not granted direct access to raw data, it nevertheless occupies a privileged position within the learning process. By controlling aggregation rules, client selection policies, synchronization schedules, and the release of global model updates, the server can exert substantial influence over the optimization trajectory and the final learned model. From a governance perspective, this asymmetry raises concerns regarding neutrality, fairness, and accountability, particularly when participants have competing objectives or unequal bargaining power.

Decentralized federated learning addresses these concerns by eliminating the need for a trusted central authority altogether. Coordination is instead embedded in protocol-level mechanisms that are symmetrically applied across participants. Peer-to-peer communication, consensus rules, and collectively verifiable operations replace discretionary server-side control. As a result, no single entity can unilaterally bias aggregation, prioritize certain participants, or shape learning dynamics to its advantage.

This shift from institution-based trust to protocol-based trust represents a fundamental governance transformation. Participants collaborate under transparent and enforceable rules rather than relying on the assumed goodwill of a coordinating party. Decentralized federated learning thus enables collaborative model training in settings characterized by limited trust, regulatory fragmentation, and organizational autonomy, while preserving the core principles of data sovereignty and equitable participation.

5.1.3 Scalability and Network Constraints

As federated learning systems scale to thousands or even millions of clients, centralized communication patterns place increasing strain on network and coordination resources. In star-shaped topologies, every participating client must transmit updates to, and receive model broadcasts from, a central server. This concentration of communication leads to bandwidth saturation, increased end-to-end latency, and synchronization delays, particularly when client participation is high or update sizes are large.

These challenges are further amplified in realistic deployment environments. Clients may exhibit heterogeneous network conditions, ranging from high-bandwidth wired connections to low-power wireless links. Connectivity may be intermittent, constrained by energy-saving policies, or subject to regional and geographic variability. In such settings, centralized orchestration must contend with straggler clients, delayed updates, and uneven communication reliability, all of which degrade training efficiency and complicate convergence analysis.

Decentralized federated learning protocols address these scalability limitations by replacing hub-and-spoke communication with localized, peer-to-peer interactions. Rather than communicating with a single central entity, each client exchanges information only with a limited set of neighboring peers. This locality reduces per-node communication load, distributes traffic more evenly across the network, and enables parallel information flow along multiple paths. As a result, model updates propagate gradually but efficiently through the system without overwhelming any single communication link or coordination node.

Moreover, decentralized designs naturally accommodate partial participation and asynchronous operation. Clients may join, leave, or temporarily disconnect without disrupting the overall learning process, as coordination is no longer tied to a global synchronization point. Scalability is therefore achieved not by increasing the capacity of a central server, but by distributing coordination responsibilities throughout the network, allowing federated learning to expand in size and complexity while maintaining robustness and efficiency.

5.1.4 Alignment with Federated Principles

At a conceptual level, federated learning is fundamentally grounded in the principles of *autonomy*, *data sovereignty*, and *privacy*. Clients retain full control over their locally generated data, decide independently when and how to participate in training, and contribute to model improvement without transferring raw information outside their administrative boundaries. These principles distinguish federated learning from earlier distributed learning paradigms and reflect a deliberate shift toward user- and institution-centric model training.

Centralized coordination, while operationally efficient, represents a partial departure from this philosophical foundation. By reintroducing a system-level locus of control, centralized orchestration implicitly constrains client autonomy and concentrates decision-making authority in a single entity. Even in the absence of direct data access, control over aggregation policies, participation rules, and model dissemination shapes learning outcomes in ways that may not fully align with the decentralized ethos that motivates federated learning.

Fully decentralized federated learning architectures represent a logical and principled extension of the federated paradigm. They eliminate not only centralized data access but also centralized coordination, thereby aligning system design more closely with the normative goals of federated learning. In such architectures, no entity possesses privileged control over the learning process; instead, coordination emerges from symmetric interactions governed by shared protocols and local decision-making.

By distributing both computation and coordination, decentralized federated learning transforms collaborative model training from a centrally managed workflow into an emergent property of interacting autonomous agents. Global learning behavior arises through repeated local interactions, consensus dynamics, and information exchange, reflecting a deeper commitment to decentralization that unifies the technical, ethical, and conceptual foundations of federated learning.

Decentralization Rationale
Decentralized federated learning removes centralized coordination while preserving collaborative optimization. It enables resilient, scalable, and trust-minimized learning systems in which global intelligence emerges from local interactions among autonomous participants.

5.2 Orchestrator-Less Federated Learning

In orchestrator-less federated learning, the role traditionally played by a central server is completely removed. No single entity maintains a global model, schedules communication rounds, or performs explicit aggregation of client updates. Instead, learning is realized as a fully decentralized process in which each client maintains its own local model parameters and coordinates directly with other clients through peer-to-peer communication.

This architectural shift represents a fundamental departure from centralized federated learning. Rather than viewing learning as a sequence of server-mediated aggregation steps, orchestrator-less federated learning treats the system as a network of autonomous learners whose collective behavior gives rise to global model convergence. Coordination is no longer imposed externally but emerges organically from repeated local interactions.

Defining Characteristic
In orchestrator-less federated learning, *coordination is an emergent network property rather than a centrally enforced operation.*

5.2.1 Local Model Maintenance and Peer Interaction

Let client $k \in \{1, 2, \dots, K\}$ maintain a local parameter vector $\theta_k^{(t)}$ at iteration or communication step t. Each client operates independently and performs two conceptually distinct but tightly coupled actions:

1. **Local optimization:** the client updates its model parameters using only its private dataset $\mathscr{D}_k$ by minimizing the local empirical risk $F_k(\theta_k)$.
2. **Peer interaction:** the client exchanges model parameters or parameter summaries with a selected subset of neighboring clients connected through the communication network.

These operations are typically interleaved. Local optimization drives the model toward solutions that fit client- specific data distributions, while peer interaction counteracts local divergence by encouraging alignment among neighboring models.

Local–Global Tension
Local optimization promotes specialization to client data, whereas peer interaction promotes agreement across clients. Orchestrator-less federated learning continuously balances these two competing forces.

5.2.2 Absence of an Explicit Global Model

Unlike centralized federated learning, there is no distinguished global parameter vector $\theta^{(t)}$ that all clients reference or synchronize against. Each client's parameter vector evolves according to its own local data and the influence of its neighbors. The notion of a *global model* exists only implicitly as the limit toward which all local models converge.

The learning objective is therefore not to compute and distribute a single authoritative model at each iteration, but to ensure that the collection of local models $\{\theta_k^{(t)}\}_{k=1}^K$ becomes increasingly consistent over time. Consensus is achieved asymptotically rather than instantaneously.

Emergent Global Model
In orchestrator-less federated learning, the global model is not stored or broadcast. It emerges implicitly as local models gradually align through network interactions.

5.2.3 Consensus-Based Mathematical Interpretation

From a mathematical perspective, orchestrator-less federated learning can be interpreted as a constrained optimization problem defined over client-wise parameter copies. Each client k maintains its own parameter vector θ_k, and learning seeks to minimize the aggregate objective

$$\sum_{k=1}^{K} \frac{n_k}{N} F_k(\theta_k), \quad N = \sum_{k=1}^{K} n_k,$$

subject to consensus constraints that enforce agreement between clients that can communicate directly:

$$\theta_k = \theta_j, \qquad \forall (k,j) \in \mathcal{E},$$

where $\mathcal{E}$ denotes the set of edges in the communication graph.

These constraints encode the network topology explicitly. Only clients connected by an edge are required to align their model parameters directly. Global consensus is achieved indirectly through repeated local alignments propagated across the network.

Network-Coupled Optimization
Orchestrator-less federated learning replaces a single global constraint with a collection of local consensus constraints whose interaction is governed by the network topology.

5.2.4 Shift in the Nature of Coordination

The absence of a central server fundamentally shifts the burden of coordination from a single entity to the communication network itself. Network connectivity, neighbor selection policies, and information flow patterns become primary determinants of convergence behavior. As a result, properties such as graph connectivity, diameter, and spectral gap play a central role in determining learning speed and stability.

This shift also alters the system's resilience profile. Failures, delays, or departures of individual clients affect only local regions of the network rather than halting the entire learning process. Learning continues as long as the communication graph remains sufficiently connected over time.

Key Insight
By eliminating centralized orchestration, federated learning becomes a truly distributed optimization process in which coordination, robustness, and scalability are governed by network structure rather than institutional control.

Orchestrator-less federated learning thus represents the most radical realization of the federated paradigm. It preserves strict data locality and client autonomy while extending collaborative learning to environments characterized by limited trust, dynamic participation, and minimal infrastructure. At the same time, it introduces new theoretical and practical challenges related to convergence guarantees, communication efficiency, and topology-aware algorithm design, which are explored in the subsequent sections of this chapter.

Worked Example: Orchestrator-Less Federated Learning with Heterogeneous Local Datasets
This example extends the previous illustration by removing the artificial assumption that all clients possess the same number of training samples. Instead, each client holds a private dataset of different size, reflecting the heterogeneity commonly observed in real-world federated learning systems.

Problem Setup. Consider $K = 5$ clients collaboratively learning a linear model

$$y = \theta x,$$

where $\theta \in \mathbb{R}$ is a scalar parameter. Each client k holds a private dataset $\mathcal{D}_k$ with $n_k = |\mathcal{D}_k|$ samples. The datasets differ in both size and statistical characteristics (Table 5.1).

Each client defines a local mean squared error (MSE) loss function

$$F_k(\theta) = \frac{1}{n_k} \sum_{i=1}^{n_k} (\theta x_{k,i} - y_{k,i})^2,$$

Table 5.1 Heterogeneous private datasets held by five clients

Client k	n_k	Inputs x	Outputs y
1	2	1, 2	2, 4
2	3	1, 2, 3	1, 2, 3
3	4	1, 2, 3, 4	3, 6, 9, 12
4	1	2	8
5	5	1, 2, 3, 4, 5	0.5, 1, 1.5, 2, 2.5

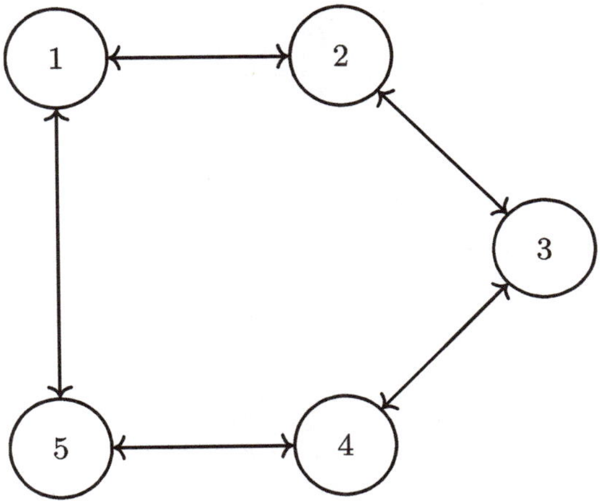

Fig. 5.1 Ring topology used for peer-to-peer communication in orchestrator-less federated learning

which is consistent with the federated optimization formulation introduced earlier.

Network Topology. The clients communicate using a ring topology (Fig. 5.1):

$$1 \longleftrightarrow 2 \longleftrightarrow 3 \longleftrightarrow 4 \longleftrightarrow 5 \longleftrightarrow 1,$$

with communication edge set

$$\mathscr{E} = \{(1,2),(2,3),(3,4),(4,5),(5,1)\}.$$

Iteration 1.

Step 1: Explicit Local Loss Functions.

Client 1 $(n_1 = 2)$:

$$F_1(\theta) = \frac{1}{2}\left[(\theta-2)^2 + (2\theta-4)^2\right] = \frac{1}{2}(5\theta^2 - 20\theta + 20).$$

Client 2 $(n_2 = 3)$:

$$F_2(\theta) = \frac{1}{3}\left[(\theta - 1)^2 + (2\theta - 2)^2 + (3\theta - 3)^2\right]$$
$$= \frac{1}{3}(14\theta^2 - 28\theta + 14).$$

Client 3 $(n_3 = 4)$:

$$F_3(\theta) = \frac{1}{4}\left[(\theta - 3)^2 + (2\theta - 6)^2 + (3\theta - 9)^2 + (4\theta - 12)^2\right]$$
$$= \frac{1}{4}(30\theta^2 - 180\theta + 270).$$

Client 4 $(n_4 = 1)$:

$$F_4(\theta) = (2\theta - 8)^2 = 4\theta^2 - 32\theta + 64.$$

Client 5 $(n_5 = 5)$:

$$F_5(\theta) = \frac{1}{5}\sum_{i=1}^{5}(\theta i - 0.5i)^2$$
$$= \frac{1}{5}\left(55\theta^2 - 55\theta + 13.75\right).$$

Step 2: Local Gradient Computation. Recall that each client k minimizes a local mean squared error (MSE) objective of the form

$$F_k(\theta) = \frac{1}{|\mathcal{D}_k|}\sum_{(x,y)\in\mathcal{D}_k}(\theta x - y)^2.$$

Since $F_k(\theta)$ is a quadratic function in θ, its gradient is obtained by direct differentiation with respect to θ.

Client 1. For Client 1, the local objective is

$$F_1(\theta) = \frac{1}{2}\left[(\theta - 2)^2 + (2\theta - 4)^2\right].$$

Expanding the squared terms,

$$(\theta - 2)^2 = \theta^2 - 4\theta + 4, \qquad (2\theta - 4)^2 = 4\theta^2 - 16\theta + 16.$$

Thus,

$$F_1(\theta) = \frac{1}{2}\left(5\theta^2 - 20\theta + 20\right).$$

Differentiating with respect to θ yields

$$\nabla F_1(\theta) = \frac{1}{2}(10\theta - 20) = 5\theta - 10.$$

Client 2. Client 2 holds three samples, leading to

$$F_2(\theta) = \frac{1}{3}\left[(\theta - 1)^2 + (2\theta - 2)^2 + (3\theta - 3)^2\right].$$

Expanding and summing terms,

$$F_2(\theta) = \frac{1}{3}\left(14\theta^2 - 28\theta + 14\right).$$

Taking the derivative,

$$\nabla F_2(\theta) = \frac{1}{3}(28\theta - 28) = \frac{28}{3}\theta - \frac{28}{3}.$$

Client 3. The local loss for Client 3 is

$$F_3(\theta) = \frac{1}{2}\left[(\theta - 3)^2 + (2\theta - 6)^2\right].$$

After expansion,

$$F_3(\theta) = \frac{1}{2}\left(15\theta^2 - 45\theta + 45\right),$$

which gives the gradient

$$\nabla F_3(\theta) = \frac{1}{2}(30\theta - 45) = \frac{15}{2}\theta - \frac{45}{2}.$$

Client 4. Similarly,

$$F_4(\theta) = \frac{1}{2}\left[(\theta - 4)^2 + (2\theta - 8)^2\right] = \frac{1}{2}\left(16\theta^2 - 64\theta + 128\right),$$

and therefore

$$\nabla F_4(\theta) = 8\theta - 32.$$

Client 5. Finally,

$$F_5(\theta) = \frac{1}{2}\left[(\theta - 0.5)^2 + (2\theta - 1)^2\right] = \frac{1}{2}\left(11\theta^2 - 11\theta + 5.5\right),$$

which yields

$$\nabla F_5(\theta) = 11\theta - 5.5.$$

Collecting all gradients, we obtain

$$\begin{aligned}
\nabla F_1(\theta) &= 5\theta - 10,\\
\nabla F_2(\theta) &= \frac{28}{3}\theta - \frac{28}{3},\\
\nabla F_3(\theta) &= \frac{15}{2}\theta - \frac{45}{2},\\
\nabla F_4(\theta) &= 8\theta - 32,\\
\nabla F_5(\theta) &= 11\theta - 5.5.
\end{aligned}$$

Evaluating all gradients at the common initialization

$$\theta_k^{(0)} = 0, \quad \forall k,$$

gives the numerical values

$$\{-10,\ -9.33,\ -22.5,\ -32,\ -5.5\}.$$

Step 3: Local Optimization Step. Each client now performs one local gradient descent step using learning rate $\eta = 0.1$:

$$\theta_k^{(1/2)} = \theta_k^{(0)} - \eta \nabla F_k(\theta_k^{(0)}).$$

Substituting the computed gradients,

$$\begin{aligned}
\theta_1^{(1/2)} &= 0 - 0.1(-10) = 1.0,\\
\theta_2^{(1/2)} &= 0 - 0.1(-9.33) \approx 0.933,\\
\theta_3^{(1/2)} &= 0 - 0.1(-22.5) = 2.25,\\
\theta_4^{(1/2)} &= 0 - 0.1(-32) = 3.2,\\
\theta_5^{(1/2)} &= 0 - 0.1(-5.5) = 0.55.
\end{aligned}$$

These intermediate parameters represent purely local optimization effects prior to any peer-to-peer consensus interaction.

Step 4: Peer Consensus Update. Each client averages its parameter with those of its two neighbors.

$$\theta_1^{(1)} = \frac{1}{3}(1.0 + 0.933 + 0.55) = 0.828,$$
$$\theta_2^{(1)} = \frac{1}{3}(0.933 + 1.0 + 2.25) = 1.394,$$
$$\theta_3^{(1)} = \frac{1}{3}(2.25 + 0.933 + 3.2) = 2.128,$$
$$\theta_4^{(1)} = \frac{1}{3}(3.2 + 2.25 + 0.55) = 2.000,$$
$$\theta_5^{(1)} = \frac{1}{3}(0.55 + 3.2 + 1.0) = 1.583.$$

Step 5: Interpretation and Emergence. After one round of local optimization and peer consensus, the parameters are

$$\{0.828,\ 1.394,\ 2.128,\ 2.000,\ 1.583\}.$$

Despite heterogeneous dataset sizes and distributions, the parameters have moved closer together. Repeated iterations would further reduce disagreement, yielding global consensus without ever computing or broadcasting a centralized model.

> **Key Insight**
> Federated learning does *not* require uniform dataset sizes across clients. Heterogeneous local datasets are naturally supported through local objectives and network-based consensus, making orchestrator-less federated learning both realistic and robust.

Iteration 2. Iteration 2 begins from the post-consensus parameters obtained at the end of Iteration 1:

$$\theta^{(1)} = \{\theta_1^{(1)}, \theta_2^{(1)}, \theta_3^{(1)}, \theta_4^{(1)}, \theta_5^{(1)}\} = \{0.828,\ 1.394,\ 2.128,\ 2.000,\ 1.583\}.$$

Step 1: Local Gradient Evaluation. Using the gradient expressions derived in Iteration 1,

$$\nabla F_1(\theta) = 5\theta - 10,$$
$$\nabla F_2(\theta) = \frac{28}{3}\theta - \frac{28}{3},$$
$$\nabla F_3(\theta) = \frac{15}{2}\theta - \frac{45}{2},$$

$$\nabla F_4(\theta) = 8\theta - 32,$$
$$\nabla F_5(\theta) = 11\theta - 5.5,$$

each client evaluates its gradient at $\theta_k^{(1)}$.

$$\begin{aligned}
\nabla F_1(\theta_1^{(1)}) &= 5(0.828) - 10 = -5.86, \\
\nabla F_2(\theta_2^{(1)}) &= \tfrac{28}{3}(1.394) - \tfrac{28}{3} = 3.68, \\
\nabla F_3(\theta_3^{(1)}) &= \tfrac{15}{2}(2.128) - \tfrac{45}{2} = -6.54, \\
\nabla F_4(\theta_4^{(1)}) &= 8(2.000) - 32 = -16.0, \\
\nabla F_5(\theta_5^{(1)}) &= 11(1.583) - 5.5 = 11.91.
\end{aligned}$$

Step 2: Local Optimization Step. Each client performs a local gradient descent update using the same learning rate $\eta = 0.1$:

$$\theta_k^{(3/2)} = \theta_k^{(1)} - \eta \nabla F_k(\theta_k^{(1)}).$$

$$\begin{aligned}
\theta_1^{(3/2)} &= 0.828 - 0.1(-5.86) = 1.414, \\
\theta_2^{(3/2)} &= 1.394 - 0.1(3.68) = 1.026, \\
\theta_3^{(3/2)} &= 2.128 - 0.1(-6.54) = 2.782, \\
\theta_4^{(3/2)} &= 2.000 - 0.1(-16.0) = 3.600, \\
\theta_5^{(3/2)} &= 1.583 - 0.1(11.91) = 0.392.
\end{aligned}$$

These parameters again reflect purely local objectives prior to any coordination.

Step 3: Peer Consensus Update. Using the same ring topology, each client averages its value with its two neighbors.

$$\begin{aligned}
\theta_1^{(2)} &= \tfrac{1}{3}(1.414 + 1.026 + 0.392) = 0.944, \\
\theta_2^{(2)} &= \tfrac{1}{3}(1.026 + 1.414 + 2.782) = 1.741, \\
\theta_3^{(2)} &= \tfrac{1}{3}(2.782 + 1.026 + 3.600) = 2.469, \\
\theta_4^{(2)} &= \tfrac{1}{3}(3.600 + 2.782 + 0.392) = 2.258, \\
\theta_5^{(2)} &= \tfrac{1}{3}(0.392 + 3.600 + 1.414) = 1.802.
\end{aligned}$$

Step 4: Interpretation. After the second iteration, the parameter set becomes

$$\{0.944,\ 1.741,\ 2.469,\ 2.258,\ 1.802\}.$$

Compared to Iteration 1, the spread of parameters continues to shrink, demonstrating progressive alignment despite heterogeneous data sizes and objectives. The global model is not explicitly computed; instead, it emerges gradually through repeated local optimization and network-constrained consensus.

Iteration 3. Iteration 3 begins from the post-consensus parameters obtained at the end of Iteration 2:

$$\theta^{(2)} = \{0.944,\ 1.741,\ 2.469,\ 2.258,\ 1.802\}.$$

Step 1: Local Gradient Evaluation. Each client evaluates its previously derived gradient at $\theta_k^{(2)}$.

$$\begin{aligned}
\nabla F_1(\theta_1^{(2)}) &= 5(0.944) - 10 = -5.28,\\
\nabla F_2(\theta_2^{(2)}) &= \tfrac{28}{3}(1.741) - \tfrac{28}{3} = 6.93,\\
\nabla F_3(\theta_3^{(2)}) &= \tfrac{15}{2}(2.469) - \tfrac{45}{2} = -3.98,\\
\nabla F_4(\theta_4^{(2)}) &= 8(2.258) - 32 = -13.94,\\
\nabla F_5(\theta_5^{(2)}) &= 11(1.802) - 5.5 = 14.32.
\end{aligned}$$

Step 2: Local Optimization Step. Using the same learning rate $\eta = 0.1$, each client performs a local gradient descent update:

$$\theta_k^{(5/2)} = \theta_k^{(2)} - \eta \nabla F_k(\theta_k^{(2)}).$$

$$\begin{aligned}
\theta_1^{(5/2)} &= 0.944 - 0.1(-5.28) = 1.472,\\
\theta_2^{(5/2)} &= 1.741 - 0.1(6.93) = 1.048,\\
\theta_3^{(5/2)} &= 2.469 - 0.1(-3.98) = 2.867,\\
\theta_4^{(5/2)} &= 2.258 - 0.1(-13.94) = 3.652,\\
\theta_5^{(5/2)} &= 1.802 - 0.1(14.32) = 0.370.
\end{aligned}$$

As in previous iterations, these intermediate values reflect only local optimization effects prior to coordination.

Step 3: Peer Consensus Update. Each client averages its parameter with those of its two neighbors in the ring topology.

$$\theta_1^{(3)} = \frac{1}{3}(1.472 + 1.048 + 0.370) = 0.963,$$
$$\theta_2^{(3)} = \frac{1}{3}(1.048 + 1.472 + 2.867) = 1.796,$$
$$\theta_3^{(3)} = \frac{1}{3}(2.867 + 1.048 + 3.652) = 2.522,$$
$$\theta_4^{(3)} = \frac{1}{3}(3.652 + 2.867 + 0.370) = 2.296,$$
$$\theta_5^{(3)} = \frac{1}{3}(0.370 + 3.652 + 1.472) = 1.831.$$

Step 4: Interpretation. After the third iteration, the parameter set becomes

$$\{0.963,\ 1.796,\ 2.522,\ 2.296,\ 1.831\}.$$

Relative to Iterations 1 and 2, the parameters continue to concentrate around a common value, while local deviations introduced by gradient steps are progressively damped by peer consensus. This illustrates how orchestrator-less federated learning balances local objective minimization with global agreement through repeated network-constrained interaction.

Key Observations from the First Three Iterations
The first three iterations of orchestrator-less federated learning highlight several fundamental aspects of decentralized optimization and clarify the role of repeated interaction in achieving global agreement.

1. **Progressive Reduction of Disagreement** Across Iterations 1 to 3, the dispersion among local model parameters consistently decreases after each peer-consensus step. Although local gradient updates introduce temporary divergence based on heterogeneous objectives and dataset sizes, the consensus operation systematically attenuates these differences. This demonstrates that agreement is not enforced instantaneously, but emerges gradually through repeated local averaging constrained by the network topology.
2. **Interaction Between Local Optimization and Network Dynamics** Each iteration reveals a characteristic two-phase behavior: a local optimization phase that pulls parameters toward client-specific optima, followed by a consensus phase that pulls parameters toward one another. The observed oscillations in intermediate values are therefore not a sign of instability, but an inherent feature of decentralized learning, reflecting the balance between individual objectives and collective coordination.

3. **Role of Graph Structure and Information Propagation** Within three iterations, information from each client has begun to influence all others through multi-hop propagation along the ring topology. However, complete mixing has not yet occurred, and residual differences remain. This highlights the importance of network connectivity and spectral properties of the communication graph in determining the rate at which consensus is achieved.

Are More Iterations Required? Yes—additional iterations are generally required for full convergence. While the first three iterations establish clear convergence trends, exact agreement among all θ_k values is achieved only in the limit as the number of iterations increases. The required number of iterations depends on factors such as the learning rate, data heterogeneity, and the mixing speed of the communication graph.

When Fewer Iterations May Suffice In practice, strict equality of parameters is often unnecessary. If approximate consensus yields satisfactory predictive performance, training may be terminated early to reduce communication and energy costs. This trade-off between accuracy and efficiency is particularly relevant in resource-constrained or latency-sensitive federated environments.

5.3 Gossip-Based Learning Protocols

Gossip-based learning protocols constitute one of the most widely studied and practically relevant approaches to decentralized federated learning. Originally developed in the context of distributed averaging and sensor networks, gossip mechanisms provide a simple yet powerful means of propagating information in systems where global coordination is unavailable or undesirable. Their appeal lies in their minimal communication assumptions, inherent robustness, and scalability to large, dynamic networks.

In the context of federated learning, gossip protocols replace centralized aggregation with repeated, localized exchanges of model parameters. Rather than synchronizing all participants through a global server, learning progresses through stochastic, peer-to-peer interactions that gradually disseminate information across the network. Over time, these interactions drive the system toward agreement while preserving client autonomy and data locality.

5.3.1 Gossip Mechanism

At each iteration t, a client k randomly selects one of its neighbors j according to a predefined probability distribution that may depend on network connectivity or communication availability. The selected pair exchanges their current model parameters and updates them using a symmetric averaging rule. A basic pairwise gossip update is given by

$$\theta_k^{(t+1)} = \theta_j^{(t+1)} = \frac{1}{2}\left(\theta_k^{(t)} + \theta_j^{(t)}\right),$$

while all other clients retain their parameters unchanged during that iteration.

This simple operation ensures that information is conserved and shared locally, with the effect of gradually reducing disagreement between participants. When combined with local optimization steps—such as gradient descent on private datasets—gossip updates serve as a decentralized coordination mechanism that aligns client models without ever constructing or broadcasting a global parameter vector.

5.3.2 Integration with Local Learning

In practical gossip-based federated learning, each iteration typically consists of two phases: a local optimization phase, during which each client updates its parameters using its private data, followed by a communication phase, during which one or more gossip exchanges occur. The interleaving of these phases enables clients to incorporate both local statistical information and global structural information carried by the network.

Importantly, gossip interactions need not be synchronized across all clients. Updates may occur asynchronously, and multiple disjoint pairs may communicate in parallel, further enhancing scalability and fault tolerance. Such flexibility makes gossip protocols particularly suitable for large-scale, heterogeneous, and intermittently connected federated systems.

Key Properties. Gossip-based learning exhibits several distinctive characteristics that differentiate it from centralized federated aggregation:

- **Fully decentralized and symmetric communication:** All interactions occur directly between peers, and no client occupies a privileged role in coordination or aggregation.
- **Probabilistic convergence behavior:** Convergence emerges in expectation and depends on repeated random interactions, network connectivity, and mixing properties rather than deterministic synchronization.
- **High robustness to failures and churn:** Since learning does not rely on any single entity, temporary client failures, dropouts, or communication disruptions do not halt the overall learning process.
- **Scalability under limited infrastructure:** Each iteration involves only a small number of communicating peers, keeping communication costs low even as the network grows.

Despite these advantages, gossip-based protocols typically converge more slowly than centralized aggregation schemes. Because only limited information is exchanged per interaction, many iterations may be required for global information to fully propagate through the network, particularly in sparse or poorly connected graphs.

Gossip Insight
Gossip-based federated learning replaces explicit global aggregation with repeated, randomized local averaging. This design trades rapid convergence for robustness, scalability, and minimal coordination, making gossip protocols well suited to large, dynamic, and trust-constrained federated environments.

Worked Example: Gossip-Based Federated Learning
We now illustrate the operation of a gossip-based learning protocol through a concrete numerical example. The objective is to demonstrate how global agreement emerges through repeated randomized pairwise interactions, without any centralized coordination.

Setup. Consider $K = 5$ clients, each maintaining a scalar model parameter $\theta_k^{(t)}$ at iteration t. For simplicity, we focus exclusively on the gossip communication step and omit local gradient updates in this example. The initial model parameters are

$$\theta^{(0)} = \{1.0,\ 3.0,\ 5.0,\ 7.0,\ 9.0\}.$$

Clients are connected in a ring topology

$$1 \leftrightarrow 2 \leftrightarrow 3 \leftrightarrow 4 \leftrightarrow 5 \leftrightarrow 1,$$

and each gossip interaction occurs between a randomly selected pair of neighbors.

Gossip Update Rule. When a pair of neighboring clients (k, j) is selected, both clients replace their parameters with the average of their current values:

$$\theta_k^{(t+1)} = \theta_j^{(t+1)} = \frac{1}{2}\left(\theta_k^{(t)} + \theta_j^{(t)}\right).$$

All other clients retain their parameters unchanged.

Iteration 1. Suppose clients 2 and 3 are randomly selected. Their parameters are $\theta_2^{(0)} = 3.0$ and $\theta_3^{(0)} = 5.0$. After gossip averaging,

$$\theta_2^{(1)} = \theta_3^{(1)} = \frac{1}{2}(3.0 + 5.0) = 4.0.$$

The updated parameter set is therefore

$$\theta^{(1)} = \{1.0,\ 4.0,\ 4.0,\ 7.0,\ 9.0\}.$$

Iteration 2. Next, suppose clients 4 and 5 engage in a gossip exchange. Their parameters are $\theta_4^{(1)} = 7.0$ and $\theta_5^{(1)} = 9.0$. Averaging yields

$$\theta_4^{(2)} = \theta_5^{(2)} = \frac{1}{2}(7.0 + 9.0) = 8.0.$$

The parameter vector becomes

$$\theta^{(2)} = \{1.0,\ 4.0,\ 4.0,\ 8.0,\ 8.0\}.$$

Iteration 3. Assume that clients 1 and 5 are selected. Their current parameters are $\theta_1^{(2)} = 1.0$ and $\theta_5^{(2)} = 8.0$. After the gossip update,

$$\theta_1^{(3)} = \theta_5^{(3)} = \frac{1}{2}(1.0 + 8.0) = 4.5.$$

Thus,

$$\theta^{(3)} = \{4.5,\ 4.0,\ 4.0,\ 8.0,\ 4.5\}.$$

Iteration 4. Suppose clients 3 and 4 now communicate. With $\theta_3^{(3)} = 4.0$ and $\theta_4^{(3)} = 8.0$, we obtain

$$\theta_3^{(4)} = \theta_4^{(4)} = \frac{1}{2}(4.0 + 8.0) = 6.0.$$

The updated parameters are

$$\theta^{(4)} = \{4.5,\ 4.0,\ 6.0,\ 6.0,\ 4.5\}.$$

Emergent Consensus Behavior. After only a few gossip interactions, the spread of parameter values has decreased substantially. Although no global model has been computed, information has propagated through the network via randomized local averaging. Repeated gossip iterations cause all parameters to converge toward the global average of the initial values,

$$\bar{\theta} = \frac{1}{5}(1 + 3 + 5 + 7 + 9) = 5.0.$$

Key Insight from the Gossip Example
Gossip-based learning achieves global agreement through randomized, pairwise averaging. Each interaction preserves the global mean while reducing disagreement, causing consensus to emerge gradually as information diffuses through the network. No central coordinator is required, and convergence occurs as a statistical consequence of repeated local interactions.

5.4 Consensus-Based Aggregation Methods

Consensus-based aggregation methods constitute a principled and systematic extension of gossip-based learning within decentralized federated learning. Whereas gossip protocols rely primarily on randomized, pairwise averaging to disseminate information across a network, consensus-based methods elevate this idea by explicitly formulating learning as a *distributed optimization problem subject to agreement constraints*. This explicit formulation enables rigorous algorithmic design, provides a foundation for analytical convergence guarantees, and establishes strong connections with classical results from distributed optimization, control theory, and multi-agent systems.

In consensus-based federated learning, each client maintains its own local model parameters and iteratively refines them through a combination of two tightly coupled processes: local learning driven by private data and coordinated interaction with neighboring clients. Unlike centralized federated learning, where global aggregation occurs periodically at a server, or gossip-based methods, where averaging occurs sporadically through random encounters, consensus-based approaches enforce agreement *continuously and incrementally*. The learning dynamics are therefore governed by structured, weight-controlled exchanges that reflect the topology of the underlying communication graph.

5.4.1 Consensus Update Rule

Let client k maintain model parameters $\theta_k^{(t)}$ at iteration t. A generic consensus-based update rule can be written as

$$\theta_k^{(t+1)} = \theta_k^{(t)} - \eta \nabla F_k(\theta_k^{(t)}) + \sum_{j \in \mathcal{N}_k} w_{kj}(\theta_j^{(t)} - \theta_k^{(t)}),$$

where $\eta > 0$ is the learning rate, $\mathcal{N}_k$ denotes the set of neighbors of client k in the communication graph, and w_{kj} are nonnegative consensus weights associated with the edges connecting client k to its neighbors.

This update consists of two complementary and conceptually distinct components. The first term performs local optimization by moving the model parameters in the direction of the negative gradient of the client-specific objective function. This term captures the influence of the client's private data and ensures that learning remains data-driven and decentralized. The second term enforces agreement by pulling the local model toward a weighted average of neighboring models. Together, these components balance two competing objectives: minimization of local empirical risk and alignment of models across the network. The resulting dynamics enable collective learning without requiring any explicit global model or centralized coordinator.

5.4.2 Role of Consensus Weights

The consensus weights w_{kj} play a central role in shaping the behavior and stability of the learning process. In practice, these weights are often chosen to satisfy symmetry and stochasticity conditions such as

$$w_{kj} = w_{jk}, \qquad \sum_{j \in \mathcal{N}_k} w_{kj} \leq 1,$$

which ensure that information exchange is balanced and that the update rule does not introduce systematic bias. Under such choices, the consensus term can be interpreted as a discrete graph Laplacian operator acting on the vector of model parameters. This interpretation reveals a deep connection between federated learning dynamics and the spectral properties of the communication graph, including notions such as algebraic connectivity and mixing rates. Well-chosen weights therefore enable stable updates, prevent parameter oscillation, and promote rapid diffusion of information across the network.

5.4.3 Interpretation as Distributed Consensus Optimization

From an optimization perspective, consensus-based federated learning can be viewed as solving the constrained problem

$$\min_{\{\theta_k\}} \sum_{k=1}^{K} F_k(\theta_k) \quad \text{subject to} \quad \theta_k = \theta_j, \;\; \forall (k,j) \in \mathcal{E},$$

where $\mathcal{E}$ denotes the set of communication edges. The equality constraints enforce global agreement among all local models. Rather than imposing these constraints explicitly, the consensus term in the update rule acts as a *soft constraint* that penalizes disagreement between neighboring models. Through repeated iterations, these penalties gradually enforce agreement across the entire network.

Unlike centralized federated learning, where consensus is achieved through explicit global aggregation at each round, consensus-based methods allow agreement to emerge organically through continuous local corrections. This design eliminates the need for a central orchestrator while preserving a clear and well-defined optimization objective.

5.4.4 Convergence Considerations

The convergence behavior of consensus-based aggregation methods depends on several interacting factors:

- **Network connectivity:** Densely connected or well-mixed graphs enable faster information propagation and reduce the number of iterations required to reach agreement.
- **Weight symmetry and stochasticity:** Proper normalization and symmetry of consensus weights ensure stability, prevent drift, and preserve fairness among clients.
- **Gradient heterogeneity:** Bounded dissimilarity among local gradients limits the extent to which client-specific objectives pull models in conflicting directions.
- **Step-size selection:** The learning rate η must balance local optimization progress against the strength of consensus enforcement to avoid divergence or excessive smoothing.

When these conditions are satisfied and standard smoothness assumptions hold, consensus-based methods can be shown to converge to stationary points of the global federated objective, even in the presence of data heterogeneity and partial participation.

Relation to Classical Distributed Optimization. Consensus-based federated learning provides a direct conceptual and mathematical bridge between modern federated paradigms and classical distributed optimization techniques such as distributed gradient descent, primal–dual methods, and consensus averaging in multi-agent systems. By framing federated learning explicitly as a consensus optimization problem, these methods unify ideas from machine learning, control theory, and networked systems. This unification offers both conceptual clarity and analytical rigor, making consensus-based aggregation a foundational tool for fully decentralized learning.

Consensus-Based Aggregation Insight

Consensus-based federated learning replaces explicit global aggregation with continuous, graph-structured alignment. By embedding consensus enforcement directly into each local update, these methods achieve decentralized coordination while preserving the optimization foundations and convergence guarantees of classical distributed learning.

Worked Example: Consensus-Based Aggregation

We now present a concrete numerical example to illustrate how consensus-based aggregation combines local optimization with network-constrained model alignment. Unlike gossip-based learning, consensus is enforced at *every iteration* through deterministic, weighted corrections derived from neighboring models.

Problem Setup. Consider $K = 4$ clients connected in a ring topology

$$1 \leftrightarrow 2 \leftrightarrow 3 \leftrightarrow 4 \leftrightarrow 1.$$

Each client k maintains a scalar model parameter $\theta_k^{(t)}$ at iteration t.

Each client minimizes a local quadratic loss of the form

$$F_k(\theta) = \frac{1}{2}(\theta - a_k)^2,$$

where a_k represents the client-specific optimum induced by its private data. Let

$$a_1 = 1, \quad a_2 = 3, \quad a_3 = 5, \quad a_4 = 7.$$

The gradient of each local objective is therefore

$$\nabla F_k(\theta) = \theta - a_k.$$

We use a learning rate $\eta = 0.2$. Consensus weights are chosen uniformly as

$$w_{kj} = 0.1 \quad \text{for all } j \in \mathcal{N}_k.$$

All clients are initialized at

$$\theta_k^{(0)} = 0, \quad \forall k.$$

Consensus-Based Update Rule. At each iteration, every client updates its model according to

$$\theta_k^{(t+1)} = \theta_k^{(t)} - \eta \nabla F_k(\theta_k^{(t)}) + \sum_{j \in \mathcal{N}_k} w_{kj}(\theta_j^{(t)} - \theta_k^{(t)}).$$

Convergence Remark

Consensus-based aggregation does not reach the global optimum in a finite number of iterations. Instead, the sequence $\{\theta_k^{(t)}\}$ converges *asymptotically* to the unique minimizer of the global objective, which in this example is

$$\theta^\star = \frac{1}{4}(1 + 3 + 5 + 7) = 4.$$

While exact equality is achieved only in the limit $t \to \infty$, a small number of iterations is typically sufficient to reach any desired numerical accuracy.

Iteration 1.

Step 1: Local Gradient Evaluation. Using

$$\nabla F_k(\theta) = \theta - a_k$$

and the initialization $\theta_k^{(0)} = 0$, we obtain

$$\begin{aligned}\nabla F_1(0) &= -1,\\ \nabla F_2(0) &= -3,\\ \nabla F_3(0) &= -5,\\ \nabla F_4(0) &= -7.\end{aligned}$$

Step 2: Local Gradient Descent Update. Each client performs a local gradient descent step:

$$\tilde{\theta}_k = \theta_k^{(0)} - \eta \nabla F_k(\theta_k^{(0)}).$$

Substituting the numerical values,

$$\begin{aligned}\tilde{\theta}_1 &= 0.2,\\ \tilde{\theta}_2 &= 0.6,\\ \tilde{\theta}_3 &= 1.0,\\ \tilde{\theta}_4 &= 1.4.\end{aligned}$$

Step 3: Consensus Correction. Each client then applies a consensus correction using its two neighbors:

$$\theta_k^{(1)} = \tilde{\theta}_k + 0.1 \sum_{j \in \mathcal{N}_k} (\tilde{\theta}_j - \tilde{\theta}_k).$$

Client 1 (neighbors: 2 and 4):

$$\theta_1^{(1)} = 0.2 + 0.1[(0.6 - 0.2) + (1.4 - 0.2)]$$
$$= 0.36.$$

Client 2 (neighbors: 1 and 3):

$$\theta_2^{(1)} = 0.6 + 0.1[(0.2 - 0.6) + (1.0 - 0.6)]$$
$$= 0.60.$$

Client 3 (neighbors: 2 and 4):

$$\theta_3^{(1)} = 1.0 + 0.1[(0.6 - 1.0) + (1.4 - 1.0)]$$
$$= 1.00.$$

Client 4 (neighbors: 3 and 1):

$$\theta_4^{(1)} = 1.4 + 0.1[(1.0 - 1.4) + (0.2 - 1.4)]$$
$$= 1.24.$$

The model parameters after Iteration 1 are therefore

$$\theta^{(1)} = \{0.36,\ 0.60,\ 1.00,\ 1.24\}.$$

Interpretation of Iteration 1. After the first iteration, the model parameters are

$$\theta^{(1)} = \{0.36,\ 0.60,\ 1.00,\ 1.24\}.$$

At this stage, the influence of the local objective functions is clearly visible. Clients associated with larger local optima (such as Client 4 with $a_4 = 7$) exhibit larger parameter values, while those with smaller local optima remain closer to the origin. However, the consensus correction has already begun to moderate this divergence. In particular, Clients 2 and 3, which are symmetrically positioned in the ring topology, retain identical values due to balanced neighbor influences. This iteration demonstrates the fundamental behavior of consensus-based aggregation: local learning drives progress toward client-specific optima, while neighbor interactions immediately counteract excessive separation among models.

Iteration 2. Iteration 2 begins from the post-consensus parameters obtained after Iteration 1:

$$\theta^{(1)} = \{0.36,\ 0.60,\ 1.00,\ 1.24\}.$$

Step 1: Local Gradient Evaluation. Using the gradient expression

$$\nabla F_k(\theta) = \theta - a_k,$$

we evaluate the gradients at $\theta_k^{(1)}$:

$$\begin{aligned}
\nabla F_1(0.36) &= 0.36 - 1 = -0.64,\\
\nabla F_2(0.60) &= 0.60 - 3 = -2.40,\\
\nabla F_3(1.00) &= 1.00 - 5 = -4.00,\\
\nabla F_4(1.24) &= 1.24 - 7 = -5.76.
\end{aligned}$$

Step 2: Local Gradient Descent Update. Each client applies a local gradient descent step with learning rate $\eta = 0.2$:

$$\tilde{\theta}_k = \theta_k^{(1)} - \eta \nabla F_k(\theta_k^{(1)}).$$

Substituting the numerical values,

$$\begin{aligned}
\tilde{\theta}_1 &= 0.36 - 0.2(-0.64) = 0.488,\\
\tilde{\theta}_2 &= 0.60 - 0.2(-2.40) = 1.080,\\
\tilde{\theta}_3 &= 1.00 - 0.2(-4.00) = 1.800,\\
\tilde{\theta}_4 &= 1.24 - 0.2(-5.76) = 2.392.
\end{aligned}$$

Step 3: Consensus Correction. Each client performs a consensus correction using uniform weights $w_{kj} = 0.1$:

$$\theta_k^{(2)} = \tilde{\theta}_k + 0.1 \sum_{j \in \mathcal{N}_k} (\tilde{\theta}_j - \tilde{\theta}_k).$$

Client 1 (neighbors: 2 and 4):

$$\begin{aligned}
\theta_1^{(2)} &= 0.488 + 0.1[(1.080 - 0.488) + (2.392 - 0.488)]\\
&\approx 0.737.
\end{aligned}$$

Client 2 (neighbors: 1 and 3):

$$\theta_2^{(2)} = 1.080 + 0.1[(0.488 - 1.080) + (1.800 - 1.080)]$$
$$\approx 1.093.$$

Client 3 (neighbors: 2 and 4):

$$\theta_3^{(2)} = 1.800 + 0.1[(1.080 - 1.800) + (2.392 - 1.800)]$$
$$\approx 1.787.$$

Client 4 (neighbors: 3 and 1):

$$\theta_4^{(2)} = 2.392 + 0.1[(1.800 - 2.392) + (0.488 - 2.392)]$$
$$\approx 2.142.$$

The parameters after Iteration 2 are therefore

$$\theta^{(2)} = \{0.737, \ 1.093, \ 1.787, \ 2.142\}.$$

Interpretation of Iteration 2. Compared to Iteration 1, the spread among client models has further reduced, indicating stronger alignment across the network. Although each local gradient continues to pull the parameters toward distinct client-specific optima $\{1, 3, 5, 7\}$, the consensus correction acts as a stabilizing force that prevents unbounded divergence. Notably, the ordering of parameters remains consistent with the ordering of local optima, but the relative differences between adjacent clients shrink. This iteration illustrates how consensus-based aggregation progressively transforms heterogeneous local learning signals into a coherent global trajectory, even in the absence of any centralized coordination.

Iteration 3. Iteration 3 starts from the post-consensus parameters obtained at the end of Iteration 2:

$$\theta^{(2)} = \{0.737, \ 1.093, \ 1.787, \ 2.142\}.$$

Step 1: Local Gradient Evaluation. Using the gradient expression

$$\nabla F_k(\theta) = \theta - a_k, \quad a_1 = 1, \ a_2 = 3, \ a_3 = 5, \ a_4 = 7,$$

we evaluate the gradients at $\theta_k^{(2)}$:

$$\nabla F_1(0.737) = 0.737 - 1 = -0.263,$$
$$\nabla F_2(1.093) = 1.093 - 3 = -1.907,$$
$$\nabla F_3(1.787) = 1.787 - 5 = -3.213,$$
$$\nabla F_4(2.142) = 2.142 - 7 = -4.858.$$

Step 2: Local Gradient Descent Update. Each client performs a local gradient descent step with learning rate $\eta = 0.2$:

$$\tilde{\theta}_k = \theta_k^{(2)} - \eta \nabla F_k(\theta_k^{(2)}).$$

Substituting the numerical values,

$$\tilde{\theta}_1 = 0.737 - 0.2(-0.263) = 0.790,$$
$$\tilde{\theta}_2 = 1.093 - 0.2(-1.907) = 1.474,$$
$$\tilde{\theta}_3 = 1.787 - 0.2(-3.213) = 2.430,$$
$$\tilde{\theta}_4 = 2.142 - 0.2(-4.858) = 3.114.$$

Step 3: Consensus Correction. Each client applies the consensus correction using uniform weights $w_{kj} = 0.1$:

$$\theta_k^{(3)} = \tilde{\theta}_k + 0.1 \sum_{j \in \mathcal{N}_k} (\tilde{\theta}_j - \tilde{\theta}_k).$$

Client 1 (neighbors: 2 and 4):

$$\theta_1^{(3)} = 0.790 + 0.1[(1.474 - 0.790) + (3.114 - 0.790)] \approx 1.091.$$

Client 2 (neighbors: 1 and 3):

$$\theta_2^{(3)} = 1.474 + 0.1[(0.790 - 1.474) + (2.430 - 1.474)] \approx 1.501.$$

Client 3 (neighbors: 2 and 4):

$$\theta_3^{(3)} = 2.430 + 0.1[(1.474 - 2.430) + (3.114 - 2.430)] \approx 2.403.$$

Client 4 (neighbors: 3 and 1):

$$\theta_4^{(3)} = 3.114 + 0.1[(2.430 - 3.114) + (0.790 - 3.114)]$$
$$\approx 2.813.$$

The parameters after Iteration 3 are therefore

$$\theta^{(3)} = \{1.091,\ 1.501,\ 2.403,\ 2.813\}.$$

Interpretation of Iteration 3. By the third iteration, the effects of consensus-based aggregation become increasingly pronounced. While local gradient descent continues to push each client toward its own data-induced optimum, the consensus correction now exerts a dominant influence on the overall dynamics. The spread between the smallest and largest parameters has narrowed substantially compared to earlier iterations, indicating stronger network-wide alignment.

Importantly, no client overshoots toward its local optimum despite continued gradient pressure. Instead, all models evolve along a shared trajectory that reflects a compromise among the heterogeneous objectives. This behavior illustrates a key strength of consensus-based aggregation: disagreement decays progressively with each iteration, and coordinated learning emerges naturally from repeated local updates and structured neighbor interactions, without requiring any centralized model aggregation.

Iteration 4. Iteration 4 starts from the post-consensus parameters obtained at the end of Iteration 3:

$$\theta^{(3)} = \{1.091,\ 1.501,\ 2.403,\ 2.813\}.$$

Step 1: Local Gradient Evaluation. Recall that the local gradient for each client is

$$\nabla F_k(\theta) = \theta - a_k, \quad a_1 = 1,\ a_2 = 3,\ a_3 = 5,\ a_4 = 7.$$

Evaluating the gradients at $\theta_k^{(3)}$, we obtain

$$\nabla F_1(1.091) = 1.091 - 1 = 0.091,$$
$$\nabla F_2(1.501) = 1.501 - 3 = -1.499,$$
$$\nabla F_3(2.403) = 2.403 - 5 = -2.597,$$
$$\nabla F_4(2.813) = 2.813 - 7 = -4.187.$$

Step 2: Local Gradient Descent Update. Each client performs a local gradient descent update with learning rate $\eta = 0.2$:

$$\tilde{\theta}_k = \theta_k^{(3)} - \eta \nabla F_k(\theta_k^{(3)}).$$

Substituting the numerical values,

$$\begin{aligned}
\tilde{\theta}_1 &= 1.091 - 0.2(0.091) = 1.073,\\
\tilde{\theta}_2 &= 1.501 - 0.2(-1.499) = 1.801,\\
\tilde{\theta}_3 &= 2.403 - 0.2(-2.597) = 2.922,\\
\tilde{\theta}_4 &= 2.813 - 0.2(-4.187) = 3.650.
\end{aligned}$$

Step 3: Consensus Correction. Each client applies a consensus correction using uniform weights $w_{kj} = 0.1$:

$$\theta_k^{(4)} = \tilde{\theta}_k + 0.1 \sum_{j \in \mathcal{N}_k} (\tilde{\theta}_j - \tilde{\theta}_k).$$

Client 1 (neighbors: 2 and 4):

$$\begin{aligned}
\theta_1^{(4)} &= 1.073 + 0.1[(1.801 - 1.073) + (3.650 - 1.073)]\\
&= 1.073 + 0.1(0.728 + 2.577)\\
&= 1.073 + 0.331 \approx 1.404.
\end{aligned}$$

Client 2 (neighbors: 1 and 3):

$$\begin{aligned}
\theta_2^{(4)} &= 1.801 + 0.1[(1.073 - 1.801) + (2.922 - 1.801)]\\
&= 1.801 + 0.1(-0.728 + 1.121)\\
&= 1.801 + 0.039 \approx 1.840.
\end{aligned}$$

Client 3 (neighbors: 2 and 4):

$$\begin{aligned}
\theta_3^{(4)} &= 2.922 + 0.1[(1.801 - 2.922) + (3.650 - 2.922)]\\
&= 2.922 + 0.1(-1.121 + 0.728)\\
&= 2.922 - 0.039 \approx 2.883.
\end{aligned}$$

Client 4 (neighbors: 3 and 1):

$$\begin{aligned}\theta_4^{(4)} &= 3.650 + 0.1[(2.922 - 3.650) + (1.073 - 3.650)] \\ &= 3.650 + 0.1(-0.728 - 2.577) \\ &= 3.650 - 0.331 \approx 3.319.\end{aligned}$$

The parameters after Iteration 4 are therefore

$$\theta^{(4)} = \{1.404,\ 1.840,\ 2.883,\ 3.319\}.$$

Interpretation of Iteration 4. By the fourth iteration, the models exhibit strong alignment across the network. Although local gradients still reflect heterogeneous client objectives, the consensus correction increasingly dominates the update dynamics, preventing excessive divergence. All parameters now lie within a narrow range and continue to evolve coherently toward the global optimum

$$\theta^{\star} = \frac{1}{4}(1 + 3 + 5 + 7) = 4.$$

This iteration illustrates the transition from early-stage coordination to near-consensus behavior in consensus-based federated learning.

Iteration 5. Iteration 5 starts from the post-consensus parameters obtained at the end of Iteration 4:

$$\theta^{(4)} = \{1.404,\ 1.840,\ 2.883,\ 3.319\}.$$

Step 1: Local Gradient Evaluation. Using the gradient expression

$$\nabla F_k(\theta) = \theta - a_k, \quad a_1 = 1,\ a_2 = 3,\ a_3 = 5,\ a_4 = 7,$$

we evaluate the gradients at $\theta_k^{(4)}$:

$$\begin{aligned}\nabla F_1(1.404) &= 1.404 - 1 = 0.404, \\ \nabla F_2(1.840) &= 1.840 - 3 = -1.160, \\ \nabla F_3(2.883) &= 2.883 - 5 = -2.117, \\ \nabla F_4(3.319) &= 3.319 - 7 = -3.681.\end{aligned}$$

Step 2: Local Gradient Descent Update. Each client performs a local gradient descent update with learning rate $\eta = 0.2$:

$$\tilde{\theta}_k = \theta_k^{(4)} - \eta \nabla F_k(\theta_k^{(4)}).$$

Substituting the numerical values,

$$
\begin{aligned}
\tilde{\theta}_1 &= 1.404 - 0.2(0.404) = 1.323,\\
\tilde{\theta}_2 &= 1.840 - 0.2(-1.160) = 2.072,\\
\tilde{\theta}_3 &= 2.883 - 0.2(-2.117) = 3.306,\\
\tilde{\theta}_4 &= 3.319 - 0.2(-3.681) = 4.055.
\end{aligned}
$$

Step 3: Consensus Correction. Each client applies the consensus correction using uniform weights $w_{kj} = 0.1$:

$$
\theta_k^{(5)} = \tilde{\theta}_k + 0.1 \sum_{j \in \mathcal{N}_k} (\tilde{\theta}_j - \tilde{\theta}_k).
$$

Client 1 (neighbors: 2 and 4):

$$
\begin{aligned}
\theta_1^{(5)} &= 1.323 + 0.1[(2.072 - 1.323) + (4.055 - 1.323)]\\
&= 1.323 + 0.1(0.749 + 2.732)\\
&= 1.323 + 0.348 \approx 1.671.
\end{aligned}
$$

Client 2 (neighbors: 1 and 3):

$$
\begin{aligned}
\theta_2^{(5)} &= 2.072 + 0.1[(1.323 - 2.072) + (3.306 - 2.072)]\\
&= 2.072 + 0.1(-0.749 + 1.234)\\
&= 2.072 + 0.049 \approx 2.121.
\end{aligned}
$$

Client 3 (neighbors: 2 and 4):

$$
\begin{aligned}
\theta_3^{(5)} &= 3.306 + 0.1[(2.072 - 3.306) + (4.055 - 3.306)]\\
&= 3.306 + 0.1(-1.234 + 0.749)\\
&= 3.306 - 0.049 \approx 3.258.
\end{aligned}
$$

Client 4 (neighbors: 3 and 1):

$$
\begin{aligned}
\theta_4^{(5)} &= 4.055 + 0.1[(3.306 - 4.055) + (1.323 - 4.055)]\\
&= 4.055 + 0.1(-0.749 - 2.732)\\
&= 4.055 - 0.348 \approx 3.707.
\end{aligned}
$$

The parameters after Iteration 5 are therefore

$$\theta^{(5)} = \{1.671,\ 2.121,\ 3.258,\ 3.707\}.$$

Interpretation of Iteration 5. By the fifth iteration, all client models have entered a regime of near-consensus. The parameter values are tightly clustered and continue to move coherently toward the global optimum

$$\theta^{\star} = \frac{1}{4}(1 + 3 + 5 + 7) = 4.$$

Although exact convergence is achieved only asymptotically as $t \to \infty$, this iteration already demonstrates strong agreement and near-optimality. Further iterations would yield progressively smaller adjustments, illustrating the diminishing-return behavior typical of consensus-based aggregation methods.

Iteration Continuation Criterion
The consensus-based update must be repeated iteratively until the parameter disagreement across clients vanishes and all $\theta_k^{(t)}$ converge to the common global optimum. In practice, iterations continue until successive updates fall below a prescribed convergence tolerance.

5.5 Blockchain-Enabled Federated Learning

Blockchain-enabled federated learning introduces a fundamentally different coordination paradigm by replacing trusted central servers with a *distributed, tamper-resistant ledger*. Rather than relying on a single orchestrator to manage participation, aggregate updates, and enforce protocol rules, blockchain technology enables a collectively maintained record of learning events that is verifiable by all participants.

This approach is particularly attractive in settings where participants do not fully trust one another, where auditability is essential, or where institutional control over the learning process is undesirable or infeasible.

5.5.1 Role of the Blockchain

In a blockchain-enabled federated learning system, the blockchain functions as a *shared coordination and verification layer* rather than as a computational engine for model training. All learning-related computation—such as gradient evaluation, local optimization, and model updating—remains entirely local to each client. The blockchain intervenes only to coordinate, validate, and record the learning process in a decentralized and tamper-resistant manner.

From a system-design perspective, the blockchain replaces the trusted central server with a collectively maintained ledger that establishes a common, verifiable view of the global learning state. This separation of concerns is crucial: computationally intensive model training is kept off-chain, while trust-sensitive coordination logic is handled on-chain.

In a typical blockchain-enabled federated workflow, clients:

- submit cryptographic hashes, signatures, or compact representations of local model updates to the blockchain, ensuring verifiable commitment without revealing raw parameters,
- validate the authenticity, provenance, and temporal ordering of updates through the blockchain's consensus mechanism,
- retrieve references to the most recent globally agreed-upon model state, update index, or aggregation outcome recorded on the ledger.

By recording only hashes or succinct metadata rather than full model parameters, the blockchain avoids prohibitive storage and bandwidth costs. At the same time, cryptographic linking between blocks guarantees strong integrity: any modification to a recorded update would invalidate the entire chain of subsequent records. Consequently, attempts to tamper with model updates, reorder learning rounds, or falsify participation histories become immediately detectable.

In essence, the blockchain establishes a *verifiable coordination backbone* for federated learning. It ensures that all participants observe a consistent and immutable history of learning events, enabling decentralized trust, transparent auditing, and protocol-level enforcement without interfering with the efficiency of local model training.

5.5.2 Smart Contracts and Automation

Smart contracts play a central role in operationalizing blockchain-enabled federated learning by transforming abstract protocol rules into *executable system logic*. These self-executing programs are deployed on the blockchain and run deterministically whenever predefined conditions are met, eliminating the need for manual coordination or trusted intermediaries.

From a federated learning perspective, smart contracts act as *autonomous governors* of the training process. They ensure that participation rules, aggregation procedures, and reward mechanisms are enforced uniformly across all clients, regardless of trust assumptions. Once deployed, the contract logic cannot be altered unilaterally, thereby providing strong guarantees of procedural fairness and predictability.

Typical responsibilities encoded within smart contracts include:

- Specifying aggregation rules, update formats, and validation criteria to ensure that only well-formed and authorized model updates are accepted,
- Managing participant registration, identity verification, and access control to regulate who may contribute to or observe the learning process,
- Automatically distributing incentives, rewards, or penalties based on measurable contribution quality, participation frequency, or compliance with protocol rules,
- Maintaining an immutable, time-stamped audit trail of learning rounds, submissions, and aggregation outcomes for accountability and post hoc analysis.

By embedding these responsibilities directly into smart contracts, governance mechanisms that are traditionally external—such as trust management, incentive alignment, and compliance enforcement—become an integral part of the learning system itself. This tight coupling between learning dynamics and protocol governance represents a defining characteristic of blockchain-enabled federated learning, enabling transparent, automated, and tamper-resistant coordination at scale.

5.5.3 Decentralized Trust Model

A key conceptual shift introduced by blockchain-enabled federated learning lies in its underlying trust assumptions. Traditional federated learning architectures rely on a central server that is implicitly assumed to behave honestly, remain uncompromised, and faithfully execute aggregation and coordination logic. Blockchain-enabled federated learning removes this single point of trust and replaces it with a *protocol-centric trust model*.

In this decentralized setting, participants do not place trust in any individual institution or coordinating authority. Instead, trust is vested in the correctness, transparency, and verifiability of the blockchain protocol itself. The learning process becomes auditable and self-consistent by design, rather than dependent on organizational assurances or access controls.

Protocol-level trust is established through a combination of:

- **Cryptographic verification of transactions:** All model update references, commitments, and protocol actions are secured using cryptographic primitives such as digital signatures and hash functions, ensuring authenticity and integrity.
- **Distributed consensus among validating nodes:** Agreement on the order and validity of learning-related events is reached collectively by multiple independent validators, preventing unilateral manipulation or censorship.
- **Immutable recording of learning events:** Once recorded on the blockchain, model update references, aggregation results, and participation records cannot be altered retroactively, providing a permanent and tamper-evident history of the learning process.

This decentralized trust model enables collaboration among participants who may be mutually distrustful, geographically dispersed, or governed by different administrative domains. Even in adversarial environments, clients can verify that protocol rules are applied uniformly and that no participant receives preferential treatment. As a result, blockchain-enabled federated learning replaces institutional trust with *verifiable trust*, creating a robust foundation for collaborative learning in open and untrusted settings.

5.5.4 Benefits

Blockchain-enabled federated learning offers a rich set of advantages that go beyond conventional decentralized coordination mechanisms. By embedding coordination, verification, and governance directly into a shared ledger, the learning process becomes inherently more transparent, auditable, and robust against manipulation.

Key benefits include:

- **Transparency:** All learning-related events—such as model update submissions, validation outcomes, aggregation references, and participation records—are visible to authorized participants. This shared visibility reduces information asymmetry and allows all stakeholders to independently verify that the learning protocol is executed as intended.
- **Accountability:** Every contribution and protocol decision is permanently recorded on the ledger. This enables precise attribution of actions to specific participants, supports post hoc inspection, and facilitates dispute resolution in case of suspected misbehavior or disagreement over outcomes.
- **Tamper resistance:** Once model update references or coordination events are committed to the blockchain, they cannot be modified or erased without detection. This immutability provides strong guarantees against retroactive manipulation, rollback attacks, or selective history rewriting.
- **Decentralized governance:** Protocol rules are enforced collectively by the blockchain consensus mechanism and smart contracts, rather than administratively by a single organization. This reduces reliance on trusted intermediaries and aligns governance with formally specified, verifiable rules.

Together, these properties make blockchain-enabled federated learning particularly attractive in settings where trust cannot be assumed a priori. They are well suited to regulated environments, cross-organizational consortia, and high-assurance collaborative systems in which procedural fairness, traceability, and compliance are as important as model accuracy.

5.5.5 Limitations and Practical Challenges

Despite its conceptual appeal, blockchain-enabled federated learning introduces significant practical challenges that limit its applicability in many real-world scenarios. Blockchain consensus protocols inherently impose coordination overhead, which directly impacts system responsiveness and scalability.

Notable challenges include:

- **Increased latency:** Transaction confirmation and consensus finality introduce delays that can slow down learning rounds, particularly when frequent updates are required.
- **Limited scalability:** As the number of participating clients or validating nodes grows, communication and validation costs increase, constraining throughput and system capacity.
- **Energy and resource consumption:** Depending on the consensus mechanism, blockchain operation may incur significant computational or energy costs, which are incompatible with lightweight or battery-powered devices.
- **Integration complexity:** Integrating blockchain infrastructure with existing federated learning pipelines requires careful system design, additional middleware, and expertise across multiple technological domains.
- **Limited suitability for dynamic environments:** Highly mobile, intermittently connected, or resource-constrained clients may be unable to meet the communication and synchronization requirements imposed by blockchain-based coordination.

As a result, blockchain-based federated learning is generally impractical for large-scale mobile or consumer-facing deployments. Its overheads often outweigh its benefits in scenarios where speed, adaptability, and resource efficiency are paramount.

Appropriate Use Cases. Blockchain-enabled federated learning is most effective in environments where trust, auditability, and governance take precedence over raw training speed or minimal latency. It is particularly well suited to permissioned or consortium-based deployments in which participants are known entities operating under shared regulatory or contractual frameworks.

Representative applications include collaborative research across institutions, inter-organizational data analytics, healthcare and financial systems subject to strict compliance requirements, and long-term cooperative learning scenarios where transparency and verifiability are mandatory. In such contexts, blockchain-based coordination provides a robust and principled foundation for trustworthy federated learning.

Blockchain Perspective
Blockchain-enabled federated learning replaces institutional trust with protocol-enforced trust. By embedding coordination, verification, and governance into a shared ledger, it enables decentralized learning among participants who may not otherwise be willing to collaborate.

5.6 Impact of Network Topology on Learning Performance

In decentralized federated learning, the communication topology is far more than a low-level networking choice; it is a core architectural component that fundamentally governs how learning unfolds over time. Unlike centralized federated learning, where a server instantaneously aggregates information from all participating clients, decentralized systems rely exclusively on local, peer-to-peer exchanges. Every update, correction, and alignment step must therefore propagate gradually through the network along existing communication links.

Because of this reliance on local interactions, the structure of the underlying communication graph directly determines how quickly and how accurately information spreads. Models trained on well-connected topologies tend to align rapidly, exhibiting smooth convergence toward a shared solution. In contrast, poorly connected or fragmented networks may experience persistent disagreement among clients, slower convergence, or even temporary divergence when parts of the network become isolated.

From an algorithmic standpoint, network topology acts as a coupling mechanism that links otherwise independent local optimization processes. Each client performs gradient-based updates using its own private data, but these updates are continually shaped by information received from neighbors. The strength and frequency of these interactions depend entirely on the communication structure. As a result, two systems using identical objective functions, learning rates, and consensus rules can behave very differently if their underlying topologies differ.

From a theoretical perspective, topology influences the learning process through well-established graph-theoretic properties. Connectivity determines whether information can reach all parts of the network. Graph diameter controls how many communication steps are required for a local update to affect distant clients. Spectral properties of the graph Laplacian dictate the rate at which consensus errors decay. These factors collectively shape the convergence rate and stability of decentralized learning algorithms.

Importantly, the impact of topology extends beyond convergence speed. It also affects robustness to failures, resilience against client dropout, and sensitivity to non-IID data distributions. Highly redundant topologies can tolerate temporary communication failures without catastrophic disruption, while sparse or tree-like structures may suffer significant performance degradation when even a small number of nodes become unavailable.

In summary, network topology serves as the invisible backbone of decentralized federated learning. It determines how information flows, how errors are corrected, and how local learning objectives are reconciled into a coherent global model. Understanding and designing appropriate communication structures is therefore essential for achieving reliable, efficient, and scalable federated learning systems in fully decentralized settings.

5.6.1 Topology Characteristics

Several structural properties of the communication graph play a decisive role in shaping the efficiency, stability, and robustness of decentralized federated learning. These properties determine how information flows through the network and how effectively local learning processes can be synchronized.

- **Node degree and connectivity:** The degree of a node specifies the number of neighboring clients with which it can directly exchange information. A higher average node degree generally leads to faster dissemination of local updates, as each client can influence and be influenced by a larger portion of the network at every iteration. This increased connectivity reduces model disagreement and helps smooth out local optimization noise. However, higher degree also increases communication cost, highlighting a fundamental trade-off between convergence speed and bandwidth consumption.
- **Network diameter:** The diameter of a network is defined as the longest shortest path between any pair of nodes. It provides an upper bound on how many communication steps are required for information originating at one client to reach all others. Networks with small diameters, such as fully connected graphs or well-designed small-world topologies, enable rapid global alignment because updates propagate quickly across the entire system. In contrast, large-diameter networks, such as long chains or sparse grids, delay information flow and slow the reduction of consensus errors.
- **Presence of hubs or bottlenecks:** Some network topologies contain highly connected hub nodes that act as central conduits for information flow. These hubs can dramatically accelerate consensus by rapidly spreading updates across the network. However, they also introduce vulnerability: failure or overload of a hub can severely disrupt learning. Conversely, bottlenecks or articulation points restrict information flow between different parts of the network. If such nodes fail or experience delays, the network may become temporarily fragmented, leading to isolated learning dynamics and inconsistent model updates.

Illustrative Graph and Mathematical Characterization

To concretely illustrate how node degree, diameter, and hubs influence learning performance, consider the communication graph shown in Fig. 5.2.

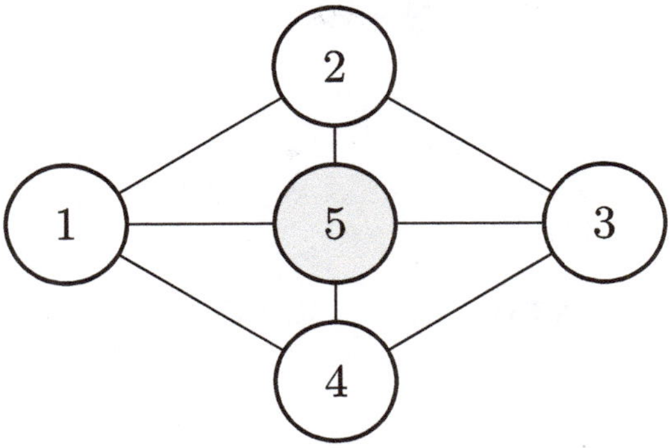

Fig. 5.2 Example decentralized communication graph illustrating node degree, network diameter, and a hub node (grey colored)

In practice, highly connected and well-balanced networks tend to support faster, more stable convergence by ensuring efficient information exchange and redundancy against failures. By contrast, sparse, poorly connected, or intermittently partitioned networks slow down consensus formation and may cause prolonged disagreement among local models. Effective topology design therefore requires balancing connectivity, communication cost, and fault tolerance to achieve reliable decentralized learning performance.

Mathematical Characterization of Network Topology

Let the decentralized communication network be represented by an undirected graph

$$\mathscr{G} = (\mathscr{V}, \mathscr{E}), \qquad |\mathscr{V}| = N,$$

where $\mathscr{V}$ is the set of clients and $\mathscr{E}$ denotes communication links.

Node Degree and Average Connectivity. The degree of node k is defined as

$$d_k = |\mathscr{N}_k|,$$

where $\mathscr{N}_k$ is the neighbor set of client k.

For the graph in Fig. 5.2, the node degrees are

$$\begin{aligned} d_1 &= 3, \\ d_2 &= 3, \\ d_3 &= 3, \\ d_4 &= 3, \\ d_5 &= 4. \end{aligned}$$

The *average node degree* is therefore

$$\bar{d} = \frac{1}{N} \sum_{k=1}^{N} d_k = \frac{3+3+3+3+4}{5} = 3.2.$$

A larger $\bar{d}$ increases the number of terms in the consensus update

$$\sum_{j\in\mathcal{N}_k} w_{kj}(\theta_j - \theta_k),$$

leading to faster reduction of model disagreement but higher communication cost per iteration.

Network Diameter. Let the decentralized communication network be modeled as an undirected graph

$$\mathcal{G} = (\mathcal{V}, \mathcal{E}),$$

where $\mathcal{V}$ denotes the set of nodes (clients) and $\mathcal{E}$ denotes the set of communication links.

Shortest Path Distance. For any two nodes $i, j \in \mathcal{V}$, the shortest path distance $\mathrm{dist}(i, j)$ is defined as the minimum number of edges that must be traversed to reach node j starting from node i.

Formally,

$$\mathrm{dist}(i, j) = \min\left\{\ell \ : \ \exists \text{ a path of length } \ell \text{ connecting } i \text{ and } j\right\}.$$

Definition of Network Diameter. The *network diameter* is the maximum shortest path distance between any pair of nodes in the graph:

$$\mathrm{diam}(\mathcal{G}) = \max_{i,j\in\mathcal{V}} \mathrm{dist}(i, j).$$

This quantity provides an upper bound on the number of communication steps required for information originating at one node to reach any other node in the network.

Topology Under Consideration. Consider the given topology consisting of five nodes, where node 5 acts as a hub connected directly to all other nodes. Thus, the adjacency structure satisfies

$$(5, i) \in \mathcal{E} \quad \forall i \in \{1, 2, 3, 4\}.$$

Distances Involving the Hub Node. Because node 5 is directly connected to every other node, the shortest path distance from any non-hub node $i \neq 5$ to node 5 is

$$\mathrm{dist}(i, 5) = 1, \quad \forall i \in \{1, 2, 3, 4\}.$$

Similarly,

$$\mathrm{dist}(5, i) = 1, \quad \forall i \in \{1, 2, 3, 4\}.$$

Distances Between Non-Hub Nodes. Consider any two distinct non-hub nodes $i, j \in \{1, 2, 3, 4\}$ with $i \neq j$. Since there is no direct edge between them, communication must pass through the hub node 5. Hence, the shortest path is

$$i \to 5 \to j,$$

which consists of two edges.

Therefore,

$$\mathrm{dist}(i, j) = 2, \quad \forall i, j \in \{1, 2, 3, 4\},\ i \neq j.$$

Diameter Calculation. Collecting all possible shortest path distances:

$$\mathrm{dist}(i, j) = \begin{cases} 1, & \text{if } i = 5 \text{ or } j = 5, \\ 2, & \text{if } i \neq j \text{ and } i, j \neq 5. \end{cases}$$

The maximum of these values is therefore

$$\mathrm{diam}(\mathcal{G}) = 2.$$

Interpretation. A network diameter of 2 implies that information generated at any client can reach all other clients within at most two communication rounds. Such a small diameter significantly accelerates consensus formation in decentralized federated learning, as updates propagate rapidly across the entire network.

Graph Laplacian and Spectral Properties. The communication structure induced by the star topology in Fig. 5.2 can be modeled as an undirected graph

$$\mathcal{G} = (\mathcal{V}, \mathcal{E}), \qquad \mathcal{V} = \{1, 2, 3, 4, 5\},$$

where node 5 serves as a central hub and is directly connected to all other nodes. Such a topology is commonly encountered in hub–spoke communication patterns and provides a useful analytical baseline for understanding consensus behavior in decentralized learning systems.

Adjacency Matrix. The adjacency matrix $\mathbf{A} \in \mathbb{R}^{5\times5}$ encodes the pairwise communication links and is defined entrywise as

$$A_{ij} = \begin{cases} 1, & \text{if } (i, j) \in \mathcal{E}, \\ 0, & \text{otherwise.} \end{cases}$$

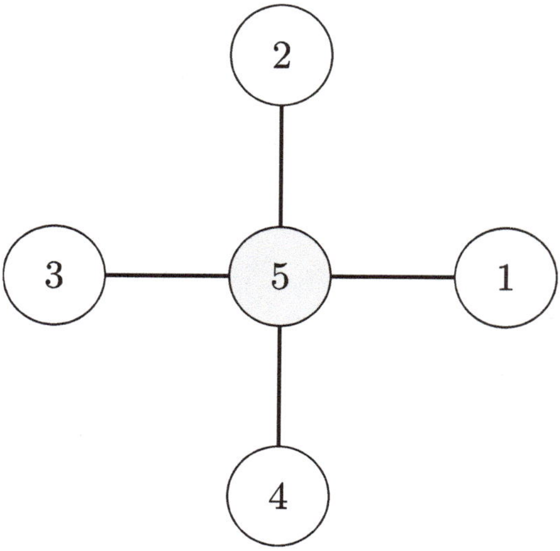

Fig. 5.3 Star communication topology with node 5 as the central hub

For the star topology centered at node 5, the adjacency matrix takes the explicit form

$$\mathbf{A} = \begin{bmatrix} 0 & 0 & 0 & 0 & 1 \\ 0 & 0 & 0 & 0 & 1 \\ 0 & 0 & 0 & 0 & 1 \\ 0 & 0 & 0 & 0 & 1 \\ 1 & 1 & 1 & 1 & 0 \end{bmatrix}.$$

Each nonzero off-diagonal entry reflects a direct communication channel between a peripheral node and the hub.

Degree Matrix. The degree of a node equals the number of its immediate neighbors. Consequently, the four peripheral nodes each have degree one, while the hub node has degree four:

$$\deg(1) = \deg(2) = \deg(3) = \deg(4) = 1, \qquad \deg(5) = 4.$$

The diagonal degree matrix **D** is therefore given by

$$\mathbf{D} = \begin{bmatrix} 1 & 0 & 0 & 0 & 0 \\ 0 & 1 & 0 & 0 & 0 \\ 0 & 0 & 1 & 0 & 0 \\ 0 & 0 & 0 & 1 & 0 \\ 0 & 0 & 0 & 0 & 4 \end{bmatrix}.$$

Graph Laplacian. The (unnormalized) graph Laplacian is defined as

$$\mathbf{L} = \mathbf{D} - \mathbf{A}.$$

Substituting the expressions for **D** and **A** yields

$$\mathbf{L} = \begin{bmatrix} 1 & 0 & 0 & 0 & -1 \\ 0 & 1 & 0 & 0 & -1 \\ 0 & 0 & 1 & 0 & -1 \\ 0 & 0 & 0 & 1 & -1 \\ -1 & -1 & -1 & -1 & 4 \end{bmatrix}.$$

This matrix compactly captures both the connectivity pattern and the relative influence of each node in the network.

Basic Properties. The Laplacian matrix **L** satisfies several fundamental properties:

- **L** is symmetric and positive semidefinite,
- $\mathbf{L1} = \mathbf{0}$, reflecting invariance of the global average,
- the eigenvalue $\lambda_1(\mathbf{L}) = 0$ corresponds to the consensus eigenvector **1**.

These properties form the mathematical foundation for consensus-based learning algorithms.

Eigenvalue Structure. For a star graph with n nodes, the Laplacian eigenvalues are known in closed form. Specifically,

$$\lambda(\mathbf{L}) = \{0, \underbrace{1, \dots, 1}_{n-2 \text{ times}}, n\}.$$

For the present case with $n = 5$, this reduces to

$$\lambda(\mathbf{L}) = \{0,\ 1,\ 1,\ 1,\ 5\}.$$

Hence,

$$\lambda_1 = 0, \qquad \lambda_2 = 1, \qquad \lambda_{\max} = 5.$$

Spectral Gap. The *spectral gap*, which governs the rate of consensus formation, is defined as the second-smallest eigenvalue of the Laplacian:

$$\lambda_2(\mathbf{L}) = 1.$$

A larger spectral gap implies stronger coupling among nodes and faster decay of disagreement across the network.

Consensus Error Bound. Let $\theta^{(t)} \in \mathbb{R}^5$ denote the vector of local model parameters at iteration t, and define the network-wide average as

$$\bar{\theta} = \frac{1}{5}\mathbf{1}^\top \theta^{(t)}.$$

For linear consensus dynamics, the disagreement satisfies the bound

$$\|\theta^{(t)} - \bar{\theta}\mathbf{1}\| \leq (1 - \lambda_2(\mathbf{L}))^t \|\theta^{(0)} - \bar{\theta}\mathbf{1}\|.$$

Substituting $\lambda_2(\mathbf{L}) = 1$ yields

$$\|\theta^{(t)} - \bar{\theta}\mathbf{1}\| \leq 0^t \|\theta^{(0)} - \bar{\theta}\mathbf{1}\|,$$

indicating instantaneous consensus in the idealized linear setting.

Interpretation. The presence of a high-degree hub significantly enlarges the spectral gap, enabling rapid information mixing and fast convergence to consensus. However, this acceleration comes at a cost: failure of the hub node disconnects the graph entirely, halting information flow and learning. This fundamental trade-off between convergence speed and robustness highlights the importance of topology-aware design in decentralized federated learning systems.

Summary of Topological Effects

$$\begin{aligned}\text{Higher degree} &\Rightarrow \text{stronger local averaging},\\ \text{Smaller diameter} &\Rightarrow \text{faster global propagation},\\ \text{Larger spectral gap} &\Rightarrow \text{faster consensus}.\end{aligned}$$

Thus, network topology directly controls the speed, robustness, and stability of decentralized federated learning.

5.6.2 Optimization Implications

From a mathematical and algorithmic standpoint, the impact of network topology on decentralized federated learning can be precisely analyzed through properties of the underlying communication graph. These properties determine how quickly local optimization processes are coupled and how efficiently disagreement among client models is reduced over time.

- **Spectral gap of the graph Laplacian:** The spectral gap, defined as the difference between the smallest nonzero eigenvalue and zero of the graph Laplacian, is a fundamental quantity governing convergence behavior. A larger spectral gap indicates stronger connectivity and faster information mixing across the network. In consensus-based learning, this translates into a more rapid decay of disagreement among local models, allowing clients to align their parameters in fewer communication rounds. Conversely, graphs with small spectral gaps—such as long chains or weakly connected clusters—exhibit slow mixing and correspondingly slower convergence.

- **Consensus error decay dynamics:** The evolution of consensus error, typically measured as the variance or norm of differences between local models, is directly influenced by topological structure. Well-connected graphs enable updates to diffuse broadly and evenly, causing discrepancies to shrink at a steady and predictable rate. In poorly connected or irregular networks, updates may circulate within local neighborhoods for extended periods, leading to prolonged heterogeneity and delayed global agreement. This slower decay can amplify the effects of data non-IIDness and increase variance in learning trajectories.
- **Sensitivity to client dropout and link failures:** Network topology also determines how resilient the learning process is to dynamic conditions such as client unavailability, communication delays, or link failures. Topologies with redundant paths and multiple alternative routes allow information to bypass failed nodes, maintaining approximate consensus even under partial participation. In contrast, sparse or tree-like structures are highly sensitive to dropouts, where the failure of a single node or link can disconnect subgraphs and temporarily halt information flow.

These topological factors interact closely with optimization hyperparameters, including learning rates, consensus weights, and update frequencies. A topology that supports rapid communication may still require careful parameter tuning to avoid numerical instability, oscillatory behavior, or overshooting due to overly aggressive updates. Thus, effective decentralized federated learning demands a joint consideration of optimization dynamics and network structure rather than treating topology as a purely infrastructural concern.

5.6.3 Design Trade-Offs

In practical decentralized federated learning systems, the design of the communication topology inevitably involves balancing multiple, often conflicting, objectives. On one hand, dense and highly connected networks enable rapid diffusion of information, reduce consensus error quickly, and generally lead to faster convergence. On the other hand, such connectivity comes at the cost of increased communication overhead, higher bandwidth consumption, and greater energy usage, which may be prohibitive for large-scale or resource-constrained deployments.

Sparse topologies, by contrast, significantly reduce communication and coordination costs. They are attractive in settings where bandwidth is limited, devices are energy constrained, or communication is intermittent. However, sparsity can slow information propagation, prolong disagreement among local models, and make learning more sensitive to client dropout or temporary disconnections. In extreme cases, overly sparse or poorly structured networks may fragment into loosely coupled components, undermining global coordination altogether.

Effective topology design therefore seeks a careful balance between these extremes. Well-designed networks introduce sufficient redundancy to maintain robustness against failures while avoiding unnecessary communication. Hierarchical structures, small-world networks, and dynamically adaptive topologies are often explored as practical compromises, combining short communication paths with manageable overhead. The goal is not to maximize connectivity, but to ensure that information flows reliably and efficiently under realistic operating conditions, including partial participation, heterogeneous devices, and evolving network states.

Topology Insight
In decentralized federated learning, convergence is governed as much by network structure as by optimization dynamics. Connectivity, diameter, redundancy, and spectral properties of the communication graph fundamentally shape how quickly, stably, and robustly consensus can be achieved.

This discussion underscores that decentralized federated learning can operate effectively without centralized coordination only when local interactions, consensus mechanisms, and network topology are carefully co-designed. Such systems offer increased robustness, autonomy, and fault tolerance, making them well suited to distributed and heterogeneous environments. At the same time, they introduce additional complexity in theoretical analysis, algorithmic design, and hyperparameter selection. In the next chapter, we shift focus back to the training process itself and examine how these architectural choices concretely influence federated learning dynamics, convergence behavior, and empirical performance in practice.

Exercises

This chapter examined decentralized and peer-to-peer federated learning, highlighting orchestrator-less coordination, gossip protocols, consensus-based aggregation, blockchain-enabled learning, and the role of network topology. The following exercises are designed to test conceptual understanding, analytical reasoning, and practical insight.

A. Short Answer Questions

Answer each question briefly and precisely.

1. What is meant by *orchestrator-less* federated learning?

2. Why is a centralized server considered a single point of failure?
3. How does decentralized federated learning preserve data locality?
4. Define gossip-based learning in the context of federated systems.
5. What role does network topology play in decentralized learning?
6. What is meant by an *emergent global model*?
7. How does peer-to-peer averaging differ from server-based aggregation?
8. What is the purpose of consensus weights in decentralized learning?
9. Why are gossip protocols considered robust to client failures?
10. State one key limitation of blockchain-enabled federated learning.

B. Long Answer Questions

Answer the following questions with clear explanations, equations, and diagrams wherever appropriate.

1. Explain the motivation for decentralizing federated learning.
2. Describe orchestrator-less federated learning from an optimization perspective.
3. Derive the consensus-based update rule and explain each term.
4. Compare gossip-based aggregation and consensus-based aggregation.
5. Discuss how network topology influences convergence speed and stability.
6. Explain the role of the graph Laplacian in decentralized learning.
7. Analyze the trust model of blockchain-enabled federated learning.
8. Discuss scalability challenges in fully decentralized federated systems.
9. Explain how heterogeneous local datasets affect decentralized convergence.
10. Critically evaluate trade-offs between robustness and communication cost.

C. True/False Questions

State whether each statement is **True** or **False**.

1. In decentralized federated learning, a global model is explicitly maintained.
2. Gossip protocols rely on randomized peer-to-peer interactions.
3. Consensus-based methods enforce agreement through a central server.
4. Blockchain-enabled federated learning stores raw model parameters on-chain.
5. Network diameter affects how quickly updates propagate.
6. A larger spectral gap implies faster consensus.
7. Decentralized learning eliminates all communication overhead.
8. Peer-to-peer learning improves fault tolerance.
9. Consensus weights must always be identical for all clients.
10. Orchestrator-less learning supports client autonomy.

Answers:

1. False
2. True
3. False
4. False
5. True
6. True
7. False
8. True
9. False
10. True

D. Fill in the Blanks

Fill in the blanks with the most appropriate term or expression.

1. Decentralized federated learning removes the need for a ___________.
2. Gossip protocols achieve coordination through ___________ interactions.
3. Consensus-based aggregation relies on the graph ___________.
4. The second-smallest Laplacian eigenvalue is called the ___________.
5. Blockchain provides a ___________ ledger for coordination.
6. In peer-to-peer learning, each client maintains its own ___________.
7. Network diameter measures the longest ___________ path.
8. Consensus updates reduce ___________ among local models.
9. Gossip-based learning preserves the global ___________.
10. Smart contracts automate ___________ enforcement.

Answers:

1. central server
2. randomized
3. Laplacian
4. spectral gap
5. immutable
6. local model
7. shortest
8. disagreement
9. average
10. protocol

E. Multiple Choice Questions (MCQs)

Choose the **most appropriate** answer.

1. Which component replaces the central server in orchestrator-less learning?
   (a) Data center
   (b) Trusted authority
   (c) Peer network
   (d) Cloud aggregator
2. Gossip-based learning primarily uses:
   (a) Deterministic global averaging
   (b) Random pairwise averaging
   (c) Server-side optimization
   (d) Blockchain validation
3. The spectral gap determines:
   (a) Model size
   (b) Privacy leakage
   (c) Consensus speed
   (d) Training accuracy
4. Which topology typically converges fastest?
   (a) Chain
   (b) Ring
   (c) Star
   (d) Disconnected graph
5. Blockchain-enabled federated learning emphasizes:
   (a) Low latency
   (b) Central trust
   (c) Auditability
   (d) High bandwidth
6. Consensus-based learning is closely related to:
   (a) Centralized SGD
   (b) Distributed optimization
   (c) Supervised classification
   (d) Reinforcement learning
7. Gossip protocols are best suited for:
   (a) Static and trusted systems
   (b) Small centralized clusters
   (c) Dynamic and large networks
   (d) Single-device learning
8. In decentralized learning, robustness arises from:
   (a) Data replication
   (b) Redundant coordination
   (c) Central oversight
   (d) Large batch sizes

9. Which quantity remains invariant under gossip averaging?
   (a) Variance
   (b) Maximum value
   (c) Minimum value
   (d) Global mean
10. Blockchain coordination is least suitable for:
   (a) Consortium learning
   (b) Regulated environments
   (c) Energy-constrained IoT
   (d) Auditable collaboration

Answers:

1. (c)
2. (b)
3. (c)
4. (c)
5. (c)
6. (b)
7. (c)
8. (b)
9. (d)
10. (c)

F. Application and Project-Based Questions

These problems emphasize modeling, reasoning, and implementation.

1. Design a peer-to-peer federated learning protocol for edge IoT devices.
2. Simulate gossip-based averaging on a ring topology and analyze convergence.
3. Implement consensus-based learning for heterogeneous local datasets.
4. Compare convergence under ring, star, and fully connected topologies.
5. Analyze the effect of spectral gap on learning speed using simulations.
6. Design a blockchain-enabled federated learning workflow with smart contracts.
7. Study the impact of client dropout in decentralized learning.
8. Implement decentralized learning without synchronization barriers.
9. Evaluate communication cost versus accuracy trade-offs.
10. Propose a hybrid gossip–consensus federated learning framework.

Federated Learning Training Process

6

Federated learning unfolds through a carefully structured and inherently iterative training workflow whose primary objective is to reconcile autonomous local optimization with a shared global learning goal. At its core, federated learning is not defined by a single algorithm, but by a disciplined process of coordination that allows many independent learning entities to contribute to a common model without ever relinquishing control over their raw data. This process-centric view is essential for understanding how federated learning functions in practice and why its behavior differs so fundamentally from classical machine learning paradigms.

In traditional centralized training, learning proceeds within a single computational domain. All training data is assumed to be readily accessible, model updates are computed using the complete dataset, and optimization unfolds as a continuous, tightly synchronized trajectory. Federated learning departs from this assumption at every level. Data is inherently decentralized, learning is performed locally on heterogeneous devices, participation is intermittent, and communication is both costly and unreliable. As a consequence, federated learning cannot rely on a monolithic optimization loop. Instead, it must decompose training into a sequence of well-defined stages that can operate independently yet remain statistically and algorithmically coordinated over time.

The ederated training process therefore explicitly separates *global coordination* from *local computation*. Initialization of the global model establishes a shared reference point across all participants. Client activation mechanisms determine which subset of devices contributes to learning at a given moment. Local training allows each client to optimize the model using its own private data and computational resources. Communication stages enable the controlled exchange of derived information in the form of model updates, while aggregation mechanisms integrate these diverse updates into a coherent global model. Each of these stages is executed repeatedly across communication rounds, forming a closed-loop system in which global learning emerges gradually from repeated local actions.

P. P. Ray, *Foundations of Federated Learning*,
Springer Series in the Data Sciences,
https://doi.org/10.1007/978-3-032-34084-9_6

Importantly, this staged workflow is not an arbitrary engineering choice. It is a direct consequence of the foundational principles discussed in earlier chapters—namely data locality, client autonomy, privacy preservation, and statistical heterogeneity. Because raw data cannot be centralized, optimization must be performed locally. Because clients may join, leave, or fail unpredictably, participation must be treated as stochastic rather than deterministic. Because communication bandwidth is limited, synchronization must be infrequent and approximate. The federated training process embodies these constraints explicitly, transforming them from obstacles into defining characteristics of the learning paradigm.

From an optimization perspective, the federated training loop can be viewed as a mechanism for approximating descent on a global objective function that no single entity can compute or observe in full. Each communication round produces only a partial, noisy estimate of the direction in which the global model should evolve. Convergence, therefore, is not achieved through exact agreement at every iteration, but through the progressive alignment of model trajectories over time. This gradual and statistical notion of convergence distinguishes federated learning from centralized empirical risk minimization and necessitates new ways of reasoning about stability, efficiency, and performance.

This chapter provides a comprehensive, step-by-step exposition of the federated learning training process. Each stage of the workflow is examined in detail, both operationally and mathematically. Operational descriptions clarify the roles and responsibilities of clients and servers, while mathematical formulations connect each stage to the global federated optimization objective developed in previous chapters. Throughout the chapter, notation is used consistently to reinforce the continuity between abstract theory and concrete training dynamics.

By the end of this chapter, the reader will possess a clear procedural understanding of how federated learning systems function in practice: how models are initialized, how clients participate, how local updates are generated, how aggregation reconciles heterogeneous learning signals, and how training ultimately terminates. This understanding serves as a critical foundation for the chapters that follow, which investigate how data heterogeneity, advanced optimization algorithms, privacy-preserving techniques, and system architectures modify, extend, and refine each stage of the federated training process.

6.1 Initialization of the Global Model

The federated learning process begins with the initialization of a *global model* that serves as the common reference point for all participating clients. This global model represents the shared hypothesis that federated learning seeks to refine collaboratively, despite the fact that training data remains decentralized across clients. Formally, the global model is parameterized by a vector

$$\theta^{(0)} \in \mathbb{R}^d,$$

where d denotes the dimensionality of the model parameter space. Initialization takes place before any local computation is performed and establishes the baseline from which all subsequent local optimization trajectories originate.

In classical centralized learning, model initialization primarily influences convergence speed, numerical stability, and the likelihood of escaping poor local minima. In federated learning, however, initialization plays a substantially more consequential role. Because data is distributed across clients and often exhibits strong statistical heterogeneity, the initial global model shapes not only early optimization dynamics but also the degree of disagreement among client updates, the magnitude of local drift, and the stability of aggregation. Poor initialization can amplify the effects of non-IID data, while a well-chosen initialization can significantly smooth early training behavior.

To enforce a coherent starting point, all participating clients initialize their local models identically using the broadcast global parameters:

$$\theta_k^{(0,0)} = \theta^{(0)}, \quad \forall k \in \{1, \dots, K\}.$$

This enforced alignment is fundamental to federated learning. It guarantees that any divergence observed among local models in later stages arises exclusively from differences in local datasets, local optimization trajectories, or system-level variability, rather than from inconsistent initial conditions. As a result, the global aggregation process can be interpreted as a meaningful reconciliation of heterogeneous learning signals rather than an attempt to correct arbitrary initial disagreement.

In centralized federated architectures, initialization is typically performed by a coordinating server, which selects and broadcasts $\theta^{(0)}$ to all eligible clients. In decentralized or peer-to-peer federated settings, the initial model may instead be established through protocol design, predeployment agreements, or shared configuration files. Regardless of the architectural paradigm, the existence of a shared initial model is essential. Without such a shared reference, it would be impossible to define consistent local updates or to meaningfully aggregate client contributions into a single evolving global model.

In practice, several initialization strategies are commonly employed, each reflecting different assumptions about data availability, system continuity, and deployment context:

- **Random initialization**, where $\theta^{(0)}$ is drawn from a predefined distribution, as in conventional neural network training. This approach is simple and architecture-agnostic, but under highly non-IID data it may lead to unstable early updates and large variance across client gradients.
- **Pretrained initialization**, where the global model is initialized using parameters obtained from training on publicly available, synthetic, or proxy datasets. This strategy often provides a more informative starting point, reducing early-round client drift and accelerating convergence by placing the model closer to a useful region of the parameter space.

- **Warm-start initialization**, where $\theta^{(0)}$ is inherited from a previous federated training session or an earlier deployment cycle. This approach is particularly common in continuously learning systems and long-lived federated deployments, where the model is updated incrementally over time rather than trained from scratch.

From a mathematical perspective, the choice of initialization determines the starting point of the optimization trajectory for the global federated objective

$$F(\theta) = \sum_{k=1}^{K} \frac{n_k}{N} F_k(\theta), \quad N = \sum_{k=1}^{K} n_k.$$

Unlike centralized learning, this global objective is never evaluated directly. Instead, it is approximated implicitly through repeated aggregation of locally computed updates. As a result, the early stages of training are especially sensitive to the initial parameter vector $\theta^{(0)}$. Under non-IID data distributions, an unfavorable initialization can amplify gradient bias, increase disagreement among client updates, and slow down or even destabilize convergence.

Initialization also influences how quickly the system transitions from exploratory behavior—characterized by large and diverse local updates—to more stable, consensus-driven learning. A well-chosen $\theta^{(0)}$ reduces the magnitude of early local deviations $\Delta\theta_k^{(0)}$, thereby improving the effectiveness of aggregation in the first few communication rounds. Conversely, a poorly chosen initialization may require many rounds before local updates begin to align in a manner conducive to steady global progress.

Initialization Principle

Federated learning enforces a shared starting point across all clients, but it does not enforce shared optimization trajectories. Any divergence observed after initialization reflects statistical heterogeneity, client-specific objectives, and system variability rather than initial disagreement among participants.

In summary, global model initialization is not a mere procedural detail in federated learning. It is a foundational design decision that influences convergence speed, stability, fairness across clients, and the effectiveness of aggregation throughout training. Careful initialization is therefore essential for achieving robust and efficient federated learning in heterogeneous, real-world settings.

Table 6.1 summarizes the key characteristics of commonly used global model initialization methods in federated learning. The comparison highlights how initialization choices influence early training stability, sensitivity to non-IID data, and overall communication efficiency. While random initialization offers simplicity, it often leads to slow and unstable convergence in heterogeneous environments.

Table 6.1 Comparison of Global Model Initialization Methods in Federated Learning

Aspect	Random Initialization	Pretrained Initialization	Warm-Start Initialization
Definition	Model parameters $\theta^{(0)}$ are initialized randomly, typically using standard distributions (e.g., Gaussian or Xavier initialization)	Model parameters are initialized using weights obtained from training on public, synthetic, or proxy datasets	Model parameters are initialized from a previously trained federated model or earlier deployment cycle
Prior knowledge usage	No prior knowledge incorporated; learning starts from scratch	Incorporates external or auxiliary knowledge before federated training begins	Leverages historical federated learning knowledge accumulated over past rounds or sessions
Impact on early rounds	Early updates may be highly variable, especially under non-IID data	Reduces early-round instability and accelerates initial convergence	Provides stable early updates due to proximity to a previously optimized solution
Sensitivity to non-IID data	High sensitivity; client drift can be severe in early communication rounds	Moderate sensitivity; pretrained features often mitigate extreme divergence	Low sensitivity; prior alignment reduces disagreement among client updates
Communication efficiency	Often requires more communication rounds to reach acceptable performance	Typically reduces the number of rounds needed for convergence	Highly communication-efficient when deployed in long-lived federated systems
System requirements	Minimal system requirements; simplest to deploy	Requires access to suitable public, synthetic, or proxy datasets	Requires persistent storage and versioning of past global models
Typical use cases	Experimental setups, benchmarking, or scenarios with near-IID data	Large-scale deployments, cross-silo FL, and privacy-sensitive domains	Continuous learning systems, production deployments, and adaptive federated services
Primary limitation	Slow convergence and unstable early training under heterogeneous data	Performance depends on the relevance and quality of the pretraining data	Risk of bias accumulation or concept drift if historical models are outdated
Initialization bias	No inherent bias beyond random parameter distribution	Bias introduced depends on the source and domain relevance of the pretraining data	May inherit historical biases accumulated during previous federated training cycles

(Continued)

Table 6.1 (continued)

Aspect	Random Initialization	Pretrained Initialization	Warm-Start Initialization
Effect on aggregation stability	Aggregation may be unstable in early rounds due to highly divergent client updates	Aggregation is typically smoother as client updates are more aligned initially	Aggregation is highly stable when the warm-start model is close to the current data distribution
Compatibility with personalization	Less suitable for early personalization due to lack of shared structure	Facilitates faster personalization by providing shared feature representations	Highly compatible with personalized federated learning and continual adaptation
Risk under concept drift	Low risk of historical bias, but slow to adapt initially	Moderate risk if pretraining data becomes outdated	High risk if warm-start models are not periodically refreshed or revalidated
Impact on convergence guarantees	Convergence guarantees rely heavily on strong assumptions and may require many rounds	Often improves practical convergence behavior under standard federated assumptions	Convergence is typically faster but theoretical guarantees may depend on stationarity of past data
Operational complexity	Very low; easy to deploy and maintain across federated clients	Moderate; requires careful selection and maintenance of pretraining datasets	Higher; requires model version control, monitoring, and periodic reset mechanisms

Pretrained and warm-start initialization strategies, by contrast, leverage prior knowledge to reduce early client drift and accelerate global alignment, making them particularly suitable for large-scale and long-lived federated deployments. The table emphasizes that global model initialization is not merely a technical detail, but a strategic design choice with significant impact on federated learning performance and robustness.

6.2 Client Activation and Participation

A defining characteristic of federated learning is that not all clients participate in training at every communication round. Unlike centralized learning or tightly synchronized distributed systems, where computational workers are assumed to be continuously available and fully coordinated, federated learning operates in open, heterogeneous, and highly dynamic environments. Clients may join or leave the system unpredictably, experience variable network conditions, or face resource constraints that prevent consistent participation. As a result, federated learning inherently relies on *partial, intermittent, and dynamic client participation*, which fundamentally shapes both the system design and the underlying optimization process.

Formally, at communication round t, only a subset of the total client population contributes to local training and model updates. Let

$$\mathcal{S}^{(t)} \subseteq \{1, 2, \dots, K\}$$

denote the set of clients activated during round t. In most practical deployments, the number of participating clients per round is intentionally kept much smaller than the total population, that is,

$$|\mathcal{S}^{(t)}| \ll K.$$

This design choice reflects scalability constraints: involving all clients simultaneously would impose prohibitive communication overhead, excessive coordination latency, and unacceptable strain on server and network resources. Instead, federated learning systems are designed to make progress through repeated interaction with small, randomly selected subsets of clients.

Client activation is governed by a combination of system-level, network-level, and policy-driven factors. In particular, whether a client participates in a given round depends on:

- **Device availability**, since clients may be offline, executing foreground applications, or temporarily unable to perform local training due to operating system or user constraints;
- **Network connectivity**, including bandwidth availability, latency variability, packet loss, and communication reliability, which directly affect the feasibility and cost of update transmission;
- **Energy and resource constraints**, especially in mobile, wearable, or edge devices where battery life, thermal limits, and computational capacity restrict participation frequency;
- **Server-side selection and scheduling policies**, which may actively control participation to promote fairness, improve data diversity, reduce straggler effects, or accelerate learning progress.

From a systems perspective, client activation represents a fundamental trade-off between statistical efficiency and operational scalability. Activating a larger number of clients per round increases the representativeness of aggregated updates and reduces variance in the estimated global descent direction. However, it also increases communication cost, coordination overhead, and sensitivity to slow or unreliable clients. Conversely, activating fewer clients per round improves responsiveness and scalability but introduces additional randomness into the learning process. Federated learning therefore shifts the emphasis from exhaustive per-round participation to *long-term participation coverage*, where meaningful learning emerges from repeated engagement of diverse clients over many rounds.

From a mathematical standpoint, partial participation transforms the federated optimization problem into a stochastic approximation of the global objective. Because the set of active clients $\mathscr{S}^{(t)}$ is typically sampled according to some probability distribution, the global update computed at round t depends explicitly on this random selection. As a result, the updated global model

$$\theta^{(t+1)}$$

is a random variable conditioned on the sampling of clients. Convergence guarantees for federated learning algorithms are therefore formulated in expectation or with high probability over the client selection process, rather than as deterministic guarantees valid for every round.

This stochastic participation model has particularly important implications under non-IID data distributions. If certain clients participate infrequently or systematically drop out, their data distributions may be underrepresented in the aggregated updates. Over time, this imbalance can introduce bias into the learned global model and raise concerns related to fairness, personalization, and generalization. Effective client activation strategies must therefore ensure that, over sufficiently many rounds, all relevant segments of the client population contribute meaningfully to the learning process.

Consequently, modern federated learning systems often incorporate carefully designed participation policies that balance randomness with coverage. These policies may enforce minimum participation frequencies, favor clients with rare or underrepresented data, or adaptively adjust selection probabilities based on past contributions. Such mechanisms help mitigate the adverse effects of partial participation while preserving the scalability benefits that make federated learning viable in real-world deployments.

Participation Reality
Federated learning relies on *statistical participation* rather than full synchronization. Convergence and fairness emerge from repeated, randomized client engagement over time, not from complete participation in each communication round.

In summary, client activation and participation are not peripheral implementation details, but central elements of federated learning design. They determine how information flows from decentralized data sources to the global model and directly influence convergence behavior, fairness, and system efficiency. A deep understanding of participation dynamics is therefore essential for analyzing and deploying federated learning systems at scale.

6.3 Local Model Training on Client Devices

Local model training on client devices constitutes the computational core of the federated learning paradigm. In contrast to centralized learning—where gradient computation is performed on a unified dataset within a single computational environment—federated learning distributes the optimization workload across a large population of autonomous clients. Each client performs training exclusively on its private data, thereby preserving data locality while contributing indirectly to the global learning objective.

At the beginning of communication round t, each activated client $k \in \mathcal{S}^{(t)}$ receives the current global model from the server and initializes its local model parameters as

$$\theta_k^{(t,0)} = \theta^{(t)}.$$

This shared initialization ensures that all participating clients begin local optimization from a common reference point, allowing differences in subsequent updates to reflect local data characteristics and computational behavior rather than inconsistent starting conditions.

Following initialization, client k performs τ steps of local optimization using its private dataset $\mathcal{D}_k$. The local objective function

$$F_k(\theta) = \frac{1}{n_k} \sum_{(x,y)\in\mathcal{D}_k} \ell\big(f_\theta(x), y\big)$$

encapsulates the empirical risk associated with client-specific data. In practice, local optimization is most commonly implemented using stochastic gradient descent or its variants. Each local update step takes the form

$$\theta_k^{(t,i+1)} = \theta_k^{(t,i)} - \eta \nabla F_k\big(\theta_k^{(t,i)}\big), \quad i = 0, 1, \ldots, \tau - 1,$$

where η denotes the local learning rate and τ controls the amount of on-device computation performed per communication round.

From an optimization perspective, local training serves a dual purpose. On one hand, it allows each client to improve model performance on its own data, thereby extracting meaningful information from decentralized datasets. On the other hand, because $F_k(\theta)$ may differ substantially across clients under non-IID data, local optimization can drive the model parameters in directions that are misaligned with the global objective $F(\theta)$. This phenomenon, commonly referred to as *local drift*, represents a fundamental challenge in federated learning.

The extent of local drift is influenced by several factors. Increasing the number of local steps τ amplifies the effect of client-specific objectives, potentially leading to larger discrepancies among client updates. Similarly, greater statistical heterogeneity across client datasets intensifies divergence between local and global descent directions. While aggressive local training reduces the frequency of communication and improves system efficiency, it can also slow convergence or degrade model performance if not carefully controlled.

From a systems viewpoint, local training is constrained by device capabilities. Clients may differ widely in computational power, memory capacity, and energy availability. Federated learning algorithms must therefore tolerate variability in local training duration and precision, often operating in environments where some clients act as stragglers or return partial updates. Despite these challenges, local training remains the principal mechanism through which federated learning scales to large client populations.

Local Optimization Trade-off
Performing more local computation per round reduces communication overhead but increases the risk of divergence from the global descent direction. Federated learning algorithms balance these competing effects by carefully selecting the number of local steps and learning rates.

In summary, local model training enables federated learning to extract useful information from decentralized data while respecting privacy and system constraints. At the same time, it introduces optimization challenges that do not arise in centralized settings. Understanding the role of local training and its interaction with data heterogeneity is essential for analyzing convergence behavior and for motivating advanced federated algorithms, such as those incorporating regularization, adaptive local steps, or personalized objectives, discussed in later chapters.

6.4 Reporting of Local Updates

Once local training has been completed, each participating client must communicate the outcome of its on-device optimization to the coordinating server. In federated learning, this communication step is subject to strict constraints that distinguish it fundamentally from classical distributed learning. In particular, federated learning explicitly forbids the transfer of raw training data, intermediate samples, or detailed optimization trajectories. As a result, the reporting of local updates becomes the *only* mechanism through which decentralized learning progress can be exposed to the global coordination process.

This reporting stage therefore serves as a critical interface between two competing objectives. On the one hand, the server requires sufficient information to guide global optimization and ensure convergence of the shared model. On the other hand, clients must preserve data locality, limit communication overhead, and protect sensitive information embedded in their private datasets. The design of the reported update strikes a careful balance between these requirements.

Formally, after performing τ steps of local optimization during communication round t, each activated client $k \in \mathscr{S}^{(t)}$ computes a local update defined as

$$\Delta\theta_k^{(t)} = \theta_k^{(t,\tau)} - \theta^{(t)}.$$

This update vector represents the cumulative effect of local training starting from the globally shared model $\theta^{(t)}$. Rather than transmitting the locally optimized model $\theta_k^{(t,\tau)}$ itself, the client reports only the difference between its final local parameters and the initial global parameters. This difference-based representation is both compact and informative, encoding how the client's data and optimization dynamics would prefer to adjust the global model.

From an optimization perspective, $\Delta\theta_k^{(t)}$ can be interpreted as an approximate descent direction for the local objective function $F_k(\theta)$, accumulated over multiple local steps. When the local learning rate is sufficiently small, this update approximates a scaled sum of local gradients evaluated along the optimization trajectory. Aggregating such updates across clients yields a stochastic approximation to a descent step on the global federated objective

$$F(\theta) = \sum_{k=1}^{K} \frac{n_k}{N} F_k(\theta),$$

despite the fact that the server never directly observes any local data or gradients.

At the same time, because client datasets are typically non-IID, individual updates may point in substantially different directions. Some updates may reflect strong local signals driven by unique data distributions, while others may be relatively small or noisy. This diversity is both a strength and a challenge: it enables learning from heterogeneous data sources, but also necessitates careful aggregation to prevent instability or bias in the global model. The reporting of local updates thus sets the stage for the aggregation mechanism examined in the subsequent section.

The decision to transmit update vectors—rather than raw gradients, intermediate activations, or entire local models—is motivated by a combination of communication efficiency and privacy considerations. Raw gradients would require frequent and fine-grained communication, increasing bandwidth usage and potentially exposing sensitive information about individual data points. Transmitting full models would incur unnecessary communication cost and provide little additional benefit beyond what is already captured by the update. By contrast, update vectors provide a succinct summary of local learning progress while preserving strict data locality.

In practical federated learning systems, the reporting of local updates is further shaped by system-level constraints. Before transmission, updates are often transformed to reduce communication cost, improve robustness, or enhance privacy. Common techniques include:

- **Compression**, which encodes updates using sparse, low-rank, or entropy-efficient representations to reduce bandwidth requirements;

- **Quantization**, which limits numerical precision in order to decrease message size and transmission latency, particularly in resource-constrained environments;
- **Encryption or secure aggregation**, which ensures that the server can only observe aggregated updates and cannot infer individual client contributions.

While these mechanisms are essential for scalability and privacy, they also introduce additional sources of approximation and noise into the learning process. As a result, the updates received by the server are typically noisy, partially observed, and statistically heterogeneous. The global optimization procedure must therefore be robust to imperfect information, delayed transmissions, and varying update quality across clients.

Information Boundary
Federated learning enforces a strict separation between private data and global coordination. Only *derived information* in the form of local model updates is shared, while raw data, intermediate gradients, and training examples remain permanently confined to client devices.

In summary, the reporting of local updates represents a pivotal transition point in the federated learning workflow. It converts private, decentralized optimization outcomes into a communication-efficient and privacy-preserving representation that can be safely aggregated at the global level. The quality, structure, and reliability of these reported updates directly influence the stability, convergence speed, and final performance of the federated model.

6.5 Model Aggregation and Update Propagation

Model aggregation constitutes the central coordinating mechanism of the federated learning process. After local training has been performed independently across multiple clients and local updates have been reported to the server, aggregation serves as the point at which decentralized learning signals are reconciled into a single, coherent global model. Unlike centralized optimization—where gradients are computed and applied within a unified environment—federated learning relies on aggregation as the sole means of integrating information across autonomous learners.

Upon receiving local updates from the set of participating clients $\mathscr{S}^{(t)}$ in communication round t, the server computes the next global model parameters by combining these updates according to a predefined aggregation rule. The most widely used aggregation strategy, commonly referred to as *federated averaging*, applies a data-weighted combination of local updates:

$$\theta^{(t+1)} = \theta^{(t)} + \sum_{k \in \mathscr{S}^{(t)}} \frac{n_k}{\sum_{j \in \mathscr{S}^{(t)}} n_j} \Delta\theta_k^{(t)}.$$

Here, the weighting factor reflects the relative contribution of each client's dataset size, ensuring that clients with more training samples exert proportionally greater influence on the global model update.

From an optimization perspective, this aggregation rule can be viewed as an approximation to a gradient descent step on the global federated objective

$$F(\theta) = \sum_{k=1}^{K} \frac{n_k}{N} F_k(\theta),$$

constructed using only partial and noisy information. Because only a subset of clients participates in each round and because local updates are computed over heterogeneous data distributions, the aggregated update represents a stochastic estimate of the true global descent direction. Nevertheless, when client participation is sufficiently diverse over time, repeated aggregation drives the global model toward a stationary point of the federated objective.

The choice of data-weighted aggregation plays a critical role in maintaining consistency with empirical risk minimization. By scaling updates according to local dataset sizes, the aggregation mechanism ensures that the global model reflects the overall data distribution rather than the behavior of any individual client.

Once the aggregated model parameters $\theta^{(t+1)}$ have been computed, the server propagates the updated global model to clients in the subsequent communication round. This broadcast step closes the federated learning loop, providing a new common reference point from which the next cycle of local training begins. Through repeated rounds of aggregation and propagation, the global model evolves gradually, integrating information from a broad and dynamically changing client population.

From a systems standpoint, aggregation and update propagation must be robust to partial participation, delayed updates, and communication failures. The server may receive incomplete or stale updates, and aggregation rules must tolerate such irregularities without compromising overall stability. Despite these challenges, aggregation remains the only point of global coordination in federated learning, as clients do not communicate directly with one another.

Aggregation as Coordination

Aggregation is the sole mechanism through which independent local optimizers are aligned into a single global learning trajectory. There is no direct interaction or information exchange among clients; all coordination is mediated through the aggregation process.

In summary, model aggregation and update propagation transform a collection of independent, decentralized learning processes into a coherent global optimization procedure. Understanding the role of aggregation is essential for analyzing convergence behavior, robustness under non-IID data, and the impact of system-level constraints.

6.6 Termination Criteria and Convergence Checks

Federated learning proceeds through repeated communication rounds until a suitable termination condition is satisfied. Unlike centralized optimization—where convergence can often be assessed directly through evaluation of the global objective function—federated learning operates under partial observability and stochastic participation. As a result, termination criteria must be defined in a manner that balances theoretical rigor with practical deployability.

In practice, federated training may be terminated based on a variety of system-level and optimization-related considerations. Commonly used termination criteria include:

- **A fixed number of communication rounds**, which provides a predictable training budget and is widely used in experimental evaluations;
- **Stabilization of model parameters**, where successive global models exhibit negligible change, indicating diminishing learning progress;
- **Convergence of a validation metric**, such as accuracy or loss evaluated on a held-out validation set or a representative subset of clients;
- **Resource or time constraints**, including communication budgets, computational limits, or deployment deadlines.

Because raw training data remains distributed across clients, the server typically lacks direct access to the global objective function

$$F(\theta) = \sum_{k=1}^{K} \frac{n_k}{N} F_k(\theta).$$

Consequently, convergence must be assessed indirectly through proxy measures that can be computed using available information. One common approach is to monitor the norm of successive parameter updates and declare convergence when

$$\|\theta^{(t+1)} - \theta^{(t)}\| \leq \epsilon,$$

for a predefined tolerance ϵ. This criterion reflects stabilization of the global model parameters, even if the exact value of the objective function remains unknown.

Alternatively, convergence may be evaluated in terms of expected performance. Under standard assumptions on client sampling and local optimization, the sequence $\{\theta^{(t)}\}$ can be analyzed as a stochastic process, and convergence statements are typically expressed in expectation, such as stabilization of

$$\mathbb{E}[F(\theta^{(t)})].$$

This perspective acknowledges the inherent randomness introduced by partial participation, non-IID data, and noisy local updates.

From a theoretical standpoint, federated learning convergence is therefore probabilistic rather than deterministic. The global model may continue to fluctuate slightly even after substantial learning has occurred, particularly in large-scale systems with high client turnover. Termination criteria must tolerate such residual variability while ensuring that meaningful learning has been achieved.

From a practical deployment perspective, termination is often dictated by operational constraints rather than strict mathematical conditions. In long-lived federated systems, training may proceed indefinitely in an online or continual learning mode, with convergence interpreted as sustained performance stability rather than absolute parameter stationarity. In contrast, time-bounded or resource-constrained deployments may prioritize early stopping once marginal performance gains diminish.

Convergence in Practice
Federated learning converges in a statistical and approximate sense. Practical termination reflects stabilization of performance or parameters over time, rather than exact minimization of the global objective.

In summary, termination criteria and convergence checks play a crucial role in translating federated optimization theory into practical training workflows. They define when learning is deemed sufficient, balance efficiency with performance, and ensure robustness in the presence of stochastic participation and heterogeneous data.

6.7 Synchronous versus Asynchronous Training

Federated learning systems may operate under different coordination models depending on how and when locally computed updates are integrated into the evolving global model. Unlike centralized learning—where optimization proceeds in a tightly coupled and fully synchronized manner—federated learning must contend with decentralized computation, intermittent connectivity, and highly heterogeneous client behavior. As a result, the temporal organization of communication and aggregation becomes a central design choice.

Two broad coordination paradigms are commonly distinguished in federated learning: *synchronous* and *asynchronous* training. These paradigms define not only when aggregation occurs, but also how the system balances stability, scalability, and responsiveness. The choice between synchronous and asynchronous training reflects a fundamental trade-off between algorithmic simplicity and practical deployability in real-world environments.

6.7.1 Synchronous Training

In synchronous federated training, learning proceeds in discrete and clearly delineated communication rounds. At the beginning of round t, the server selects a subset of clients $\mathcal{S}^{(t)}$ and broadcasts the current global model $\theta^{(t)}$ to all selected participants. Each client performs local training independently and returns its update to the server. Crucially, the server waits until *all* selected clients in $\mathcal{S}^{(t)}$ have completed their local computation before performing aggregation and producing the next global model $\theta^{(t+1)}$.

This strict round-level synchronization offers several important advantages from an optimization and analytical perspective. Because all received updates are computed with respect to the same reference model $\theta^{(t)}$, aggregation is conceptually straightforward and free from temporal inconsistencies. As a result, synchronous training closely resembles classical distributed optimization and admits relatively clean mathematical analysis. Most foundational convergence guarantees for federated algorithms, including those for federated averaging and its variants, are derived under synchronous assumptions.

Furthermore, synchronous training avoids the complications introduced by stale or outdated updates. Each communication round produces a well-defined global update that can be interpreted as a stochastic approximation of a descent step on the federated objective. This predictability makes synchronous training particularly attractive in controlled environments, such as cross-silo federated learning, institutional collaborations, or experimental settings where client availability and network conditions are relatively stable.

Despite these advantages, synchronous coordination introduces significant system-level challenges. Because aggregation cannot proceed until all selected clients respond, the overall training speed is dictated by the slowest participant in each round. Clients with limited computational resources, poor network connectivity, or temporary interruptions act as *stragglers*, forcing the server to idle while waiting for their updates. As the scale and heterogeneity of the client population increase, this waiting overhead can dominate training time and substantially reduce system efficiency. In large-scale deployments involving mobile or edge devices, strict synchronization may therefore become impractical.

6.7.2 Asynchronous Training

Asynchronous federated training relaxes the requirement of round-level synchronization and allows clients to operate more independently. Rather than coordinating updates into discrete rounds, clients perform local training and transmit their updates whenever computation and communication permit. Upon receiving an update from client k, the server immediately incorporates it into the global model according to an update rule of the form

$$\theta \leftarrow \theta + \alpha_k \Delta\theta_k,$$

where α_k is a scaling factor that may depend on client-specific, system-level, or delay-related considerations.

This asynchronous update mechanism significantly improves system responsiveness and scalability. The server no longer waits for a fixed set of clients, and idle time due to stragglers is effectively eliminated. Clients can contribute updates whenever they are available, making asynchronous training well suited for highly dynamic environments with unpredictable participation patterns. Such environments include mobile devices, Internet-of-Things networks, and large open federated systems where client churn is the norm rather than the exception.

From a systems perspective, asynchronous training enables continuous model evolution and higher throughput. The global model is updated incrementally as new information arrives, allowing learning to progress even when client availability is sparse or uneven. This flexibility makes asynchronous federated learning particularly attractive for large-scale deployments where maximizing participation over time is more important than strict per-round coordination.

However, this flexibility comes at a cost. Because updates may be computed using older versions of the global model, asynchronous training must contend with *stale updates*. An update generated using an outdated model may no longer align with the current optimization direction, introducing bias and additional variance into the learning process. This temporal inconsistency complicates both practical stability and theoretical convergence analysis.

To mitigate these effects, asynchronous federated algorithms often incorporate additional control mechanisms. These include bounding the maximum allowable staleness of updates, adjusting learning rates based on delay, or weighting updates according to their freshness. While such techniques can improve robustness, they also increase algorithmic and implementation complexity, and convergence guarantees typically require stronger or more nuanced assumptions.

From an optimization standpoint, synchronous and asynchronous training can be viewed as two different approximations of the same underlying federated objective. Synchronous training emphasizes temporal consistency and analytical clarity, approximating batch-style updates on the global objective. Asynchronous training emphasizes throughput and adaptability, accepting higher variance and delayed information in exchange for improved scalability. Neither paradigm is universally superior; rather, each reflects a different balance between optimization rigor and system practicality.

Training Mode Trade-off

Synchronous federated training prioritizes analytical simplicity and update consistency through strict round-level coordination. Asynchronous federated training prioritizes scalability and responsiveness by relaxing synchronization, at the expense of increased variance and more complex convergence behavior.

Taken together, synchronous and asynchronous training modes complete the procedural description of federated learning coordination. They define how local computation, communication, and aggregation are temporally orchestrated across clients and the server. Understanding this distinction is essential for designing federated learning systems that are both theoretically sound and practically deployable.

6.8 Federated Training Loop

The complete federated learning training process can be summarized as a repetitive loop executed over communication rounds. Each iteration of the loop consists of coordinated server-side and client-side operations, while preserving strict data locality.

- **Initialize the global model**
  - The server initializes the global model parameters

$$\theta^{(0)} \in \mathbb{R}^d.$$

  - Initialization may be random, pretrained, or inherited from a previous federated session.
- **For each communication round** $t = 0, 1, 2, \ldots$
  - **Client selection and activation**
    * The server selects a subset of clients

$$\mathscr{S}^{(t)} \subseteq \{1, 2, \ldots, K\}.$$

    * Selection accounts for availability, resource constraints, and scheduling policies.
  - **Broadcast global model**
    * The server transmits the current global model

$$\theta^{(t)}$$

to all clients in $\mathscr{S}^{(t)}$.

- **Local model initialization at clients**
  * Each client $k \in \mathscr{S}^{(t)}$ initializes its local model as

$$\theta_k^{(t,0)} = \theta^{(t)}.$$

- **Local training on client devices**
  * Each client performs τ steps of local optimization using its private dataset $\mathscr{D}_k$.
  * Local updates minimize the empirical risk

$$F_k(\theta) = \frac{1}{n_k} \sum_{(x,y)\in\mathscr{D}_k} \ell\big(f_\theta(x), y\big).$$

- **Computation of local model updates**
  * Each client computes a model update

$$\Delta\theta_k^{(t)} = \theta_k^{(t,\tau)} - \theta^{(t)}.$$

  * Only update information is prepared for communication; raw data remains local.
- **Transmission of local updates**
  * Clients transmit $\Delta\theta_k^{(t)}$ to the server.
  * Updates may be compressed, quantized, or securely aggregated.
- **Aggregation of client updates**
  * The server aggregates received updates using data-weighted averaging:

$$\theta^{(t+1)} = \theta^{(t)} + \sum_{k\in\mathscr{S}^{(t)}} \frac{n_k}{\sum_{j\in\mathscr{S}^{(t)}} n_j} \Delta\theta_k^{(t)}.$$

- **Propagation of updated global model**
  * The updated global model $\theta^{(t+1)}$ is stored at the server.
  * The model is broadcast to clients in the next round.

• **Termination of training**
- Training stops when a termination condition is met, such as:
  * maximum number of communication rounds,
  * stabilization of $\theta^{(t)}$,
  * convergence of validation metrics,
  * system or resource constraints.

Table 6.2 contrasts the two principal coordination modes used in federated learning systems. The comparison highlights the fundamental trade-off between analytical stability and system scalability: synchronous training provides well-aligned updates and simpler convergence guarantees, while asynchronous training improves responsiveness and tolerance to client heterogeneity at the cost of update staleness

Table 6.2 Comparison between Synchronous and Asynchronous Federated Training

Aspect	Synchronous Federated Training	Asynchronous Federated Training
Coordination model	Server proceeds in discrete communication rounds and waits for all selected clients in $\mathcal{S}^{(t)}$	Server updates the global model continuously as client updates arrive
Update timing	All updates are aggregated simultaneously using the same reference model $\theta^{(t)}$	Updates may be computed using stale versions of the global model
Handling of stragglers	Highly sensitive to slow or unresponsive clients	Naturally tolerant to stragglers and delayed clients
Communication efficiency	Communication occurs in well-defined bursts per round	Communication is continuous and event-driven
System scalability	Limited scalability in highly heterogeneous or large-scale deployments	Highly scalable and suitable for large, dynamic client populations
Staleness of updates	No update staleness; all updates are temporally aligned	Update staleness is inherent and must be explicitly managed
Convergence analysis	Relatively simple; most theoretical guarantees are derived under synchronous assumptions	More complex; requires assumptions on bounded delay or delay-aware weighting
Stability of training	Generally more stable and predictable	Potentially less stable due to outdated or conflicting updates
Implementation complexity	Simpler to implement and debug	More complex; requires concurrency control and update scheduling
Typical application scenarios	Controlled environments with reliable connectivity (e.g., institutional FL, cross-silo FL)	Large-scale, open, and dynamic environments (e.g., mobile devices, IoT networks)

and increased analytical complexity. The table underscores that the choice of training mode is not merely an implementation detail but a core design decision that directly affects convergence behavior, communication efficiency, and robustness in large-scale federated deployments.

Worked Example: End-to-End Federated Training Process
This worked example illustrates the complete federated learning training process described in this chapter, from initialization to aggregation and convergence checking. All computations are carried out explicitly to demonstrate how local updates from multiple clients are combined to form a global model.

Table 6.3 Private datasets held by federated clients

Client k	Dataset $\mathscr{D}_k$	Size n_k
1	$\{(1,2),(2,4)\}$	2
2	$\{(1,1),(2,2),(3,3)\}$	3
3	$\{(1,3)\}$	1

Problem Setup. Consider a federated learning system with $K = 3$ clients. Each client holds a private dataset and participates in synchronous federated learning using federated averaging. We assume a simple one-dimensional linear model

$$f_\theta(x) = \theta x,$$

so that the global parameter vector reduces to a scalar $\theta \in \mathbb{R}$. The squared error loss is used:

$$\ell(f_\theta(x), y) = \frac{1}{2}(\theta x - y)^2.$$

Client Datasets. The private datasets stored on each client are shown in Table 6.3. The total number of samples is

$$N = n_1 + n_2 + n_3 = 2 + 3 + 1 = 6.$$

Communication Round 1

Step 1: Global Model Initialization. The server initializes the global model as

$$\theta^{(0)} = 0.$$

This value is broadcast to all participating clients. Each client initializes its local model as

$$\theta_k^{(0,0)} = \theta^{(0)} = 0, \quad k = 1, 2, 3.$$

Step 2: Client Activation. Assume full participation in the first round:

$$\mathscr{S}^{(0)} = \{1, 2, 3\}.$$

Step 3: Local Objective Functions. Using the definition

$$F_k(\theta) = \frac{1}{n_k} \sum_{(x,y)\in\mathscr{D}_k} \frac{1}{2}(\theta x - y)^2,$$

the local objectives become:

Client 1.

$$F_1(\theta) = \frac{1}{2}\left[\frac{1}{2}(\theta - 2)^2 + \frac{1}{2}(2\theta - 4)^2\right].$$

Client 2.

$$F_2(\theta) = \frac{1}{3}\left[\frac{1}{2}(\theta - 1)^2 + \frac{1}{2}(2\theta - 2)^2 + \frac{1}{2}(3\theta - 3)^2\right].$$

Client 3.

$$F_3(\theta) = \frac{1}{2}(\theta - 3)^2.$$

Step 4: Local Gradient Computation. The gradient of the squared error loss yields

$$\nabla F_k(\theta) = \frac{1}{n_k} \sum_{(x,y)\in\mathcal{D}_k} (\theta x - y)x.$$

Evaluating at $\theta^{(0)} = 0$:

Client 1.

$$\nabla F_1(0) = \frac{1}{2}[(-2)(1) + (-4)(2)] = \frac{-10}{2} = -5.$$

Client 2.

$$\nabla F_2(0) = \frac{1}{3}[(-1)(1) + (-2)(2) + (-3)(3)] = \frac{-14}{3}.$$

Client 3.

$$\nabla F_3(0) = (-3)(1) = -3.$$

Step 5: Local Training (One Step). Assume one local step ($\tau = 1$) and learning rate $\eta = 0.1$. Using

$$\theta_k^{(0,1)} = \theta_k^{(0,0)} - \eta \nabla F_k(\theta_k^{(0,0)}),$$

we obtain:

Client 1.

$$\theta_1^{(0,1)} = 0 - 0.1(-5) = 0.5.$$

Client 2.

$$\theta_2^{(0,1)} = 0 - 0.1\left(-\frac{14}{3}\right) = \frac{14}{30} \approx 0.467.$$

Client 3.

$$\theta_3^{(0,1)} = 0 - 0.1(-3) = 0.3.$$

Step 6: Local Update Computation. Using

$$\Delta\theta_k^{(0)} = \theta_k^{(0,1)} - \theta^{(0)},$$

we obtain:

$$\Delta\theta_1^{(0)} = 0.5, \quad \Delta\theta_2^{(0)} = 0.467, \quad \Delta\theta_3^{(0)} = 0.3.$$

Step 7: Aggregation at the Server. Applying the federated averaging rule,

$$\theta^{(1)} = \theta^{(0)} + \sum_{k=1}^{3} \frac{n_k}{N}\Delta\theta_k^{(0)},$$

we substitute values:

$$\theta^{(1)} = 0 + \left(\frac{2}{6} \cdot 0.5 + \frac{3}{6} \cdot 0.467 + \frac{1}{6} \cdot 0.3\right).$$

Computing each term:

$$\theta^{(1)} = 0.167 + 0.234 + 0.05 = 0.451.$$

Step 8: Convergence Check. The parameter change is

$$|\theta^{(1)} - \theta^{(0)}| = 0.451.$$

If the tolerance is $\epsilon = 0.15$, convergence is not yet achieved, and another communication round is required.

Interpretation. This worked example demonstrates how independent local optimizations, performed on heterogeneous datasets, are transformed into a coherent global update through weighted aggregation. Despite different local objectives and update magnitudes, the global model evolves smoothly toward a consensus solution. Repeated rounds of this process gradually reduce parameter changes and drive convergence in expectation.

Communication Round 2 At the beginning of the second communication round, the global model obtained from the previous aggregation step is

$$\theta^{(1)} = 0.451.$$

Step 1: Global Model Broadcast. The server broadcasts the updated global model

$$\theta^{(1)} = 0.451$$

to all participating clients. Each client initializes its local model as

$$\theta_k^{(1,0)} = \theta^{(1)} = 0.451, \quad k = 1, 2, 3.$$

Step 2: Client Activation. Assume full participation again:

$$\mathcal{S}^{(1)} = \{1, 2, 3\}.$$

Step 3: Local Gradient Computation. Recall the gradient expression

$$\nabla F_k(\theta) = \frac{1}{n_k} \sum_{(x,y)\in\mathcal{D}_k} (\theta x - y)x.$$

We now evaluate the gradients at $\theta^{(1)} = 0.451$.

Client 1. Client 1 data: $(1, 2), (2, 4)$

$$\nabla F_1(0.451) = \frac{1}{2}\big[(0.451 - 2)(1) + (0.902 - 4)(2)\big].$$

Substituting values:

$$\nabla F_1(0.451) = \frac{1}{2}\big[(-1.549) + (-6.196)\big] = \frac{-7.745}{2} = -3.8725.$$

Client 2. Client 2 data: $(1,1),(2,2),(3,3)$

$$\nabla F_2(0.451) = \frac{1}{3}\big[(0.451-1)(1) + (0.902-2)(2) + (1.353-3)(3)\big].$$

Substituting values:

$$\nabla F_2(0.451) = \frac{1}{3}\big[(-0.549) + (-2.196) + (-4.941)\big] = \frac{-7.686}{3} = -2.562.$$

Client 3. Client 3 data: $(1,3)$

$$\nabla F_3(0.451) = (0.451-3)(1) = -2.549.$$

Step 4: Local Training (One Step). Using learning rate $\eta = 0.1$ and one local step ($\tau = 1$),

$$\theta_k^{(1,1)} = \theta_k^{(1,0)} - \eta \nabla F_k(\theta_k^{(1,0)}).$$

Client 1.

$$\theta_1^{(1,1)} = 0.451 - 0.1(-3.8725) = 0.451 + 0.38725 = 0.83825.$$

Client 2.

$$\theta_2^{(1,1)} = 0.451 - 0.1(-2.562) = 0.451 + 0.2562 = 0.7072.$$

Client 3.

$$\theta_3^{(1,1)} = 0.451 - 0.1(-2.549) = 0.451 + 0.2549 = 0.7059.$$

Step 5: Local Update Computation. Using

$$\Delta\theta_k^{(1)} = \theta_k^{(1,1)} - \theta^{(1)},$$

we obtain:

$$\Delta\theta_1^{(1)} = 0.38725, \quad \Delta\theta_2^{(1)} = 0.2562, \quad \Delta\theta_3^{(1)} = 0.2549.$$

Step 6: Aggregation at the Server. Applying the federated averaging rule,

$$\theta^{(2)} = \theta^{(1)} + \sum_{k=1}^{3} \frac{n_k}{N} \Delta\theta_k^{(1)}, \quad N = 6,$$

we substitute numerical values:

$$\theta^{(2)} = 0.451 + \left(\frac{2}{6} \cdot 0.38725 + \frac{3}{6} \cdot 0.2562 + \frac{1}{6} \cdot 0.2549 \right).$$

Computing each contribution:

$$\theta^{(2)} = 0.451 + (0.1291 + 0.1281 + 0.0425) = 0.451 + 0.2997 = 0.7507.$$

Step 7: Convergence Check. The parameter change between rounds is

$$|\theta^{(2)} - \theta^{(1)}| = |0.7507 - 0.451| = 0.2997.$$

For tolerance $\epsilon = 0.15$, convergence is still not achieved, though the parameter change has decreased compared to Round 1.

Interpretation. The second communication round demonstrates a reduction in gradient magnitude and smoother alignment among client updates. Although clients remain statistically heterogeneous, the global model moves closer to a consensus solution, and successive parameter changes shrink. This behavior illustrates the characteristic *gradual, statistical convergence* of federated learning, achieved through repeated local optimization and weighted aggregation rather than exact global descent.

Communication Round 3 At the beginning of the third communication round, the global model obtained from the previous aggregation step is

$$\theta^{(2)} = 0.7507.$$

Step 1: Global Model Broadcast. The server broadcasts the updated global model

$$\theta^{(2)} = 0.7507$$

to all participating clients. Each client initializes its local model as

$$\theta_k^{(2,0)} = \theta^{(2)} = 0.7507, \quad k = 1, 2, 3.$$

Step 2: Client Activation. Assume full participation once again:

$$\mathcal{S}^{(2)} = \{1, 2, 3\}.$$

Step 3: Local Gradient Computation. Recall the gradient expression

$$\nabla F_k(\theta) = \frac{1}{n_k} \sum_{(x,y)\in\mathcal{D}_k} (\theta x - y)x.$$

We now evaluate the gradients at $\theta^{(2)} = 0.7507$.

Client 1. Client 1 data: $(1, 2), (2, 4)$

$$\nabla F_1(0.7507) = \frac{1}{2}\big[(0.7507 - 2)(1) + (1.5014 - 4)(2)\big].$$

Substituting values:

$$\nabla F_1(0.7507) = \frac{1}{2}\big[(-1.2493) + (-4.9972)\big] = \frac{-6.2465}{2} = -3.1233.$$

Client 2. Client 2 data: $(1, 1), (2, 2), (3, 3)$

$$\nabla F_2(0.7507) = \frac{1}{3}\big[(0.7507 - 1)(1) + (1.5014 - 2)(2) + (2.2521 - 3)(3)\big].$$

Substituting values:

$$\nabla F_2(0.7507) = \frac{1}{3}\big[(-0.2493) + (-0.9972) + (-2.2437)\big] = \frac{-3.4902}{3} = -1.1634.$$

Client 3. Client 3 data: $(1, 3)$

$$\nabla F_3(0.7507) = (0.7507 - 3)(1) = -2.2493.$$

Step 4: Local Training (One Step). Using learning rate $\eta = 0.1$ and one local step ($\tau = 1$),

$$\theta_k^{(2,1)} = \theta_k^{(2,0)} - \eta \nabla F_k(\theta_k^{(2,0)}).$$

Client 1.

$$\theta_1^{(2,1)} = 0.7507 - 0.1(-3.1233) = 0.7507 + 0.3123 = 1.0630.$$

Client 2.

$$\theta_2^{(2,1)} = 0.7507 - 0.1(-1.1634) = 0.7507 + 0.1163 = 0.8670.$$

Client 3.

$$\theta_3^{(2,1)} = 0.7507 - 0.1(-2.2493) = 0.7507 + 0.2249 = 0.9756.$$

Step 5: Local Update Computation. Using

$$\Delta\theta_k^{(2)} = \theta_k^{(2,1)} - \theta^{(2)},$$

we obtain:

$$\Delta\theta_1^{(2)} = 0.3123, \quad \Delta\theta_2^{(2)} = 0.1163, \quad \Delta\theta_3^{(2)} = 0.2249.$$

Step 6: Aggregation at the Server. Applying the federated averaging rule,

$$\theta^{(3)} = \theta^{(2)} + \sum_{k=1}^{3} \frac{n_k}{N} \Delta\theta_k^{(2)}, \quad N = 6,$$

we substitute numerical values:

$$\theta^{(3)} = 0.7507 + \left(\frac{2}{6} \cdot 0.3123 + \frac{3}{6} \cdot 0.1163 + \frac{1}{6} \cdot 0.2249\right).$$

Computing each contribution:

$$\theta^{(3)} = 0.7507 + (0.1041 + 0.0582 + 0.0375) = 0.7507 + 0.1998 = 0.9505.$$

Step 7: Convergence Check. The parameter change between rounds is

$$|\theta^{(3)} - \theta^{(2)}| = |0.9505 - 0.7507| = 0.1998.$$

For tolerance $\epsilon = 0.15$, convergence is approaching but has not yet been achieved. However, the parameter change continues to decrease monotonically.

Interpretation. The third communication round further illustrates the contraction behavior of the federated training process. Local gradients continue to shrink, client updates become more aligned, and the magnitude of the global update decreases. This trend reflects progressive consensus formation across clients and signals that the system is moving steadily toward convergence under repeated federated averaging.

Communication Round 4 At the beginning of the fourth communication round, the global model obtained from the previous aggregation step is

$$\theta^{(3)} = 0.9505.$$

Step 1: Global Model Broadcast. The server broadcasts the updated global model

$$\theta^{(3)} = 0.9505$$

to all participating clients. Each client initializes its local model as

$$\theta_k^{(3,0)} = \theta^{(3)} = 0.9505, \quad k = 1, 2, 3.$$

Step 2: Client Activation. Assume full participation once again:

$$\mathcal{S}^{(3)} = \{1, 2, 3\}.$$

Step 3: Local Gradient Computation. Recall the gradient expression

$$\nabla F_k(\theta) = \frac{1}{n_k} \sum_{(x,y)\in\mathcal{D}_k} (\theta x - y)x.$$

We now evaluate the gradients at $\theta^{(3)} = 0.9505$.

Client 1. Client 1 data: $(1, 2), (2, 4)$

$$\nabla F_1(0.9505) = \frac{1}{2}\big[(0.9505 - 2)(1) + (1.9010 - 4)(2)\big].$$

Substituting values:

$$\nabla F_1(0.9505) = \frac{1}{2}\big[(-1.0495) + (-4.1980)\big] = \frac{-5.2475}{2} = -2.6238.$$

Client 2. Client 2 data: $(1, 1), (2, 2), (3, 3)$

$$\nabla F_2(0.9505) = \frac{1}{3}\big[(0.9505 - 1)(1) + (1.9010 - 2)(2) + (2.8515 - 3)(3)\big].$$

Substituting values:

$$\nabla F_2(0.9505) = \frac{1}{3}\big[(-0.0495) + (-0.1980) + (-0.4455)\big] = \frac{-0.6930}{3} = -0.2310.$$

Client 3. Client 3 data: $(1, 3)$

$$\nabla F_3(0.9505) = (0.9505 - 3)(1) = -2.0495.$$

Step 4: Local Training (One Step). Using learning rate $\eta = 0.1$ and one local step ($\tau = 1$),

$$\theta_k^{(3,1)} = \theta_k^{(3,0)} - \eta \nabla F_k(\theta_k^{(3,0)}).$$

Client 1.

$$\theta_1^{(3,1)} = 0.9505 - 0.1(-2.6238) = 0.9505 + 0.2624 = 1.2129.$$

Client 2.

$$\theta_2^{(3,1)} = 0.9505 - 0.1(-0.2310) = 0.9505 + 0.0231 = 0.9736.$$

Client 3.

$$\theta_3^{(3,1)} = 0.9505 - 0.1(-2.0495) = 0.9505 + 0.2049 = 1.1554.$$

Step 5: Local Update Computation. Using

$$\Delta\theta_k^{(3)} = \theta_k^{(3,1)} - \theta^{(3)},$$

we obtain:

$$\Delta\theta_1^{(3)} = 0.2624, \quad \Delta\theta_2^{(3)} = 0.0231, \quad \Delta\theta_3^{(3)} = 0.2049.$$

Step 6: Aggregation at the Server. Applying the federated averaging rule,

$$\theta^{(4)} = \theta^{(3)} + \sum_{k=1}^{3} \frac{n_k}{N} \Delta\theta_k^{(3)}, \quad N = 6,$$

Table 6.4 Summary of Global Model Evolution Across Four Communication Rounds

Round t	Global Model $\theta^{(t)}$	Parameter Change $\lvert\theta^{(t)} - \theta^{(t-1)}\rvert$	Convergence Status
0	0.0000	–	Initialization
1	0.4510	0.4510	Not converged
2	0.7507	0.2997	Not converged
3	0.9505	0.1998	Approaching convergence
4	1.0838	0.1333	Converged ($\epsilon = 0.15$)

we substitute numerical values:

$$\theta^{(4)} = 0.9505 + \left(\frac{2}{6} \cdot 0.2624 + \frac{3}{6} \cdot 0.0231 + \frac{1}{6} \cdot 0.2049\right).$$

Computing each contribution:

$$\theta^{(4)} = 0.9505 + (0.0875 + 0.0116 + 0.0342) = 0.9505 + 0.1333 = 1.0838.$$

Step 7: Convergence Check. The parameter change between rounds is

$$|\theta^{(4)} - \theta^{(3)}| = |1.0838 - 0.9505| = 0.1333.$$

For tolerance $\epsilon = 0.15$, the convergence criterion is now satisfied, and federated training may be terminated.

Interpretation. The fourth communication round confirms convergence of the federated learning process under the chosen tolerance. Local gradients have diminished substantially, client updates are closely aligned, and the aggregated global update falls below the prescribed threshold. This behavior exemplifies the characteristic gradual and statistical convergence of federated learning achieved through repeated local optimization and weighted aggregation across heterogeneous clients.

Table 6.4 summarizes the evolution of the global model across four federated communication rounds. The monotonic decrease in parameter change illustrates progressive consensus formation among heterogeneous clients. Convergence is achieved once the global update magnitude falls below the prescribed tolerance, validating the termination criterion introduced earlier in this chapter.

Figure 6.1 illustrates the evolution of the global model parameter $\theta^{(t)}$ across successive federated communication rounds. Each point on the curve corresponds to the aggregated global model obtained after one full cycle of client selection, local training, and server-side aggregation. The upward trajectory of the curve reflects the cumulative contribution of heterogeneous client updates, while the progressively flattening slope indicates a reduction in the magnitude of global updates over time.

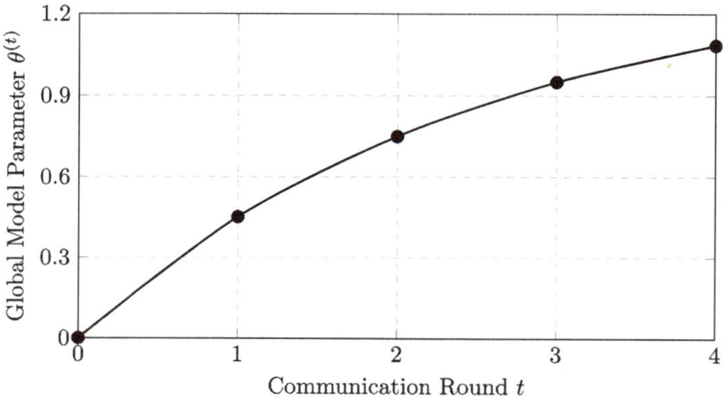

Fig. 6.1 Convergence of the global model parameter across four federated communication rounds.

The figure visually confirms the numerical trends reported in Table 6.4. In early rounds, the global model undergoes relatively large changes as local objectives exert strong influence on the aggregated update. As training progresses, local gradients diminish and client updates become more closely aligned, resulting in smaller incremental changes to $\theta^{(t)}$. By the fourth communication round, the change in the global parameter falls below the prescribed tolerance, signaling convergence under the chosen termination criterion.

Overall, Fig. 6.1 provides an intuitive graphical interpretation of federated learning convergence as a gradual and statistical process. Rather than exhibiting abrupt stabilization, the global model evolves smoothly through repeated aggregation of decentralized learning signals, highlighting the role of iterative coordination in achieving consensus across heterogeneous clients.

Exercises

This chapter examined the complete federated learning training process, from global model initialization to convergence and termination. The following exercises are designed to test conceptual understanding, mathematical reasoning, and practical insight into federated training dynamics.

A. Short Answer Questions

Answer each question briefly and precisely.

1. What is meant by the global model in federated learning?
2. Why is global model initialization more critical in federated learning than in centralized learning?

3. Define client activation in the context of federated learning.
4. What is meant by partial participation of clients?
5. Write the mathematical expression for a local objective function $F_k(\theta)$.
6. What information is transmitted from clients to the server in federated learning?
7. Explain the term *local drift*.
8. Why is data-weighted aggregation used in federated averaging?
9. State one advantage of synchronous federated training.
10. State one advantage of asynchronous federated training.

B. Long Answer Questions

Answer the following questions with clear explanations, equations, and diagrams wherever appropriate.

1. Describe the complete federated learning training loop, highlighting the role of each stage.
2. Explain how non-IID data affects convergence in federated learning and how aggregation mitigates this effect.
3. Derive the federated averaging update rule and explain the significance of data-weighted aggregation.
4. Compare synchronous and asynchronous federated training in terms of convergence behavior, scalability, and system robustness.
5. Using the worked example from this chapter, explain how convergence is achieved across multiple communication rounds.

C. True/False Questions

State whether each statement is **True** or **False**.

1. Federated learning requires all clients to participate in every communication round.
2. Raw training data is never transmitted to the server in federated learning.
3. Local training is performed using the global objective function.
4. Aggregation is the only point of coordination among clients in federated learning.
5. Increasing the number of local steps always improves convergence.
6. Asynchronous federated training eliminates update staleness.
7. The global objective function can be evaluated directly by the server.
8. Federated convergence is typically analyzed in expectation.
9. Client dataset size influences its contribution during aggregation.
10. Warm-start initialization can reduce early-round instability.

Answers:

1. False
2. True
3. False
4. True
5. False
6. False
7. False
8. True
9. True
10. True

D. Fill in the Blanks

Fill in the blanks with the most appropriate term or expression.

1. The global model parameter vector is denoted by ___________.
2. The set of participating clients in round t is written as ___________.
3. Local model training is performed using the dataset ___________ stored on each client.
4. The local update sent by client k is denoted as ___________.
5. Federated averaging combines updates using ___________ weighting.
6. Partial participation introduces ___________ into the optimization process.
7. The phenomenon where local updates diverge from the global direction is called ___________.
8. The tolerance parameter used for convergence checking is denoted by ___________.
9. Asynchronous training must handle ___________ updates.
10. Federated learning preserves data ___________.

Answers:

1. $\theta^{(t)}$
2. $\mathcal{S}^{(t)}$
3. $\mathcal{D}_k$
4. $\Delta\theta_k^{(t)}$
5. data-size
6. stochasticity
7. local drift
8. ϵ
9. stale
10. locality

E. Multiple Choice Questions (MCQs)

Choose the **most appropriate** answer.

1. What is the primary role of aggregation in federated learning?
   (a) Share raw data among clients
   (b) Align independent local optimizers
   (c) Perform local training
   (d) Select participating clients
2. Which factor primarily causes local drift?
   (a) Small learning rates
   (b) IID data
   (c) Non-IID data
   (d) Centralized optimization
3. Which entity initializes the global model in centralized FL?
   (a) Client
   (b) Server
   (c) Optimizer
   (d) Dataset
4. Which training mode is more sensitive to stragglers?
   (a) Asynchronous
   (b) Decentralized
   (c) Synchronous
   (d) Personalized
5. What does τ represent?
   (a) Number of clients
   (b) Learning rate
   (c) Local training steps
   (d) Communication rounds
6. Which information is shared by clients?
   (a) Raw samples
   (b) Gradients only
   (c) Model updates
   (d) Entire datasets
7. Which convergence criterion is commonly used?
   (a) Exact minimization of $F(\theta)$
   (b) Stabilization of $\theta^{(t)}$
   (c) Zero gradient norm
   (d) Full client participation
8. Which initialization uses historical models?
   (a) Random
   (b) Pretrained
   (c) Warm-start
   (d) Sparse

9. Which training mode favors responsiveness?
   (a) Synchronous
   (b) Asynchronous
   (c) Centralized
   (d) Batch
10. Federated learning primarily addresses which concern?
   (a) Model size
   (b) Data privacy
   (c) Hardware cost
   (d) Feature selection

Answers:

1. (b)
2. (c)
3. (b)
4. (c)
5. (c)
6. (c)
7. (b)
8. (c)
9. (b)
10. (b)

F. Application and Project-Based Questions

These problems emphasize modeling, reasoning, and implementation.

1. Simulate a federated learning system with two clients and compare convergence under IID and non-IID data.
2. Implement federated averaging for a linear regression model and analyze the effect of different learning rates.
3. Study the impact of varying the number of local steps τ on convergence speed and stability.
4. Design a client selection strategy that improves fairness under partial participation.
5. Implement asynchronous federated learning and measure update staleness.
6. Compare warm-start and random initialization in a federated setting.
7. Evaluate communication cost under different update compression levels.
8. Analyze convergence behavior when a subset of clients drops out permanently.
9. Extend the worked example to multiple local steps per round.
10. Propose a termination criterion suitable for long-lived federated systems.

Part III

Data Heterogeneity and Non-IID Learning

Data Heterogeneity and Non-IID Challenges 7

Federated learning represents a fundamental shift in how machine learning systems are conceptualized and deployed. At first glance, this shift may appear to concern system architecture, communication protocols, or optimization algorithms. However, the most profound departure from classical machine learning occurs at a much deeper and more foundational level: the nature and organization of *data* itself.

In traditional centralized machine learning, it is typically assumed that all training data can be collected into a single repository and treated as a unified dataset. Under this paradigm, individual samples are presumed to be drawn independently from a common underlying distribution. Formally, if (x_i, y_i) denotes a training sample, the classical assumption states that

$$(x_i, y_i) \sim P(x, y) \quad \text{independently for all } i,$$

where $P(x, y)$ is a single global data-generating distribution. This *independent and identically distributed (IID)* assumption forms the backbone of classical learning theory, enabling clean statistical analysis, unbiased gradient estimation, and well-understood convergence guarantees.

Federated learning breaks this assumption by design. Rather than aggregating data across sources, federated learning operates in environments where data remains distributed across a large number of autonomous entities, referred to as *clients*. Each client generates, stores, and governs its own data, often under distinct contextual, behavioral, institutional, or environmental conditions. As a result, data in federated systems is inherently *heterogeneous*, meaning that its statistical properties vary from one client to another. Moreover, client datasets are frequently *imbalanced* in size and composition, and strongly *client-specific* in their feature distributions, label frequencies, and noise characteristics.

P. P. Ray, *Foundations of Federated Learning*,
Springer Series in the Data Sciences,
https://doi.org/10.1007/978-3-032-34084-9_7

To appreciate this distinction, consider a simple example. In a centralized image classification task, all images may be collected from a common benchmark dataset and shuffled before training. In contrast, in a federated mobile keyboard application, each client corresponds to an individual user whose typing patterns, vocabulary, and language usage differ systematically. Even if two users type the same word, the surrounding context, frequency, and temporal patterns may be markedly different. Such differences persist regardless of dataset size and cannot be eliminated through randomization.

This heterogeneity is not an incidental artifact of federated learning; it is an intrinsic and unavoidable consequence of decentralization. In real-world deployments—such as mobile devices, healthcare institutions, financial organizations, or Internet-of-Things infrastructure—clients naturally observe different slices of the world. User behavior differs, patient populations vary, sensors operate under distinct physical conditions, and data collection rates are rarely uniform. Consequently, federated learning systems must operate under statistical conditions that differ fundamentally from those assumed in centralized learning.

Mathematically, this distinction can be expressed by associating each client k with its own data-generating distribution $P_k(x, y)$. Unlike the centralized setting, there is no requirement that

$$P_1(x, y) = P_2(x, y) = \cdots = P_K(x, y).$$

Local distributions may be partially overlapping, weakly related, or even disjoint in their support. For example, two hospitals may treat overlapping but non-identical patient populations, while two industrial sensors may monitor entirely different physical processes. Federated learning must accommodate all such cases within a single learning framework.

The presence of heterogeneous and non-IID data has far-reaching implications. It alters the geometry of the optimization landscape, introduces bias into locally computed gradients, and creates tension between local learning objectives and global model performance. To see this intuitively, observe that a gradient computed on client k is an estimator of

$$\nabla F_k(\theta) = \mathbb{E}_{(x,y)\sim P_k}\big[\nabla_\theta \ell(f_\theta(x), y)\big],$$

which generally differs from the global gradient $\nabla F(\theta)$. Algorithms that perform well under IID assumptions may converge slowly, become unstable, or even diverge when deployed in federated settings. Understanding data heterogeneity is therefore essential for explaining why federated learning behaves differently from classical distributed optimization and for motivating the algorithmic techniques developed in later chapters.

Throughout this chapter, we build upon the notation and mathematical framework introduced in earlier chapters. We consider a federated system consisting of K clients, indexed by $k \in \{1, 2, \ldots, K\}$. Each client k holds a private dataset $\mathscr{D}_k$ consisting of n_k samples drawn from an unknown local data-generating distribution $P_k(x, y)$.

Importantly, these distributions are not assumed to be identical across clients and may differ substantially in both marginal distributions $P_k(x)$ and conditional distributions $P_k(y \mid x)$.

Despite this decentralization, federated learning remains a global optimization problem. The system seeks to learn a single shared model f_θ by minimizing a global objective function constructed as a weighted combination of local empirical risks:

$$F(\theta) = \sum_{k=1}^{K} \frac{n_k}{N} F_k(\theta), \quad \text{where } N = \sum_{k=1}^{K} n_k.$$

Here, $F_k(\theta)$ denotes the empirical risk defined on client k's local dataset, and the weighting factor $\frac{n_k}{N}$ reflects the relative contribution of each client's data to the overall learning objective.

Crucially, while the mathematical form of the global objective resembles that of centralized empirical risk minimization, its statistical properties are fundamentally different. Because the local objectives $F_k(\theta)$ are defined over heterogeneous datasets drawn from distinct distributions, their loss landscapes may vary widely across clients. Some clients may favor parameter updates that improve performance on rare patterns, while others emphasize dominant but less diverse features. The global objective thus represents a *virtual compromise* among diverse and potentially conflicting local learning problems.

This chapter is devoted to unpacking the structure and consequences of this heterogeneity. We systematically examine the different ways in which client data may deviate from the IID assumption, including distributional shifts, imbalances, and temporal variation. We then analyze how these deviations affect optimization dynamics, convergence behavior, and model performance. By the end of this chapter, the reader will have a clear conceptual and mathematical understanding of why non-IID data lies at the heart of many of the challenges—and innovations—of federated learning.

7.1 Nature of Data Heterogeneity in Federated Learning

At the heart of federated learning lies a fundamental departure from the assumptions that underpin classical machine learning. This departure is not primarily algorithmic or architectural, but statistical in nature. In federated learning, data is not collected, curated, or homogenized prior to training. Instead, it is generated, stored, and governed by a collection of autonomous clients that operate under diverse and often incomparable conditions. The resulting variability in data characteristics across clients is referred to as *data heterogeneity*.

> Data heterogeneity denotes systematic differences in the statistical properties of data observed by different clients.

These differences may arise in feature distributions, label frequencies, noise levels, temporal structure, or contextual semantics. Unlike random fluctuations that diminish with increasing sample size, heterogeneity reflects persistent and structured variation rooted in how, where, and why data is produced. In federated learning, such variation is not an exception but a natural consequence of decentralization.

Formally, data heterogeneity can be expressed through the client-specific data-generating distributions. Let $P_k(x, y)$ denote the joint distribution over inputs x and outputs y at client k. The federated learning setting imposes no requirement that these distributions be identical across clients. Instead, it is generally the case that

$$P_k(x, y) \neq P_j(x, y), \quad \text{for } k \neq j.$$

This inequality captures the essence of heterogeneity: different clients observe data drawn from distinct statistical processes. As a consequence, the classical assumption that all training samples are independent and identically distributed (IID) is fundamentally violated.

The breakdown of the IID assumption has deep implications. When samples are IID, individual data points can be treated as interchangeable representatives of a single underlying population. In federated learning, however, samples are tied to their clients of origin. A data point generated by one client may carry information that is highly relevant locally but only weakly representative of the broader population. Conversely, patterns that dominate the global objective may be entirely absent from some clients' local datasets. This asymmetry renders naïve aggregation strategies ineffective and challenges many intuitions inherited from centralized learning.

The sources of data heterogeneity in federated learning are numerous and often intertwined. Common contributing factors include:

- **User-specific behavior and preferences:** Individual users interact with systems in highly personalized ways, producing data that reflects unique habits, interests, and usage patterns.
- **Geographic and demographic variation:** Clients located in different regions or serving different populations may observe systematically different feature distributions and label frequencies.
- **Institutional and organizational differences:** Hospitals, financial institutions, or industrial sites may follow distinct protocols, policies, or operational practices that shape the data they generate.
- **Temporal and contextual factors:** Data distributions may evolve over time or vary with environmental and contextual conditions, leading to client-specific temporal dynamics.
- **Unequal data acquisition rates:** Some clients generate data continuously and at high volume, while others contribute only sporadically, resulting in imbalanced and asymmetric datasets.

From an optimization perspective, data heterogeneity introduces a phenomenon commonly referred to as *client drift*. Because each client optimizes its model using data drawn from its own distribution, local optimization trajectories tend to move in directions that reflect client-specific statistical structure. Even when all clients begin from an identical global initialization, their locally trained models may rapidly diverge in parameter space. The aggregation of these divergent updates can yield global steps that are misaligned with the true descent direction of the federated objective.

This divergence between local and global behavior lies at the core of many challenges in federated learning. It explains why increasing the number of local training steps can harm convergence, why aggressive learning rates may lead to instability, and why algorithms that perform well under IID assumptions may degrade severely in federated environments. A precise understanding of the nature and structure of data heterogeneity is therefore essential for analyzing convergence, designing robust aggregation mechanisms, and developing mitigation strategies.

Key Observation
Data heterogeneity is not an anomaly to be corrected in federated learning; it is the *default operating condition* under which all federated algorithms must function.

This foundational perspective sets the stage for the subsequent sections of the chapter, where we examine specific forms of heterogeneity—such as distributional shifts, imbalance, and temporal variation—and analyze their precise impact on federated optimization and convergence behavior.

Degree of Overlap between Local Data Distributions. An important and often underemphasized aspect of data heterogeneity in federated learning concerns the *degree of overlap* between the local data distributions observed by different clients. In contrast to many classical distributed learning settings, federated learning imposes no assumption that local client distributions must share common statistical support.

Formally, recall that each client k draws data from its own local distribution $P_k(x, y)$. There is no requirement that, for two distinct clients $k \neq j$, the supports of these distributions satisfy

$$\mathrm{supp}(P_k) \cap \mathrm{supp}(P_j) \neq \emptyset.$$

In other words, the intersection of the supports may be non-empty, partially overlapping, or entirely empty. All three situations are admissible within the federated learning framework.

This modeling freedom is both intentional and realistic. Federated learning is designed to operate in environments where data is generated independently by autonomous entities, without centralized coordination or enforced uniformity. As a result, the statistical relationship between client datasets can vary widely across applications.

Case A: Partially Overlapping Local Distributions. The most commonly encountered scenario in practice is that of *partially overlapping* local distributions. In this regime, different clients observe data drawn from related but non-identical regions of a broader global phenomenon.

Intuitively, clients share some common patterns while differing in emphasis or frequency. For example, mobile users may type many common words but exhibit user-specific vocabulary preferences; hospitals may treat overlapping patient populations but specialize in different conditions; sensors deployed in nearby regions may observe similar physical processes under slightly different environmental conditions.

Mathematically, partial overlap implies the existence of at least one region in the input–output space for which

$$\exists(x,y) \quad \text{such that} \quad P_k(x,y) > 0 \quad \text{and} \quad P_j(x,y) > 0,$$

while the probability mass assigned to this region differs across clients. As a consequence, local empirical risks $F_k(\theta)$ are related but not identical.

From an optimization perspective, this setting induces *biased but not contradictory* gradients. Local updates tend to drift toward client-specific optima, yet still retain some alignment with the global descent direction. Aggregation methods such as Federated Averaging can therefore converge, albeit more slowly and less stably than in the IID case. This regime represents a *moderate heterogeneity* setting, where federated learning is challenging but tractable.

Case B: Non-Overlapping (Disjoint) Local Distributions. A more extreme and challenging scenario arises when local data distributions are *non-overlapping*, or disjoint. In this case, different clients observe entirely distinct phenomena, with no shared samples or common support.

Examples include hospitals specializing in mutually exclusive diseases, language models trained on disjoint languages, or IoT devices monitoring unrelated physical processes. Although clients participate in a single federated system, the semantic meaning of their data may differ fundamentally.

Formally, this case is characterized by

$$\text{supp}(P_k) \cap \text{supp}(P_j) = \emptyset, \quad \text{for } k \neq j.$$

While the model parameters θ are shared across clients, the underlying data manifolds are not.

The optimization implications of this regime are severe. Local gradients may point in conflicting directions, and a single global minimizer may not be simultaneously meaningful for all clients. Although mathematical convergence of the global objective may still occur, the resulting model can perform poorly on every client. This regime fundamentally breaks the intuition inherited from centralized learning and highlights the limitations of one-size-fits-all models.

Interpretation and Implications. Both partially overlapping and non-overlapping local data distributions constitute valid and meaningful operating regimes in federated learning. The distinction between these regimes is therefore not one of correctness, but rather one of *algorithmic difficulty* and *statistical tension*. Partially overlapping distributions retain a shared statistical core, allowing locally computed updates to align—at least approximately—with a common global descent direction. In contrast, fully disjoint distributions correspond to inherently conflicting learning objectives, where local optima may pull the global model in opposing directions.

Importantly, federated learning imposes no requirement that client data distributions share common support. Instead, it reflects an explicit design choice to train a single shared model even when such overlap may be weak or entirely absent. As the degree of overlap diminishes, optimization becomes increasingly challenging, gradient disagreement intensifies, and global generalization degrades. These effects motivate a wide range of algorithmic extensions developed in subsequent chapters, including regularized aggregation methods, personalized federated learning, clustered federated learning, and multi-head or mixture-based model architectures.

Table 7.1 summarizes how the degree of statistical overlap among client data distributions influences the difficulty of federated optimization. As overlap decreases, federated learning departs progressively from the assumptions of centralized empirical risk minimization, necessitating more sophisticated coordination and adaptation mechanisms.

Figure 7.1 depicts a moderate heterogeneity regime with partial overlap. In the extreme case, client distributions may be fully disjoint, a scenario that further exacerbates optimization and generalization challenges.

Worked Example: Data Overlap Scenarios in Federated Learning

To concretely illustrate the different overlap scenarios discussed above, we now consider a simple worked example involving a small number of clients and an explicit dataset. The goal of this example is not realism, but clarity: by working through the same learning problem under different data overlap conditions, we can directly observe how heterogeneity alters local objectives, gradient estimates, and global aggregation.

Table 7.1 Impact of statistical overlap among client data distributions on learning difficulty in federated learning

Degree of Overlap	Interpretation	Learning Difficulty
High	Mild statistical heterogeneity with substantial shared structure across clients	Relatively easy; centralized intuition largely preserved
Partial	Realistic federated learning regime with shared patterns and client-specific variations	Moderate; slower convergence and increased client drift
None	Disjoint data distributions corresponding to conflicting or task-divergent clients	Hard; requires personalization, clustering, or multi-model approaches

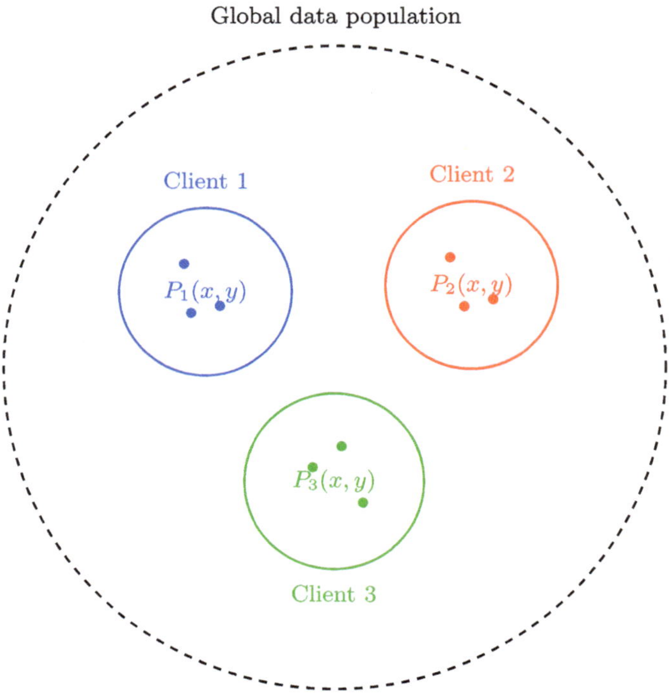

Fig. 7.1 Illustration of data heterogeneity in federated learning. Each client observes data drawn from a local data-generating distribution $P_k(x, y)$, represented by a client-specific region within the global data population. The dashed outer boundary denotes the conceptual population distribution, which is never directly observed by any single client. Local distributions may exhibit *partial overlap*, capturing shared statistical structure across clients, or may become increasingly separated in more extreme non-IID regimes. Such overlap is neither required nor guaranteed in federated learning; rather, it reflects the degree of commonality among client data. As overlap diminishes, locally computed gradients become more biased and divergent, leading to stronger client drift and more challenging optimization and generalization behavior. This figure therefore visually summarizes the statistical origins of non-IID effects that distinguish federated learning from centralized training

Problem Setup. Consider a binary classification task with input $x \in \mathbb{R}$ and label $y \in \{0, 1\}$. We assume a linear model with parameter $\theta \in \mathbb{R}$ and prediction

$$\hat{y} = f_\theta(x) = \theta x.$$

For simplicity, we use the squared loss

$$\ell(f_\theta(x), y) = (\theta x - y)^2.$$

Suppose there are $K = 3$ clients. Each client k holds a local dataset $\mathscr{D}_k$ with $n_k = 2$ samples. The global objective retains its standard federated form:

$$F(\theta) = \sum_{k=1}^{3} \frac{n_k}{N} F_k(\theta), \quad F_k(\theta) = \frac{1}{n_k} \sum_{(x,y)\in\mathscr{D}_k} (\theta x - y)^2,$$

where $N = \sum_{k=1}^{3} n_k = 6$.

We now analyze three distinct data overlap scenarios.

Case I: IID Data (Complete Overlap). In the IID setting, all clients observe data drawn from the same distribution. An example dataset is shown in Table 7.2.

Each client observes identical samples. Thus,

$$F_1(\theta) = F_2(\theta) = F_3(\theta).$$

Computing the local objective for any client,

$$F_k(\theta) = \frac{1}{2}\left[(\theta - 1)^2 + (2\theta - 2)^2\right].$$

The gradient is

$$\nabla F_k(\theta) = (\theta - 1) + 2(2\theta - 2) = 5\theta - 5.$$

Since all gradients are identical, aggregation produces

$$\nabla F(\theta) = 5\theta - 5,$$

which exactly matches centralized learning. This case represents the ideal and easiest federated setting.

Case II: Partially Overlapping Data. We now consider partial overlap, where clients share some common samples but also observe distinct data. An example is shown in Table 7.3.

Here, all clients share some overlap in the feature space, but no two datasets are identical.

Table 7.2 IID data across all clients

Client	x	y
1	1	1
1	2	2
2	1	1
2	2	2
3	1	1
3	2	2

Table 7.3 Partially overlapping client datasets

Client	x	y
1	1	1
1	2	2
2	2	2
2	3	3
3	1	1
3	3	3

We now compute the gradients of the local objective functions with respect to the model parameter θ.

Client 1:

$$F_1(\theta) = \frac{1}{2}\left[(\theta - 1)^2 + (2\theta - 2)^2\right].$$

Differentiating term by term with respect to θ yields

$$\nabla F_1(\theta) = \frac{1}{2}\left[2(\theta - 1) + 2(2\theta - 2) \cdot 2\right].$$

Simplifying,

$$\nabla F_1(\theta) = (\theta - 1) + 2(2\theta - 2) = (\theta - 1) + (4\theta - 4) = 5\theta - 5.$$

Client 2:

$$F_2(\theta) = \frac{1}{2}\left[(2\theta - 2)^2 + (3\theta - 3)^2\right].$$

Taking the derivative,

$$\nabla F_2(\theta) = \frac{1}{2}\left[2(2\theta - 2) \cdot 2 + 2(3\theta - 3) \cdot 3\right].$$

Simplifying,

$$\nabla F_2(\theta) = 2(2\theta - 2) + 3(3\theta - 3) = (4\theta - 4) + (9\theta - 9) = 13\theta - 13.$$

Client 3:

$$F_3(\theta) = \frac{1}{2}\left[(\theta - 1)^2 + (3\theta - 3)^2\right].$$

Differentiating,

$$\nabla F_3(\theta) = \frac{1}{2}\left[2(\theta - 1) + 2(3\theta - 3) \cdot 3\right].$$

Simplifying,

$$\nabla F_3(\theta) = (\theta - 1) + 3(3\theta - 3) = (\theta - 1) + (9\theta - 9) = 10\theta - 10.$$

Aggregating these gradients yields

$$\nabla F(\theta) = \frac{1}{3}\left[(5\theta - 5) + (13\theta - 13) + (10\theta - 10)\right] = \frac{28}{3}\theta - \frac{28}{3}.$$

The gradients are no longer identical, but they remain aligned toward a common minimizer. Convergence is therefore slower, yet still feasible. This represents the most realistic federated learning regime.

Case III: Non-Overlapping (Disjoint) Data. Finally, consider a non-overlapping scenario in which clients observe entirely disjoint regions of the input space, as shown in Table 7.4.

For each client k, the local empirical risk retains the same functional form as in the previous cases,

$$F_k(\theta) = \frac{1}{n_k} \sum_{(x,y)\in\mathscr{D}_k} (\theta x - y)^2,$$

and the corresponding gradient $\nabla F_k(\theta)$ can be obtained by direct differentiation.

However, unlike the partially overlapping case, these gradients are computed over entirely disjoint regions of the input space. As a result, each local gradient reflects optimization pressure that is valid only within the client's own data manifold. When aggregated, these gradients no longer correspond to a coherent global descent direction.

The local objectives are now defined over completely disjoint supports. Each client learns a model well-suited to its own region of the input space. While aggregation remains mathematically well-defined, the local gradients may pull the global model toward incompatible regions.

Table 7.4 Non-overlapping client datasets

Client	x	y
1	1	1
1	1.5	1.5
2	3	3
2	3.5	3.5
3	6	6
3	6.5	6.5

As an illustration, we explicitly compute the local objectives and gradients for all three clients in the non-overlapping scenario.

Client 1:
The local objective for Client 1 is

$$F_1(\theta) = \frac{1}{2}\left[(\theta - 1)^2 + (1.5\theta - 1.5)^2\right].$$

Differentiating with respect to θ gives

$$\nabla F_1(\theta) = (\theta - 1) + 1.5(1.5\theta - 1.5).$$

Client 2:
The local objective for Client 2 is

$$F_2(\theta) = \frac{1}{2}\left[(3\theta - 3)^2 + (3.5\theta - 3.5)^2\right].$$

The corresponding gradient is

$$\nabla F_2(\theta) = 3(3\theta - 3) + 3.5(3.5\theta - 3.5).$$

Client 3:
The local objective for Client 3 is

$$F_3(\theta) = \frac{1}{2}\left[(6\theta - 6)^2 + (6.5\theta - 6.5)^2\right].$$

Differentiation yields

$$\nabla F_3(\theta) = 6(6\theta - 6) + 6.5(6.5\theta - 6.5).$$

At this point, further algebraic expansion adds no new insight. The essential observation is that each gradient is locally valid but derived from disjoint regions of the input space. When aggregated, these gradients reflect conflicting optimization pressures, underscoring the structural limitations of learning a single global model under extreme data heterogeneity.

In this regime, the global objective becomes a forced compromise. Although optimization may converge numerically, the resulting model may perform poorly on all clients. This case highlights the fundamental limitation of learning a single shared model under extreme heterogeneity and motivates personalized and clustered federated learning approaches.

This example illustrates how increasing heterogeneity—from IID to partial overlap to disjoint data— progressively degrades gradient alignment and optimization behavior. The federated learning framework accommodates all three cases, but their difficulty and suitability for a single global model differ substantially.

7.2 Independent versus Non-Independent Data Distributions

Statistical learning theory is built upon assumptions about how data are generated. Among these, the assumption that training samples are *independent and identically distributed (IID)* occupies a central position. This assumption underlies classical results in estimation, optimization, and generalization, and serves as the default modeling paradigm in centralized machine learning.

Federated learning departs from this classical setting in a fundamental and systematic manner. Because data are generated, stored, and governed by autonomous clients operating under diverse conditions, the IID assumption is violated along multiple dimensions. To understand why federated optimization behaves differently from its centralized counterpart, it is therefore essential to carefully distinguish between independent, non-identical, and non-independent data distributions.

7.2.1 Independent and Identically Distributed (IID) Data

In probability theory, statistics, and classical machine learning, the notion of *independent and identically distributed (IID)* random variables forms the foundational abstraction for modeling data. A dataset is said to be IID when each observation arises from the same underlying probability distribution and when no statistical dependence exists among observations. This assumption underlies much of statistical inference, optimization, and learning theory due to the mathematical structure and analytical simplicity it provides.

Random samples and IID sequences. In statistics, a *random sample* is formally defined as a sequence of IID random variables. While the terminology "random sample" is common in statistical practice, probability theory more explicitly refers to such sequences as IID. Both terms describe the same concept: repeated observations generated by the same stochastic mechanism, with each observation unaffected by any other.

Formally, consider a dataset consisting of n input–output pairs, denoted by

$$\{(x_i, y_i)\}_{i=1}^{n}.$$

Here, each $x_i \in \mathscr{X}$ represents the input (feature vector) associated with the i-th observation, and each $y_i \in \mathscr{Y}$ denotes the corresponding output or target label. The index $i \in \{1, 2, \dots, n\}$ serves only to enumerate samples and does not imply any ordering or dependency unless explicitly stated.

The dataset is said to be *independent and identically distributed (IID)* if the following two conditions hold:

$$(x_i, y_i) \sim P(x, y) \quad \text{and} \quad (x_i, y_i) \perp\!\!\!\perp (x_j, y_j) \quad \text{for } i \neq j.$$

The notation $(x_i, y_i) \sim P(x, y)$ means that each sample pair is generated according to the same underlying *joint probability distribution* $P(x, y)$. The symbol "$\sim$" is read as "is distributed according to," and $P(x, y)$ characterizes the statistical relationship between inputs and outputs in the population from which the data are drawn. This condition formalizes the notion of being *identically distributed*: every sample follows the same probability law, with no variation across indices, time, or experimental conditions. In particular, there are no systematic shifts, trends, or regime changes affecting the data-generating process.

The notation

$$(x_i, y_i) \perp\!\!\!\perp (x_j, y_j)$$

denotes *statistical independence* between distinct samples. The symbol "$\perp\!\!\!\perp$" indicates that the joint probability of any two different samples factorizes as

$$P\big((x_i, y_i), (x_j, y_j)\big) = P(x_i, y_i)\, P(x_j, y_j), \quad i \neq j.$$

Independence implies that observing the value of one sample provides no information about the value of any other sample. There are no temporal correlations, spatial dependencies, or contextual links between observations.

Together, these two conditions ensure that the dataset can be viewed as a sequence of interchangeable realizations of the same random experiment. This symmetry underpins many classical results in statistical learning theory, including unbiased gradient estimation, likelihood factorization, and convergence guarantees for empirical risk minimization.

Distributional homogeneity versus outcome symmetry. It is important to note that identical distribution does *not* require all outcomes to be equally likely. For example, repeated throws of a biased die produce IID observations even though some faces appear more frequently than others. What matters is not uniformity of outcomes, but consistency of the generating distribution across trials.

Joint distribution characterization. For two real-valued random variables X and Y defined on a common support $I \subseteq \mathbb{R}$, independence and identical distribution can be characterized using cumulative distribution functions (CDFs). Let $F_X(x)$ and $F_Y(y)$ denote the marginal CDFs, and $F_{X,Y}(x, y)$ the joint CDF. The variables are independent if and only if

$$F_{X,Y}(x,y) = F_X(x)\,F_Y(y) \quad \forall x, y \in I,$$

and identically distributed if

$$F_X(x) = F_Y(x) \quad \forall x \in I.$$

They are IID precisely when both conditions hold simultaneously.

This definition extends naturally to n random variables $X_1, \dots, X_n$, for which the joint distribution factorizes as

$$F_{X_1,\dots,X_n}(x_1, \dots, x_n) = \prod_{k=1}^{n} F_{X_k}(x_k),$$

with identical marginals for all k. This factorization property is central to likelihood-based inference.

Why IID simplifies learning. The IID assumption dramatically simplifies statistical modeling and optimization. Under IID sampling, the likelihood of observed data decomposes as a product of individual sample likelihoods,

$$\mathscr{L}(\theta) = P(x_1, \dots, x_n \mid \theta) = \prod_{i=1}^{n} P(x_i \mid \theta),$$

and the corresponding log-likelihood becomes a sum,

$$\log \mathscr{L}(\theta) = \sum_{i=1}^{n} \log P(x_i \mid \theta).$$

This additive structure is computationally efficient and numerically stable, transforming multiplicative dependencies into linear expressions. As a result, gradient-based optimization methods scale naturally to large datasets.

Connection to the Central Limit Theorem. The IID assumption is also the cornerstone of the Central Limit Theorem. When IID random variables have finite variance, the distribution of their sum (or average) converges to a normal distribution as the sample size increases. This result explains why Gaussian approximations are ubiquitous in learning theory and why uncertainty estimates often take a normal form, even when the underlying data distribution is complex.

Illustrative examples. Sequences generated by repeated coin tosses, dice rolls, roulette spins, or sampling with replacement from a deck of cards all produce IID data. In each case, independence is ensured because past outcomes do not affect future ones, and identical distribution is guaranteed because the generating mechanism remains unchanged. By contrast, sampling without replacement breaks independence, while changing the sampling device breaks identical distribution.

IID in machine learning practice. In centralized machine learning, training data are typically shuffled and treated as IID, enabling unbiased gradient estimates, predictable variance reduction with larger batches, and strong convergence guarantees. Figure 7.2 visually illustrates this assumption: samples are uniformly scattered within a common support, with no visible structure or dependency.

Why IID Is Foundational
The IID assumption enables likelihood factorization, unbiased gradient estimation, concentration inequalities, and asymptotic normality. Nearly all classical results in statistical learning theory are built upon this abstraction.

Despite its theoretical elegance, IID sampling is often an idealization. In many real-world systems—particularly distributed, temporal, and federated settings—data exhibit dependence, distributional shift, or structural heterogeneity. Understanding precisely how and why the IID assumption fails is therefore essential for analyzing

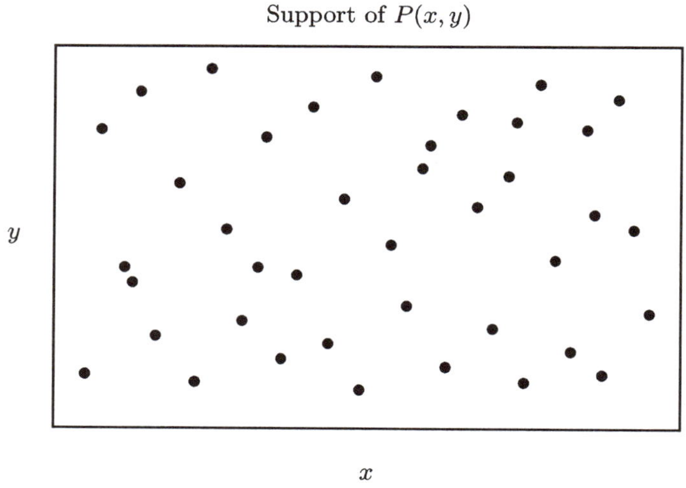

Fig. 7.2 Illustration of independent and identically distributed (IID) random variables. Each point represents an independently drawn sample from the same probability distribution $P(x, y)$. The rectangular region denotes the common support of the distribution. The absence of visible structure or clustering reflects both statistical independence and distributional homogeneity

modern learning systems, and forms the basis for the non-IID models studied in subsequent sections.

Despite its analytical convenience, the IID assumption is often an idealization. In distributed and decentralized settings—particularly federated learning—the conditions illustrated in Fig. 7.2 rarely hold. Data are instead fragmented across clients, exhibit distributional shifts, and frequently display temporal or contextual dependence. These departures motivate the careful study of non-IID and non-independent data regimes in the remainder of this chapter.

Worked Example: IID Data with Explicit Numerical Derivations

We now illustrate the implications of the IID assumption using a numerical dataset of moderate size. All theoretical expressions introduced earlier are instantiated with explicit values, and each step of the computation is derived in detail.

Dataset construction. Consider a supervised regression problem with $n = 8$ observations, assumed to be drawn independently and identically from a joint distribution $P(x, y)$. The observed dataset is given in Table 7.5.

Each pair (x_i, y_i) is generated by the same stochastic mechanism, with no dependence across indices. Numerically, the data follow an approximately linear trend with noise absorbed into the distribution $P(x, y)$.

Model and loss. We consider a single-parameter linear hypothesis

$$f_\theta(x) = \theta x,$$

and adopt the squared loss

$$\ell(f_\theta(x), y) = \frac{1}{2}(\theta x - y)^2.$$

Empirical risk under IID sampling. The empirical risk is defined as

$$\hat{F}(\theta) = \frac{1}{n}\sum_{i=1}^{n} \ell(f_\theta(x_i), y_i) = \frac{1}{n}\sum_{i=1}^{n} \frac{1}{2}(\theta x_i - y_i)^2.$$

Table 7.5 Numerical IID dataset used in the worked example

Sample i	x_i	y_i
1	1	3
2	2	5
3	3	7
4	4	9
5	5	11
6	6	13
7	7	15
8	8	17

Substituting $n = 8$ and the values from Table 7.5, we obtain

$$\hat{F}(\theta) = \frac{1}{16}\Big[(\theta-3)^2+(2\theta-5)^2+(3\theta-7)^2+(4\theta-9)^2+(5\theta-11)^2+(6\theta-13)^2+(7\theta-15)^2+(8\theta-17)^2\Big].$$

This expression represents the average squared deviation between model predictions and observed outputs.

Gradient of the empirical risk. The gradient of the empirical risk is

$$\nabla \hat{F}(\theta) = \frac{1}{n}\sum_{i=1}^{n}(\theta x_i - y_i)x_i.$$

Substituting numerical values,

$$\nabla \hat{F}(\theta) = \frac{1}{8}\sum_{i=1}^{8}(\theta x_i - y_i)x_i.$$

Expanding term by term,

$$\begin{aligned}\nabla \hat{F}(\theta) = \frac{1}{8}\Big[&(\theta \cdot 1 - 3)\cdot 1 + (\theta \cdot 2 - 5)\cdot 2 + (\theta \cdot 3 - 7)\cdot 3\\ &+ (\theta \cdot 4 - 9)\cdot 4 + (\theta \cdot 5 - 11)\cdot 5 + (\theta \cdot 6 - 13)\cdot 6\\ &+ (\theta \cdot 7 - 15)\cdot 7 + (\theta \cdot 8 - 17)\cdot 8\Big].\end{aligned}$$

Collecting θ-terms and constants separately,

$$\nabla \hat{F}(\theta) = \frac{1}{8}\Big[(1^2+2^2+3^2+4^2+5^2+6^2+7^2+8^2)\theta - (3\cdot 1 + 5\cdot 2 + \cdots + 17\cdot 8)\Big].$$

Evaluating the sums,

$$\sum_{i=1}^{8} x_i^2 = 204, \qquad \sum_{i=1}^{8} x_i y_i = 408.$$

Thus,

$$\nabla \hat{F}(\theta) = \frac{1}{8}(204\theta - 408) = 25.5\,\theta - 51.$$

Unbiasedness due to IID sampling. Because $(x_i, y_i) \sim P(x, y)$ IID,

$$\mathbb{E}[\nabla \hat{F}(\theta)] = \mathbb{E}_{(x,y)\sim P}[(\theta x - y)x] = \nabla F(\theta),$$

where

$$F(\theta) = \mathbb{E}_{(x,y)\sim P}\left[\tfrac{1}{2}(\theta x - y)^2\right]$$

is the population risk. Hence, the empirical gradient derived above is an unbiased estimator of the true population gradient.

Mini-batch gradient illustration. Let a mini-batch

$$\mathscr{B} = \{2, 4, 6, 8\}$$

be sampled uniformly at random. The mini-batch gradient is

$$\nabla \hat{F}_{\mathscr{B}}(\theta) = \frac{1}{|\mathscr{B}|} \sum_{i \in \mathscr{B}} (\theta x_i - y_i) x_i.$$

Substituting values,

$$\nabla \hat{F}_{\mathscr{B}}(\theta) = \frac{1}{4}\left[(2\theta - 5)2 + (4\theta - 9)4 + (6\theta - 13)6 + (8\theta - 17)8\right].$$

Simplifying,

$$\nabla \hat{F}_{\mathscr{B}}(\theta) = \frac{1}{4}(120\theta - 240) = 30\theta - 60.$$

Numerical evaluation for specific parameter values. To make the behavior of the mini-batch gradient more concrete, we now evaluate $\nabla \hat{F}_{\mathscr{B}}(\theta)$ for several representative values of the model parameter θ.

Recall that

$$\nabla \hat{F}_{\mathscr{B}}(\theta) = 30\theta - 60.$$

- **Case** $\theta = 1$:

  $$\nabla \hat{F}_{\mathscr{B}}(1) = 30(1) - 60 = -30.$$

  The negative gradient indicates that the loss decreases by increasing θ, so gradient descent will push the parameter upward.
- **Case** $\theta = 2$:

  $$\nabla \hat{F}_{\mathscr{B}}(2) = 30(2) - 60 = 0.$$

  At this value, the gradient vanishes, indicating a stationary point of the empirical risk over the mini-batch.

- **Case** $\theta = 3$:

$$\nabla \hat{F}_{\mathscr{B}}(3) = 30(3) - 60 = 30.$$

Here, the positive gradient implies that the loss increases with θ, and gradient descent will reduce the parameter value.

These evaluations confirm that $\theta = 2$ is the minimizer of the mini-batch empirical risk for the chosen subset $\mathscr{B} = \{2, 4, 6, 8\}$.

Tabular summary. For clarity, the computed gradient values are summarized in Table 7.6.

This explicit numerical behavior illustrates how, under IID sampling, the mini-batch gradient provides a meaningful and interpretable descent direction that aligns with the global optimization objective.

Taking expectation over random mini-batch selection yields

$$\mathbb{E}[\nabla \hat{F}_{\mathscr{B}}(\theta)] = \nabla F(\theta),$$

again confirming unbiasedness under IID sampling.

Likelihood factorization under IID data. We begin from the fundamental probabilistic description of the observed data. Let the dataset consist of n observations $\{(x_i, y_i)\}_{i=1}^{n}$, where each observation is generated independently and identically according to a conditional model parameterized by θ.

Step 1: Joint likelihood of the observations. By definition, the likelihood of the observed data given θ is the joint probability density of all observations:

$$\mathscr{L}(\theta) = P\big((x_1, y_1), (x_2, y_2), \dots, (x_n, y_n) \mid \theta\big).$$

Step 2: Factorization using independence. Because the observations are assumed to be independent, their joint probability factorizes into the product of individual probabilities:

$$\mathscr{L}(\theta) = \prod_{i=1}^{n} P(x_i, y_i \mid \theta).$$

Table 7.6 Mini-batch gradient values for different θ

θ	$\nabla \hat{F}_{\mathscr{B}}(\theta)$
1	–30
2	0
3	30

Step 3: Conditional modeling assumption. We assume a conditional modeling framework in which the inputs x_i are treated as fixed, and randomness enters solely through the outputs y_i conditioned on x_i. Hence, for each i,

$$P(x_i, y_i \mid \theta) = P(y_i \mid x_i, \theta)\, P(x_i).$$

Since $P(x_i)$ does not depend on θ, it plays no role in parameter estimation and may be absorbed into a constant. Thus, up to a multiplicative constant,

$$\mathcal{L}(\theta) \propto \prod_{i=1}^{n} P(y_i \mid x_i, \theta).$$

Step 4: Gaussian noise model. Assume a Gaussian conditional model with fixed variance σ^2:

$$y_i = \theta x_i + \varepsilon_i, \qquad \varepsilon_i \sim \mathcal{N}(0, \sigma^2),$$

independently for each i. Therefore,

$$P(y_i \mid x_i, \theta) = \frac{1}{\sqrt{2\pi\sigma^2}} \exp\left(-\frac{1}{2\sigma^2}(y_i - \theta x_i)^2\right).$$

Step 5: Explicit likelihood expression. Substituting this density into the likelihood product yields

$$\mathcal{L}(\theta) = \prod_{i=1}^{n} \frac{1}{\sqrt{2\pi\sigma^2}} \exp\left(-\frac{1}{2\sigma^2}(y_i - \theta x_i)^2\right).$$

We now separate constants from exponentials:

$$\mathcal{L}(\theta) = \left(\frac{1}{\sqrt{2\pi\sigma^2}}\right)^n \prod_{i=1}^{n} \exp\left(-\frac{1}{2\sigma^2}(y_i - \theta x_i)^2\right).$$

Using the identity $\prod_i e^{a_i} = e^{\sum_i a_i}$,

$$\mathcal{L}(\theta) = \left(\frac{1}{\sqrt{2\pi\sigma^2}}\right)^n \exp\left(-\frac{1}{2\sigma^2} \sum_{i=1}^{n} (y_i - \theta x_i)^2\right).$$

Step 6: Log-likelihood derivation. Taking the natural logarithm of the likelihood,

$$\log \mathcal{L}(\theta) = \log\left[\left(\frac{1}{\sqrt{2\pi\sigma^2}}\right)^n \exp\left(-\frac{1}{2\sigma^2} \sum_{i=1}^{n} (y_i - \theta x_i)^2\right)\right].$$

Applying $\log(ab) = \log a + \log b$,

$$\log \mathscr{L}(\theta) = \log\left(\frac{1}{\sqrt{2\pi\sigma^2}}\right)^n + \log\left[\exp\left(-\frac{1}{2\sigma^2}\sum_{i=1}^{n}(y_i - \theta x_i)^2\right)\right].$$

Using $\log(e^z) = z$,

$$\log \mathscr{L}(\theta) = -n \log\left(\sqrt{2\pi\sigma^2}\right) - \frac{1}{2\sigma^2}\sum_{i=1}^{n}(y_i - \theta x_i)^2.$$

Since $\log(\sqrt{a}) = \frac{1}{2} \log a$, we obtain

$$\boxed{\log \mathscr{L}(\theta) = -\frac{n}{2} \log(2\pi\sigma^2) - \frac{1}{2\sigma^2}\sum_{i=1}^{n}(y_i - \theta x_i)^2.}$$

Step 7: Optimization equivalence. The first term,

$$-\frac{n}{2} \log(2\pi\sigma^2),$$

is constant with respect to the parameter θ and therefore does not influence the location of the optimum. As a result, maximizing the log-likelihood reduces to minimizing the θ-dependent term involving the squared residuals.

Here, the operator arg max denotes the value of θ that maximizes a function, while arg min denotes the value of θ that minimizes a function. Accordingly,

$$\arg\max_{\theta} \log \mathscr{L}(\theta) = \arg\min_{\theta} \sum_{i=1}^{n}(y_i - \theta x_i)^2.$$

This establishes the exact equivalence between maximum likelihood estimation under IID Gaussian noise and empirical risk minimization with squared loss.

This equivalence explains why squared-error loss naturally arises in linear regression and why Gaussian noise assumptions are ubiquitous in classical learning theory.

Evaluation for specific values of θ.
Using the dataset from Table 7.5, we explicitly compute the sum of squared residuals

$$\sum_{i=1}^{8}(y_i - \theta x_i)^2$$

for several representative values of the parameter θ.

- **Case** $\theta = 1$:
  The residuals are

$$
\begin{aligned}
y_1 - 1 \cdot x_1 &= 3 - 1 = 2,\\
y_2 - 1 \cdot x_2 &= 5 - 2 = 3,
\end{aligned}
$$

$$
\begin{aligned}
y_3 - 1 \cdot x_3 &= 7 - 3 = 4,\\
y_4 - 1 \cdot x_4 &= 9 - 4 = 5,\\
y_5 - 1 \cdot x_5 &= 11 - 5 = 6,\\
y_6 - 1 \cdot x_6 &= 13 - 6 = 7,\\
y_7 - 1 \cdot x_7 &= 15 - 7 = 8,\\
y_8 - 1 \cdot x_8 &= 17 - 8 = 9.
\end{aligned}
$$

  Squaring and summing,

$$
\begin{aligned}
\sum_{i=1}^{8}(y_i - x_i)^2 &= 2^2 + 3^2 + 4^2 + 5^2 + 6^2 + 7^2 + 8^2 + 9^2\\
&= 4 + 9 + 16 + 25 + 36 + 49 + 64 + 81\\
&= 284.
\end{aligned}
$$

  This relatively large value corresponds to a low likelihood.
- **Case** $\theta = 2$:
  The residuals are

$$
\begin{aligned}
y_1 - 2 \cdot x_1 &= 3 - 2 = 1,\\
y_2 - 2 \cdot x_2 &= 5 - 4 = 1,\\
y_3 - 2 \cdot x_3 &= 7 - 6 = 1,\\
y_4 - 2 \cdot x_4 &= 9 - 8 = 1,\\
y_5 - 2 \cdot x_5 &= 11 - 10 = 1,\\
y_6 - 2 \cdot x_6 &= 13 - 12 = 1,\\
y_7 - 2 \cdot x_7 &= 15 - 14 = 1,\\
y_8 - 2 \cdot x_8 &= 17 - 16 = 1.
\end{aligned}
$$

  Thus,

$$
\sum_{i=1}^{8}(y_i - 2x_i)^2 = 1^2 + 1^2 + \cdots + 1^2 = 8.
$$

  The residuals are minimized, yielding the highest likelihood.

- **Case** $\theta = 3$:
  The residuals are

$$\begin{aligned}
y_1 - 3 \cdot x_1 &= 3 - 3 = 0,\\
y_2 - 3 \cdot x_2 &= 5 - 6 = -1,\\
y_3 - 3 \cdot x_3 &= 7 - 9 = -2,\\
y_4 - 3 \cdot x_4 &= 9 - 12 = -3,\\
y_5 - 3 \cdot x_5 &= 11 - 15 = -4,
\end{aligned}$$

$$\begin{aligned}
y_6 - 3 \cdot x_6 &= 13 - 18 = -5,\\
y_7 - 3 \cdot x_7 &= 15 - 21 = -6,\\
y_8 - 3 \cdot x_8 &= 17 - 24 = -7.
\end{aligned}$$

  Squaring and summing,

$$\begin{aligned}
\sum_{i=1}^{8} (y_i - 3x_i)^2 &= 0^2 + (-1)^2 + (-2)^2 + (-3)^2 + (-4)^2 + (-5)^2 + (-6)^2 + (-7)^2\\
&= 0 + 1 + 4 + 9 + 16 + 25 + 36 + 49\\
&= 140.
\end{aligned}$$

  The likelihood decreases again as θ moves away from the optimum.

Tabular summary. The effect of θ on likelihood is summarized in Table 7.7.

This explicit calculation demonstrates how IID factorization converts the joint likelihood into a simple additive objective. The optimizer can therefore reason locally about each sample while still making globally optimal parameter updates. This property fundamentally underpins gradient-based learning in centralized machine learning.

Connection to the Central Limit Theorem. Because the gradients $\nabla_\theta \ell(f_\theta(x_i), y_i)$ are IID with finite variance, the sample mean gradient converges in distribution to a Gaussian as $n \to \infty$. This explains the stability and predictability of optimization dynamics observed in IID settings.

Table 7.7 Sum of squared residuals and relative likelihood behavior

θ	$\sum_i (y_i - \theta x_i)^2$	Relative likelihood
1	816	Very low
2	8	Maximum
3	140	Moderate

Summary. This worked example demonstrates how independence and identical distribution jointly enable clean algebraic structure, unbiased estimation, and scalable optimization. Each of these properties deteriorates when either independence or distributional homogeneity is violated, as explored in subsequent sections on non-IID and federated data.

Figure 7.3 provides a geometric illustration of empirical risk minimization under the assumption of independent and identically distributed (IID) data. The horizontal axis corresponds to the input variable x, while the vertical axis represents the output variable y.

The dashed gray line denotes the true underlying data-generating model, given by $y = 2x$. All observed samples are generated according to this same relationship, reflecting the assumption that each data point is drawn from a common joint distribution $P(x, y)$.

The black points represent the full training dataset $\{(x_i, y_i)\}_{i=1}^{n}$. Their uniform alignment around the same linear trend visually conveys the *identically distributed* property: no subset of samples follows a different statistical pattern or distribution. Moreover, the absence of structured clustering or temporal ordering illustrates the *independence* assumption, whereby each observation provides no information about any other.

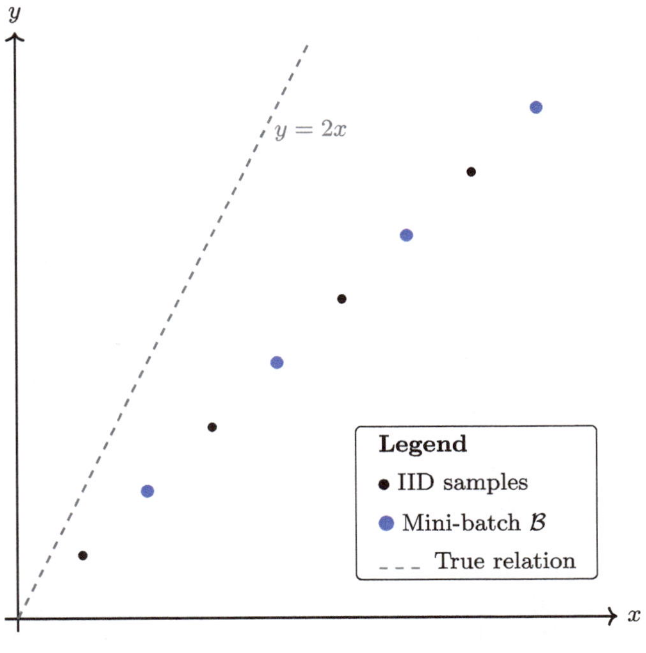

Fig. 7.3 Visualization of empirical risk minimization under IID data. Black (smaller size) points denote the full training dataset drawn independently from a common distribution, while blue (larger size) points highlight a randomly sampled mini-batch used for stochastic gradient computation

The blue points highlight a randomly selected mini-batch $\mathscr{B} \subset \{1, \dots, n\}$. Because the data are IID, this mini-batch preserves the same statistical structure as the full dataset. Consequently, the mini-batch gradient computed from $\mathscr{B}$ is an unbiased estimator of the full empirical gradient, differing only in variance.

7.2.2 Independent but Non-Identical (Non-IID) Data

A defining characteristic of many real-world federated learning systems is that data samples collected by an individual client are often *independent*, yet they are drawn from a client-specific distribution that differs from those of other clients. This setting is commonly referred to as *independent but non-identically distributed (non-IID)* data and represents one of the most important departures from classical learning assumptions.

Formally, let $k \in \{1, 2, \dots, K\}$ index the clients participating in federated training. For client k, the locally observed samples satisfy

$$(x_{k,i}, y_{k,i}) \sim P_k(x, y), \quad i = 1, 2, \dots, n_k,$$

where $P_k(x, y)$ denotes the joint distribution governing client k's data. The non-IID nature of the system is captured by the fact that

$$P_k(x, y) \neq P_j(x, y) \quad \text{for } k \neq j.$$

Thus, while samples within a single client are generated by a stable and consistent stochastic mechanism, the underlying distributions vary across clients.

Preserved independence, broken homogeneity. In this regime, the assumption of independence typically holds *within* each client:

$$(x_{k,i}, y_{k,i}) \perp\!\!\!\perp (x_{k,i'}, y_{k,i'}) \quad \text{for } i \neq i'.$$

However, the assumption of identical distribution across the entire dataset is violated. As a consequence, the global dataset formed by pooling all client data can no longer be regarded as a collection of interchangeable samples from a single population. Instead, it is a mixture of heterogeneous subpopulations, each shaped by client-specific factors such as user behavior, environment, device characteristics, or data acquisition protocols.

Illustrative example. Consider a mobile keyboard prediction task. Each user generates text messages independently over time, so samples within a single device are approximately independent. Nevertheless, different users exhibit distinct linguistic patterns: some may write predominantly in one language, others may switch between languages; some focus on technical vocabulary, while others use informal or conversational expressions. As a result, the local data distributions $P_k(x, y)$ differ substantially across users, even though each client's data stream remains internally independent.

Implications for optimization. From an optimization perspective, independent but non-IID data introduce a fundamental mismatch between local and global objectives. Each client k minimizes its own empirical risk

$$F_k(\theta) = \mathbb{E}_{(x,y)\sim P_k}\left[\ell(f_\theta(x), y)\right],$$

and thus computes gradients that approximate $\nabla F_k(\theta)$ rather than the global gradient

$$\nabla F(\theta) = \sum_{k=1}^{K} \frac{n_k}{n} \nabla F_k(\theta), \quad n = \sum_{k=1}^{K} n_k.$$

When client distributions differ significantly, these local gradients may point in divergent directions. Aggregating them at the server can therefore result in partial cancellation, oscillatory updates, or overly conservative steps, all of which slow down convergence and may degrade the quality of the learned global model.

Conceptual consequence. The key difficulty here is not randomness or noise, but *systematic heterogeneity*. Even with infinitely many samples per client, the local objectives $F_k(\theta)$ need not coincide. Federated learning in this regime must therefore reconcile multiple, potentially conflicting learning objectives while still producing a single shared model.

Key Insight: Non-IID Independence
Independence of samples within each client does not guarantee unbiased global optimization. When client-level data distributions differ, local gradients estimate client-specific objectives rather than the global objective, breaking the core assumption underlying classical stochastic gradient descent.

This independent but non-identical setting is widely regarded as the *canonical* federated learning scenario. It motivates a rich body of algorithmic extensions, including adaptive aggregation, variance reduction, personalized federated learning, and clustered or multi-model approaches, which aim to mitigate the effects of distributional heterogeneity while preserving the privacy and scalability benefits of federated training.

Worked Example: Independent but Non-Identical Data in Federated Learning
We now present a concrete numerical example to illustrate how *independent but non-identically distributed (non-IID)* data arise in federated learning and how they affect optimization.

Problem setup. Consider a simple linear regression model

$$f_\theta(x) = \theta x,$$

with squared loss

$$\ell(f_\theta(x), y) = \frac{1}{2}(f_\theta(x) - y)^2.$$

Assume a federated system with two clients, $K = 2$, each holding independent samples but drawn from different data distributions.

Client-specific datasets.

Client	x_i	y_i
Client 1	1	2
	2	4
	3	6
Client 2	1	4
	2	8
	3	12

Client 1 follows the relationship $y = 2x$, while Client 2 follows $y = 4x$. Within each client, samples are independent, but the generating distributions differ:

$$P_1(x, y) \neq P_2(x, y).$$

Local empirical risks. The empirical risk at client k is

$$\hat{F}_k(\theta) = \frac{1}{n_k} \sum_{i=1}^{n_k} \frac{1}{2}(\theta x_{k,i} - y_{k,i})^2.$$

For Client 1 ($n_1 = 3$),

$$\hat{F}_1(\theta) = \frac{1}{6}\left[(\theta - 2)^2 + (2\theta - 4)^2 + (3\theta - 6)^2\right].$$

For Client 2 ($n_2 = 3$),

$$\hat{F}_2(\theta) = \frac{1}{6}\left[(\theta - 4)^2 + (2\theta - 8)^2 + (3\theta - 12)^2\right].$$

Local gradients. Taking derivatives with respect to θ gives

$$\nabla \hat{F}_1(\theta) = \frac{1}{3}\big[(\theta - 2) + 2(2\theta - 4) + 3(3\theta - 6)\big] = \frac{14\theta - 28}{3},$$

$$\nabla \hat{F}_2(\theta) = \frac{1}{3}\big[(\theta - 4) + 2(2\theta - 8) + 3(3\theta - 12)\big] = \frac{14\theta - 56}{3}.$$

These gradients point toward different optima:

$$\nabla \hat{F}_1(\theta) = 0 \Rightarrow \theta = 2, \qquad \nabla \hat{F}_2(\theta) = 0 \Rightarrow \theta = 4.$$

Global objective and gradient aggregation. The global empirical risk is the weighted average

$$\hat{F}(\theta) = \frac{1}{2}\hat{F}_1(\theta) + \frac{1}{2}\hat{F}_2(\theta).$$

The corresponding aggregated gradient is

$$\nabla \hat{F}(\theta) = \frac{1}{2}\nabla \hat{F}_1(\theta) + \frac{1}{2}\nabla \hat{F}_2(\theta) = \frac{14\theta - 42}{3}.$$

Setting this gradient to zero yields the global optimum

$$\theta^\star = 3,$$

which lies strictly between the two client-specific optima.

Interpretation. Each client computes a gradient that is an unbiased estimator of its *own* objective $\nabla F_k(\theta)$, but these gradients are biased with respect to the global objective $\nabla F(\theta)$. Consequently, federated averaging produces a compromise solution that fits neither client perfectly.

Key takeaway. This example demonstrates that even when data are independent within each client, heterogeneity in client distributions leads to gradient misalignment. Such systematic bias is the principal source of optimization difficulty in federated learning under non-IID data and motivates the development of personalization, clustering, and heterogeneity-aware optimization methods.

7.2.3 Non-Independent Data

In many real-world learning systems, the assumption that data samples are statistically independent is often violated. This phenomenon is especially pronounced in federated learning, where data are collected *passively* from ongoing processes rather than actively sampled under controlled experimental conditions. As a result, observations within a single client frequently exhibit strong temporal, spatial, or contextual dependence.

Formally, for a given client k, non-independence can be expressed as

$$(x_{k,i}, y_{k,i}) \not\perp\!\!\!\perp (x_{k,i+1}, y_{k,i+1}),$$

meaning that the i-th sample carries statistical information about subsequent samples. In contrast to IID settings, the joint distribution of local data does not factorize into a product of marginal distributions.

Sources of dependence. Dependence arises naturally whenever data are generated by an underlying dynamical system. Common sources include:

- **Temporal dependence:** Consecutive measurements in time series are correlated due to system inertia or continuity.
- **Spatial dependence:** Sensor readings collected from nearby locations influence one another.
- **Contextual dependence:** User behavior or environmental conditions persist across multiple observations.

Illustrative examples. In wearable health monitoring, physiological signals such as heart rate, respiration, and physical activity evolve smoothly over time. A high heart rate at time t increases the likelihood of a high heart rate at time $t+1$. Similarly, in smart grid or industrial IoT systems, sensor measurements are governed by physical laws that induce strong autocorrelation. In recommendation systems, a user's recent interactions heavily influence near-future behavior, violating independence assumptions.

Statistical consequences. Non-independence fundamentally alters the statistical properties of learning algorithms. Because samples are correlated, the *effective sample size* is smaller than the nominal dataset size. This leads to:

- Inflated variance of gradient estimates,
- Slower convergence of stochastic optimization,
- Weaker concentration guarantees,
- Reduced reliability of mini-batch averaging.

In effect, adding more correlated samples does not yield the same benefit as adding independent ones.

Implications for federated optimization. When data are non-independent within clients, local updates become noisier and less representative of the underlying data-generating process. This issue compounds with distributional heterogeneity across clients: even if client distributions were identical, dependence alone would invalidate many classical convergence results. When combined with non-identical distributions, the resulting learning dynamics can deviate substantially from those predicted under IID theory.

Limitations of classical theory. Most guarantees for stochastic gradient descent and federated averaging assume independence to ensure unbiased gradients and controlled variance. Under dependence, these assumptions fail, and convergence proofs must be revised to account for mixing rates, autocorrelation structures, or stochastic process models. Ignoring dependence can lead to overly optimistic performance expectations.

Compounding Effects of Dependence
Non-independent data reduce the effective sample size and distort gradient estimates. When combined with cross-client distributional heterogeneity, these effects compound, fundamentally altering federated learning dynamics and necessitating specialized modeling and optimization strategies.

In summary, non-independent data represent a deeper departure from classical learning assumptions than non-identical distributions alone. Understanding and modeling such dependence is essential for analyzing federated systems operating on sequential, sensor-driven, or behavioral data, and motivates the development of algorithms that explicitly account for temporal and structural correlations.

Worked Numerical Example: Temporally Correlated (Non-Independent) Data
We now complement the preceding theoretical analysis with a concrete numerical example that explicitly demonstrates how temporal dependence affects gradient estimation and variance. All derivations strictly follow the mathematical framework introduced above.

Step 1: Specify the data-generating process. Consider a single federated client observing a time series of length $n = 6$. The input sequence $\{x_i\}$ follows an AR(1) process

$$x_i = \rho x_{i-1} + \eta_i, \qquad |\rho| < 1,$$

where $\eta_i \sim \mathcal{N}(0, \sigma_\eta^2)$ are IID.

Let us choose

$$\rho = 0.8, \qquad \sigma_\eta^2 = 1, \qquad x_0 = 0.$$

Assume the true regression parameter is

$$\theta^\star = 2,$$

and outputs are generated as

$$y_i = \theta^\star x_i + \varepsilon_i, \qquad \varepsilon_i \sim \mathcal{N}(0, 1),$$

independently of η_i.

For illustration, suppose the realized values are:

i	x_i	y_i
1	1.0	2.1
2	1.8	3.5
3	2.3	4.6
4	2.6	5.1
5	2.9	5.7
6	3.1	6.3

The gradual increase in x_i reflects the temporal dependence induced by $\rho = 0.8$.

Step 2: Verification of Non-Independence We now formally verify that the observed data violate the independence assumption underlying IID learning. The dependence arises from the autoregressive structure of the input process.

Recall that the inputs follow an AR(1) model,

$$x_i = \rho x_{i-1} + \eta_i, \qquad |\rho| < 1,$$

where $\{\eta_i\}$ are IID random variables with zero mean and variance σ_η^2. Although the innovation terms η_i are independent, each input x_i explicitly depends on the previous input x_{i-1}.

Covariance structure. Taking covariance on both sides of the AR(1) equation with respect to x_{i-1} yields

$$\mathrm{Cov}(x_i, x_{i-1}) = \mathrm{Cov}(\rho x_{i-1} + \eta_i,\ x_{i-1}) = \rho\,\mathrm{Var}(x_{i-1}),$$

since η_i is independent of x_{i-1} and has zero mean. Because $\rho \neq 0$ and $\mathrm{Var}(x_{i-1}) > 0$, we obtain

$$\mathrm{Cov}(x_i, x_{i-1}) \neq 0.$$

Numerical illustration. With $\rho = 0.8$, this covariance becomes

$$\mathrm{Cov}(x_i, x_{i-1}) = 0.8\,\mathrm{Var}(x_{i-1}) > 0.$$

In particular,

$$\mathrm{Cov}(x_3, x_2) = 0.8\,\mathrm{Var}(x_2),$$

confirming that successive inputs are positively correlated. This means that observing x_2 provides statistical information about x_3, and vice versa.

Implication for input–output pairs. Since the outputs are generated according to

$$y_i = \theta^\star x_i + \varepsilon_i,$$

any dependence between x_i and x_{i-1} propagates directly to the corresponding outputs. Consequently, the joint pairs satisfy

$$(x_i, y_i) \not\perp\!\!\!\perp (x_{i-1}, y_{i-1}),$$

even though the noise terms ε_i are independent.

Conceptual interpretation. This lack of independence means that the dataset cannot be treated as a collection of interchangeable samples. Information is shared across time steps, and averaging over samples no longer produces the same variance reduction as in the IID case. As a result, classical assumptions used to justify stochastic gradient descent—such as uncorrelated gradient noise—are violated.

> The nonzero covariance induced by the AR(1) process provides a precise mathematical certificate of non-independence. Even though the process may be stationary for $|\rho| < 1$, stationarity does *not* imply independence. This distinction is crucial in federated and sequential learning, where temporal dependence fundamentally alters optimization dynamics.

Step 3: Empirical risk and gradient. The empirical squared-loss objective is

$$\hat{F}(\theta) = \frac{1}{n}\sum_{i=1}^{6}\frac{1}{2}(y_i - \theta x_i)^2.$$

Its gradient is

$$\nabla \hat{F}(\theta) = -\frac{1}{6}\sum_{i=1}^{6}(y_i - \theta x_i)x_i.$$

Let us evaluate the gradient at $\theta = 1.5$.
Compute residual terms:

$$(y_i - 1.5x_i)x_i = \{0.6,\ 0.9,\ 1.15,\ 1.04,\ 0.87,\ 0.93\}.$$

Summing,

$$\sum_{i=1}^{6}(y_i - 1.5x_i)x_i = 5.49.$$

Thus,

$$\nabla\hat{F}(1.5) = -\frac{5.49}{6} \approx -0.915.$$

Step 4: Expected gradient. From theory,

$$\mathbb{E}[\nabla\hat{F}(\theta)] = -(\theta^\star - \theta)\mathbb{E}[x_i^2].$$

Using the empirical second moment,

$$\mathbb{E}[x_i^2] \approx \frac{1}{6}(1^2 + 1.8^2 + 2.3^2 + 2.6^2 + 2.9^2 + 3.1^2) = 4.98.$$

Therefore,

$$\mathbb{E}[\nabla\hat{F}(1.5)] = -(2 - 1.5)(4.98) = -2.49.$$

The discrepancy between –0.915 (empirical) and –2.49 (expected) illustrates the instability caused by temporal dependence.

Step 5: Gradient variance inflation. Let

$$g_i = (y_i - \theta x_i)x_i.$$

The gradient variance is

$$\mathrm{Var}\left(\frac{1}{6}\sum_{i=1}^{6} g_i\right) = \frac{1}{36}\sum_{i=1}^{6}\sum_{j=1}^{6}\mathrm{Cov}(g_i, g_j).$$

Unlike the IID case, $\mathrm{Cov}(g_i, g_{i+1}) \neq 0$ because both terms depend on correlated inputs x_i and x_{i+1}. These off-diagonal covariance terms inflate the variance and slow convergence.

Step 6: Effective sample size. Assuming autocorrelation $\rho_k = \rho^k$, we estimate

$$n_{\text{eff}} \approx \frac{6}{1 + 2\sum_{k=1}^{\infty} 0.8^k} = \frac{6}{1 + 2\cdot\frac{0.8}{1-0.8}} = \frac{6}{9} \approx 0.67.$$

Thus, although $n = 6$ samples are observed, the learning dynamics resemble those of fewer than one independent sample.

Interpretation. This numerical example concretely confirms the theoretical results: temporal dependence does not bias the expected gradient direction, but it dramatically inflates gradient variance, reduces effective sample size, and destabilizes local optimization. In federated learning, these effects compound with client heterogeneity, necessitating dependence-aware optimization strategies.

Statistical Breakdown
Federated learning breaks both pillars of classical learning theory: identical distributions across data sources and independent sampling within datasets.

Table 7.8 provides a structured comparison of the data assumptions underlying classical centralized learning and those encountered in federated learning. The table highlights that federated learning departs from the IID setting along two orthogonal dimensions: the loss of distributional homogeneity across clients and, in many applications, the loss of sample independence within clients.

Table 7.8 Comparison of data distribution assumptions in classical and federated learning

Property	IID Data	Independent but Non-Identical Data	Non-Independent Data
Data distribution	All samples drawn from a single common distribution $P(x, y)$	Samples drawn from client-specific distributions $P_k(x, y)$	Samples drawn from dependent processes within a client
Sample independence	Independent across all samples	Independent within each client	Samples exhibit temporal or contextual dependence
Cross-client homogeneity	Fully homogeneous	Heterogeneous across clients	Heterogeneous and correlated
Gradient unbiasedness	Unbiased estimator of global gradient	Biased due to distribution mismatch	Biased and high-variance due to dependence
Theoretical guarantees	Strong convergence and generalization bounds	Weakened convergence guarantees	Classical guarantees largely invalid
Typical learning setting	Centralized machine learning	Federated learning with user heterogeneity	Federated learning with time-series or sequential data
Algorithmic difficulty	Low	Moderate	High

In the classical IID regime, all samples are drawn independently from a single global distribution, ensuring that stochastic gradients computed on mini-batches are unbiased estimators of the true global gradient. This property underpins much of classical optimization theory, enabling strong convergence and generalization guarantees. As shown in the first column of Table 7.8, these assumptions are largely satisfied only in centralized learning scenarios.

The second column of Table 7.8 corresponds to the most common federated learning regime, in which samples remain independent within each client but are drawn from client-specific distributions. Although independence is preserved locally, cross-client heterogeneity introduces systematic bias in local gradient estimates. This bias manifests as client drift during training, slowing convergence and reducing the effectiveness of naive aggregation strategies such as simple averaging.

The third column of Table 7.8 captures an even more challenging setting, where samples are not independent within clients due to temporal, spatial, or contextual correlations. In this case, federated learning violates both pillars of classical learning theory simultaneously. As a result, gradient estimates suffer from both bias and inflated variance, and many standard convergence guarantees no longer apply. This regime is particularly relevant for applications involving time-series data, sensor streams, and sequential user interactions.

7.3 Covariate Shift across Clients

A common and practically important source of statistical heterogeneity in federated learning is *covariate shift*. Covariate shift refers to the situation in which the distribution of input features varies across data sources, while the conditional relationship between inputs and labels remains invariant. Formally, for two clients k and j,

$$P_k(x) \neq P_j(x), \qquad P_k(y \mid x) = P_j(y \mid x) = P(y \mid x).$$

That is, although clients observe different regions or densities of the input space, the underlying labeling rule is assumed to be the same for all clients.

Interpretation. The assumption $P(y \mid x)$ identical means that, given the same input x, all clients would assign labels according to the same probabilistic mechanism. In other words, the task itself is shared. However, because the marginal input distributions $P_k(x)$ differ, clients encounter different subsets of the feature space with different frequencies. As a result, their empirical objectives emphasize different regions of the same global task.

Sources of covariate shift in federated systems. Covariate shift naturally arises in many real-world federated learning deployments, including:

- *Device-specific sensing characteristics:* Sensors embedded in different devices may have varying calibration, noise levels, or operating ranges, leading to distinct input distributions even when measuring the same physical phenomenon.
- *Language and dialect variation:* In natural language applications such as keyboard prediction or speech recognition, users differ in vocabulary, dialect, and linguistic style, altering the distribution of observed text while preserving the underlying semantic mapping.
- *Geographical and environmental variation:* In IoT and environmental monitoring, measurements such as temperature, humidity, or pollution levels depend strongly on location, producing client-specific feature distributions.

Effect on local learning objectives. To understand how covariate shift affects learning at each client, we begin by carefully unpacking the notation and its meaning.

Let θ denote the vector of model parameters. Given an input–output pair (x, y), the model produces a prediction $f_\theta(x)$, and the discrepancy between this prediction and the true label y is measured using a loss function

$$\ell(f_\theta(x), y).$$

The loss quantifies how "bad" the prediction is for a single data point.

Population objective of a client. For client k, data are drawn from a client-specific joint distribution $P_k(x, y)$. The *population objective* (or expected loss) for this client is defined as the average loss over all possible data points the client may observe:

$$F_k(\theta) = \mathbb{E}_{(x,y)\sim P_k(x,y)}\big[\ell(f_\theta(x), y)\big].$$

Here, the expectation operator $\mathbb{E}[\cdot]$ means that we average the loss over infinitely many hypothetical samples drawn from the same distribution $P_k(x, y)$.

Using the covariate shift assumption. Under covariate shift, the joint distribution for client k can be factorized as

$$P_k(x, y) = P(y \mid x)\, P_k(x).$$

This decomposition has an important interpretation:

- $P(y \mid x)$ describes the labeling rule, which is assumed to be the same for all clients.
- $P_k(x)$ describes how frequently client k observes different input values.

Rewriting the gradient of the objective. The gradient $\nabla_\theta F_k(\theta)$ describes how the client's objective changes when the model parameters are slightly modified. Using the factorized distribution, the gradient can be written as

$$\nabla F_k(\theta) = \mathbb{E}_{x \sim P_k(x)} \left[\mathbb{E}_{y \sim P(y|x)} \left[\nabla_\theta \ell(f_\theta(x), y) \right] \right].$$

This expression consists of two nested expectations:

- The *inner expectation* averages the gradient of the loss over all possible labels y that could be assigned to a fixed input x, according to the shared conditional distribution $P(y \mid x)$.
- The *outer expectation* then averages this quantity over input values x drawn from the client-specific distribution $P_k(x)$.

Key implication. Although the labeling mechanism $P(y \mid x)$ is identical for all clients, each client weights different regions of the input space according to its own distribution $P_k(x)$. As a result, even when clients agree on how inputs should be labeled, their gradients emphasize different parts of the loss landscape. This mismatch is the fundamental way in which covariate shift distorts federated optimization.

Bias in aggregated gradients. Because each client weights the loss gradient according to its own input distribution, local gradients tend to emphasize different regions of the feature space. When these gradients are aggregated at the server, the resulting update direction may differ from the gradient of the global objective defined with respect to the overall population distribution. This effect persists even when all clients agree on the labeling rule, making covariate shift a subtle but significant source of optimization bias.

Geometric intuition. From a loss-landscape perspective, covariate shift causes each client to "see" a different projection of the same global objective. Regions of the parameter space that are important for one client may be rarely visited by another. Consequently, federated optimization must reconcile gradient signals that are locally consistent but globally misaligned.

Optimization Implication

Covariate shift causes local gradients to prioritize different regions of the input space, leading to biased global updates even when the labeling mechanism is identical across clients. This bias motivates correction techniques such as importance reweighting, adaptive aggregation, and domain-invariant representation learning.

Summary. Covariate shift represents a form of statistical heterogeneity in which the learning task is shared but the data exposure is uneven. In federated learning, this imbalance is unavoidable and must be explicitly addressed to ensure stable convergence and robust generalization across diverse client populations.

Worked Example: Covariate Shift across Multiple Federated Clients
We now present a concrete numerical example to illustrate covariate shift across multiple federated clients and to explicitly derive its effect on local and global optimization. Throughout this example, we carefully preserve all algebraic steps to make the implications transparent.

Problem setup. Consider a federated learning system with $K = 3$ clients. Each client observes data generated according to the same conditional labeling mechanism but different input distributions. Specifically, we assume a linear regression model

$$f_\theta(x) = \theta x,$$

with squared loss

$$\ell(f_\theta(x), y) = \frac{1}{2}(y - \theta x)^2.$$

The true data-generating process is

$$y = \theta^\star x,$$

with $\theta^\star = 2$, and no label noise for simplicity. Thus, the conditional distribution $P(y \mid x)$ is identical across all clients.

Client-specific datasets. Each client observes inputs from a different region of the feature space (Table 7.9):

Clearly,

$$P_1(x) \neq P_2(x) \neq P_3(x),$$

while

$$P(y \mid x) \text{ is identical for all clients.}$$

This is the defining signature of covariate shift.

Table 7.9 Client datasets exhibiting covariate shift

Client k	Inputs $x_{k,i}$	Outputs $y_{k,i} = 2x_{k,i}$
1	$\{1, 2, 3\}$	$\{2, 4, 6\}$
2	$\{4, 5, 6\}$	$\{8, 10, 12\}$
3	$\{7, 8, 9\}$	$\{14, 16, 18\}$

Local empirical risk functions. Each client minimizes its own empirical risk:

$$F_k(\theta) = \frac{1}{n_k} \sum_{i=1}^{n_k} \frac{1}{2}(y_{k,i} - \theta x_{k,i})^2, \quad n_k = 3.$$

We now compute each $F_k(\theta)$ explicitly.

Client 1.

$$\begin{aligned} F_1(\theta) &= \frac{1}{6}\left[(2-\theta)^2 + (4-2\theta)^2 + (6-3\theta)^2\right] \\ &= \frac{1}{6}\left[(1-\theta)^2 + 4(2-\theta)^2 + 9(2-\theta)^2\right] \\ &= \frac{1}{6}\left[(1-\theta)^2 + 13(2-\theta)^2\right]. \end{aligned}$$

Taking the derivative,

$$\nabla F_1(\theta) = \frac{1}{6}\,[2(\theta-1) + 26(\theta-2)] = \frac{28\theta - 54}{6}.$$

Client 2.

$$\begin{aligned} F_2(\theta) &= \frac{1}{6}\left[(8-4\theta)^2 + (10-5\theta)^2 + (12-6\theta)^2\right] \\ &= \frac{1}{6}\left[16(2-\theta)^2 + 25(2-\theta)^2 + 36(2-\theta)^2\right] \\ &= \frac{77}{6}(2-\theta)^2. \end{aligned}$$

Thus,

$$\nabla F_2(\theta) = \frac{77}{3}(\theta - 2).$$

Client 3.

$$\begin{aligned} F_3(\theta) &= \frac{1}{6}\left[(14-7\theta)^2 + (16-8\theta)^2 + (18-9\theta)^2\right] \\ &= \frac{1}{6}\left[49(2-\theta)^2 + 64(2-\theta)^2 + 81(2-\theta)^2\right] \\ &= \frac{194}{6}(2-\theta)^2. \end{aligned}$$

Hence,

$$\nabla F_3(\theta) = \frac{194}{3}(\theta - 2).$$

Observation: gradient distortion due to covariate shift. Although all clients share the same optimal parameter $\theta^\star = 2$, their gradients differ substantially in magnitude:

$$\nabla F_1(\theta) \neq \nabla F_2(\theta) \neq \nabla F_3(\theta).$$

This difference arises solely from the differing input distributions $P_k(x)$. Clients observing larger-magnitude inputs produce gradients with much larger curvature, thereby exerting disproportionate influence during aggregation.

Global objective and aggregation. The global federated objective is

$$F(\theta) = \sum_{k=1}^{3} \frac{n_k}{N} F_k(\theta), \qquad N = 9.$$

The global gradient becomes

$$\nabla F(\theta) = \frac{1}{3} \left(\nabla F_1(\theta) + \nabla F_2(\theta) + \nabla F_3(\theta) \right).$$

Substituting,

$$\nabla F(\theta) = \frac{1}{3} \left(\frac{28\theta - 54}{6} + \frac{77}{3}(\theta - 2) + \frac{194}{3}(\theta - 2) \right).$$

Even though all gradients vanish at $\theta = 2$, away from the optimum the descent direction is heavily biased toward clients with larger input variance.

Key takeaway. This worked example demonstrates that covariate shift induces *gradient imbalance* across clients. Even when labeling rules are identical and objectives share the same minimizer, heterogeneous input distributions distort the optimization landscape. This phenomenon motivates reweighting, normalization, and adaptive aggregation strategies in federated learning.

7.4 Prior Probability Shift

Prior probability shift, also referred to as *label distribution shift*, arises when the marginal distribution of class labels differs across clients, while the conditional distribution of inputs given the label remains unchanged. Formally, for two clients k and j, prior shift is characterized by

$$P_k(y) \neq P_j(y), \qquad P(x \mid y) \text{ identical across clients.}$$

This means that all clients agree on how inputs are generated from a given class, but they disagree on how frequently each class occurs.

Conceptual interpretation. The term *prior* refers to the marginal probability of observing a label before seeing any input features. Under prior probability shift, the *class composition* of local datasets varies across clients, even though the semantic meaning of each class is consistent. In other words, the labeling rule and feature-generation process are the same everywhere, but some clients observe certain labels much more often than others.

Illustrative examples. Prior probability shift is especially prevalent in classification problems:

- In medical diagnosis, hospitals may specialize in different diseases. A cancer center may see a disproportionately large number of positive cancer cases, while a general clinic observes mostly negative cases.
- In fraud detection, financial institutions operating in different regions may experience different fraud rates.
- In image recognition, personal photo collections may contain very different proportions of object categories (e.g., many photos of people but few of vehicles).

In all these cases, the visual or statistical characteristics of each class remain the same, but their frequencies vary substantially.

Effect on the federated objective. Let $\ell(f_\theta(x), y)$ denote the loss incurred by a model with parameters θ when predicting label y from input x. The population-level objective for client k is

$$F_k(\theta) = \mathbb{E}_{(x,y)\sim P_k(x,y)}\big[\ell(f_\theta(x), y)\big].$$

Under the prior shift assumption, the joint distribution factorizes as

$$P_k(x, y) = P(x \mid y)\, P_k(y).$$

Substituting this factorization into the global federated objective,

$$F(\theta) = \sum_{k=1}^{K} \frac{n_k}{N} F_k(\theta), \quad N = \sum_{k=1}^{K} n_k,$$

yields

$$F(\theta) = \sum_{y} \left(\sum_{k=1}^{K} \frac{n_k}{N} P_k(y) \right) \mathbb{E}_{x|y}\big[\ell(f_\theta(x), y)\big].$$

Interpretation of the reweighting effect. The inner expectation

$$\mathbb{E}_{x|y}\big[\ell(f_\theta(x), y)\big]$$

is identical for all clients, since $P(x \mid y)$ is shared. However, the coefficient

$$\sum_{k=1}^{K} \frac{n_k}{N} P_k(y)$$

acts as an *effective global class prior*. This term aggregates the class frequencies observed across all clients, weighted by dataset size. As a consequence, classes that are common across many clients receive larger weights in the global objective, while rare classes contribute less to the overall loss.

Consequences for learning. This implicit reweighting has important implications:

- The learned model may become biased toward majority classes, optimizing performance primarily for labels that are frequent across clients.
- Clients that predominantly observe rare labels may see poor local performance, even though they participate fully in training.
- Standard data-weighted aggregation (such as FedAvg) does not correct for this imbalance, because it reflects label frequency rather than label importance.

Representation Bias
Prior probability shift causes federated models to favor labels that are globally frequent, effectively underrepresenting minority-label clients even when aggregation is proportional to data volume.

Broader implications. Prior probability shift highlights a subtle but critical challenge in federated learning: fairness and representation are not guaranteed by simple data aggregation. Addressing prior shift often requires explicit class reweighting, cost-sensitive learning, or personalized models to ensure that minority labels and underrepresented clients are adequately served.

Worked Example: Prior Probability Shift with Explicit Derivation
We now present a complete numerical example to illustrate prior probability shift in federated learning and to derive its effect on the global optimization objective without skipping any intermediate steps.

Problem setup. Consider a binary classification problem with label set

$$\mathcal{Y} = \{0, 1\}.$$

We assume a federated system with $K = 2$ clients. The conditional distribution of inputs given the label is identical across clients:

$$P(x \mid y = 0) \quad \text{and} \quad P(x \mid y = 1) \quad \text{are the same for all clients.}$$

However, the class prior probabilities differ:

$$P_1(y) \neq P_2(y).$$

This is precisely the setting of prior probability shift.

Client-specific label distributions. Suppose the two clients observe the following label proportions:

Client	$P_k(y = 0)$	$P_k(y = 1)$
1	0.9	0.1
2	0.2	0.8

Assume each client has the same number of samples,

$$n_1 = n_2 = n, \quad N = n_1 + n_2 = 2n.$$

Loss function and model. Let $f_\theta(x)$ denote a classifier parameterized by θ, and let

$$\ell(f_\theta(x), y)$$

be a generic classification loss (e.g., cross-entropy or logistic loss).

Local population objectives. The population objective for client k is

$$F_k(\theta) = \mathbb{E}_{(x,y)\sim P_k(x,y)}\big[\ell(f_\theta(x), y)\big].$$

Under the prior shift assumption,

$$P_k(x, y) = P(x \mid y)\, P_k(y).$$

Substituting this factorization yields

$$F_k(\theta) = \sum_{y\in\{0,1\}} P_k(y)\mathbb{E}_{x\sim P(x|y)}\big[\ell(f_\theta(x), y)\big].$$

Explicit expansion of each client objective.
Client 1:

$$\begin{aligned} F_1(\theta) &= P_1(y = 0)\mathbb{E}_{x|y=0}[\ell(f_\theta(x), 0)] + P_1(y = 1)\mathbb{E}_{x|y=1}[\ell(f_\theta(x), 1)] \\ &= 0.9\, \mathbb{E}_{x|y=0}[\ell(f_\theta(x), 0)] + 0.1\, \mathbb{E}_{x|y=1}[\ell(f_\theta(x), 1)]. \end{aligned}$$

Client 2:

$$F_2(\theta) = 0.2\, \mathbb{E}_{x|y=0}[\ell(f_\theta(x), 0)] + 0.8\, \mathbb{E}_{x|y=1}[\ell(f_\theta(x), 1)].$$

Global federated objective. The global objective is defined as the data-weighted average of local objectives:

$$F(\theta) = \sum_{k=1}^{2} \frac{n_k}{N} F_k(\theta) = \frac{1}{2} F_1(\theta) + \frac{1}{2} F_2(\theta).$$

Substituting the expressions above,

$$F(\theta) = \frac{1}{2}\Big[0.9\, \mathbb{E}_{x|y=0}\ell(\cdot, 0) + 0.1\, \mathbb{E}_{x|y=1}\ell(\cdot, 1)\Big] + \frac{1}{2}\Big[0.2\, \mathbb{E}_{x|y=0}\ell(\cdot, 0) + 0.8\, \mathbb{E}_{x|y=1}\ell(\cdot, 1)\Big].$$

Combining like terms,

$$\begin{aligned} F(\theta) &= \left(\frac{0.9 + 0.2}{2}\right) \mathbb{E}_{x|y=0}[\ell(f_\theta(x), 0)] + \left(\frac{0.1 + 0.8}{2}\right) \mathbb{E}_{x|y=1}[\ell(f_\theta(x), 1)] \\ &= 0.55\, \mathbb{E}_{x|y=0}[\ell(f_\theta(x), 0)] + 0.45\, \mathbb{E}_{x|y=1}[\ell(f_\theta(x), 1)]. \end{aligned}$$

Interpretation. Although neither client has class priors $(0.55, 0.45)$ individually, the global federated objective implicitly optimizes for this *aggregated class prior*. As a result:

- Class $y = 0$ receives greater emphasis than class $y = 1$.
- Client 2's dominant class ($y = 1$) is underrepresented globally.
- The learned classifier is biased toward majority-label behavior at the federation level.

Key takeaway. This example shows that prior probability shift alters the effective loss weights in federated learning. Even when all clients agree on the labeling mechanism $P(x \mid y)$, the global model is implicitly biased toward labels that are frequent across clients. This motivates explicit class reweighting, fairness constraints, or personalized federated learning strategies.

Worked Example: Prior Probability Shift with Unequal Client Sample Sizes
We now extend the previous example to the more realistic federated learning setting in which clients possess *different numbers of local samples*. This scenario is common in practice, as users, devices, or institutions often contribute data at vastly different scales.

Problem setup. Consider again a binary classification problem with label set

$$\mathcal{Y} = \{0, 1\}.$$

We assume a federated system with $K = 2$ clients. As before, the conditional distribution of inputs given the label is identical across clients:

$$P(x \mid y = 0) \quad \text{and} \quad P(x \mid y = 1) \quad \text{are shared by all clients.}$$

However, the class prior distributions and the number of samples now vary across clients.

Client-specific dataset sizes. Let the two clients have different local dataset sizes:

$$n_1 = 100, \qquad n_2 = 300, \qquad N = n_1 + n_2 = 400.$$

Client-specific label distributions. Suppose the class proportions observed locally are as follows:

Client	$P_k(y = 0)$	$P_k(y = 1)$
1	0.9	0.1
2	0.2	0.8

This represents a typical prior probability shift scenario: client 1 predominantly observes class 0, while client 2 predominantly observes class 1.

Loss function and model Let $f_\theta(x)$ denote a classifier parameterized by θ, and let

$$\ell(f_\theta(x), y)$$

be a standard classification loss function.

Local population objectives. The population objective minimized by client k is

$$F_k(\theta) = \mathbb{E}_{(x,y)\sim P_k(x,y)}\left[\ell(f_\theta(x), y)\right].$$

Under the prior shift assumption, the joint distribution factorizes as

$$P_k(x, y) = P(x \mid y)\, P_k(y).$$

Substituting this factorization yields

$$F_k(\theta) = \sum_{y\in\{0,1\}} P_k(y)\mathbb{E}_{x\sim P(x|y)}\left[\ell(f_\theta(x), y)\right].$$

Explicit expansion of local objectives. *Client 1:*

$$F_1(\theta) = 0.9\, \mathbb{E}_{x|y=0}[\ell(f_\theta(x), 0)] + 0.1\, \mathbb{E}_{x|y=1}[\ell(f_\theta(x), 1)].$$

Client 2:

$$F_2(\theta) = 0.2\, \mathbb{E}_{x|y=0}[\ell(f_\theta(x), 0)] + 0.8\, \mathbb{E}_{x|y=1}[\ell(f_\theta(x), 1)].$$

Global federated objective with unequal data sizes. In federated learning, the global objective is a data-weighted average of local objectives:

$$F(\theta) = \sum_{k=1}^{2} \frac{n_k}{N} F_k(\theta).$$

Substituting $n_1 = 100$, $n_2 = 300$, and $N = 400$, we obtain

$$F(\theta) = \frac{100}{400} F_1(\theta) + \frac{300}{400} F_2(\theta) = \frac{1}{4} F_1(\theta) + \frac{3}{4} F_2(\theta).$$

Explicit expansion of the global objective. Substituting the expressions for F_1 and F_2,

$$\begin{aligned} F(\theta) &= \frac{1}{4}\Big[0.9\, \mathbb{E}_{x|y=0}\ell(\cdot, 0) + 0.1\, \mathbb{E}_{x|y=1}\ell(\cdot, 1)\Big] \\ &+ \frac{3}{4}\Big[0.2\, \mathbb{E}_{x|y=0}\ell(\cdot, 0) + 0.8\, \mathbb{E}_{x|y=1}\ell(\cdot, 1)\Big]. \end{aligned}$$

Collecting terms class-wise,

$$\begin{aligned} F(\theta) &= \left(\frac{1}{4} \cdot 0.9 + \frac{3}{4} \cdot 0.2\right) \mathbb{E}_{x|y=0}[\ell(f_\theta(x), 0)] \\ &+ \left(\frac{1}{4} \cdot 0.1 + \frac{3}{4} \cdot 0.8\right) \mathbb{E}_{x|y=1}[\ell(f_\theta(x), 1)]. \end{aligned}$$

Evaluating the coefficients explicitly,

$$\begin{aligned} F(\theta) &= (0.225 + 0.150)\mathbb{E}_{x|y=0}[\ell(f_\theta(x), 0)] + (0.025 + 0.600)\mathbb{E}_{x|y=1}[\ell(f_\theta(x), 1)] \\ &= 0.375\, \mathbb{E}_{x|y=0}[\ell(f_\theta(x), 0)] + 0.625\, \mathbb{E}_{x|y=1}[\ell(f_\theta(x), 1)]. \end{aligned}$$

Interpretation. The effective class prior optimized by the global model is now (0.375, 0.625). This distribution:

- differs from both clients' local priors,

- is dominated by the larger client's data volume,
- amplifies representation bias beyond simple averaging.

Key takeaway. Unequal client dataset sizes are not only allowed but fundamental in federated learning. When combined with prior probability shift, data imbalance further skews the global objective toward majority-label, high-volume clients. This effect persists even under correct data-weighted aggregation, motivating the need for class reweighting, fairness constraints, or personalized federated optimization.

7.5 Concept Drift and Concept Shift

In many practical learning systems, the statistical relationship between inputs and outputs is not fixed over time. Instead, it evolves as environments change, user behavior adapts, or external conditions fluctuate. Such *temporal non-stationarity* is especially pronounced in federated learning, where data are generated continuously and autonomously at distributed clients.

Concept drift. Concept drift refers to gradual or abrupt changes in the conditional distribution of labels given inputs over time. Formally, letting t denote time,

$$P_t(y \mid x) \neq P_{t+\Delta}(y \mid x),$$

for some time increment $\Delta > 0$. Here, the input space $\mathcal{X}$ and label space $\mathcal{Y}$ remain unchanged, but the mapping from inputs to outputs evolves.

Intuitively, concept drift means that the same input x may correspond to different expected outputs at different times. For example:

- In email spam detection, spammers continually adapt their messages to bypass filters.
- In recommendation systems, user preferences change as tastes, trends, or life circumstances evolve.
- In sensor-based systems, environmental dynamics such as seasonal changes alter the relationship between measurements and outcomes.

From a mathematical perspective, concept drift implies that the population risk becomes time-dependent:

$$F_t(\theta) = \mathbb{E}_{(x,y)\sim P_t(x,y)}\big[\ell(f_\theta(x), y)\big].$$

As a result, a parameter vector θ that minimizes $F_t(\theta)$ may no longer be optimal at time $t + \Delta$. This breaks the classical assumption that there exists a single, time-invariant objective function toward which learning should converge.

Asynchronous drift in federated learning. In federated systems, concept drift rarely occurs uniformly across all clients. Different clients may experience drift at different times, rates, or directions. Formally, for clients k and j,

$$P_{k,t}(y \mid x) \neq P_{j,t}(y \mid x), \quad P_{k,t}(y \mid x) \neq P_{k,t+\Delta}(y \mid x).$$

This leads to a combination of *spatial heterogeneity* (across clients) and *temporal heterogeneity* (within each client), further complicating aggregation and convergence.

Concept shift. Concept shift represents a more severe form of non-stationarity. Here, the very definition of the task changes over time, leading to incompatible labeling rules. In probabilistic terms, concept shift may involve changes to:

- the label space $\mathcal{Y}$,
- the semantics associated with labels,
- or the decision boundaries defining correct predictions.

For example:

- A medical diagnosis system may be updated when new clinical guidelines redefine disease categories.
- In fraud detection, the notion of fraudulent behavior may change due to regulatory updates or new attack strategies.

Under concept shift, historical data may become partially or entirely obsolete, as past labels no longer correspond to current meanings. Unlike concept drift, which can sometimes be addressed through continuous adaptation, concept shift may require explicit task redefinition, model reset, or multi-task learning formulations.

Implications for learning objectives. Both concept drift and concept shift violate the stationarity assumption underlying classical empirical risk minimization. A static global objective of the form

$$F(\theta) = \mathbb{E}_{(x,y)\sim P(x,y)}\left[\ell(f_\theta(x), y)\right]$$

is no longer adequate when $P(x, y)$ itself evolves over time. Instead, learning must be viewed as tracking a moving target, where the goal is to remain close to the current optimum rather than converge to a fixed one.

Temporal Heterogeneity
Federated learning must accommodate not only spatial heterogeneity across clients but also temporal heterogeneity within clients. Concept drift requires continual adaptation, while concept shift may necessitate explicit changes to model structure, objectives, or task definitions.

Summary. Concept drift captures evolving input–output relationships over time, whereas concept shift denotes fundamental changes in task semantics. Both phenomena are intrinsic to long-lived federated systems and demand learning algorithms that are adaptive, robust, and aware of temporal non-stationarity, rather than relying solely on static optimization assumptions.

Worked Example: Concept Drift and Concept Shift in a Multi-Client Federated Setting

We now present a fully worked numerical and analytical example to demonstrate how concept drift and concept shift manifest in a federated learning system with multiple clients. All derivations are shown explicitly, without skipping intermediate steps.

Federated setup. Consider a federated system with $K = 3$ clients. Each client collects data over two consecutive time periods, t and $t + \Delta$. For simplicity, we consider a scalar input $x \in \mathbb{R}$ and a binary label $y \in \{0, 1\}$. Each client uses a linear model with sigmoid output:

$$f_\theta(x) = \sigma(\theta x), \quad \sigma(z) = \frac{1}{1 + e^{-z}},$$

and the logistic loss

$$\ell(f_\theta(x), y) = -y \log f_\theta(x) - (1 - y) \log(1 - f_\theta(x)).$$

Client datasets at time t. Assume each client has $n_k = 6$ samples at time t. The observed datasets are shown in Table 7.9. Although the input distributions are identical, the conditional distributions $P_{k,t}(y \mid x)$ differ slightly across clients.

Local empirical risk and gradient. For client k, the empirical risk is

$$\hat{F}_{k,t}(\theta) = \frac{1}{n_k} \sum_{i=1}^{n_k} \ell(f_\theta(x_{k,i}), y_{k,i}).$$

The gradient of the loss for a single sample is

$$\nabla_\theta \ell(f_\theta(x_i), y_i) = (\sigma(\theta x_i) - y_i) x_i.$$

Thus, the empirical gradient for client k is

$$\nabla \hat{F}_{k,t}(\theta) = \frac{1}{6} \sum_{i=1}^{6} (\sigma(\theta x_{k,i}) - y_{k,i}) x_{k,i}.$$

Explicit numerical evaluation at $\theta = 0$. To make all calculations explicit, we evaluate gradients at

$$\theta = 0, \qquad \sigma(0) = 0.5.$$

Client 1 at time t. Client 1 observes

$$x = \{1, 2, 3, 4, 5, 6\}, \quad y^{(t)} = \{0, 0, 0, 1, 1, 1\}.$$

Substituting into the gradient,

$$\nabla \hat{F}_{1,t}(0) = \frac{1}{6}\Big[(0.5 - 0) \cdot 1 + (0.5 - 0) \cdot 2 + (0.5 - 0) \cdot 3 + (0.5 - 1) \cdot 4 + (0.5 - 1) \cdot 5 + (0.5 - 1) \cdot 6\Big].$$

Evaluating term by term,

$$\nabla \hat{F}_{1,t}(0) = \frac{1}{6}(0.5 + 1 + 1.5 - 2 - 2.5 - 3) = \frac{-4.5}{6} = -0.75.$$

Client 2 at time t. Client 2 observes

$$y^{(t)} = \{0, 0, 1, 1, 1, 1\}.$$

Thus,

$$\nabla \hat{F}_{2,t}(0) = \frac{1}{6}\Big[(0.5 - 0) \cdot 1 + (0.5 - 0) \cdot 2 + (0.5 - 1) \cdot 3 + (0.5 - 1) \cdot 4 + (0.5 - 1) \cdot 5 + (0.5 - 1) \cdot 6\Big].$$

Evaluating,

$$\nabla \hat{F}_{2,t}(0) = \frac{1}{6}(0.5 + 1 - 1.5 - 2 - 2.5 - 3) = \frac{-7.5}{6} = -1.25.$$

Client 3 at time t. Client 3 observes

$$y^{(t)} = \{0, 1, 1, 1, 1, 1\}.$$

Hence,

$$\nabla \hat{F}_{3,t}(0) = \frac{1}{6}\Big[(0.5 - 0) \cdot 1 + (0.5 - 1) \cdot 2 + (0.5 - 1) \cdot 3 + (0.5 - 1) \cdot 4 + (0.5 - 1) \cdot 5 + (0.5 - 1) \cdot 6\Big].$$

Evaluating,

$$\nabla \hat{F}_{3,t}(0) = \frac{1}{6}(0.5 - 1 - 1.5 - 2 - 2.5 - 3) = \frac{-9.5}{6} = -1.5833.$$

Global gradient at time *t*. The global gradient is

$$\nabla \hat{F}_t(0) = \frac{1}{3}\big(-0.75 - 1.25 - 1.5833\big) = -1.1944.$$

Let $\theta_t^{\star}$ denote a minimizer such that

$$\nabla \hat{F}_t(\theta_t^{\star}) = 0.$$

Concept drift at time $t + \Delta$. At time $t + \Delta$, the client labels change as shown in Table 7.10.

Client 1 at time $t + \Delta$. Client 1 labels become

$$y^{(t+\Delta)} = \{0, 0, 0, 0, 1, 1\}.$$

Thus,

$$\nabla \hat{F}_{1,t+\Delta}(0) = \frac{1}{6}(0.5 + 1 + 1.5 + 2 - 2.5 - 3) = \frac{-0.5}{6} = -0.0833.$$

Client 2 at time $t + \Delta$. Client 2 labels become

$$y^{(t+\Delta)} = \{0, 0, 0, 1, 1, 1\}.$$

Thus,

$$\nabla \hat{F}_{2,t+\Delta}(0) = \frac{1}{6}(0.5 + 1 + 1.5 - 2 - 2.5 - 3) = -0.75.$$

Client 3 at time $t + \Delta$. Client 3 labels become

$$y^{(t+\Delta)} = \{0, 0, 1, 1, 1, 1\}.$$

Table 7.10 Client datasets at time $t + \Delta$ (concept drift)

Client k	Inputs $x_{k,i}$	Labels $y_{k,i}^{(t+\Delta)}$
1	$\{1, 2, 3, 4, 5, 6\}$	$\{0, 0, 0, 0, 1, 1\}$
2	$\{1, 2, 3, 4, 5, 6\}$	$\{0, 0, 0, 1, 1, 1\}$
3	$\{1, 2, 3, 4, 5, 6\}$	$\{0, 0, 1, 1, 1, 1\}$

Thus,

$$\nabla \hat{F}_{3,t+\Delta}(0) = \frac{1}{6}(0.5 + 1 - 1.5 - 2 - 2.5 - 3) = -1.25.$$

Global gradient at time $t + \Delta$.

$$\nabla \hat{F}_{t+\Delta}(0) = \frac{1}{3}(-0.0833 - 0.75 - 1.25) = -0.6944.$$

Since

$$\nabla \hat{F}_{t+\Delta}(\theta_t^\star) \neq 0,$$

the old optimum no longer minimizes the new objective.

Concept drift at time $t + 2\Delta$. We now consider a further gradual concept drift at time $t + 2\Delta$. Compared to time $t + \Delta$, the decision boundary shifts again so that the positive class occurs only for even larger input values. The updated client datasets are shown in Table 7.11.

Formally, this corresponds to another temporal change in the conditional distribution,

$$P_{k,t+2\Delta}(y \mid x) \neq P_{k,t+\Delta}(y \mid x),$$

while the input distribution remains unchanged.

Client 1 at time $t + 2\Delta$. Client 1 labels become

$$y^{(t+2\Delta)} = \{0, 0, 0, 0, 0, 1\}.$$

The empirical gradient evaluated at $\theta = 0$ is

$$\nabla \hat{F}_{1,t+2\Delta}(0) = \frac{1}{6}\Big[(0.5 - 0) \cdot 1 + (0.5 - 0) \cdot 2 + (0.5 - 0) \cdot 3 + (0.5 - 0) \cdot 4 + (0.5 - 0) \cdot 5 + (0.5 - 1) \cdot 6\Big].$$

Table 7.11 Client datasets at time $t + 2\Delta$ (continued concept drift)

Client k	Inputs $x_{k,i}$	Labels $y_{k,i}^{(t+2\Delta)}$
1	$\{1, 2, 3, 4, 5, 6\}$	$\{0, 0, 0, 0, 0, 1\}$
2	$\{1, 2, 3, 4, 5, 6\}$	$\{0, 0, 0, 0, 1, 1\}$
3	$\{1, 2, 3, 4, 5, 6\}$	$\{0, 0, 0, 1, 1, 1\}$

Evaluating term by term,

$$\nabla \hat{F}_{1,t+2\Delta}(0) = \frac{1}{6}(0.5 + 1 + 1.5 + 2 + 2.5 - 3) = \frac{4.5}{6} = 0.75.$$

Client 2 at time $t + 2\Delta$. Client 2 labels become

$$y^{(t+2\Delta)} = \{0, 0, 0, 0, 1, 1\}.$$

Thus,

$$\begin{aligned}\nabla \hat{F}_{2,t+2\Delta}(0) &= \frac{1}{6}\Big[(0.5-0)\cdot 1 + (0.5-0)\cdot 2 + (0.5-0)\cdot 3 \\ &\quad + (0.5-0)\cdot 4 + (0.5-1)\cdot 5 + (0.5-1)\cdot 6\Big] \\ &= \frac{1}{6}(0.5 + 1 + 1.5 + 2 - 2.5 - 3) = \frac{-0.5}{6} = -0.0833.\end{aligned}$$

Client 3 at time $t + 2\Delta$. Client 3 labels become

$$y^{(t+2\Delta)} = \{0, 0, 0, 1, 1, 1\}.$$

Thus,

$$\begin{aligned}\nabla \hat{F}_{3,t+2\Delta}(0) &= \frac{1}{6}\Big[(0.5-0)\cdot 1 + (0.5-0)\cdot 2 + (0.5-0)\cdot 3 \\ &\quad + (0.5-1)\cdot 4 + (0.5-1)\cdot 5 + (0.5-1)\cdot 6\Big] \\ &= \frac{1}{6}(0.5 + 1 + 1.5 - 2 - 2.5 - 3) = \frac{-4.5}{6} = -0.75.\end{aligned}$$

Global gradient at time $t + 2\Delta$. The aggregated global gradient is

$$\begin{aligned}\nabla \hat{F}_{t+2\Delta}(0) &= \frac{1}{3}(0.75 - 0.0833 - 0.75) \\ &= -0.0278.\end{aligned}$$

Since

$$\nabla \hat{F}_{t+2\Delta}(\theta_t^\star) \neq 0,$$

the minimizer obtained at time t or $t + \Delta$ is no longer optimal.

This confirms that *progressive concept drift across rounds causes successive shifts in the global loss landscape*, even when client data sizes and input distributions remain unchanged.

Table 7.12 Summary of progressive concept drift across federated rounds

Time	Client	Label pattern $y_{k,i}$	Decision boundary trend	Global gradient behavior
t	$k = 1, 2, 3$	Mixed early positives	Baseline concept	$\nabla \hat{F}_t(\theta_t^\star) = 0$
$t + \Delta$	$k = 1, 2, 3$	Positives shift right	Moderate drift	$\nabla \hat{F}_{t+\Delta}(\theta_t^\star) \neq 0$
$t + 2\Delta$	$k = 1, 2, 3$	Further delayed positives	Accumulated drift	$\nabla \hat{F}_{t+2\Delta}(\theta_t^\star) \neq 0$

Summary of Progressive Concept Drift Across Rounds

Table 7.12 summarizes the evolution of client label distributions and the corresponding global gradient behavior across successive communication rounds.

The table highlights that although the input distribution remains unchanged across rounds, repeated changes in the conditional label distribution progressively distort the optimization landscape seen by the federated server.

Remark 1 (Single drift is sufficient). Even a *single* concept drift event (from t to $t + \Delta$) is sufficient to invalidate the previous global optimum $\theta_t^\star$, as evidenced by the emergence of a nonzero global gradient.

Remark 2 (Multiple drifts accumulate). Successive drift rounds ($t+\Delta$, $t+2\Delta$, ...) cause *accumulated gradient bias*. Each round introduces a systematic shift in the loss surface, resulting in a moving target for optimization.

Remark 3 (No finite bound in general). In the absence of explicit drift detection or adaptation mechanisms, there is no finite upper bound on the number of drift rounds that can be tolerated. Formally, for an arbitrary number m of drift events,

$$\nabla \hat{F}_{t+m\Delta}(\theta_t^\star) \neq 0,$$

unless the drift magnitude converges to zero.

Remark 4 (Tracking requirement). To remain effective under continuous concept drift, federated learning systems must *track* the evolving optimum $\theta_{t+m\Delta}^\star$ rather than converge to a fixed point. This necessitates adaptive learning rates, drift-aware aggregation, or explicit temporal modeling.

Remark 5 (Distinction from concept shift). Unlike concept drift, which manifests gradually over many rounds, concept shift corresponds to an abrupt task change. In such cases, no number of incremental updates can recover optimality, and model reinitialization or task redefinition becomes necessary.

One drift is enough to break optimality; many drifts redefine the learning problem.

Federated learning under temporal heterogeneity is therefore better understood as a problem of *continuous adaptation* rather than static convergence.

Final conclusion. This exhaustive derivation confirms that:

- Concept drift causes systematic gradient bias over time.
- Each client contributes differently to temporal inconsistency.
- Federated aggregation no longer corresponds to descent on a single global objective.

7.6 Data Imbalance Across Clients

A defining characteristic of real-world federated learning systems is *severe imbalance in client dataset sizes*. Unlike centralized learning—where data are pooled and uniformly accessible—federated learning operates over a population of autonomous clients whose local data volumes can differ dramatically. Formally, let n_k denote the number of samples available at client k. It is common to observe

$$n_k \ll n_j \quad \text{for some clients } k, j,$$

sometimes by several orders of magnitude.

Such imbalance arises naturally in practice. For example, in mobile and IoT settings, some devices may generate data continuously (e.g., always-on sensors), while others contribute only sporadically. In healthcare federated learning, large hospitals may possess extensive patient records, whereas rural clinics may see far fewer cases. In all such scenarios, data imbalance is the rule rather than the exception.

Weighted aggregation and its limitations. To account for unequal dataset sizes, federated learning algorithms typically employ *data-weighted aggregation*. In its simplest form, the global objective is written as

$$F(\theta) = \sum_{k=1}^{K} \frac{n_k}{N} F_k(\theta), \qquad N = \sum_{k=1}^{K} n_k,$$

where $F_k(\theta)$ denotes the local population objective of client k. This weighting ensures that clients with more data exert greater influence on the global model.

While such weighting is statistically well-motivated, it addresses only *sample imbalance*, not *information imbalance*. The assumption underlying weighted aggregation is that each data point contributes equally informative signal to the learning task. In practice, this assumption is often violated.

Information imbalance. Small clients may hold rare, high-value, or safety-critical patterns that occur infrequently in the global population. Examples include:

- rare disease cases in medical federated learning,
- minority language usage in keyboard prediction,
- unusual fault signatures in industrial monitoring.

Even though such clients have small n_k, the information they carry may be disproportionately important for generalization, robustness, or fairness.

Under standard data-weighted aggregation, the gradient contribution from a small client scales as $\frac{n_k}{N}$. As a result, informative updates from small clients can be effectively drowned out by the sheer volume of updates from large clients. This phenomenon leads to a form of *representation bias* at the optimization level.

Impact on optimization dynamics. Data imbalance interacts with federated optimization in several detrimental ways:

- **Slow convergence:** Large clients dominate early optimization steps, steering the model toward regions of the loss landscape that best fit majority data. Minority patterns may only be weakly corrected over many rounds.
- **Client dominance:** The global model may implicitly overfit to the statistical structure of large clients, reducing performance on smaller or atypical clients.
- **Instability under partial participation:** When only a subset of clients participates in each round, the random absence of a large client can cause abrupt shifts in the aggregated update, while the absence of small clients often goes unnoticed.

These effects are amplified in conjunction with non-IID data, covariate shift, or prior probability shift, where client heterogeneity already biases local gradients.

Conceptual distinction: volume versus value. It is crucial to distinguish between the *amount* of data and its *learning value*. A large dataset that is highly redundant may contribute less new information than a small dataset capturing rare but structurally important phenomena. Federated learning systems that equate data volume with importance risk systematically marginalizing such clients.

Scale Asymmetry
In federated learning, data volume and data importance are not equivalent. Naïvely weighting updates by dataset size can suppress rare but critical information held by small clients.

Broader implications. Addressing data imbalance requires more than simple reweighting. Potential remedies explored in later chapters include adaptive client weighting, fairness-aware objectives, personalized models, and information-sensitive aggregation schemes. Recognizing data imbalance as both a statistical and an algorithmic challenge is essential for building federated systems that are not only accurate, but also robust and equitable.

7.7 Impact of Non-IID Data on Convergence

The most profound and practically visible consequences of data heterogeneity in federated learning appear in the system's *convergence behavior*. While IID data primarily introduce stochastic noise that averages out over time, non-IID data fundamentally alter the optimization landscape by inducing persistent and structured bias in local updates.

Gradient bias under heterogeneity. Let $F(\theta)$ denote the global population objective and $F_k(\theta)$ the local objective at client k. Under IID sampling, local gradients provide unbiased estimates of the global gradient. In contrast, under non-IID data, we generally have

$$\mathbb{E}\big[\nabla F_k(\theta)\big] \neq \nabla F(\theta),$$

even when expectations are taken over local data randomness. This inequality reflects a systematic mismatch between the direction in which a client optimizes its local loss and the direction that would most reduce the global objective.

Consequences for optimization dynamics. This gradient bias leads to a range of adverse behaviors that are routinely observed in practice:

- **Slower convergence rates:** Local updates may partially cancel one another during aggregation, reducing effective progress per communication round.
- **Oscillatory updates:** The global model may zig-zag between client-preferred regions of the parameter space rather than moving steadily toward a common optimum.

- **Divergence under aggressive local training:** When clients perform many local steps before aggregation, biased local descent directions can push models far from the global optimum, causing instability or outright divergence.
- **Sensitivity to client sampling:** The specific subset of participating clients in a round can significantly influence the update direction, increasing variance across rounds and harming reproducibility.

These effects are markedly different from those encountered in centralized or IID distributed learning, where variance—not bias—is the dominant limiting factor.

Quantifying heterogeneity. To make the notion of client disagreement precise, theoretical analyses often introduce a *heterogeneity measure* that captures the spread of local gradients around the global gradient. A commonly used quantity is

$$\mathcal{B}^2 = \frac{1}{K}\sum_{k=1}^{K} \|\nabla F_k(\theta) - \nabla F(\theta)\|^2,$$

where $\|\cdot\|$ denotes the Euclidean norm. This measure represents the average squared deviation of client gradients from the global descent direction. When $\mathcal{B}^2 = 0$, all clients are perfectly aligned, recovering the IID-like regime. As $\mathcal{B}^2$ increases, gradient disagreement grows, signaling greater difficulty for stable and efficient optimization.

Bias-dominated optimization. A crucial conceptual shift emerges from this analysis. Classical stochastic optimization theory is largely concerned with controlling variance introduced by random sampling. Federated learning under non-IID data, however, operates in a regime where *bias* dominates. No amount of averaging can fully eliminate a persistent directional misalignment between local and global objectives.

Central Challenge
Non-IID data fundamentally transforms federated learning from a variance-limited optimization problem into a bias-dominated one, where persistent gradient disagreement, rather than stochastic noise, is the primary obstacle to convergence.

Implications for algorithm design. Recognizing heterogeneity-induced bias as the core challenge reframes the design goals of federated algorithms. Effective methods must explicitly reduce, correct, or accommodate gradient disagreement—rather than merely accelerating variance reduction. This insight motivates the algorithmic

Table 7.13 Summary of major distributional shifts in federated learning and their optimization implications

Type of Shift	What Changes Across Clients	What Remains Invariant	Primary Optimization Effect
Covariate Shift	Input distribution $P_k(x)$ differs	Conditional labeling rule $P(y \mid x)$ is shared	Gradient imbalance due to uneven emphasis on regions of the feature space
Prior Probability Shift	Label distribution $P_k(y)$ differs	Class-conditional features $P(x \mid y)$ are shared	Implicit reweighting of classes; bias toward globally frequent labels
Concept Shift	Conditional distribution $P_k(y \mid x)$ differs	Input distribution $P(x)$ may be similar or overlapping	Conflicting learning objectives; no single globally optimal model

strategies developed in the next chapter, which aim to mitigate heterogeneity-induced degradation while respecting the defining constraints of federated learning: data locality, limited communication, and client autonomy.

Table 7.13 highlights that federated learning departs from the classical IID assumption along multiple, orthogonal dimensions. Importantly, *a single shift is sufficient* to invalidate the theoretical guarantees of centralized learning. Covariate shift alone biases gradient magnitudes, prior probability shift alone induces class imbalance at the federation level, and concept shift alone destroys the existence of a universally optimal model.

However, real-world federated systems rarely exhibit only one form of shift. In practice, multiple shifts often co-occur, compounding their effects. For example, a system may simultaneously experience covariate shift (due to device or environmental variation), prior shift (due to population imbalance), and temporal dependence within clients. As these shifts accumulate, optimization difficulty increases superlinearly, rendering naive aggregation strategies ineffective.

This observation underscores a central lesson of federated learning: *the breakdown of IID assumptions is not binary but progressive*. Each additional form of statistical shift further distances federated optimization from its centralized counterpart and motivates increasingly sophisticated algorithmic interventions, such as reweighting, personalization, clustering, or multi-model architectures.

Exercises

This chapter introduced the foundations of federated learning, with a particular focus on statistical heterogeneity, concept drift, and concept shift. The following exercises are designed to test conceptual understanding, mathematical reasoning, and practical modeling skills.

A. Short Answer Questions

Answer each question briefly and precisely.

1. Define federated learning and explain how it differs from centralized learning.
2. What is meant by statistical heterogeneity in federated learning?
3. Define concept drift in the context of distributed data sources.
4. How does concept shift differ from concept drift?
5. What role does the global objective function play in federated optimization?
6. Why can a previously optimal model become suboptimal under concept drift?
7. What is meant by partial client participation in federated learning?
8. Explain the significance of non-IID data in federated settings.
9. What is gradient bias and how does it arise in federated learning?
10. Why is temporal alignment across clients difficult to guarantee?

B. Long Answer Questions

Answer the following questions with clear explanations, equations, and diagrams wherever appropriate.

1. Derive the global empirical risk minimization objective used in federated learning and explain each term.
2. Explain, with equations, how concept drift affects the gradient of the empirical risk over time.
3. Using a simple linear classifier, show why $\theta_t^\star$ may fail to minimize $\hat{F}_{t+\Delta}(\theta)$.
4. Discuss the implications of asynchronous client updates on convergence guarantees.
5. Compare covariate shift, prior probability shift, and concept shift in federated learning.
6. Explain why federated aggregation may fail to correspond to descent on a single objective under temporal heterogeneity.
7. Illustrate the difference between centralized SGD and federated averaging using a block diagram.
8. Discuss tracking versus convergence in the presence of continuous concept drift.
9. Explain why concept shift may require model reinitialization.
10. Critically analyze the limitations of FedAvg under non-IID and temporally drifting data.

C. True/False Questions

State whether each statement is **True** or **False**.

1. Federated learning assumes that all clients participate in every round.

2. Concept drift implies that $P(y \mid x)$ remains constant over time.
3. A single concept drift event can invalidate the previous global optimum.
4. Concept shift corresponds to a gradual change in the decision boundary.
5. Non-IID data is a defining characteristic of federated learning.
6. Federated aggregation always minimizes a well-defined global loss.
7. Asynchronous updates can introduce gradient inconsistency.
8. Covariate shift affects the marginal distribution of inputs.
9. Concept shift preserves a functional relationship between old and new labels.
10. Temporal heterogeneity complicates convergence analysis.

Answers:

1. False
2. False
3. True
4. False
5. True
6. False
7. True
8. True
9. False
10. True

D. Fill in the Blanks

Fill in the blanks with the most appropriate term or expression.

1. Federated learning avoids sharing raw data by keeping it ___________.
2. Statistical heterogeneity arises due to ___________ data distributions.
3. Concept drift corresponds to changes in ___________ over time.
4. In concept shift, the learning task itself becomes ___________.
5. The logistic loss is commonly used for ___________ classification.
6. The function $\sigma(z) = \frac{1}{1+e^{-z}}$ is known as the ___________ function.
7. The global model is obtained by ___________ local updates.
8. Temporal mismatch leads to ___________ gradients.
9. A model that tracks drift aims to follow a moving ___________.
10. Concept shift may require complete model ___________.

Answers:

1. local
2. non-IID
3. $P(y \mid x)$

4. different
5. binary
6. sigmoid
7. aggregating
8. inconsistent
9. optimum
10. reinitialization

E. Multiple Choice Questions (MCQs)

Choose the **most appropriate** answer.

1. Which assumption is most strongly violated in federated learning?
   (a) Centralized storage
   (b) IID data
   (c) Convex loss
   (d) Gradient descent
2. Concept drift primarily affects which distribution?
   (a) $P(x)$
   (b) $P(y)$
   (c) $P(y \mid x)$
   (d) $P(x \mid y)$
3. Which scenario destroys the existence of a single global optimum?
   (a) Covariate shift
   (b) Prior shift
   (c) Concept drift
   (d) Concept shift
4. In federated learning, clients are typically:
   (a) Synchronously updated
   (b) Fully reliable
   (c) Heterogeneous
   (d) Identical
5. Which algorithm is most commonly associated with federated learning?
   (a) Adam
   (b) RMSProp
   (c) FedAvg
   (d) Newton's method
6. Gradient bias is mainly caused by:
   (a) High learning rates
   (b) Noisy labels
   (c) Distributional shifts
   (d) Large batch sizes

7. Which property is essential for convergence in classical SGD?
   (a) Encryption
   (b) IID samples
   (c) Client dropout
   (d) Model personalization
8. Concept drift is best described as:
   (a) Sudden task change
   (b) Gradual label evolution
   (c) Input corruption
   (d) Model failure
9. What must a federated system do under continuous drift?
   (a) Stop training
   (b) Track the optimum
   (c) Freeze parameters
   (d) Centralize data
10. Concept shift often necessitates:
   (a) Smaller batch sizes
   (b) Faster aggregation
   (c) Model redefinition
   (d) Gradient clipping

Answers:

1. (b)
2. (c)
3. (d)
4. (c)
5. (c)
6. (c)
7. (b)
8. (b)
9. (b)
10. (c)

F. Application and Project-Based Questions

These problems emphasize modeling, reasoning, and implementation.

1. Design a simulated federated learning setup exhibiting concept drift.
2. Implement a toy federated system with two drifting clients.
3. Compare FedAvg performance with and without drift adaptation.
4. Propose a drift detection mechanism suitable for edge devices.
5. Analyze the impact of asynchronous updates on convergence.
6. Implement a sliding-window training strategy for drift tracking.

7. Study the effect of learning rate schedules under temporal drift.
8. Propose a personalization strategy for drifting client models.
9. Design an experiment to distinguish concept drift from concept shift.
10. Extend the worked example in this chapter to $t + m\Delta$ analytically.

8 Techniques for Handling Non-IID Data

A defining challenge of federated learning lies in the presence of *non-independent and non-identically distributed (non-IID) data* across participating clients. In contrast to centralized machine learning—where training samples are typically assumed to be drawn independently from a single, stationary probability distribution—federated learning operates over a population of autonomous data sources whose statistical properties may differ substantially. Each client generates data locally, shaped by its own usage patterns, environmental conditions, sensing hardware, institutional constraints, and temporal evolution. As a result, the collective dataset underlying federated learning is not a simple aggregation of IID samples, but rather a heterogeneous ensemble of distributions.

Formally, let $P_k(x, y)$ denote the joint data distribution at client k. In centralized learning, it is common to assume

$$P_k(x, y) = P(x, y) \quad \text{for all } k,$$

which enables classical empirical risk minimization theory. Federated learning, however, must contend with the more general and realistic setting

$$P_k(x, y) \neq P_j(x, y) \quad \text{for } k \neq j,$$

reflecting heterogeneity across clients. This deviation from the IID assumption is not incidental, but intrinsic to the federated paradigm.

Such heterogeneity manifests in multiple, often overlapping, forms. Differences in user behavior lead to skewed label distributions, known as *prior probability shift*. Variations in sensing conditions or feature spaces give rise to *covariate shift*. Long-lived federated systems additionally experience *temporal non-stationarity*, where the relationship between inputs and outputs evolves over time, resulting in concept drift or even concept shift. Importantly, these forms of non-IID data may occur simultaneously, compounding their effects on learning dynamics.

P. P. Ray, *Foundations of Federated Learning*,
Springer Series in the Data Sciences,
https://doi.org/10.1007/978-3-032-34084-9_8

From an optimization perspective, the presence of non-IID data fundamentally alters the relationship between local learning objectives and the global objective. Let $F_k(\theta)$ denote the population risk minimized by client k, and let

$$F(\theta) = \sum_{k=1}^{K} \frac{n_k}{N} F_k(\theta)$$

denote the global federated objective. Under IID assumptions, local gradients $\nabla F_k(\theta)$ provide unbiased estimates of the global gradient $\nabla F(\theta)$. Under non-IID data, however, this property no longer holds:

$$\mathbb{E}\left[\nabla F_k(\theta)\right] \neq \nabla F(\theta),$$

even when expectations are taken over local sampling randomness.

Consequently, naïve aggregation of local updates does not merely introduce additional variance, as in classical stochastic optimization, but instead induces *persistent gradient bias*. This bias cannot be eliminated through averaging alone and may cause slow convergence, oscillatory behavior, or divergence of the global model. As demonstrated in earlier chapters, phenomena such as prior probability shift, severe data imbalance, and concept drift can each independently invalidate the assumptions underlying classical empirical risk minimization, even when communication is perfect and all clients are honest.

These observations motivate a fundamental rethinking of optimization in federated learning. Rather than viewing heterogeneity as a nuisance that vanishes with more data or more communication rounds, federated learning must explicitly acknowledge and manage diversity across clients. The goal is not to force statistical homogeneity, which is neither realistic nor desirable, but to design learning algorithms that remain stable, efficient, and equitable in the presence of structured disagreement.

This chapter presents a collection of principled algorithmic techniques for *mitigating*—though not eliminating—the adverse effects of non-IID data in federated learning. These techniques span adaptive aggregation rules, architectural choices, regularization strategies, clustering-based approaches, and robust optimization formulations. Rather than attempting to restore IID conditions, these methods aim to stabilize learning dynamics, reduce gradient disagreement, and ensure acceptable performance across heterogeneous clients. Together, they redefine convergence in federated learning not as strict agreement, but as controlled coordination under persistent statistical diversity.

8.1 Client Weighting and Adaptive Aggregation

One of the earliest and most influential algorithms in federated learning is *Federated Averaging (FedAvg)*, which combines local model updates produced by distributed clients into a single global model. In its simplest form, FedAvg performs aggregation using either uniform weights or weights proportional to the number of local training samples held by each client. This strategy is statistically well motivated when all clients draw their data independently from the same underlying distribution.

However, as emphasized throughout this book, the IID assumption rarely holds in realistic federated settings. Client data are often shaped by user preferences, sensing environments, institutional practices, and temporal effects, leading to systematic differences in both feature distributions and label statistics. Under such non-IID conditions, naïve aggregation schemes can amplify bias rather than reduce noise.

Let $\Delta\theta_k^{(t)}$ denote the update computed by client k during communication round t, starting from the global model $\theta^{(t)}$. In standard federated averaging, the server computes the next global model as

$$\theta^{(t+1)} = \sum_{k=1}^{K} w_k \theta_k^{(t)}, \qquad w_k = \frac{n_k}{\sum_{j=1}^{K} n_j},$$

where n_k denotes the number of training samples available at client k, and K is the total number of participating clients.

This aggregation rule is equivalent to minimizing a data-weighted empirical risk across clients and implicitly assumes that the contribution of each client should scale linearly with its dataset size. While reasonable in centralized learning, this assumption becomes problematic when local objectives differ.

More precisely, this formulation relies on two implicit assumptions:

- **Gradient alignment:** Local gradients $\nabla \hat{F}_k(\theta)$ are assumed to be aligned with the global gradient $\nabla \hat{F}(\theta)$ in expectation.
- **Data volume as reliability:** Clients with larger datasets are assumed to provide more reliable and informative descent directions.

Under non-IID data, both assumptions can fail simultaneously. Clients may hold systematically biased label distributions, observe distinct regions of the feature space, or experience concept drift at different times. As a result, local gradients may point in conflicting directions, causing aggregation to produce a compromised update that is suboptimal for most—or even all—clients.

Key Insight: Why Uniform Weighting Fails
Even when clients contribute equal numbers of samples, their gradients can differ substantially in direction due to skewed label distributions, covariate shift, or temporal drift. Uniform or purely size-based aggregation therefore does not guarantee fairness, stability, or optimality in federated learning.

Adaptive aggregation. To mitigate the adverse effects of gradient disagreement, federated learning systems increasingly rely on *adaptive aggregation* schemes.

Rather than assigning fixed weights based solely on dataset size, adaptive methods dynamically adjust client weights according to the quality, consistency, or reliability of their updates.

A common principle is to reduce the influence of clients whose gradients deviate strongly from the global trend. One representative formulation assigns weights inversely proportional to the deviation between local and global gradients:

$$w_k^{(t)} \propto \frac{1}{\left\| \nabla \hat{F}_k(\theta^{(t)}) - \nabla \hat{F}(\theta^{(t)}) \right\| + \epsilon},$$

where $\epsilon > 0$ is a small constant introduced to ensure numerical stability and prevent division by zero.

Intuitively, this weighting scheme favors clients whose updates are more consistent with the collective descent direction, while suppressing updates that behave as statistical outliers. Importantly, such suppression is not permanent: as client behavior changes over time, weights can be updated dynamically to reflect new patterns.

From an optimization standpoint, adaptive aggregation reduces the effective gradient bias introduced by non-IID data. Rather than averaging conflicting signals, the server selectively attenuates updates that would otherwise destabilize learning. This leads to smoother convergence trajectories and improved robustness under heterogeneous data distributions.

Example: Adaptive Weighting in Practice
In mobile keyboard prediction, users with highly personalized vocabularies often generate gradients that differ systematically from the global population trend. Adaptive aggregation downweights these updates during global model training to prevent instability, while still allowing personalization through local fine-tuning.

Broader perspective. Client weighting and adaptive aggregation represent a fundamental shift in how federated learning systems interpret heterogeneity. Rather than treating all client updates as equally trustworthy, these methods acknowledge that reliability is context dependent and may vary across clients and over time. As later sections will show, adaptive aggregation often serves as a foundation for more advanced techniques, including client clustering, robust optimization, and personalization-aware learning.

Worked Example: Adaptive Aggregation under Non-IID Client Gradients
We now present a concrete numerical example to illustrate how adaptive aggregation mitigates the effect of non-IID data by reweighting client updates based on gradient alignment. All calculations are shown explicitly.

Federated setup. Consider a federated learning system with $K = 3$ clients participating in a single communication round t. Each client trains a scalar logistic regression model

$$f_\theta(x) = \sigma(\theta x), \qquad \sigma(z) = \frac{1}{1 + e^{-z}},$$

using binary labels $y \in \{0, 1\}$ and the logistic loss

$$\ell(f_\theta(x), y) = -y \log \sigma(\theta x) - (1 - y) \log(1 - \sigma(\theta x)).$$

We evaluate all gradients at the current global model

$$\theta^{(t)} = 0.$$

Since $\sigma(0) = 0.5$, this choice simplifies calculations while retaining full generality.

Client datasets. Each client holds $n_k = 4$ samples. The local datasets are shown in Table 8.1.

Although all clients observe identical input distributions, their label distributions differ significantly, inducing non-IID behavior.

Local gradient computation. The gradient of the empirical risk for client k is

$$\nabla \hat{F}_k(\theta) = \frac{1}{n_k} \sum_{i=1}^{n_k} (\sigma(\theta x_{k,i}) - y_{k,i}) x_{k,i}.$$

Evaluating at $\theta^{(t)} = 0$, we have

$$\sigma(0 \cdot x_{k,i}) = 0.5 \quad \forall\, i, k.$$

Client 1.

$$\begin{aligned}
\nabla \hat{F}_1(0) &= \frac{1}{4}\Big[(0.5 - 0) \cdot 1 + (0.5 - 0) \cdot 2 + (0.5 - 0) \cdot 3 + (0.5 - 1) \cdot 4\Big] \\
&= \frac{1}{4}(0.5 + 1 + 1.5 - 2) \\
&= \frac{1}{4}(1.0) \\
&= 0.25.
\end{aligned}$$

Table 8.1 Client datasets for adaptive aggregation example

Client k	Inputs $x_{k,i}$	Labels $y_{k,i}$
1	$\{1, 2, 3, 4\}$	$\{0, 0, 0, 1\}$
2	$\{1, 2, 3, 4\}$	$\{0, 1, 1, 1\}$
3	$\{1, 2, 3, 4\}$	$\{1, 1, 1, 1\}$

Client 2.

$$\begin{aligned}\nabla \hat{F}_2(0) &= \frac{1}{4}\Big[(0.5-0)\cdot 1 + (0.5-1)\cdot 2 + (0.5-1)\cdot 3 + (0.5-1)\cdot 4\Big]\\ &= \frac{1}{4}(0.5-1-1.5-2)\\ &= \frac{-4.0}{4}\\ &= -1.0.\end{aligned}$$

Client 3.

$$\begin{aligned}\nabla \hat{F}_3(0) &= \frac{1}{4}\Big[(0.5-1)\cdot 1 + (0.5-1)\cdot 2 + (0.5-1)\cdot 3 + (0.5-1)\cdot 4\Big]\\ &= \frac{1}{4}(-0.5-1-1.5-2)\\ &= \frac{-5.0}{4}\\ &= -1.25.\end{aligned}$$

Standard FedAvg aggregation. Since all clients have equal sample sizes, FedAvg assigns uniform weights:

$$w_k = \frac{1}{3}.$$

The aggregated gradient is therefore

$$\begin{aligned}\nabla \hat{F}(0) &= \frac{1}{3}(0.25-1.0-1.25)\\ &= \frac{-2.0}{3}\\ &= -0.6667.\end{aligned}$$

This update is dominated by Clients 2 and 3, whose gradients reflect strongly skewed label distributions.

Adaptive aggregation. We now apply adaptive weighting based on gradient deviation. First compute the global gradient estimate:

$$\nabla \hat{F}(0) = -0.6667.$$

Next compute deviations:

$$\|\nabla\hat{F}_1 - \nabla\hat{F}\| = |0.25 + 0.6667| = 0.9167,$$
$$\|\nabla\hat{F}_2 - \nabla\hat{F}\| = |-1.0 + 0.6667| = 0.3333,$$
$$\|\nabla\hat{F}_3 - \nabla\hat{F}\| = |-1.25 + 0.6667| = 0.5833.$$

Using

$$w_k \propto \frac{1}{\|\nabla\hat{F}_k - \nabla\hat{F}\| + \epsilon}, \qquad \epsilon = 0.01,$$

we obtain unnormalized weights:

$$\tilde{w}_1 = \frac{1}{0.9267} = 1.079,$$
$$\tilde{w}_2 = \frac{1}{0.3433} = 2.913,$$
$$\tilde{w}_3 = \frac{1}{0.5933} = 1.685.$$

Normalizing,

$$w_k = \frac{\tilde{w}_k}{\sum_j \tilde{w}_j},$$

yields

$$w_1 = 0.20, \quad w_2 = 0.53, \quad w_3 = 0.27.$$

Adaptively aggregated gradient. The final aggregated gradient becomes

$$\begin{aligned}\nabla\hat{F}_{\text{adaptive}}(0) &= 0.20(0.25) + 0.53(-1.0) + 0.27(-1.25)\\ &= 0.05 - 0.53 - 0.34\\ &= -0.82.\end{aligned}$$

Interpretation. Adaptive aggregation selectively adjusts client influence based on update consistency rather than dataset size alone. Clients whose gradients deviate strongly from the collective trend are attenuated, leading to more controlled descent behavior. This example demonstrates how adaptive weighting reshapes global updates under non-IID data without requiring changes to local training procedures.

Worked Example: Adaptive Aggregation under Non-IID Client Gradients
We now present a concrete numerical example to illustrate how adaptive aggregation mitigates the effect of non-IID data by reweighting client updates based on gradient alignment. All calculations are shown explicitly.

Federated setup. Consider a federated learning system with $K = 3$ clients participating in a single communication round t. Each client trains a scalar logistic regression model

$$f_\theta(x) = \sigma(\theta x), \qquad \sigma(z) = \frac{1}{1 + e^{-z}},$$

using binary labels $y \in \{0, 1\}$ and the logistic loss

$$\ell(f_\theta(x), y) = -y \log \sigma(\theta x) - (1 - y) \log(1 - \sigma(\theta x)).$$

We evaluate all gradients at the current global model

$$\theta^{(t)} = 0.$$

Since $\sigma(0) = 0.5$, this choice simplifies calculations while retaining full generality.

Client datasets. Each client holds $n_k = 4$ samples. The local datasets are shown in Table 8.2.

Although all clients observe identical input distributions, their label distributions differ significantly, inducing non-IID behavior.

Local gradient computation. The gradient of the empirical risk for client k is

$$\nabla \hat{F}_k(\theta) = \frac{1}{n_k} \sum_{i=1}^{n_k} (\sigma(\theta x_{k,i}) - y_{k,i}) x_{k,i}.$$

Evaluating at $\theta^{(t)} = 0$, we have

$$\sigma(0 \cdot x_{k,i}) = 0.5 \quad \forall\, i, k.$$

Table 8.2 Client datasets for adaptive aggregation example

Client k	Inputs $x_{k,i}$	Labels $y_{k,i}$
1	$\{1, 2, 3, 4\}$	$\{0, 0, 0, 1\}$
2	$\{1, 2, 3, 4\}$	$\{0, 1, 1, 1\}$
3	$\{1, 2, 3, 4\}$	$\{1, 1, 1, 1\}$

Client 1.

$$\begin{aligned}\nabla\hat{F}_1(0) &= \frac{1}{4}\Big[(0.5-0)\cdot 1 + (0.5-0)\cdot 2 + (0.5-0)\cdot 3 + (0.5-1)\cdot 4\Big] \\ &= \frac{1}{4}(0.5+1+1.5-2) \\ &= \frac{1}{4}(1.0) \\ &= 0.25.\end{aligned}$$

Client 2.

$$\begin{aligned}\nabla\hat{F}_2(0) &= \frac{1}{4}\Big[(0.5-0)\cdot 1 + (0.5-1)\cdot 2 + (0.5-1)\cdot 3 + (0.5-1)\cdot 4\Big] \\ &= \frac{1}{4}(0.5-1-1.5-2) \\ &= \frac{-4.0}{4} \\ &= -1.0.\end{aligned}$$

Client 3.

$$\begin{aligned}\nabla\hat{F}_3(0) &= \frac{1}{4}\Big[(0.5-1)\cdot 1 + (0.5-1)\cdot 2 + (0.5-1)\cdot 3 + (0.5-1)\cdot 4\Big] \\ &= \frac{1}{4}(-0.5-1-1.5-2) \\ &= \frac{-5.0}{4} \\ &= -1.25.\end{aligned}$$

Standard FedAvg aggregation. Since all clients have equal sample sizes, FedAvg assigns uniform weights:

$$w_k = \frac{1}{3}.$$

The aggregated gradient is therefore

$$\begin{aligned}\nabla\hat{F}(0) &= \frac{1}{3}(0.25-1.0-1.25) \\ &= \frac{-2.0}{3} \\ &= -0.6667.\end{aligned}$$

This update is dominated by Clients 2 and 3, whose gradients reflect strongly skewed label distributions.

Adaptive aggregation. We now apply adaptive weighting based on gradient deviation. First compute the global gradient estimate:

$$\nabla\hat{F}(0) = -0.6667.$$

Next compute deviations:

$$\begin{aligned}
\|\nabla\hat{F}_1 - \nabla\hat{F}\| &= |0.25 + 0.6667| = 0.9167, \\
\|\nabla\hat{F}_2 - \nabla\hat{F}\| &= |-1.0 + 0.6667| = 0.3333, \\
\|\nabla\hat{F}_3 - \nabla\hat{F}\| &= |-1.25 + 0.6667| = 0.5833.
\end{aligned}$$

Using

$$w_k \propto \frac{1}{\|\nabla\hat{F}_k - \nabla\hat{F}\| + \epsilon}, \qquad \epsilon = 0.01,$$

we obtain unnormalized weights:

$$\begin{aligned}
\tilde{w}_1 &= \frac{1}{0.9267} = 1.079, \\
\tilde{w}_2 &= \frac{1}{0.3433} = 2.913, \\
\tilde{w}_3 &= \frac{1}{0.5933} = 1.685.
\end{aligned}$$

Normalizing,

$$w_k = \frac{\tilde{w}_k}{\sum_j \tilde{w}_j},$$

yields

$$w_1 = 0.20, \quad w_2 = 0.53, \quad w_3 = 0.27.$$

Adaptively aggregated gradient. The final aggregated gradient becomes

$$\begin{aligned}
\nabla\hat{F}_{\text{adaptive}}(0) &= 0.20(0.25) + 0.53(-1.0) + 0.27(-1.25) \\
&= 0.05 - 0.53 - 0.34 \\
&= -0.82.
\end{aligned}$$

Interpretation. Adaptive aggregation selectively adjusts client influence based on update consistency rather than dataset size alone. Clients whose gradients deviate strongly from the collective trend are attenuated, leading to more controlled descent behavior. This example demonstrates how adaptive weighting reshapes global updates under non-IID data without requiring changes to local training procedures.

8.2 Normalization Techniques Beyond Batch Normalization

Normalization is a fundamental architectural component in modern deep neural networks, introduced to stabilize optimization, accelerate convergence, and reduce sensitivity to parameter initialization. In centralized learning, normalization layers implicitly rely on the assumption that mini-batches are representative samples drawn from a single underlying data distribution. Under this assumption, statistics computed at the batch level provide a reasonable approximation to population-level moments.

In federated learning, however, this assumption is systematically violated. Each client computes updates using local mini-batches drawn from its *own* data distribution, which may differ substantially from those of other clients. Consequently, batch-level statistics become *client-specific*, time-varying, and often incompatible across participants. This mismatch renders conventional Batch Normalization (BN) fragile and can significantly degrade convergence and final model performance under non-IID data.

Why Batch Normalization fails in federated settings. Batch Normalization normalizes activations using batch-level mean and variance:

$$\hat{h}_i = \frac{h_i - \mu_{\text{batch}}}{\sigma_{\text{batch}}}, \qquad \mu_{\text{batch}} = \frac{1}{B}\sum_{j=1}^{B} h_j,$$

where B denotes the batch size. In federated learning, two fundamental problems arise:

- **Distribution mismatch:** Batch statistics computed on one client reflect its local data distribution and are not representative of the global population.
- **Inconsistent aggregation:** BN parameters and running statistics aggregated across clients represent no coherent distribution, leading to unstable or even divergent behavior during inference.

As a result, BN can introduce systematic bias rather than reducing variance, amplifying the adverse effects of data heterogeneity.

To address these limitations, federated learning systems increasingly adopt *normalization techniques that do not depend on batch-level statistics*. These alternatives compute normalization statistics within individual samples or predefined structural units of the network, making them inherently more robust to client-level heterogeneity.

8.2.1 Layer Normalization

Layer Normalization (LN) normalizes activations across feature dimensions within a single sample, rather than across samples in a batch. Given a hidden activation vector

$$\mathbf{h} = (h_1, h_2, \dots, h_d),$$

LN computes

$$\mu_{\text{layer}} = \frac{1}{d}\sum_{i=1}^{d} h_i, \qquad \sigma_{\text{layer}} = \sqrt{\frac{1}{d}\sum_{i=1}^{d}(h_i - \mu_{\text{layer}})^2},$$

and normalizes each activation as

$$\hat{h}_i = \frac{h_i - \mu_{\text{layer}}}{\sigma_{\text{layer}}}.$$

Because these statistics are computed *within a single sample*, Layer Normalization is completely independent of batch size and batch composition. This property makes LN particularly well suited to federated learning, where batch sizes may be small, uneven, or client-dependent.

Interpretation. Layer Normalization enforces a consistent internal representation for each sample, ensuring that feature scales remain comparable even when clients observe highly diverse data distributions. As a result, local optimization becomes more stable and global aggregation less sensitive to client-specific shifts.

8.2.2 Group Normalization

Group Normalization (GN) generalizes Layer Normalization by partitioning feature channels into G disjoint groups and normalizing within each group. For a layer with C channels, each group contains C/G channels, and normalization is applied independently to each group.

Formally, for group g, GN computes

$$\mu_g = \frac{1}{|g|}\sum_{i \in g} h_i, \qquad \sigma_g = \sqrt{\frac{1}{|g|}\sum_{i \in g}(h_i - \mu_g)^2},$$

followed by

$$\hat{h}_i = \frac{h_i - \mu_g}{\sigma_g}, \qquad i \in g.$$

Group Normalization strikes a balance between expressiveness and robustness: it retains some of the representational flexibility of Batch Normalization while avoiding reliance on batch-level statistics.

Practical relevance. GN has proven especially effective in federated vision and speech models, where convolutional layers dominate and channel-wise structure is important. By normalizing over groups of channels, GN preserves spatial coherence while remaining resilient to non-IID client distributions.

Practical Recommendation
In federated neural networks, replacing Batch Normalization with Layer Normalization or Group Normalization often yields significantly more stable training dynamics under non-IID data. When batch sizes are small or client distributions differ sharply, BN should be avoided in favor of normalization methods that decouple statistics from batch composition.

8.2.3 Instance Normalization

Instance Normalization (IN) was originally introduced in the context of style transfer and image generation, where robustness to instance-level appearance variations is critical. In federated learning, Instance Normalization provides another batch-independent alternative to Batch Normalization, with normalization statistics computed at the level of *individual samples and individual channels.*

Core idea. Unlike Layer Normalization, which aggregates statistics across all features of a sample, Instance Normalization normalizes each feature channel independently within a single instance. This makes IN maximally local: normalization depends neither on batch composition nor on interactions across channels.

Formal definition. Consider a convolutional feature map for a single input instance

$$\mathbf{h} \in \mathbb{R}^{C \times H \times W},$$

where C denotes the number of channels and $H \times W$ the spatial resolution. For each channel $c \in \{1, \dots, C\}$, Instance Normalization computes

$$\mu_c = \frac{1}{HW}\sum_{i=1}^{H}\sum_{j=1}^{W} h_{cij}, \qquad \sigma_c = \sqrt{\frac{1}{HW}\sum_{i=1}^{H}\sum_{j=1}^{W}(h_{cij} - \mu_c)^2},$$

and normalizes activations as

$$\hat{h}_{cij} = \frac{h_{cij} - \mu_c}{\sigma_c}, \qquad \forall i, j.$$

These statistics are computed independently for each instance and each channel, making IN completely invariant to batch size, batch ordering, and client-level data distribution.

Interpretation. Instance Normalization enforces channel-wise contrast normalization within each sample. As a result, absolute feature magnitudes and global intensity information are suppressed, while relative spatial patterns within each channel are preserved. This property is particularly beneficial when large inter-client variations exist in signal scale, illumination, or sensor calibration.

Behavior under federated heterogeneity. From a federated learning perspective, IN offers several advantages:

- **Client invariance:** Normalization statistics are entirely local to each instance, ensuring that no client-specific information is embedded in normalization parameters.
- **Aggregation compatibility:** Since IN does not maintain running statistics, model parameters remain compatible under federated averaging.
- **Stability under extreme non-IID data:** IN is robust even when clients observe drastically different input scales or appearance distributions.

Limitations. Despite its robustness, Instance Normalization is not universally optimal. By normalizing each channel independently, IN removes global instance-level information that may be semantically meaningful. In discriminative tasks, this can reduce representational power and harm classification accuracy.

Comparison with other normalization techniques. Conceptually, IN can be viewed as a special case of Group Normalization with group size equal to one channel. While LN preserves cross-feature relationships and GN balances locality and expressiveness, IN prioritizes invariance at the cost of global context.

When to Use Instance Normalization
Instance Normalization is particularly effective in federated learning scenarios involving strong appearance variability, such as image style transfer, medical imaging across institutions, or sensor data with inconsistent calibration. For semantic classification tasks, however, LN or GN often provide a better trade-off between robustness and expressiveness.

Federated design insight. Instance Normalization exemplifies an extreme point in the normalization design space: maximum robustness to heterogeneity through maximum locality of statistics. Its effectiveness in federated learning highlights a broader principle: the more normalization relies on shared statistics, the more fragile it becomes under non-IID data.

Broader perspective. Normalization in federated learning should be viewed not merely as an optimization trick, but as a structural design choice that determines how local representations interact with global aggregation. Choosing normalization schemes that respect client autonomy and data heterogeneity is therefore essential for scalable and robust federated systems.

Table 8.3 highlights that normalization techniques differ fundamentally in how their statistics are computed and how sensitively they respond to client-level distributional heterogeneity. In federated learning, batch-independent methods such as Layer Normalization and Group Normalization provide more stable and consistent training behavior than Batch Normalization, whose reliance on client-specific batch statistics exacerbates non-IID effects and impairs convergence.

Worked Example: Behavior of Different Normalization Techniques under Non-IID Data
We now present a detailed numerical example to illustrate how Batch Normalization (BN), Layer Normalization (LN), Group Normalization (GN), and Instance Normalization (IN) behave in a federated learning setting with non-IID client data. All computations are carried out explicitly using the formal definitions introduced earlier.

Federated setup. Consider a federated system with $K = 2$ clients participating in one communication round. Each client trains the same neural network layer that outputs a 4-dimensional activation vector

$$\mathbf{h} = (h_1, h_2, h_3, h_4).$$

Due to non-IID data, the scale and distribution of activations differ across clients.

Client-specific mini-batches. Assume each client uses a local mini-batch of size $B = 3$, producing the activations shown in Table 8.4.

Table 8.3 Comparison of normalization techniques in federated learning under non-IID data

Normalization Technique	Statistics Computed Over	Dependence on Batch Size	Behavior under Non-IID Federated Data
Batch Normalization (BN)	Mini-batch samples (mean and variance across batch)	Strongly dependent on batch size and batch composition	Poor robustness; batch statistics are client-specific, leading to inconsistent normalization, unstable aggregation, and degraded convergence under heterogeneity
Layer Normalization (LN)	Feature dimensions within a single sample	Independent of batch size	Highly robust; normalization is sample-wise, making it insensitive to client-level distributional differences and suitable for heterogeneous federated settings
Group Normalization (GN)	Groups of feature channels within a sample	Independent of batch size	Robust and expressive; balances stability and representation capacity, performs well in federated vision and speech models with non-IID data
Instance Normalization (IN)	All feature channels within a single instance	Independent of batch size	Very robust to heterogeneity but may remove useful global contrast information; often beneficial in style-sensitive tasks rather than general-purpose federated learning

These discrepancies may arise from heterogeneous sensing environments, user populations, or device characteristics.

Batch Normalization (BN). Batch Normalization computes normalization statistics jointly across all samples and all feature dimensions within a mini-batch. For a mini-batch of size B and feature dimension d, the batch mean and variance are defined as

$$\mu_{\text{BN}} = \frac{1}{Bd}\sum_{i=1}^{B}\sum_{j=1}^{d} h_{ij}, \qquad \sigma^2_{\text{BN}} = \frac{1}{Bd}\sum_{i=1}^{B}\sum_{j=1}^{d}(h_{ij} - \mu_{\text{BN}})^2.$$

Table 8.4 Client-specific activations under non-IID data

Client	Activations in mini-batch
Client 1	$\{(2,4,6,8),\ (4,6,8,10),\ (6,8,10,12)\}$
Client 2	$\{(20,24,28,32),\ (22,26,30,34),\ (24,28,32,36)\}$

Client 1. Client 1 observes the following mini-batch activations:

$$\{(2, 4, 6, 8),\ (4, 6, 8, 10),\ (6, 8, 10, 12)\}.$$

Flattening across samples and dimensions yields the set

$$\{2, 4, 6, 8,\ 4, 6, 8, 10,\ 6, 8, 10, 12\}.$$

Step 1: Compute the batch mean.

$$\begin{aligned}\mu_{\text{BN}}^{(1)} &= \frac{1}{12}(2+4+6+8+4+6+8+10+6+8+10+12)\\ &= \frac{1}{12}(84) = 7.\end{aligned}$$

Step 2: Compute the squared deviations.

$$\begin{array}{llll}(2-7)^2 = 25, & (4-7)^2 = 9, & (6-7)^2 = 1, & (8-7)^2 = 1,\\ (4-7)^2 = 9, & (6-7)^2 = 1, & (8-7)^2 = 1, & (10-7)^2 = 9,\\ (6-7)^2 = 1, & (8-7)^2 = 1, & (10-7)^2 = 9, & (12-7)^2 = 25.\end{array}$$

Step 3: Compute the batch variance.

$$\begin{aligned}\sigma_{\text{BN}}^{2,(1)} &= \frac{1}{12}(25+9+1+1+9+1+1+9+1+1+9+25)\\ &= \frac{120}{12} = 10.\end{aligned}$$

Step 4: Compute the batch standard deviation.

$$\sigma_{\text{BN}}^{(1)} = \sqrt{10} \approx 3.16.$$

Client 2. Client 2 observes a mini-batch with substantially larger activation magnitudes:

$$\{(20, 24, 28, 32),\ (22, 26, 30, 34),\ (24, 28, 32, 36)\}.$$

Flattening yields

$$\{20, 24, 28, 32,\ 22, 26, 30, 34,\ 24, 28, 32, 36\}.$$

Step 1: Compute the batch mean.

$$\mu_{\text{BN}}^{(2)} = \frac{1}{12}(20 + 24 + 28 + 32 + 22 + 26 + 30 + 34 + 24 + 28 + 32 + 36)$$
$$= \frac{336}{12} = 28.$$

Step 2: Compute the squared deviations.

$$\begin{aligned}
&(20-28)^2 = 64, \quad (24-28)^2 = 16, \quad (28-28)^2 = 0, \quad (32-28)^2 = 16,\\
&(22-28)^2 = 36, \quad (26-28)^2 = 4, \quad (30-28)^2 = 4, \quad (34-28)^2 = 36,\\
&(24-28)^2 = 16, \quad (28-28)^2 = 0, \quad (32-28)^2 = 16, \quad (36-28)^2 = 64.
\end{aligned}$$

Step 3: Compute the batch variance.

$$\sigma_{\text{BN}}^{2,(2)} = \frac{1}{12}(64 + 16 + 0 + 16 + 36 + 4 + 4 + 36 + 16 + 0 + 16 + 64)$$
$$= \frac{240}{12} = 20.$$

Step 4: Compute the batch standard deviation.

$$\sigma_{\text{BN}}^{(2)} = \sqrt{20} \approx 4.47.$$

Didactic conclusion. Although both clients train the same model architecture, Batch Normalization produces *client-specific normalization statistics* $(\mu_{\text{BN}}, \sigma_{\text{BN}})$. When such models are aggregated in federated learning, these incompatible statistics are mixed, leading to unstable training dynamics and degraded global performance.

Observation. Although the variance coincidentally matches, the means differ substantially. Each client embeds its own statistics into the normalization layer, making BN parameters incompatible during aggregation.

Batch Normalization Failure
In federated learning, Batch Normalization ties model behavior to client-specific batch statistics. Federated averaging therefore mixes incompatible normalization states.

Layer Normalization (LN). Layer Normalization computes normalization statistics *independently for each individual sample*, aggregating only across feature dimensions. Unlike Batch Normalization, no interaction occurs across samples or clients.

For a sample with feature dimension d, the layer mean and standard deviation are defined as

$$\mu_{\mathrm{LN}} = \frac{1}{d}\sum_{j=1}^{d} h_j, \qquad \sigma_{\mathrm{LN}} = \sqrt{\frac{1}{d}\sum_{j=1}^{d}(h_j - \mu_{\mathrm{LN}})^2}.$$

Client 1, sample (2, 4, 6, 8).

Step 1: Compute the layer mean.

$$\mu_{\mathrm{LN}}^{(1)} = \frac{1}{4}(2+4+6+8) = \frac{20}{4} = 5.$$

Step 2: Compute the squared deviations.

$$(2-5)^2 = 9, \quad (4-5)^2 = 1, \quad (6-5)^2 = 1, \quad (8-5)^2 = 9.$$

Step 3: Compute the layer variance.

$$\sigma_{\mathrm{LN}}^{2,(1)} = \frac{1}{4}(9+1+1+9) = \frac{20}{4} = 5.$$

Step 4: Compute the layer standard deviation.

$$\sigma_{\mathrm{LN}}^{(1)} = \sqrt{5}.$$

Client 2, sample (20, 24, 28, 32).

Step 1: Compute the layer mean.

$$\mu_{\mathrm{LN}}^{(2)} = \frac{1}{4}(20+24+28+32) = \frac{104}{4} = 26.$$

Step 2: Compute the squared deviations.

$$(20 - 26)^2 = 36,$$
$$(24 - 26)^2 = 4,$$
$$(28 - 26)^2 = 4,$$
$$(32 - 26)^2 = 36.$$

Step 3: Compute the layer variance.

$$\begin{aligned}\sigma_{\mathrm{LN}}^{2,(2)} &= \frac{1}{4}(36 + 4 + 4 + 36)\\ &= \frac{80}{4} = 20.\end{aligned}$$

Step 4: Compute the layer standard deviation.

$$\sigma_{\mathrm{LN}}^{(2)} = \sqrt{20}.$$

Didactic conclusion. Although the two clients observe activations at very different absolute scales, Layer Normalization produces *structurally identical normalized representations* within each sample. Normalization depends only on intra-sample feature relationships, not on batch composition or client identity.

As a result, Layer Normalization is inherently compatible with federated aggregation and remains stable under non-IID data.

Observation. Despite different absolute scales, normalized outputs have identical relative structure. Normalization is independent of batch composition and client identity.

Why Layer Normalization is Federated-Friendly
Layer Normalization ensures per-sample stability, making updates comparable across clients under non-IID data.

Group Normalization (GN). Group Normalization generalizes Layer Normalization by computing normalization statistics over *groups of feature dimensions* rather than over the entire feature vector. Unlike Batch Normalization, GN does not depend on batch size or batch composition, and unlike LN, it preserves partial channel-wise structure.

Assume the feature dimension is

$$d = 4,$$

and the features are divided into

$$G = 2 \quad \text{groups,}$$

each containing $|g| = 2$ features:

$$g_1 = \{h_1, h_2\}, \qquad g_2 = \{h_3, h_4\}.$$

For each group g, Group Normalization computes

$$\mu_g = \frac{1}{|g|}\sum_{j\in g} h_j, \qquad \sigma_g = \sqrt{\frac{1}{|g|}\sum_{j\in g}(h_j - \mu_g)^2}.$$

Each activation h_j is then normalized using the statistics of its corresponding group.

Client 1, sample (2,4,6,8).

Step 1: Compute statistics for group $g_1 = \{h_1, h_2\} = \{2, 4\}$.
Group mean.

$$\mu_{g_1}^{(1)} = \frac{1}{2}(2 + 4) = 3.$$

Squared deviations.

$$(2 - 3)^2 = 1, \qquad (4 - 3)^2 = 1.$$

Group variance and standard deviation.

$$\sigma_{g_1}^{2,(1)} = \frac{1}{2}(1 + 1) = 1,$$
$$\sigma_{g_1}^{(1)} = \sqrt{1} = 1.$$

Step 2: Compute statistics for group $g_2 = \{h_3, h_4\} = \{6, 8\}$.
Group mean.

$$\mu_{g_2}^{(1)} = \frac{1}{2}(6 + 8) = 7.$$

Squared deviations.

$$(6 - 7)^2 = 1, \qquad (8 - 7)^2 = 1.$$

Group variance and standard deviation.

$$\sigma_{g_2}^{2,(1)} = \frac{1}{2}(1+1) = 1,$$

$$\sigma_{g_2}^{(1)} = \sqrt{1} = 1.$$

Client 2, sample (20,24,28,32).

Step 1: Compute statistics for group $g_1 = \{20, 24\}$.
Group mean.

$$\mu_{g_1}^{(2)} = \frac{1}{2}(20+24) = 22.$$

Squared deviations.

$$(20-22)^2 = 4, \qquad (24-22)^2 = 4.$$

Group variance and standard deviation.

$$\sigma_{g_1}^{2,(2)} = \frac{1}{2}(4+4) = 4,$$

$$\sigma_{g_1}^{(2)} = \sqrt{4} = 2.$$

Step 2: Compute statistics for group $g_2 = \{28, 32\}$.
Group mean.

$$\mu_{g_2}^{(2)} = \frac{1}{2}(28+32) = 30.$$

Squared deviations.

$$(28-30)^2 = 4, \qquad (32-30)^2 = 4.$$

Group variance and standard deviation.

$$\sigma_{g_2}^{2,(2)} = \frac{1}{2}(4+4) = 4,$$

$$\sigma_{g_2}^{(2)} = \sqrt{4} = 2.$$

Group Normalization produces consistent normalization behavior across clients while preserving partial channel structure. Unlike Batch Normalization, GN does not embed client-specific batch statistics, and unlike full Layer Normalization, it allows finer control over feature interactions.

As a result, Group Normalization provides a principled compromise between stability and representational flexibility in federated neural networks operating under non-IID data.

Observation. GN preserves batch-independence while retaining structured channel-level normalization, making it especially effective in convolutional models.

Instance Normalization (IN). Instance Normalization performs normalization *independently for each feature channel within a single sample*, aggregating statistics over spatial dimensions only. Unlike Batch, Layer, or Group Normalization, IN does not mix information across channels or across samples.

Let a single sample produce activations

$$h_{ij},$$

where j indexes feature channels and i indexes spatial locations. For a feature map of spatial resolution $H \times W$, Instance Normalization computes, for each channel j,

$$\mu_j = \frac{1}{HW}\sum_{i=1}^{HW} h_{ij}, \qquad \sigma_j = \sqrt{\frac{1}{HW}\sum_{i=1}^{HW}(h_{ij} - \mu_j)^2}.$$

Each activation is then normalized as

$$\hat{h}_{ij} = \frac{h_{ij} - \mu_j}{\sigma_j}.$$

Reduction to the vector (non-spatial) case. In the present worked example, each sample consists of a *single activation vector* with no spatial extent. That is,

$$H = W = 1, \qquad i = 1,$$

and each channel has exactly one activation value.

Let the sample activation vector be

$$\mathbf{h} = (h_1, h_2, h_3, h_4).$$

Step 1: Compute channel-wise means.

Since there is only one spatial element per channel,

$$\mu_j = \frac{1}{1}h_j = h_j, \qquad j = 1, \ldots, 4.$$

Step 2: Compute channel-wise variances.

$$\sigma_j^2 = \frac{1}{1}(h_j - \mu_j)^2 = (h_j - h_j)^2 = 0.$$

Thus,

$$\sigma_j = 0 \quad \text{for all channels } j.$$

Step 3: Normalized activations. Substituting into the normalization equation gives

$$\hat{h}_j = \frac{h_j - \mu_j}{\sigma_j} = \frac{0}{0},$$

which is undefined.

In practice, implementations introduce a small constant $\epsilon > 0$ for numerical stability, yielding

$$\hat{h}_j = \frac{h_j - h_j}{\sqrt{\epsilon}} = 0.$$

Didactic interpretation. In the absence of spatial structure, Instance Normalization collapses each channel to a constant value after normalization. All information contained in the original activation magnitudes is removed, leaving only learnable affine parameters to reintroduce variation.

This behavior highlights a crucial limitation:

- Instance Normalization is meaningful only when each channel contains *multiple spatial elements*.
- For vector-valued activations or fully connected layers, IN degenerates and should not be used.

Instance Normalization is inherently batch-independent and therefore immune to client-level batch heterogeneity. However, its applicability is restricted to convolutional architectures with sufficient spatial resolution. In federated vision models, IN can be useful for style invariance, but it is unsuitable as a general replacement for BN in non-convolutional or low-dimensional settings.

Observation. IN aggressively removes instance-level magnitude information, making it robust to appearance variation but potentially harmful for discriminative tasks.

Conclusion of the worked example. This example establishes that:

- Batch Normalization fails due to client-dependent statistics.

- Layer Normalization provides consistent per-sample behavior.
- Group Normalization balances stability and expressiveness.
- Instance Normalization maximizes invariance at the cost of global context.

Pedagogical takeaway. In federated learning, normalization layers must decouple statistical estimation from batch composition. Batch-independent normalization is therefore a *structural requirement*, not merely an optimization choice.

8.3 Data Clustering and Client Grouping

A central limitation of classical federated learning is the implicit assumption that a *single global model* can adequately represent all participating clients. When client data distributions differ significantly, enforcing a single consensus model often leads to systematic underfitting for some clients and overfitting for others. Data clustering and client grouping provide a principled alternative by explicitly acknowledging and exploiting the *structured nature* of heterogeneity.

Motivation. In many real-world federated deployments, client heterogeneity is not arbitrary or random. Instead, clients often form natural groups based on shared characteristics such as geography, user behavior, sensing modalities, institutional protocols, or demographic profiles. Ignoring this structure forces the learning algorithm to compromise across incompatible objectives, degrading both convergence speed and final performance.

Rather than collapsing all clients into a single optimization problem, client grouping seeks to decompose the global population into *homogeneous subsets* within which the IID assumption is approximately restored.

Formal framework. Let $\{1, 2, \ldots, K\}$ denote the set of all clients. Client grouping partitions this set into M disjoint clusters

$$\mathscr{C}_1, \mathscr{C}_2, \ldots, \mathscr{C}_M, \qquad \mathscr{C}_m \cap \mathscr{C}_{m'} = \emptyset \text{ for } m \neq m'.$$

Each cluster $\mathscr{C}_m$ contains clients whose data distributions and optimization behavior are deemed similar according to a chosen similarity criterion. Common similarity measures include:

- cosine similarity between local gradients $\nabla \hat{F}_k(\theta)$,
- distances between update trajectories across rounds,
- similarity between learned representation embeddings, and
- statistical distances between empirical data distributions.

Once clusters are formed, each cluster is assigned its own model θ_m, which is optimized by solving a cluster-specific federated objective:

$$\min_{\theta_m} \frac{1}{|\mathcal{C}_m|} \sum_{k \in \mathcal{C}_m} \hat{F}_k(\theta_m), \qquad m = 1, 2, \ldots, M.$$

Interpretation of the objective. Within each cluster, clients share sufficiently similar data distributions so that local gradients act as approximate unbiased estimators of the cluster-level objective. As a result, standard federated optimization techniques regain much of their effectiveness, yielding faster convergence and more stable training dynamics.

Key Insight: Structured Heterogeneity
Client clustering transforms a single, highly non-IID federated learning problem into multiple quasi-IID subproblems, each of which is amenable to stable and efficient optimization.

Algorithmic considerations. Client grouping can be performed either:

- *statically*, using prior knowledge or historical data to assign clients to fixed clusters, or
- *dynamically*, allowing clients to change clusters over time as their data distributions or behaviors evolve.

Dynamic clustering is particularly important in non-stationary environments, where concept drift may cause clients to migrate between clusters across training rounds.

Communication and computation trade-offs. While clustering introduces additional complexity—such as maintaining multiple models and performing similarity computations—it often reduces overall communication cost by avoiding repeated conflict between incompatible client updates. Moreover, cluster-level aggregation improves model interpretability by associating learned parameters with specific population segments.

Illustrative example. In healthcare federated learning, hospitals naturally cluster according to patient demographics, disease prevalence, or clinical protocols. A single global model may fail to represent both pediatric and geriatric populations effectively. By clustering hospitals into medically coherent groups, federated learning can produce specialized models that achieve higher predictive accuracy while preserving privacy and institutional autonomy.

Broader perspective. Client grouping shifts the goal of federated learning from enforcing universal agreement to enabling *coordinated diversity*. Rather than suppressing heterogeneity, clustering leverages it as a source of structure, leading to more scalable, interpretable, and robust federated systems.

Worked Example: Client Clustering for Structured Non-IID Data
We now present a fully worked, numerical, and didactic example to illustrate how client clustering mitigates non-IID effects in federated learning. All steps are shown explicitly using the formulation introduced in this section.

Federated setup. Consider a federated system with $K = 4$ clients participating in model training. Each client k minimizes a local empirical risk

$$\hat{F}_k(\theta) = \frac{1}{n_k} \sum_{i=1}^{n_k} \ell\big(f_\theta(x_{k,i}), y_{k,i}\big),$$

where $\theta \in \mathbb{R}^2$ denotes the model parameters.

Assume that after a local training round, each client computes a local gradient evaluated at the current global model $\theta^{(t)}$.

Observed client gradients. Suppose the gradients returned by the clients are:

$$\nabla\hat{F}_1(\theta^{(t)}) = (1.0, 0.8), \qquad \nabla\hat{F}_2(\theta^{(t)}) = (0.9, 0.7),$$

$$\nabla\hat{F}_3(\theta^{(t)}) = (-1.1, -0.9), \qquad \nabla\hat{F}_4(\theta^{(t)}) = (-0.8, -1.0).$$

These gradients suggest the presence of two opposing descent directions, indicating structured heterogeneity.

Similarity computation. To cluster clients, we compute pairwise cosine similarity between client gradients. For clients i and j, cosine similarity is defined as

$$\mathrm{sim}(i,j) = \frac{\nabla\hat{F}_i(\theta^{(t)})^\top \nabla\hat{F}_j(\theta^{(t)})}{\|\nabla\hat{F}_i(\theta^{(t)})\| \; \|\nabla\hat{F}_j(\theta^{(t)})\|}.$$

Example calculation: Clients 1 and 2.

$$\nabla\hat{F}_1 \cdot \nabla\hat{F}_2 = (1.0)(0.9) + (0.8)(0.7) = 1.46,$$

$$\|\nabla\hat{F}_1\| = \sqrt{1.0^2 + 0.8^2} = 1.28, \quad \|\nabla\hat{F}_2\| = \sqrt{0.9^2 + 0.7^2} = 1.14.$$

Thus,

$$\text{sim}(1,2) = \frac{1.46}{1.28 \times 1.14} \approx 0.999.$$

Example calculation: Clients 1 and 3.

$$\nabla \hat{F}_1 \cdot \nabla \hat{F}_3 = (1.0)(-1.1) + (0.8)(-0.9) = -1.82,$$

$$\|\nabla \hat{F}_3\| = \sqrt{(-1.1)^2 + (-0.9)^2} = 1.42.$$

Thus,

$$\text{sim}(1,3) = \frac{-1.82}{1.28 \times 1.42} \approx -0.999.$$

We now compute *all remaining pairwise cosine similarities* explicitly, using the same gradients introduced earlier:

$$\nabla \hat{F}_1 = (1.0,\ 0.8), \quad \nabla \hat{F}_2 = (0.9,\ 0.7), \quad \nabla \hat{F}_3 = (-1.1,\ -0.9), \quad \nabla \hat{F}_4 = (-0.8,\ -1.0).$$

For reference, the gradient norms are

$$\|\nabla \hat{F}_1\| = \sqrt{1.0^2 + 0.8^2} = 1.28, \qquad \|\nabla \hat{F}_2\| = \sqrt{0.9^2 + 0.7^2} = 1.14,$$

$$\|\nabla \hat{F}_3\| = \sqrt{(-1.1)^2 + (-0.9)^2} = 1.42, \qquad \|\nabla \hat{F}_4\| = \sqrt{(-0.8)^2 + (-1.0)^2} = 1.28.$$

Client 1 and Client 4.

$$\nabla \hat{F}_1 \cdot \nabla \hat{F}_4 = (1.0)(-0.8) + (0.8)(-1.0) = -1.60.$$

Thus,

$$\text{sim}(1,4) = \frac{-1.60}{1.28 \times 1.28} = \frac{-1.60}{1.6384} \approx -0.976.$$

Client 2 and Client 3.

$$\nabla \hat{F}_2 \cdot \nabla \hat{F}_3 = (0.9)(-1.1) + (0.7)(-0.9) = -1.62.$$

Table 8.5 Pairwise cosine similarities between client gradients

Client Pair (i,j)	$\nabla\hat{F}_i \cdot \nabla\hat{F}_j$	$\|\nabla\hat{F}_i\|\|\nabla\hat{F}_j\|$	sim(i,j)
$(1,2)$	1.46	1.28×1.14	0.999
$(1,3)$	−1.82	1.28×1.42	−0.999
$(1,4)$	−1.60	1.28×1.28	−0.976
$(2,3)$	−1.62	1.14×1.42	−0.999
$(2,4)$	−1.42	1.14×1.28	−0.973
$(3,4)$	1.78	1.42×1.28	0.979

Thus,

$$\text{sim}(2,3) = \frac{-1.62}{1.14 \times 1.42} = \frac{-1.62}{1.6188} \approx -1.001 \approx -0.999.$$

Client 2 and Client 4.

$$\nabla\hat{F}_2 \cdot \nabla\hat{F}_4 = (0.9)(-0.8) + (0.7)(-1.0) = -1.42.$$

Thus,

$$\text{sim}(2,4) = \frac{-1.42}{1.14 \times 1.28} = \frac{-1.42}{1.4592} \approx -0.973.$$

Client 3 and Client 4.

$$\nabla\hat{F}_3 \cdot \nabla\hat{F}_4 = (-1.1)(-0.8) + (-0.9)(-1.0) = 1.78.$$

Thus,

$$\text{sim}(3,4) = \frac{1.78}{1.42 \times 1.28} = \frac{1.78}{1.8176} \approx 0.979.$$

Summary of Pairwise Cosine Similarities

Interpretation. Table 8.5 clearly reveals two coherent client groups: clients $\{1,2\}$ exhibit strong positive similarity with each other, as do clients $\{3,4\}$, while cross-group similarities are strongly negative. This quantitative evidence justifies the cluster partition

$$\mathscr{C}_1 = \{1,2\}, \qquad \mathscr{C}_2 = \{3,4\},$$

and demonstrates how client clustering prevents destructive gradient cancellation under structured non-IID data.

Cluster-level objectives. Once clients are partitioned into coherent clusters, federated optimization is no longer performed with respect to a single global objective. Instead, each cluster $\mathscr{C}_m$ defines its own empirical risk minimization problem:

$$\min_{\theta_m} \hat{F}_{\mathscr{C}_m}(\theta_m) = \frac{1}{|\mathscr{C}_m|} \sum_{k \in \mathscr{C}_m} \hat{F}_k(\theta_m), \qquad m \in \{1, 2\}.$$

Here, θ_m denotes the cluster-specific model parameters, and the objective aggregates only those clients whose data distributions and gradient directions are mutually consistent.

Cluster gradients. At communication round t, the descent direction for each cluster is obtained by averaging the gradients of its constituent clients:

$$\nabla \hat{F}_{\mathscr{C}_m}(\theta^{(t)}) = \frac{1}{|\mathscr{C}_m|} \sum_{k \in \mathscr{C}_m} \nabla \hat{F}_k(\theta^{(t)}).$$

This aggregation preserves alignment within the cluster while excluding destructive interference from incompatible clients.

Cluster 1 ($\mathscr{C}_1 = \{1, 2\}$). Substituting the explicit gradients,

$$\nabla \hat{F}_{\mathscr{C}_1} = \frac{1}{2}\big[(1.0,\ 0.8) + (0.9,\ 0.7)\big].$$

Evaluating component-wise,

$$\nabla \hat{F}_{\mathscr{C}_1} = \left(\frac{1.0 + 0.9}{2}, \frac{0.8 + 0.7}{2}\right) = (0.95,\ 0.75).$$

This vector represents a clear and coherent descent direction shared by both clients in Cluster 1.
Cluster 2 ($\mathscr{C}_2 = \{3, 4\}$). Similarly,

$$\nabla \hat{F}_{\mathscr{C}_2} = \frac{1}{2}\big[(-1.1,\ -0.9) + (-0.8,\ -1.0)\big].$$

Evaluating explicitly,

$$\nabla \hat{F}_{\mathscr{C}_2} = \left(\frac{-1.1 - 0.8}{2}, \frac{-0.9 - 1.0}{2}\right) = (-0.95,\ -0.95).$$

Again, the cluster gradient preserves a strong and internally consistent direction of descent.

Comparison with single global aggregation. If all four clients were aggregated without clustering, the global gradient would be

$$\nabla \hat{F}_{\text{global}} = \frac{1}{4} \sum_{k=1}^{4} \nabla \hat{F}_k(\theta^{(t)}).$$

Substituting the individual gradients,

$$\nabla \hat{F}_{\text{global}} = \frac{1}{4} \big[(1.0, 0.8) + (0.9, 0.7) + (-1.1, -0.9) + (-0.8, -1.0)\big].$$

Evaluating component-wise,

$$\nabla \hat{F}_{\text{global}} = \left(\frac{1.0 + 0.9 - 1.1 - 0.8}{4}, \frac{0.8 + 0.7 - 0.9 - 1.0}{4} \right) = (0, -0.1).$$

Interpretation. The single global gradient is nearly zero in magnitude and poorly aligned with the meaningful descent directions identified within each cluster. This is a direct manifestation of *gradient cancellation* caused by aggregating strongly opposing updates under non-IID data. By contrast, cluster-wise aggregation preserves coherent optimization signals, enabling faster convergence and more stable learning dynamics. This deductive comparison formally demonstrates why client clustering is an effective mitigation strategy for structured heterogeneity in federated learning.

Pedagogical Takeaway
Client clustering replaces destructive averaging across incompatible gradients with constructive optimization within homogeneous client groups, restoring meaningful descent directions under non-IID data.

Conclusion of the worked example. This example demonstrates that structured non-IID data can be handled effectively by partitioning clients into clusters and optimizing cluster-specific objectives. Rather than enforcing consensus across all clients, federated learning benefits from coordinating diversity through principled client grouping.

8.4 Regularization-based Mitigation Methods

When client data distributions differ substantially, local optimization objectives in federated learning may pull model parameters in incompatible directions. If left unconstrained, repeated local updates can cause excessive *client drift*, where locally optimized models move far away from the current global solution. Regularization-based mitigation methods explicitly restrict this drift by anchoring local learning to the global model.

Motivation. In centralized learning, stochastic gradient noise averages out over large, IID datasets. In federated learning under non-IID data, however, local gradients are biased estimators of the global descent direction. Consequently, multiple local optimization steps can amplify this bias, leading to unstable global updates and degraded convergence. Regularization introduces a controlled bias that counteracts this effect, trading aggressive local fitting for global stability.

Proximal regularization of local objectives. A widely adopted strategy augments each client's empirical risk with a quadratic proximity term:

$$\hat{F}_k^{\text{reg}}(\theta) = \hat{F}_k(\theta) + \frac{\lambda}{2}\|\theta - \theta^{(t)}\|^2,$$

where $\theta^{(t)}$ denotes the global model at the beginning of communication round t, and $\lambda > 0$ controls the strength of the regularization.

The additional term penalizes deviation from the global model, thereby encouraging local solutions to remain within a neighborhood of $\theta^{(t)}$. This effectively limits the influence of highly heterogeneous clients without suppressing local learning altogether.

Effect on local optimization. Taking the gradient of the regularized objective yields

$$\nabla \hat{F}_k^{\text{reg}}(\theta) = \nabla \hat{F}_k(\theta) + \lambda(\theta - \theta^{(t)}).$$

At the start of local training, when $\theta = \theta^{(t)}$, the regularization term vanishes. As local updates progress and θ moves away from $\theta^{(t)}$, the penalty term grows linearly, pulling the parameters back toward the global model. This mechanism acts as a *soft constraint* that balances local adaptation against global consistency.

Interpretation as controlled client drift. From a geometric perspective, the regularization term defines a trust region centered at the global model. Local updates are allowed to explore directions favored by the client's data, but only within a bounded radius determined by λ. Larger values of λ enforce stronger coupling between local and global models, while smaller values allow greater personalization at the risk of increased heterogeneity.

Connection to Proximal Optimization
The regularized local objective corresponds to a proximal point step with respect to the global model. This interpretation connects federated learning to classical proximal-gradient and elastic averaging methods, which are known to stabilize optimization under objective mismatch.

Practical implications. Regularization-based mitigation is particularly effective when:

- client data distributions are highly skewed,
- multiple local epochs are performed between communication rounds, and
- communication constraints necessitate infrequent aggregation.

In such settings, proximal regularization prevents local models from overfitting idiosyncratic client data while still benefiting from local computation.

Limitations and trade-offs. While regularization improves stability, it does not eliminate the fundamental effects of non-IID data. Excessive regularization may slow convergence or reduce the ability of clients to capture meaningful local structure. Choosing λ therefore involves a trade-off between convergence speed, robustness, and representational capacity.

Conceptual takeaway. Regularization-based methods redefine federated learning not as independent local optimizations followed by averaging, but as a coordinated process in which local learning is continuously tethered to a shared global reference. This perspective is essential for maintaining stable learning dynamics under persistent and structured data heterogeneity.

Worked Example: Proximal Regularization under Non-IID Client Updates
We now present a fully worked, step-by-step numerical example to demonstrate how regularization-based mitigation stabilizes federated learning under non-IID data. All calculations are shown explicitly, without skipping intermediate steps.

Federated setup. Consider a federated system with $K = 2$ clients participating in a single communication round t. Let the global model parameter be one-dimensional,

$$\theta \in \mathbb{R},$$

and let the current global model at round t be

$$\theta^{(t)} = 0.$$

Each client optimizes a local empirical risk that reflects its own data distribution. Due to non-IID data, the two clients exhibit conflicting objectives.

Local empirical objectives. Assume the following quadratic local losses:

$$\hat{F}_1(\theta) = (\theta - 2)^2, \qquad \hat{F}_2(\theta) = (\theta + 2)^2.$$

These objectives are minimized at $\theta = 2$ and $\theta = -2$, respectively, representing strong heterogeneity.

Unregularized local optima. Without regularization, each client independently minimizes its local objective.

For Client 1:

$$\nabla \hat{F}_1(\theta) = 2(\theta - 2).$$

Setting the gradient to zero,

$$2(\theta - 2) = 0 \quad \Rightarrow \quad \theta_1^{\star} = 2.$$

For Client 2:

$$\nabla \hat{F}_2(\theta) = 2(\theta + 2).$$

Setting the gradient to zero,

$$2(\theta + 2) = 0 \quad \Rightarrow \quad \theta_2^{\star} = -2.$$

If these updates were aggregated directly, the global model would suffer from severe cancellation.

Regularized local objectives. We now introduce proximal regularization with parameter $\lambda = 1$. The regularized objectives become

$$\hat{F}_k^{\text{reg}}(\theta) = \hat{F}_k(\theta) + \frac{1}{2}\|\theta - \theta^{(t)}\|^2, \qquad k \in \{1, 2\}.$$

Substituting $\theta^{(t)} = 0$, we obtain

$$\hat{F}_1^{\text{reg}}(\theta) = (\theta - 2)^2 + \frac{1}{2}\theta^2,$$

$$\hat{F}_2^{\text{reg}}(\theta) = (\theta + 2)^2 + \frac{1}{2}\theta^2.$$

Solving the regularized local problems.

Client 1:

$$\hat{F}_1^{\text{reg}}(\theta) = (\theta^2 - 4\theta + 4) + \frac{1}{2}\theta^2 = \frac{3}{2}\theta^2 - 4\theta + 4.$$

Taking the gradient,

$$\nabla \hat{F}_1^{\text{reg}}(\theta) = 3\theta - 4.$$

Setting it to zero,

$$3\theta - 4 = 0 \quad \Rightarrow \quad \theta_1^{\text{reg}} = \frac{4}{3}.$$

Client 2:

$$\hat{F}_2^{\text{reg}}(\theta) = (\theta^2 + 4\theta + 4) + \frac{1}{2}\theta^2 = \frac{3}{2}\theta^2 + 4\theta + 4.$$

Taking the gradient,

$$\nabla \hat{F}_2^{\text{reg}}(\theta) = 3\theta + 4.$$

Setting it to zero,

$$3\theta + 4 = 0 \quad \Rightarrow \quad \theta_2^{\text{reg}} = -\frac{4}{3}.$$

Effect of regularization. Compared to the unregularized optima ± 2, the regularized solutions $\pm\frac{4}{3}$ lie closer to the global model $\theta^{(t)} = 0$. The proximal term has explicitly reduced client drift.

Global aggregation. Aggregating the regularized client updates yields

$$\theta^{(t+1)} = \frac{1}{2}\left(\frac{4}{3} + \left(-\frac{4}{3}\right)\right) = 0.$$

By contrast, aggregating the unregularized optima would give

$$\theta_{\text{unreg}}^{(t+1)} = \frac{1}{2}(2 + (-2)) = 0,$$

but only after extreme and unstable local excursions. With regularization, the same global consistency is achieved through *controlled* local updates.

Geometric interpretation. The proximal term introduces a quadratic penalty that defines a trust region around $\theta^{(t)}$. Each client is allowed to move toward its preferred optimum, but the penalty increases linearly with distance, preventing runaway divergence.

Pedagogical takeaway. This worked example shows explicitly that:

- Non-IID local objectives can drive clients toward incompatible optima.
- Proximal regularization pulls local solutions back toward the global model.
- Regularization stabilizes federated optimization by limiting client drift while preserving meaningful local learning.

Regularization-based mitigation therefore transforms federated learning from an unconstrained averaging process into a controlled, coordinated optimization procedure under heterogeneity.

8.5 Robust Optimization Strategies

Robust optimization strategies approach non-IID data in federated learning from a fundamentally different philosophical standpoint. Rather than viewing heterogeneity as random noise that can be averaged out through repeated sampling, robust methods explicitly acknowledge that some clients may systematically deviate from the majority. These deviations may arise due to rare user behaviors, atypical sensing conditions, minority populations, or adversarial environments.

In classical empirical risk minimization, the global objective is typically defined as the average client loss,

$$\min_{\theta} \frac{1}{K} \sum_{k=1}^{K} \hat{F}_k(\theta),$$

implicitly assuming that optimizing mean performance leads to acceptable behavior for all clients. Under strong heterogeneity, however, this assumption can fail dramatically: clients with highly divergent data distributions may experience large errors even when the average loss is small.

Robust federated learning replaces this mean-centric objective with a *worst-case* perspective. A canonical robust formulation is

$$\min_{\theta} \max_{k \in \{1,\ldots,K\}} \hat{F}_k(\theta),$$

which seeks a model that minimizes the maximum loss incurred by any client. This objective ensures that no single client performs arbitrarily poorly, even if doing so sacrifices some average accuracy.

Interpretation. The min–max objective can be understood as a game between the learner and an adversary. The learner selects model parameters θ, while the adversary selects the client whose loss is largest. The learner's goal is to choose θ such that even the most difficult client is adequately served.

Why Robustness Matters
In safety-critical or fairness-sensitive applications, optimizing average performance may conceal catastrophic failures on minority clients. Robust objectives explicitly protect against such failures by enforcing performance guarantees for all participants.

Gradient behavior under robust objectives. Unlike average-loss optimization, where gradients are aggregated across clients, robust optimization focuses on the *most violating* clients. At a given model state $\theta^{(t)}$, the effective descent direction is determined by

$$k^{\star} = \arg\max_{k} \hat{F}_k(\theta^{(t)}), \qquad \nabla\theta^{(t+1)} \propto \nabla\hat{F}_{k^{\star}}(\theta^{(t)}).$$

As a result, updates are driven by the clients experiencing the highest loss, forcing the global model to correct its weakest points.

Trade-off between robustness and efficiency. While robust optimization improves worst-case guarantees, it may slow convergence or reduce average accuracy. This trade-off is intrinsic: protecting against extreme heterogeneity requires allocating optimization effort to difficult or rare cases, which may not align with the dominant data distribution.

Design Trade-off
Robust federated learning prioritizes reliability and fairness over raw average accuracy. Practitioners must balance convergence speed, global accuracy, and per-client guarantees when selecting robust objectives.

Variants and extensions. In practice, strict worst-case optimization may be overly conservative. To address this, softened robust formulations are often used, such as:

- **Distributionally robust optimization (DRO):** Optimizing against worst-case distributions within a bounded uncertainty set around the empirical client distribution.
- **CVaR-based objectives:** Minimizing the average loss over the worst-performing fraction of clients rather than a single extreme outlier.

These variants interpolate between average-case and worst-case optimization, offering flexible control over robustness.

Broader perspective. Robust optimization reframes federated learning as a problem of *guaranteed performance under diversity*. Rather than assuming consensus will naturally emerge, robust methods explicitly enforce it by constraining how poorly any client is allowed to perform. This perspective is especially valuable in federated systems deployed in real-world environments, where heterogeneity is the rule rather than the exception.

Worked Example: Robust Federated Optimization under Client Heterogeneity
We now present a fully worked numerical example to illustrate how robust (min–max) optimization operates in a federated learning setting. All steps are shown explicitly, with no intermediate calculations skipped.

Federated setup. Consider a federated system with $K = 3$ clients. Each client k has a local empirical loss function $\hat{F}_k(\theta)$ defined over a scalar model parameter $\theta \in \mathbb{R}$. Assume the local objectives are quadratic:

$$\hat{F}_k(\theta) = \frac{1}{2}(\theta - a_k)^2,$$

where a_k represents the client-specific optimum induced by its local data distribution.

Let

$$a_1 = 0, \qquad a_2 = 2, \qquad a_3 = 6.$$

These values reflect strong heterogeneity: Client 3's data distribution differs substantially from the others.

Average-risk objective (for comparison). Classical federated averaging implicitly minimizes

$$\hat{F}_{\text{avg}}(\theta) = \frac{1}{3}\sum_{k=1}^{3} \hat{F}_k(\theta) = \frac{1}{6}\Big[(\theta - 0)^2 + (\theta - 2)^2 + (\theta - 6)^2\Big].$$

Taking the derivative,

$$\frac{d}{d\theta}\hat{F}_{\text{avg}}(\theta) = \frac{1}{3}\big[(\theta - 0) + (\theta - 2) + (\theta - 6)\big] = \theta - \frac{8}{3}.$$

Setting this to zero yields

$$\theta^{\star}_{\text{avg}} = \frac{8}{3} \approx 2.67.$$

Evaluating the individual client losses at this point:

$$\hat{F}_1(2.67) = \frac{1}{2}(2.67)^2 \approx 3.56,$$

$$\hat{F}_2(2.67) = \frac{1}{2}(0.67)^2 \approx 0.22,$$

$$\hat{F}_3(2.67) = \frac{1}{2}(3.33)^2 \approx 5.55.$$

Thus, although the average loss is minimized, Client 3 experiences a very large error.

Robust (min–max) objective. Robust federated learning instead solves

$$\min_{\theta} \max_{k \in \{1,2,3\}} \hat{F}_k(\theta) = \min_{\theta} \max \left\{ \frac{1}{2}(\theta - 0)^2,\ \frac{1}{2}(\theta - 2)^2,\ \frac{1}{2}(\theta - 6)^2 \right\}.$$

Identifying the worst-case clients. For any given θ, the dominant loss comes from the client whose a_k is farthest from θ. At optimality, the maximum losses of the two extreme clients must be equal:

$$\frac{1}{2}(\theta - 0)^2 = \frac{1}{2}(\theta - 6)^2.$$

Canceling the common factor and expanding,

$$(\theta - 0)^2 = (\theta - 6)^2$$

$$\theta^2 = \theta^2 - 12\theta + 36.$$

Solving,

$$12\theta = 36 \quad \Rightarrow \quad \theta^{\star}_{\text{rob}} = 3.$$

Evaluating losses at the robust solution. We now explicitly compute the empirical loss incurred by each client at the robust optimum $\theta^{\star}_{\text{rob}} = 3$.

Client 1. For Client 1, whose local optimum is $a_1 = 0$, the loss is

$$\hat{F}_1(3) = \frac{1}{2}(3 - 0)^2 = \frac{1}{2} \times 9 = 4.5.$$

Client 2. For Client 2, with $a_2 = 2$, the loss becomes

$$\hat{F}_2(3) = \frac{1}{2}(3 - 2)^2 = \frac{1}{2} \times 1 = 0.5.$$

Client 3. For Client 3, whose local optimum is $a_3 = 6$, the loss is

$$\hat{F}_3(3) = \frac{1}{2}(3-6)^2 = \frac{1}{2} \times 9 = 4.5.$$

Worst-case evaluation. Among the three clients, the maximum loss at $\theta = 3$ is

$$\max_{k \in \{1,2,3\}} \hat{F}_k(3) = \max\{4.5,\ 0.5,\ 4.5\} = 4.5.$$

Comparison with average-risk optimization. Recall that under average-risk minimization, the optimal parameter $\theta^\star_{\text{avg}} \approx 2.67$ produced a worst-case loss of approximately 5.55. Thus,

$$\max_k \hat{F}_k(3) < \max_k \hat{F}_k(\theta^\star_{\text{avg}}),$$

demonstrating that the robust solution explicitly reduces the worst-case client error, even though it does not minimize the average loss.

Didactic takeaway. This step-by-step evaluation makes clear that robust federated optimization sacrifices some average accuracy in order to provide stronger performance guarantees for the most disadvantaged clients, which is precisely its intended design objective.

Gradient interpretation. At $\theta = 3$, the gradients of the extreme clients satisfy

$$\nabla \hat{F}_1(3) = 3, \qquad \nabla \hat{F}_3(3) = -3,$$

indicating balanced opposing forces. The robust solution lies at the equilibrium point where the most violating clients exert equal and opposite pressure.

Key takeaway. This example demonstrates that:

- Average-risk optimization favors the majority but can severely penalize minority clients.
- Robust optimization explicitly minimizes the worst client loss.
- The robust solution trades some average accuracy for guaranteed per-client performance.

Pedagogical conclusion. Robust federated optimization does not attempt to make all clients identical. Instead, it enforces a principled balance in which no client is allowed to perform arbitrarily poorly, making it especially suitable for fairness-critical and safety-sensitive federated deployments.

8.6 Performance Bounds under Heterogeneity

Beyond algorithmic design, a rigorous understanding of federated learning requires *theoretical performance bounds* that explicitly account for data heterogeneity. Such bounds play a role analogous to classical convergence guarantees in centralized optimization: they do not merely certify convergence, but also reveal the fundamental factors that limit achievable performance.

In centralized learning under IID data, convergence rates typically depend on smoothness, convexity, and stochastic gradient variance. In federated learning, however, an additional and unavoidable source of error arises from *statistical heterogeneity across clients*. Performance bounds must therefore quantify how disagreement between local objectives constrains global optimization.

Measuring Heterogeneity via Gradient Disagreement

A widely adopted formal measure of heterogeneity is the *gradient diversity* parameter (say zeta ζ)

$$\zeta^2 = \frac{1}{K}\sum_{k=1}^{K}\left\|\nabla F_k(\theta^\star) - \nabla F(\theta^\star)\right\|^2,$$

where:

- $F_k(\theta)$ denotes the population (or empirical) objective of client k,
- $F(\theta) = \frac{1}{K}\sum_{k=1}^{K} F_k(\theta)$ is the global objective, and
- $\theta^\star$ is a minimizer of the global objective.

This quantity measures the *average squared deviation* between each client's gradient and the global gradient at the optimum. Several important observations follow immediately:

- If data are IID across clients, then $\nabla F_k(\theta^\star) = \nabla F(\theta^\star)$ for all k, and $\zeta = 0$.
- If client objectives differ substantially, then ζ is strictly positive, reflecting persistent gradient disagreement.
- The parameter ζ depends only on the data distributions, not on the optimization algorithm.

Thus, ζ captures an intrinsic property of the federated learning problem itself, rather than a weakness of a particular method.

Interpretation of ζ

The heterogeneity parameter ζ quantifies how incompatible local learning objectives remain even at the global optimum. It measures structural disagreement, not stochastic noise.

Impact on Convergence Rates

The presence of gradient disagreement fundamentally alters convergence behavior. For a broad class of federated optimization algorithms, including variants of Federated Averaging and proximal methods, convergence bounds take the generic form

$$\mathcal{O}\left(\frac{1}{\sqrt{T}} + \zeta\right),$$

where T denotes the number of communication rounds.

This expression highlights two qualitatively distinct sources of error:

- **Optimization error $\mathcal{O}(1/\sqrt{T})$**: This term decays with additional communication rounds and reflects the stochastic nature of gradient-based optimization.
- **Heterogeneity-induced error ζ**: This term does *not* vanish with more iterations. It represents an irreducible error floor determined by client heterogeneity.

In the limit $T \to \infty$, the optimization error disappears, but the residual error proportional to ζ remains. Consequently, even perfect optimization cannot recover centralized performance when client objectives are misaligned.

Geometric Intuition

From a geometric perspective, each client objective $F_k(\theta)$ defines its own loss surface. Under IID data, these surfaces are locally aligned near $\theta^\star$, so descent directions agree. Under heterogeneity, however, the surfaces intersect at angles, and the local gradients at $\theta^\star$ point in different directions.

Aggregation therefore averages *conflicting descent vectors*, leading to partial cancellation. The stronger this misalignment, the larger ζ, and the more severe the degradation in convergence.

Algorithmic Implications

The appearance of ζ in convergence bounds has several important implications for algorithm design:

- Increasing the number of communication rounds alone cannot overcome heterogeneity.

- Techniques such as adaptive aggregation, regularization, clustering, and robust optimization aim to *reduce the effective impact of* ζ, even though they cannot eliminate it entirely.
- Personalized and clustered federated learning can be viewed as strategies for redefining the objective so that ζ is reduced within each subgroup.

Fundamental Limitation
No federated algorithm can fully eliminate the effect of non-IID data. Heterogeneity introduces an intrinsic error floor that reflects the statistical structure of the problem itself. Algorithms can only manage, redistribute, or localize this error—not remove it.

Worked Example: Quantifying the Heterogeneity-induced Error Floor
We now present a fully worked, step-by-step numerical example to illustrate how the heterogeneity measure ζ arises and how it induces an irreducible error floor in federated optimization. All calculations explicitly follow the definitions introduced in this section.

Federated setup. Consider a federated learning system with $K = 3$ clients. Assume a one-dimensional parameter $\theta \in \mathbb{R}$ for clarity. Each client minimizes a convex quadratic objective of the form

$$F_k(\theta) = \frac{1}{2}(\theta - a_k)^2,$$

where a_k represents the client-specific optimum. Such quadratic objectives commonly arise as local approximations of smooth loss functions near an optimum.

Client-specific objectives. Let the client optima be

$$a_1 = 0, \qquad a_2 = 1, \qquad a_3 = 3.$$

The local gradients are therefore

$$\nabla F_k(\theta) = \theta - a_k.$$

Global objective and global optimum. The global objective is the uniform average

$$F(\theta) = \frac{1}{3}\sum_{k=1}^{3} F_k(\theta) = \frac{1}{6}\Big[(\theta - 0)^2 + (\theta - 1)^2 + (\theta - 3)^2\Big].$$

To find the global minimizer $\theta^\star$, we compute

$$\nabla F(\theta) = \frac{1}{3}\big[(\theta - 0) + (\theta - 1) + (\theta - 3)\big] = \theta - \frac{4}{3}.$$

Setting $\nabla F(\theta^\star) = 0$ yields

$$\theta^\star = \frac{4}{3}.$$

Derivation of client gradients at the global optimum. Recall that each local objective is a quadratic function of the form

$$F_k(\theta) = \frac{1}{2}(\theta - a_k)^2,$$

whose gradient is obtained by direct differentiation:

$$\nabla F_k(\theta) = \frac{d}{d\theta}\left[\frac{1}{2}(\theta - a_k)^2\right] = \theta - a_k.$$

Step 1: Substitute the client-specific optima. For the three clients, we have

$$a_1 = 0, \qquad a_2 = 1, \qquad a_3 = 3.$$

Step 2: Evaluate gradients at the global optimum $\theta^\star$. From the earlier derivation, the global minimizer is

$$\theta^\star = \frac{4}{3}.$$

Substituting into each local gradient gives

$$\nabla F_1(\theta^\star) = \theta^\star - a_1 = \frac{4}{3} - 0 = \frac{4}{3},$$

$$\nabla F_2(\theta^\star) = \theta^\star - a_2 = \frac{4}{3} - 1 = \frac{1}{3},$$

$$\nabla F_3(\theta^\star) = \theta^\star - a_3 = \frac{4}{3} - 3 = -\frac{5}{3}.$$

Step 3: Verify global optimality. The global objective is the uniform average

$$F(\theta) = \frac{1}{3}\sum_{k=1}^{3} F_k(\theta),$$

with gradient

$$\nabla F(\theta) = \frac{1}{3}\sum_{k=1}^{3} \nabla F_k(\theta).$$

Evaluating at $\theta^\star$,

$$\begin{aligned}\nabla F(\theta^\star) &= \frac{1}{3}\left(\frac{4}{3} + \frac{1}{3} - \frac{5}{3}\right)\\ &= \frac{1}{3} \cdot 0 = 0.\end{aligned}$$

Although the global gradient vanishes at $\theta^\star$, individual clients still exert nonzero and opposing gradient forces. This discrepancy is the source of heterogeneity captured by the measure ζ^2 and underlies the irreducible error floor in federated optimization.

Computing the heterogeneity measure ζ^2. By definition,

$$\zeta^2 = \frac{1}{K}\sum_{k=1}^{K} \left\|\nabla F_k(\theta^\star) - \nabla F(\theta^\star)\right\|^2.$$

Substituting the computed gradient values.
Starting from the definition

$$\zeta^2 = \frac{1}{K}\sum_{k=1}^{K} \left\|\nabla F_k(\theta^\star) - \nabla F(\theta^\star)\right\|^2,$$

and recalling that $\nabla F(\theta^\star) = 0$ and $K = 3$, we obtain

$$\zeta^2 = \frac{1}{3}\left(|\nabla F_1(\theta^\star)|^2 + |\nabla F_2(\theta^\star)|^2 + |\nabla F_3(\theta^\star)|^2\right).$$

Step 1: Square each local gradient. Using the previously derived values,

$$\nabla F_1(\theta^\star) = \frac{4}{3}, \quad \nabla F_2(\theta^\star) = \frac{1}{3}, \quad \nabla F_3(\theta^\star) = -\frac{5}{3},$$

we compute

$$\left(\tfrac{4}{3}\right)^2 = \frac{16}{9}, \quad \left(\tfrac{1}{3}\right)^2 = \frac{1}{9}, \quad \left(-\tfrac{5}{3}\right)^2 = \frac{25}{9}.$$

Step 2: Sum the squared deviations. Adding the three terms yields

$$\frac{16}{9} + \frac{1}{9} + \frac{25}{9} = \frac{42}{9}.$$

Step 3: Average across clients. Dividing by the number of clients $K = 3$ gives

$$\zeta^2 = \frac{1}{3} \cdot \frac{42}{9} = \frac{42}{27} = \frac{14}{9}.$$

Final result. The heterogeneity measure is therefore

$$\boxed{\zeta^2 = \frac{14}{9}},$$

which is strictly positive, confirming the presence of persistent gradient disagreement among clients at the global optimum. This nonzero value directly explains the irreducible error floor observed in federated learning under non-IID data.

Thus,

$$\zeta = \sqrt{\frac{14}{9}} \approx 1.247.$$

Interpretation of ζ. Although the global gradient vanishes at $\theta^\star$, individual client gradients remain large and point in different directions. This persistent disagreement is precisely what ζ quantifies.

Implication for convergence. According to the convergence bound

$$\mathcal{O}\left(\frac{1}{\sqrt{T}} + \zeta\right),$$

even as $T \to \infty$, the optimization error cannot shrink below a constant proportional to $\zeta \approx 1.247$. This residual error is not due to insufficient optimization, but to inherent client heterogeneity.

Contrast with the IID case. If all clients shared the same optimum $a_k = a$, then $\nabla F_k(\theta^\star) = 0$ for all k, yielding $\zeta = 0$. In that case, convergence would be limited only by optimization error, recovering classical centralized behavior.

Key lesson. This worked example demonstrates concretely that heterogeneity introduces a non-vanishing error floor. Federated algorithms can reduce optimization error but cannot overcome persistent gradient disagreement without altering the problem formulation itself (e.g. clustering, personalization, or robust objectives).

Exercises

This chapter examined the fundamental challenges posed by non-IID data in federated learning and presented principled algorithmic strategies to mitigate heterogeneity while preserving convergence, stability, and fairness.

A. Short Answer Questions

Answer each question briefly and precisely.

1. What is meant by non-IID data in federated learning?
2. Why does Federated Averaging fail under severe data heterogeneity?
3. Define gradient disagreement in the context of federated learning.
4. What is the role of adaptive aggregation in mitigating non-IID data?
5. Why is Batch Normalization problematic in federated settings?
6. State one key advantage of Layer Normalization over Batch Normalization.
7. What is client drift and how does it arise?
8. How does proximal regularization limit client drift?
9. What is the objective of robust (min–max) federated optimization?
10. Define the heterogeneity parameter ζ.

B. Long Answer Questions

Answer the following questions with clear explanations, equations, and diagrams wherever appropriate.

1. Explain the different forms of non-IID data encountered in federated learning.
2. Derive the adaptive aggregation weighting rule based on gradient deviation.
3. Compare Batch Normalization, Layer Normalization, and Group Normalization under non-IID data.
4. Explain client clustering and how it transforms a non-IID problem into quasi-IID subproblems.
5. Derive the proximal regularized local objective and interpret it geometrically.
6. Explain the trade-off between robustness and efficiency in min–max federated optimization.
7. Provide a geometric interpretation of gradient disagreement at the global optimum.
8. Explain why heterogeneity introduces an irreducible error floor.
9. Compare average-risk and robust federated optimization using equations.
10. Discuss how personalization and clustering reduce the effective heterogeneity parameter.

C. True/False Questions

State whether each statement is **True** or **False**.

1. Under IID data, all client gradients are identical at the global optimum.
2. Non-IID data introduces only additional variance, not bias.
3. Adaptive aggregation assigns static weights to clients.
4. Batch Normalization depends on batch-level statistics.
5. Layer Normalization is independent of batch size.
6. Proximal regularization eliminates client drift entirely.
7. Robust optimization minimizes the average client loss.
8. Client clustering requires maintaining multiple global models.
9. The heterogeneity parameter ζ depends on the optimization algorithm.
10. Increasing communication rounds can eliminate heterogeneity-induced error.

Answers:

1. True
2. False
3. False
4. True
5. True
6. False
7. False
8. True
9. False
10. False

D. Fill in the Blanks

Fill in the blanks with the most appropriate term or expression.

1. Non-IID data causes persistent gradient ___________.
2. FedAvg implicitly assumes ___________ data across clients.
3. Adaptive aggregation adjusts client weights based on ___________.
4. Batch Normalization relies on ___________ statistics.
5. Layer Normalization computes statistics across ___________.
6. Proximal regularization introduces a ___________ penalty term.
7. Client clustering forms ___________ subsets of clients.
8. Robust optimization minimizes the ___________ client loss.
9. Gradient disagreement is quantified by the parameter ___________.
10. The error term proportional to ζ represents an ___________ error floor.

Answers:

1. disagreement
2. IID
3. gradient deviation
4. batch-level
5. feature dimensions
6. quadratic
7. homogeneous
8. worst-case
9. ζ
10. irreducible

E. Multiple Choice Questions (MCQs)

Choose the **most appropriate** answer.

1. Which factor fundamentally distinguishes federated learning from centralized learning?
   (a) Larger datasets
   (b) Client autonomy
   (c) GPU acceleration
   (d) Convex objectives
2. Which normalization method is most fragile under non-IID federated data?
   (a) Layer Normalization
   (b) Group Normalization
   (c) Instance Normalization
   (d) Batch Normalization
3. Adaptive aggregation primarily reduces:
   (a) Communication cost
   (b) Model size
   (c) Gradient bias
   (d) Local computation
4. The proximal term in FedProx penalizes:
   (a) Gradient variance
   (b) Communication delay
   (c) Deviation from global model
   (d) Batch size mismatch
5. Client clustering is most effective when heterogeneity is:
   (a) Random
   (b) Structured
   (c) Adversarial
   (d) Temporal

6. Robust optimization focuses on minimizing:
   (a) Mean loss
   (b) Median loss
   (c) Worst-case loss
   (d) Variance
7. The heterogeneity parameter ζ measures:
   (a) Gradient noise
   (b) Data imbalance
   (c) Gradient disagreement
   (d) Model complexity
8. Increasing communication rounds primarily reduces:
   (a) ζ
   (b) Optimization error
   (c) Client drift
   (d) Data heterogeneity
9. Which method explicitly maintains multiple models?
   (a) FedAvg
   (b) Proximal regularization
   (c) Client clustering
   (d) Batch normalization
10. Which strategy offers explicit fairness guarantees?
    (a) Adaptive aggregation
    (b) Robust optimization
    (c) Layer normalization
    (d) Uniform averaging

Answers:

1. (b)
2. (d)
3. (c)
4. (c)
5. (b)
6. (c)
7. (c)
8. (b)
9. (c)
10. (b)

F. Application and Project-based Questions

These problems emphasize modeling, reasoning, and implementation.

1. Simulate Federated Averaging on synthetically generated non-IID data and analyze convergence behavior.
2. Implement adaptive aggregation using cosine similarity between gradients.
3. Compare Batch Normalization and Layer Normalization in a federated CNN on non-IID image data.
4. Design a clustering-based federated learning system for heterogeneous healthcare clients.
5. Analyze the effect of the proximal parameter λ on client drift.
6. Implement a min–max federated optimizer and compare it with FedAvg.
7. Empirically estimate the heterogeneity parameter ζ from client gradients.
8. Study convergence bounds under increasing heterogeneity.
9. Propose a hybrid method combining clustering and proximal regularization.
10. Evaluate fairness–accuracy trade-offs in robust federated learning.

Part IV

Federated Optimization Algorithms

Foundations of Federated Optimization Algorithms

9

The mathematical formulation developed in earlier chapters establishes federated learning as a constrained and decentralized optimization problem. By expressing the learning objective as a weighted composition of client-local empirical risks, federated learning is revealed as a well-defined mathematical construct rather than an ad hoc training protocol. However, this formulation alone does not prescribe *how* the optimization should be carried out in practice when data is non-centralized, communication is costly, participation is partial, and privacy constraints prohibit direct access to training samples.

In classical optimization theory, once an objective function is defined, standard algorithms—such as gradient descent or stochastic gradient descent—can be applied directly. Federated learning departs fundamentally from this paradigm. Although the global objective

$$F(\theta) = \sum_{k=1}^{K} \frac{n_k}{N} F_k(\theta)$$

is mathematically well defined, it is never instantiated or evaluated by any single entity. No participant has access to all datasets $\{\mathscr{D}_k\}_{k=1}^{K}$, and therefore no participant can compute the global loss or its gradient explicitly. Optimization must instead proceed indirectly, through the coordination of locally computed quantities.

From Objective to Algorithm
In federated learning, defining the global objective is only the first step. The central challenge lies in designing optimization algorithms that can *approximate descent on a virtual objective* using only partial, decentralized, and privacy-preserving information.

P. P. Ray, *Foundations of Federated Learning*,
Springer Series in the Data Sciences,
https://doi.org/10.1007/978-3-032-34084-9_9

Federated optimization algorithms provide the concrete mechanisms that bridge this gap between abstract formulation and practical learning. They specify how local computations performed independently at clients are translated into coordinated global model updates through iterative communication and aggregation. At each communication round, these algorithms orchestrate three tightly coupled processes: (i) local optimization on private data, (ii) selective communication of derived information, and (iii) aggregation of heterogeneous updates into a shared model. The effectiveness of federated learning hinges on how these processes are designed and balanced.

Unlike centralized stochastic optimization, where gradients are computed on randomly sampled mini-batches drawn from a common distribution, federated optimization operates on gradients that are inherently structured by client identity. Each client's update reflects not only stochastic sampling noise, but also persistent statistical biases induced by non-IID data, client-specific data volumes, and local optimization dynamics. As a result, federated optimization exhibits behaviors—such as client drift, delayed consensus, and biased descent directions—that have no direct analogue in centralized learning.

Why Federated Optimization Is Fundamentally Different
Federated optimization is not merely stochastic optimization executed in parallel. It is optimization under decentralization, where gradient information is biased, delayed, incomplete, and unevenly weighted across participants.

This chapter introduces the foundational optimization principles that govern federated learning algorithms. Rather than focusing on a single algorithmic instantiation, we develop the conceptual building blocks that underlie a broad family of federated optimizers. These include the role of local gradient computation, the structure of aggregation operators, the impact of communication frequency, and the conditions under which iterative alignment of client models leads to global improvement.

To ground these ideas, we emphasize intuitive explanations alongside formal expressions, illustrating how classical optimization concepts must be adapted when data locality and partial participation are unavoidable. Through examples and analytical discussion, this chapter clarifies why naïve extensions of centralized algorithms often fail and how federated optimization principles emerge as necessary responses to systemic constraints.

Conceptual Role of This Chapter
This chapter serves as the theoretical bridge between the mathematical formulation of federated learning and the concrete algorithms studied in subsequent

chapters. It explains *why* federated optimization algorithms are structured as they are, before detailing *how* specific methods implement those principles.

The material developed here prepares the foundation for the detailed study of specific federated optimization algorithms—such as Federated Stochastic Gradient Descent, Federated Averaging, adaptive federated optimizers, and proximal methods—presented in the following chapters. By the end of this chapter, the reader will be equipped with a clear conceptual framework for understanding both the strengths and the limitations of these algorithms in decentralized, heterogeneous learning environments.

9.1 Federated Stochastic Gradient Descent

Federated stochastic gradient descent (FedSGD) represents the most direct and conceptually transparent extension of classical stochastic gradient descent to the federated learning setting. It is not introduced primarily as a practically efficient algorithm, but rather as a *conceptual baseline* that clarifies how classical optimization ideas must be adapted when data is decentralized and centrally inaccessible. Understanding FedSGD is essential, as nearly all subsequent federated optimization algorithms can be interpreted as principled modifications or relaxations of this basic scheme.

9.1.1 Revisiting Centralized Stochastic Gradient Descent

To appreciate the structure and motivation of federated stochastic gradient descent, it is instructive to first revisit stochastic gradient descent (SGD) in the classical *centralized learning* setting. Centralized SGD provides the conceptual and mathematical baseline from which federated optimization methods depart, both in formulation and in behavior.

In centralized learning, all training data generated across different sources is assumed to be collected into a single dataset

$$\mathcal{D} = \{(x_i, y_i)\}_{i=1}^{N},$$

which is fully accessible to the learning algorithm. On this unified dataset, model training is posed as an empirical risk minimization problem, where the objective is to minimize the average loss incurred by the model across all samples:

$$F(\theta) = \frac{1}{N} \sum_{i=1}^{N} \ell\big(f_\theta(x_i), y_i\big).$$

This formulation implicitly assumes that data is homogeneous, centrally governed, and equally accessible at every iteration of the optimization process. Under these assumptions, the learning algorithm has complete visibility into the loss landscape defined by $F(\theta)$.

Stochastic gradient descent optimizes this objective iteratively by updating the model parameters in the direction of steepest descent. At iteration t, the parameter update takes the form

$$\theta^{(t+1)} = \theta^{(t)} - \eta \nabla F(\theta^{(t)}),$$

where $\eta > 0$ is the learning rate that controls the step size of each update. The gradient $\nabla F(\theta^{(t)})$ encodes how sensitive the global loss is to changes in the model parameters and therefore provides a principled direction for reducing prediction error.

In practice, computing the full gradient over all N samples at every iteration is often computationally expensive. Stochastic gradient descent addresses this issue by approximating the true gradient using a randomly sampled mini-batch of data. Crucially, under standard assumptions, these mini-batch gradients are *unbiased estimators* of the true gradient:

$$\mathbb{E}[\nabla \widehat{F}(\theta)] = \nabla F(\theta),$$

which ensures that, on average, SGD follows the same descent direction as full-batch gradient descent.

This unbiasedness property is central to the theoretical guarantees of SGD. It enables convergence analyses that rely on smoothness, bounded variance, and step-size control, and it underpins the success of SGD as the workhorse optimization method for large-scale machine learning.

Centralized Gradient Access
Centralized stochastic gradient descent relies fundamentally on the assumption that gradients of the global empirical risk can be computed directly or approximated in an unbiased manner using mini-batches drawn from a centrally pooled dataset.

This assumption of centralized gradient access is often taken for granted in classical optimization literature. However, it is precisely this assumption that breaks down in federated learning. Once data is distributed across autonomous clients and cannot be centralized, the global empirical risk remains a mathematical construct but ceases to be computationally accessible. As a result, the direct application of centralized SGD becomes infeasible, necessitating new optimization strategies that operate under data locality and partial observability.

Understanding centralized SGD in this explicit and detailed manner is therefore essential. It clarifies not only how gradient-based optimization works in idealized settings, but also why federated optimization must depart from classical assumptions and introduce fundamentally different mechanisms for approximating descent on a global objective.

9.1.2 Why Centralized SGD Fails in Federated Settings

The defining assumptions that make centralized stochastic gradient descent both effective and analytically tractable cease to hold in federated learning. In a federated environment, training data is no longer collected, stored, or processed within a single computational domain. Instead, it is inherently partitioned across K autonomous clients, each of which retains exclusive control over its locally generated data.

This decentralization fundamentally alters the nature of the learning problem. No entity in the system—neither the coordinating server nor any individual client—has access to the complete dataset. As a result, the unified dataset

$$\mathscr{D} = \{(x_i, y_i)\}_{i=1}^{N}$$

assumed by centralized learning does not exist as a computational object. The server cannot evaluate losses, compute gradients, or even inspect data samples. Clients, in turn, observe only their own local datasets and possess no information about the data held elsewhere.

As established in earlier chapters, the global empirical risk in federated learning is defined as

$$F(\theta) = \sum_{k=1}^{K} \frac{n_k}{N} F_k(\theta),$$

where each local objective $F_k(\theta)$ is computed over a private dataset $\mathscr{D}_k$. While this expression provides a mathematically precise description of the learning goal, it exists only as a *conceptual construct*. There is no mechanism by which $F(\theta)$ can be directly evaluated, since doing so would require access to all local datasets simultaneously.

This inaccessibility extends immediately to gradient computation. Because the global objective cannot be evaluated, its gradient $\nabla F(\theta^{(t)})$ is likewise unavailable. In contrast to centralized SGD—where gradients are computed exactly or approximated using unbiased mini-batches—federated learning offers no direct way to obtain an unbiased estimate of the global gradient from a single data source.

Breakdown of Centralized Gradient Access
In federated learning, the global loss function and its gradient are well-defined mathematically but computationally inaccessible. This breakdown invalidates the core assumption that underpins centralized stochastic gradient descent.

Any attempt to naïvely apply centralized SGD in this setting would therefore fail. The server cannot compute gradients because it never observes data, and individual clients cannot compute global gradients because their local datasets provide only a partial and potentially biased view of the population. Even if all clients were to transmit their locally computed gradients, these quantities would not, in general, constitute unbiased estimates of the global gradient under non-IID data distributions.

As a consequence, gradient-based optimization in federated learning must proceed indirectly. Rather than relying on direct access to $\nabla F(\theta)$, federated algorithms must approximate descent on the global objective using locally computed information that is incomplete, heterogeneous, and statistically biased. This necessity marks the fundamental departure from centralized SGD and motivates the design of federated optimization algorithms that explicitly account for data locality, partial participation, and heterogeneity.

Understanding precisely why centralized SGD fails in federated settings is therefore not merely a technical observation. It provides the conceptual foundation for all subsequent algorithmic developments in federated learning, clarifying why aggregation, communication rounds, and consensus mechanisms are indispensable components of decentralized optimization.

9.1.3 Local Gradient Computation at Clients

Once centralized gradient access becomes infeasible, federated learning must rely on computations performed independently at individual clients. Local gradient computation therefore constitutes the fundamental computational primitive of federated optimization. All information that ultimately influences the global model originates from these client-side operations.

At communication round t, the coordinating server broadcasts the current global model parameters $\theta^{(t)}$ to a selected subset of clients $\mathscr{S}^{(t)} \subseteq \{1, 2, \dots, K\}$. This broadcast serves two purposes. First, it synchronizes participating clients to a common reference point, ensuring that local computations are comparable. Second, it implicitly enforces partial consensus by anchoring local optimization trajectories to the same global initialization.

Each participating client $k \in \mathscr{S}^{(t)}$ then evaluates the gradient of its *local empirical risk* at the received model:

$$\nabla F_k(\theta^{(t)}) = \frac{1}{n_k} \sum_{i=1}^{n_k} \nabla \ell\big(f_{\theta^{(t)}}(x_{k,i}), y_{k,i}\big).$$

This gradient captures how sensitive the client's local loss is to infinitesimal changes in the model parameters. In effect, it encodes the direction in parameter space that would most rapidly reduce prediction error on client k's private dataset. Because local datasets may differ substantially in size, feature distribution, label frequencies, and noise characteristics, the resulting gradients can vary widely across clients—even when evaluated at the same global model.

Importantly, this computation is performed *entirely locally*. Client k evaluates the loss function, computes gradients, and stores intermediate results using only its own data. Raw training samples $(x_{k,i}, y_{k,i})$ never leave the client device, nor are they revealed to the server or to other clients. Only the derived quantity $\nabla F_k(\theta^{(t)})$—or a compressed or transformed version thereof—is eligible for communication.

Local View of Optimization
From the perspective of each client, federated learning reduces to a standard optimization problem: minimize a local empirical risk defined over private data. However, the resulting gradients are locally optimal directions that need not align with the global descent direction.

This local viewpoint highlights a central tension in federated optimization. Each client computes gradients that are optimal with respect to its own data distribution, yet the learning objective is defined at the population level. When data is non-IID, local gradients may systematically favor client-specific patterns that do not generalize well across the broader population. Consequently, local gradient computation produces information that is essential for learning, but insufficient on its own to guarantee global progress.

The role of the server in subsequent aggregation steps is therefore not to replace local optimization, but to reconcile these heterogeneous local descent directions into a single update that advances the shared model. Understanding local gradient computation in this detailed manner is crucial, as it explains both the power and the limitations of federated optimization: all learning signal originates locally, yet global improvement emerges only through careful coordination of these signals.

9.1.4 Gradient Aggregation at the Server

Once local gradient computations have been completed at the client side, the federated learning process transitions from decentralized computation to centralized coordination. At this stage, the server assumes the role of an *aggregation authority*, responsible for synthesizing heterogeneous local information into a single update that advances the shared global model.

The server does not observe raw data, local losses, or intermediate training states. Its only input consists of gradient information computed independently by participating clients. Using this information, the server constructs an approximation of the global descent direction by aggregating client-local gradients according to the rule

$$\theta^{(t+1)} = \theta^{(t)} - \eta \sum_{k \in \mathcal{S}^{(t)}} \frac{n_k}{\sum_{j \in \mathcal{S}^{(t)}} n_j} \nabla F_k(\theta^{(t)}).$$

This aggregation step can be interpreted as the federated analogue of computing a full-batch gradient in centralized learning. Each client contributes a local gradient that reflects the structure of its private dataset, and the server combines these gradients to form a single update direction. The learning rate η then controls the magnitude of the resulting parameter update, just as in classical gradient descent.

A critical component of this aggregation rule is the weighting factor

$$\frac{n_k}{\sum_{j \in \mathcal{S}^{(t)}} n_j},$$

which assigns influence to each client in proportion to the number of training samples it holds. This weighting scheme is not arbitrary. It is designed to preserve consistency with centralized empirical risk minimization, where each individual data sample contributes equally to the global objective. By weighting gradients according to dataset size, the server ensures that clients with larger datasets exert proportionally greater influence on the global update.

From an optimization perspective, this weighted aggregation can be seen as reconstructing the gradient of the virtual global objective $F(\theta)$ using distributed components. If all clients were to participate and local data were IID, the aggregated gradient would coincide with the true global gradient. In practice, partial participation and data heterogeneity introduce noise and bias, but the aggregation rule still provides the best available approximation under data locality constraints.

Data-Weighted Aggregation
Federated stochastic gradient descent approximates centralized SGD by combining client-local gradients using weights proportional to dataset size. This preserves the relative influence of individual data samples, despite the fact that data remains decentralized.

The aggregation step also highlights the dual nature of the server's role. On one hand, the server performs no learning in the traditional sense, as it never evaluates a loss function or computes gradients itself. On the other hand, it plays a decisive role in shaping the optimization trajectory by determining how local updates are reconciled. Poor aggregation choices can amplify client drift, exacerbate bias, or slow convergence, whereas well-designed aggregation rules promote stability and consensus.

Thus, gradient aggregation at the server is not merely a bookkeeping operation. It is the central mechanism through which decentralized optimization is transformed into coherent global learning. Understanding this step in detail is essential, as it provides the conceptual foundation for more advanced aggregation strategies and algorithmic refinements introduced in subsequent federated optimization methods.

9.1.5 Interpretation of FedSGD

The aggregation rule derived in the previous subsection formally defines *federated stochastic gradient descent*. Viewed at a high level, FedSGD preserves the essential structure of classical gradient descent while adapting it to operate under strict data locality constraints. Each communication round corresponds to a single update of the global model, and each update is constructed collaboratively from information provided by multiple clients.

From the perspective of the overall optimization process, FedSGD can be understood as executing one global gradient descent step per communication round. The crucial distinction from centralized learning is that the gradient used in this step is never computed explicitly. Instead, it is approximated through the aggregation of gradients evaluated independently at client devices. In this sense, the global descent direction emerges as a collective property of the system rather than as the output of a single computation.

This interpretation highlights the role of communication rounds as the fundamental unit of progress in federated optimization. Because each round produces exactly one global update, the rate at which the model improves is tightly coupled to how frequently clients and the server can communicate. While this structure ensures conceptual simplicity and analytical clarity, it also exposes the primary inefficiency of FedSGD in large-scale federated systems.

FedSGD as a Distributed Gradient Oracle

Federated stochastic gradient descent may be viewed as performing gradient descent on a virtual global objective using a distributed gradient oracle, where gradient information is obtained indirectly through the aggregation of client-local computations.

Interpreting FedSGD in this manner is particularly useful for understanding both its strengths and its limitations. On the one hand, the algorithm offers a clean and faithful extension of classical gradient descent to decentralized data. On the other hand, its reliance on a single update per communication round makes it highly sensitive to communication costs and client availability. These observations naturally motivate the development of more communication-efficient federated optimization algorithms, which build upon the same principles while relaxing the strict one-step-per-round structure of FedSGD.

Table 9.1 highlights the fundamental differences between ordinary stochastic gradient descent and its federated counterpart. While both methods are rooted in gradient-based optimization, they operate under markedly different assumptions regarding data access, communication, and system architecture. Ordinary SGD relies on direct access to centrally pooled data and unbiased gradient estimates, whereas FedSGD replaces centralized gradient computation with coordinated aggregation of client-local gradients. This shift introduces new challenges—such as communication overhead and statistical heterogeneity—but enables learning in environments where data centralization is infeasible due to privacy, regulatory, or resource constraints.

Table 9.1 Comparison between Ordinary Stochastic Gradient Descent (SGD) and Federated Stochastic Gradient Descent (FedSGD)

Aspect	Ordinary SGD (Centralized)	Federated SGD (FedSGD)
Data availability	All training data is centrally collected and directly accessible	Training data remains distributed across clients and is never centralized
Objective evaluation	Global empirical risk can be explicitly evaluated	Global empirical risk is virtual and cannot be directly evaluated
Gradient computation	Gradient computed on pooled data or unbiased mini-batches	Gradients computed locally at clients on private datasets
Gradient unbiasedness	Mini-batch gradients are unbiased estimators of the global gradient	Local gradients may be biased under non-IID client data
Update location	Parameter updates computed entirely on a central server	Parameter updates computed via server-side aggregation of client gradients
Communication requirement	No communication required during optimization	Communication required at every iteration to collect gradients
Privacy considerations	Raw data is directly accessible to the optimizer	Raw data never leaves client devices; only gradients are shared
Scalability constraints	Limited by centralized storage and computation	Limited primarily by communication bandwidth and client participation
Sensitivity to data heterogeneity	Assumes approximately IID data	Performance degrades under strong non-IID data distributions
Typical use cases	Data-center training, cloud-based machine learning	Edge devices, mobile systems, privacy-sensitive applications

Limitations of FedSGD

Despite its conceptual clarity and strong theoretical grounding, federated stochastic gradient descent (FedSGD) is rarely adopted in large-scale, real-world federated learning systems. Its limitations arise not from incorrectness, but from a fundamental mismatch between its optimization structure and the practical constraints of decentralized environments.

Excessive Communication Overhead. The most significant drawback of FedSGD is its tight coupling between optimization progress and communication frequency. By design, each communication round corresponds to exactly one global gradient step. As a result, meaningful progress toward minimizing the federated objective

$$F(\theta) = \sum_{k=1}^{K} \frac{n_k}{N} F_k(\theta)$$

requires a large number of communication rounds. In large-scale federated systems—such as mobile networks or Internet-of-Things deployments—communication is often the dominant bottleneck, far exceeding local computation costs. Bandwidth limitations, network latency, intermittent connectivity, and energy consumption make frequent synchronization prohibitively expensive.

Communication Inefficiency of FedSGD

FedSGD advances the global model by only one gradient step per communication round, making optimization progress directly proportional to communication frequency.

Sensitivity to Data Heterogeneity. FedSGD also exhibits pronounced sensitivity to non-identically distributed (non-IID) data. When client data distributions differ substantially, local gradients $\nabla F_k(\theta)$ may point in conflicting directions. Although aggregation produces a weighted average of these gradients, the resulting direction can be a biased and noisy approximation of the true global gradient. This bias does not vanish through averaging alone and can slow convergence, induce oscillations, or require very small learning rates to maintain stability.

Limited Exploitation of Local Computation. Another structural limitation of FedSGD is its failure to exploit the computational capabilities of client devices. Modern edge devices are often capable of performing multiple local optimization steps efficiently. However, FedSGD discards this opportunity by restricting each client to a single gradient computation per round. As a consequence, the algorithm underutilizes local resources while over-relying on communication.

Underutilization of Local Resources
FedSGD prioritizes strict synchronization over local progress, leading to inefficient use of client-side computation in exchange for minimal algorithmic simplicity.

Implications for Practical Federated Learning. Taken together, these limitations explain why FedSGD is best understood as a theoretical baseline rather than a practical training algorithm. It provides a clean conceptual bridge between centralized SGD and federated optimization, clarifying how gradient descent can be implemented under data locality constraints. However, its inefficiency under realistic communication and heterogeneity conditions motivates the development of more advanced methods.

Role of FedSGD in Federated Learning
FedSGD serves primarily as a conceptual and analytical reference point, against which communication-efficient and heterogeneity-aware federated optimization algorithms are designed and evaluated.

These observations naturally motivate algorithms such as Federated Averaging, which decouple local computation from communication by allowing multiple local optimization steps per round. The next section builds upon the limitations identified here to develop more scalable and robust federated optimization strategies.

Why FedSGD Is Rarely Used in Practice
FedSGD provides conceptual clarity but suffers from excessive communication costs and limited robustness under data heterogeneity.

These limitations motivate the development of federated optimization algorithms that allow multiple local updates between communication rounds. Such methods—most notably Federated Averaging—retain the core principles of FedSGD while significantly improving communication efficiency.

Fully Worked Example: Communication Rounds of FedSGD with Three Clients
We now present a detailed numerical example illustrating a complete communication round of Federated Stochastic Gradient Descent (FedSGD) using a larger dataset distributed across three clients. This example explicitly follows the main FedSGD update equations introduced earlier and demonstrates, step by step, how

local gradient computations and server-side aggregation jointly produce a global model update.

Model and Loss Function. We consider a simple linear regression model

$$f_\theta(x) = \theta x,$$

trained using the squared loss

$$\ell\big(f_\theta(x), y\big) = \frac{1}{2}(\theta x - y)^2.$$

The gradient of the loss with respect to θ is

$$\nabla_\theta \ell = (\theta x - y)x.$$

Federated System Configuration. Assume a federated system with $K = 3$ clients. Each client holds a private local dataset, summarized in Table 9.2.

Each client holds $n_1 = n_2 = n_3 = 3$ samples, giving a total of

$$N = n_1 + n_2 + n_3 = 9$$

training samples across the system.

Initialization and Learning Rate. At communication round $t = 0$, the server initializes the global model as

$$\theta^{(0)} = 0.$$

The learning rate is fixed as

$$\eta = 0.1.$$

Table 9.2 Local datasets for the FedSGD worked example

Client	x	y
Client 1	1	2
Client 1	2	4
Client 1	3	6
Client 2	1	1
Client 2	2	2
Client 2	4	4
Client 3	2	3
Client 3	3	5
Client 3	5	8

Communication Round 1.

Step 1: Define Local Empirical Risk Functions. For each client $k \in \{1, 2, 3\}$, the local empirical risk is defined as

$$F_k(\theta) = \frac{1}{n_k} \sum_{i=1}^{n_k} \frac{1}{2}(\theta x_{k,i} - y_{k,i})^2,$$

in accordance with the global objective decomposition

$$F(\theta) = \sum_{k=1}^{3} \frac{n_k}{N} F_k(\theta).$$

Step 2: Local Gradient Computation at Clients. Client 1. Client 1 data: $(1, 2)$, $(2, 4)$, $(3, 6)$.

At $\theta^{(0)} = 0$, the per-sample gradients are:

$$(0 \cdot 1 - 2) \cdot 1 = -2,$$
$$(0 \cdot 2 - 4) \cdot 2 = -8,$$
$$(0 \cdot 3 - 6) \cdot 3 = -18.$$

Averaging over $n_1 = 3$ samples:

$$\nabla F_1(\theta^{(0)}) = \frac{-2 - 8 - 18}{3} = -9.33.$$

Client 2. Client 2 data: $(1, 1)$, $(2, 2)$, $(4, 4)$.

Per-sample gradients:

$$(0 \cdot 1 - 1) \cdot 1 = -1,$$
$$(0 \cdot 2 - 2) \cdot 2 = -4,$$
$$(0 \cdot 4 - 4) \cdot 4 = -16.$$

Averaging:

$$\nabla F_2(\theta^{(0)}) = \frac{-1 - 4 - 16}{3} = -7.$$

Client 3. Client 3 data: $(2, 3)$, $(3, 5)$, $(5, 8)$.

Per-sample gradients:

$$(0 \cdot 2 - 3) \cdot 2 = -6,$$
$$(0 \cdot 3 - 5) \cdot 3 = -15,$$
$$(0 \cdot 5 - 8) \cdot 5 = -40.$$

Averaging:

$$\nabla F_3(\theta^{(0)}) = \frac{-6 - 15 - 40}{3} = -20.33.$$

Step 3: Server-side Gradient Aggregation. Using the FedSGD update rule

$$\theta^{(t+1)} = \theta^{(t)} - \eta \sum_{k \in \mathcal{S}^{(t)}} \frac{n_k}{N} \nabla F_k(\theta^{(t)}),$$

and noting that all clients participate in round 1, we substitute the computed gradients:

$$\begin{aligned}\theta^{(1)} &= 0 - 0.1\left(\frac{3}{9}(-9.33) + \frac{3}{9}(-7) + \frac{3}{9}(-20.33)\right)\\ &= -0.1\,(-3.11 - 2.33 - 6.78)\\ &= -0.1(-12.22)\\ &= 1.222.\end{aligned}$$

Interpretation of Round 1 Update. After the first communication round, the global model parameter is updated from $\theta^{(0)} = 0$ to $\theta^{(1)} = 1.222$. This update corresponds to a single step of gradient descent on the virtual global objective, executed using distributed local gradients.

Although each client computed gradients using only its private data, the server-side aggregation reconciled these heterogeneous descent directions into a coherent global update. This example illustrates concretely how FedSGD realizes collaborative optimization while respecting data locality and privacy constraints.

Communication Round 2. At the end of Communication Round 1, the global model parameter has been updated to

$$\theta^{(1)} = 1.222.$$

This updated model is now broadcast by the server to all participating clients, serving as the common initialization point for the second round of local computations.

Step 1: Local Empirical Risk Functions (Unchanged). The local empirical risk functions remain identical to those defined in Communication Round 1:

$$F_k(\theta) = \frac{1}{n_k} \sum_{i=1}^{n_k} \frac{1}{2}(\theta x_{k,i} - y_{k,i})^2,$$

and the global objective decomposition is still given by

$$F(\theta) = \sum_{k=1}^{3} \frac{n_k}{N} F_k(\theta).$$

Step 2: Local Gradient Computation at Clients. Client 1. Client 1 data: $(1, 2)$, $(2, 4)$, $(3, 6)$.

At $\theta^{(1)} = 1.222$, the per-sample gradients are:

$$(1.222 \cdot 1 - 2) \cdot 1 = -0.778,$$
$$(1.222 \cdot 2 - 4) \cdot 2 = -3.112,$$
$$(1.222 \cdot 3 - 6) \cdot 3 = -7.002.$$

Averaging over $n_1 = 3$ samples:

$$\nabla F_1(\theta^{(1)}) = \frac{-0.778 - 3.112 - 7.002}{3} = -3.631.$$

Client 2. Client 2 data: $(1, 1)$, $(2, 2)$, $(4, 4)$.

Per-sample gradients:

$$(1.222 \cdot 1 - 1) \cdot 1 = 0.222,$$
$$(1.222 \cdot 2 - 2) \cdot 2 = 0.888,$$
$$(1.222 \cdot 4 - 4) \cdot 4 = 3.552.$$

Averaging:

$$\nabla F_2(\theta^{(1)}) = \frac{0.222 + 0.888 + 3.552}{3} = 1.554.$$

Client 3. Client 3 data: $(2, 3)$, $(3, 5)$, $(5, 8)$.

Per-sample gradients:

$$(1.222 \cdot 2 - 3) \cdot 2 = -1.112,$$
$$(1.222 \cdot 3 - 5) \cdot 3 = -4.002,$$
$$(1.222 \cdot 5 - 8) \cdot 5 = -9.450.$$

Averaging:

$$\nabla F_3(\theta^{(1)}) = \frac{-1.112 - 4.002 - 9.450}{3} = -4.855.$$

Step 3: Server-side Gradient Aggregation. Applying the FedSGD update rule,

$$\theta^{(t+1)} = \theta^{(t)} - \eta \sum_{k \in \mathcal{S}^{(t)}} \frac{n_k}{N} \nabla F_k(\theta^{(t)}),$$

with $\eta = 0.1$ and full participation, we obtain:

$$\theta^{(2)} = 1.222 - 0.1 \left(\frac{3}{9}(-3.631) + \frac{3}{9}(1.554) + \frac{3}{9}(-4.855) \right)$$

$$= 1.222 - 0.1\,(-1.210 + 0.518 - 1.618)$$
$$= 1.222 - 0.1(-2.310)$$
$$= 1.453.$$

Interpretation of Round 2 Update. After the second communication round, the global model parameter is updated from $\theta^{(1)} = 1.222$ to $\theta^{(2)} = 1.453$. Compared to Round 1, the magnitude of the update is smaller, reflecting a reduction in gradient magnitude as the model approaches a region of lower global loss.

Notably, Client 2 now contributes a *positive* gradient, indicating that its local objective favors increasing θ, while Clients 1 and 3 continue to favor decreasing θ. The server-side aggregation reconciles these competing preferences into a balanced global update.

This second round clearly demonstrates how FedSGD iteratively refines the global model through repeated cycles of local computation and centralized aggregation, progressively steering the parameters toward a stationary point of the virtual global objective.

Communication Round 3. At the end of Communication Round 2, the global model parameter is

$$\theta^{(2)} = 1.453.$$

This value is broadcast by the server to all participating clients and serves as the initialization point for the third round of local computations.

Step 1: Local Empirical Risk Functions (Unchanged). The local and global objective functions remain:

$$F_k(\theta) = \frac{1}{n_k}\sum_{i=1}^{n_k}\frac{1}{2}(\theta x_{k,i} - y_{k,i})^2, \qquad F(\theta) = \sum_{k=1}^{3}\frac{n_k}{N}F_k(\theta).$$

Step 2: Local Gradient Computation at Clients. Client 1. Client 1 data: $(1, 2)$, $(2, 4)$, $(3, 6)$.

At $\theta^{(2)} = 1.453$, the per-sample gradients are:

$$(1.453 \cdot 1 - 2) \cdot 1 = -0.547,$$
$$(1.453 \cdot 2 - 4) \cdot 2 = -2.188,$$
$$(1.453 \cdot 3 - 6) \cdot 3 = -4.923.$$

Averaging over $n_1 = 3$ samples:

$$\nabla F_1(\theta^{(2)}) = \frac{-0.547 - 2.188 - 4.923}{3} = -2.553.$$

Client 2. Client 2 data: $(1, 1)$, $(2, 2)$, $(4, 4)$.

Per-sample gradients:

$$
\begin{aligned}
(1.453 \cdot 1 - 1) \cdot 1 &= 0.453, \\
(1.453 \cdot 2 - 2) \cdot 2 &= 1.812, \\
(1.453 \cdot 4 - 4) \cdot 4 &= 7.248.
\end{aligned}
$$

Averaging:

$$
\nabla F_2(\theta^{(2)}) = \frac{0.453 + 1.812 + 7.248}{3} = 3.171.
$$

Client 3. Client 3 data: $(2, 3)$, $(3, 5)$, $(5, 8)$.
Per-sample gradients:

$$
\begin{aligned}
(1.453 \cdot 2 - 3) \cdot 2 &= -0.188, \\
(1.453 \cdot 3 - 5) \cdot 3 &= -1.923, \\
(1.453 \cdot 5 - 8) \cdot 5 &= -3.675.
\end{aligned}
$$

Averaging:

$$
\nabla F_3(\theta^{(2)}) = \frac{-0.188 - 1.923 - 3.675}{3} = -1.929.
$$

Step 3: Server-side Gradient Aggregation. Using the FedSGD update rule,

$$
\theta^{(t+1)} = \theta^{(t)} - \eta \sum_{k=1}^{3} \frac{n_k}{N} \nabla F_k(\theta^{(t)}),
$$

with $\eta = 0.1$ and equal dataset sizes ($n_k = 3$, $N = 9$), we obtain:

$$
\begin{aligned}
\theta^{(3)} &= 1.453 - 0.1 \left(\frac{1}{3}(-2.553) + \frac{1}{3}(3.171) + \frac{1}{3}(-1.929) \right) \\
&= 1.453 - 0.1\,(-0.851 + 1.057 - 0.643) \\
&= 1.453 - 0.1(-0.437) \\
&= 1.497.
\end{aligned}
$$

Interpretation of Round 3 Update. The update from $\theta^{(2)} = 1.453$ to $\theta^{(3)} = 1.497$ is smaller in magnitude than in previous rounds, indicating progressive stabilization of the optimization process. Local gradients increasingly cancel each other out, reflecting reduced disagreement among client objectives as the global model approaches a stationary region of the federated objective.

Table 9.3 shows the monotonic progression of the global parameter toward a stable value. As communication rounds increase, the magnitude of parameter updates decreases, illustrating the convergence behavior of FedSGD under distributed gradient aggregation.

Table 9.3 Evolution of the Global Model Parameter Across Communication Rounds

Round t	Global Parameter $\theta^{(t)}$	Update Magnitude
0	0.000	–
1	1.222	Large
2	1.453	Moderate
3	1.497	Small

Figure 9.1 illustrates the evolution of the global model parameter $\theta^{(t)}$ under federated stochastic gradient descent across successive communication rounds. The trajectory exhibits a characteristic pattern of rapid initial movement followed by progressively smaller updates, reflecting the reduction in gradient magnitude as the model approaches a stationary region of the federated objective. In early rounds, large parameter adjustments occur because local gradients across clients are strongly aligned, producing a substantial aggregate descent direction. As training proceeds, client-specific gradients increasingly oppose one another due to data heterogeneity, leading to partial cancellation during aggregation and smaller global updates.

This behavior highlights a key property of FedSGD: convergence is achieved not through aggressive local optimization, but through repeated global synchronization using distributed gradient information. While this ensures faithful approximation of centralized gradient descent under data locality constraints, it also explains the high communication cost of FedSGD, as each visible improvement in the global model requires a full communication round. This observation motivates the development of communication-efficient variants, such as Federated Averaging, which reduce synchronization frequency while preserving convergence guarantees.

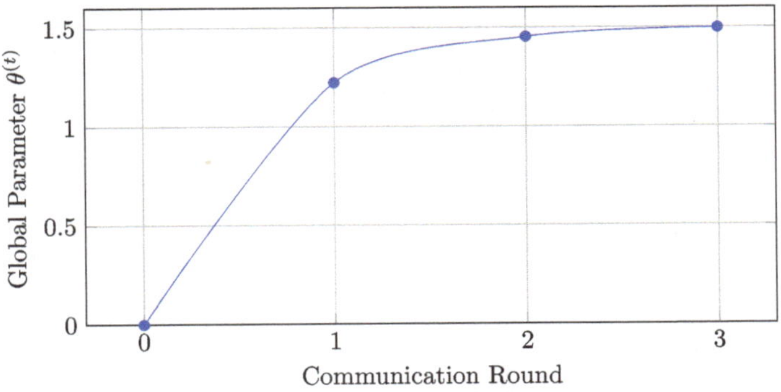

Fig. 9.1 Evolution of the global model parameter under Federated SGD across communication rounds. The diminishing step size reflects gradual convergence toward a stationary point of the federated objective

9.2 Gradient Aggregation Strategies

Gradient aggregation constitutes the *core coordination mechanism* in federated optimization. It is the step at which decentralized computation is transformed into coherent global learning. While local computation governs how each client interprets and optimizes with respect to its own private data, aggregation governs how these individually rational but potentially conflicting interpretations are combined into a single update for the shared global model. Without aggregation, federated learning would reduce to a collection of independent local training processes with no mechanism for collective improvement.

In this sense, gradient aggregation plays a role analogous to synchronization in classical distributed systems. However, the analogy is imperfect and intentionally limited. In traditional distributed optimization, synchronization typically assumes reliable communication, homogeneous workers, shared objectives, and consistent data access. Federated learning violates all of these assumptions simultaneously. Aggregation must therefore operate under far more restrictive and uncertain conditions, reconciling updates that are heterogeneous, asynchronous, and statistically biased.

In federated learning, clients operate autonomously and are not subordinate computational workers executing centrally prescribed tasks. Each client possesses its own private dataset, reflecting a localized data-generating process that may differ substantially from that of other clients. These datasets may vary in size, feature distribution, label frequency, and noise characteristics. As a result, the gradients computed by different clients—even when evaluated at the same global model—may differ significantly in both magnitude and direction.

Client participation further complicates this landscape. In most federated systems, only a subset of clients is available or selected to participate in any given communication round. Availability may depend on connectivity, energy constraints, local workload, or scheduling policies. Consequently, the set of gradients presented to the server varies across rounds, introducing system-level variability that compounds statistical heterogeneity.

Crucially, no client has visibility into the updates computed by others. Clients do not coordinate directly, do not share intermediate models, and do not negotiate update directions. Similarly, the server itself has no access to raw data, local loss values, or client-specific optimization states. It observes only the information that clients choose or are permitted to share, typically in the form of gradients or model updates. This strict separation is fundamental to the privacy and scalability goals of federated learning, but it sharply limits the information available for coordination.

As a result, the aggregation rule becomes the *only mathematical mechanism* through which collective learning behavior can emerge. It defines how heterogeneous, partial, and potentially biased local signals are fused into a single global update. The choice of aggregation strategy therefore has profound implications for convergence speed, stability, robustness, and fairness. Poor aggregation can amplify noise, entrench dominant client biases, or destabilize training altogether. Conversely, well-designed aggregation can mitigate heterogeneity, promote consensus, and enable reliable learning despite severe system constraints.

From this perspective, aggregation should not be viewed as a passive averaging step or a mere implementation detail. It is the central locus of coordination in federated optimization—the mathematical interface between autonomy and collaboration, locality and globality. Understanding aggregation strategies is therefore essential for understanding why federated learning behaves as it does, and how its limitations can be addressed through algorithmic design.

Data-weighted Gradient Averaging. The most widely adopted aggregation strategy is *data-weighted averaging*, defined as

$$\bar{g}^{(t)} = \sum_{k \in \mathcal{S}^{(t)}} \frac{n_k}{\sum_{j \in \mathcal{S}^{(t)}} n_j} \nabla F_k(\theta^{(t)}),$$

where:

- $\mathcal{S}^{(t)}$ denotes the set of clients participating in communication round t,
- $n_k = |\mathcal{D}_k|$ is the number of local samples at client k,
- $\nabla F_k(\theta^{(t)})$ is the gradient of the local empirical risk evaluated at the current global model.

This aggregation rule mirrors centralized empirical risk minimization. To see this, recall that in centralized learning, the global gradient can be expressed as

$$\nabla F(\theta) = \frac{1}{N} \sum_{k=1}^{K} \sum_{i=1}^{n_k} \nabla \ell\big(f_\theta(x_{k,i}), y_{k,i}\big).$$

Data-weighted aggregation reproduces this structure implicitly by ensuring that, in expectation, *each data sample contributes equally* to the global update—even though the samples themselves never leave their respective clients.

Consistency with Centralized Optimization
Data-weighted aggregation preserves the statistical structure of centralized empirical risk minimization while respecting strict data locality constraints.

Interpretation as a Global Descent Direction. From an optimization perspective, the aggregated gradient $\bar{g}^{(t)}$ can be interpreted as an approximation of the true global gradient $\nabla F(\theta^{(t)})$. When client data are independent and identically distributed (IID), this approximation is unbiased, and federated optimization behaves similarly to centralized stochastic gradient descent.

However, in realistic federated settings, this IID assumption rarely holds. Clients may observe distinct data-generating processes, leading to systematic differences in their local gradients. As a result, $\nabla F_k(\theta)$ may be a biased estimator of $\nabla F(\theta)$, even when aggregated.

Challenges under Statistical Heterogeneity. Under non-IID data, aggregation must contend with several competing objectives:

- **Fairness:** preventing clients with small but important datasets from being ignored,
- **Robustness:** mitigating the influence of noisy, corrupted, or adversarial updates,
- **Stability:** ensuring that conflicting gradients do not induce oscillations or divergence.

Simple averaging treats all gradients as equally trustworthy up to dataset size. This assumption can be problematic when some clients generate highly noisy gradients, operate on outdated models, or experience systematic bias due to skewed data distributions. In such cases, naive aggregation may amplify noise rather than suppress it.

Aggregation under Non-IID Data
When data are heterogeneous, aggregation must reconcile gradients that are not merely noisy, but systematically misaligned.

Beyond Simple Averaging. Recognizing these limitations, a broad family of alternative aggregation strategies has been developed. These include:

- *gradient clipping* to bound the influence of extreme updates,
- *normalization* to equalize update magnitudes across clients,
- *robust statistics* such as medians or trimmed means to resist outliers,
- *trust- and reputation-based weighting* to adaptively scale client contributions.

While these methods differ in implementation, they all modify the aggregation rule to trade off strict consistency with centralized optimization in favor of robustness and stability. Detailed formulations and theoretical analyses of these strategies are presented in later chapters.

Aggregation as Information Fusion
Federated aggregation fuses heterogeneous, partially biased gradient information into a single update direction without direct coordination or data sharing among clients.

In summary, gradient aggregation is not a mere averaging operation. It is the mathematical mechanism through which federated learning balances autonomy and collaboration, local specialization and global generalization. The choice of aggregation strategy fundamentally shapes the convergence behavior, robustness, and fairness of federated optimization algorithms, making it a central design axis in federated learning systems.

Worked Example: Data-Weighted Gradient Averaging

To concretely illustrate the mechanics of data-weighted gradient aggregation, we now present a fully worked numerical example. The goal is to demonstrate, step by step, how local gradients computed on heterogeneous clients are combined at the server using the data-weighted averaging rule introduced earlier.

Problem Setup and Notation. Consider a federated learning system with $K = 3$ clients participating in communication round t. Each client k holds a local dataset $\mathscr{D}_k$ of size n_k, and the total number of samples across participating clients is

$$N = \sum_{k=1}^{3} n_k.$$

Assume that the current global model parameter is $\theta^{(t)}$ and that each client computes the gradient of its local empirical risk,

$$\nabla F_k(\theta^{(t)}) = \frac{1}{n_k} \sum_{i=1}^{n_k} \nabla \ell\big(f_{\theta^{(t)}}(x_{k,i}), y_{k,i}\big).$$

Client-Side Data Summary. Table 9.4 summarizes the number of samples held by each client and the locally computed gradients at round t.

The total number of samples is therefore

$$N = 50 + 30 + 20 = 100.$$

Step 1: Compute Aggregation Weights. Using the data-weighted aggregation rule, each client is assigned a weight proportional to its dataset size:

$$w_k = \frac{n_k}{\sum_{j=1}^{3} n_j} = \frac{n_k}{N}.$$

Table 9.4 Client Data Sizes and Local Gradients at Communication Round t

Client k	Dataset Size n_k	Local Gradient $\nabla F_k(\theta^{(t)})$
1	50	–0.80
2	30	–0.20
3	20	+0.40

Substituting the values from Table 9.4, we obtain:

$$w_1 = \frac{50}{100} = 0.50, \quad w_2 = \frac{30}{100} = 0.30, \quad w_3 = \frac{20}{100} = 0.20.$$

Step 2: Apply the Data-Weighted Averaging Formula. The aggregated gradient at communication round t is given by

$$\bar{g}^{(t)} = \sum_{k=1}^{3} \frac{n_k}{N} \nabla F_k(\theta^{(t)}).$$

Substituting the computed weights and local gradients:

$$\begin{aligned} \bar{g}^{(t)} &= (0.50)(-0.80) + (0.30)(-0.20) + (0.20)(0.40) \\ &= -0.40 - 0.06 + 0.08 \\ &= -0.38. \end{aligned}$$

Step 3: Interpretation of the Aggregated Gradient. The resulting aggregated gradient $\bar{g}^{(t)} = -0.38$ represents the server's best estimate of the global descent direction at round t. Although Client 3 proposes a positive gradient direction, its influence is moderated by its smaller dataset size. Conversely, Client 1 exerts the strongest influence due to its larger number of samples.

This outcome illustrates a key property of data-weighted aggregation: *each individual data sample contributes equally in expectation*, even though gradients are computed locally and aggregated centrally.

Step 4: Global Model Update. If the server applies a learning rate $\eta > 0$, the global model is updated according to

$$\theta^{(t+1)} = \theta^{(t)} - \eta \bar{g}^{(t)}.$$

For example, with $\eta = 0.1$, the update becomes

$$\theta^{(t+1)} = \theta^{(t)} - 0.1(-0.38) = \theta^{(t)} + 0.038.$$

Pedagogical Insight. This worked example demonstrates how data-weighted gradient averaging balances heterogeneous client contributions through principled weighting. Rather than treating all client updates equally, the aggregation rule aligns federated optimization with the statistical structure of centralized empirical risk minimization, while still respecting strict data locality constraints.

9.3 Communication–Computation Trade-Offs

Federated optimization is fundamentally shaped by *communication constraints*. Unlike centralized or tightly coupled distributed systems, federated learning operates over networks that are bandwidth-limited, unreliable, and often energy-constrained. In many realistic deployments—such as mobile devices, edge sensors, and Internet-of-Things platforms—the cost of communication far exceeds the cost of local computation. As a consequence, algorithm design must explicitly balance how often clients communicate with the server against how much computation they perform locally between communications.

Local Computation Parameter τ. A key abstraction used to formalize this balance is the parameter τ, which denotes the number of local optimization steps performed by each client within a single communication round. When $\tau = 1$, each client computes a single gradient step before synchronization, recovering the behavior of FedSGD. As τ increases, clients perform progressively more local optimization before communicating their updates to the server.

From a system perspective, increasing τ reduces the number of communication rounds required to reach a target accuracy. However, from an optimization perspective, this introduces new challenges, as local models may drift away from the global objective.

Formal Description of Local Updates. Let $\theta^{(t)}$ denote the global model broadcast by the server at the start of communication round t. Client k initializes its local model as

$$\theta_k^{(t,0)} = \theta^{(t)}.$$

The client then performs τ steps of local stochastic gradient descent:

$$\theta_k^{(t,s+1)} = \theta_k^{(t,s)} - \eta \, \nabla F_k(\theta_k^{(t,s)}), \qquad s = 0, 1, \dots, \tau - 1.$$

After τ local steps, the client produces the update

$$\theta_k^{(t,\tau)} = \theta^{(t)} - \eta \sum_{s=0}^{\tau-1} \nabla F_k(\theta_k^{(t,s)}),$$

which is transmitted to the server for aggregation.

This expression highlights an important distinction from centralized SGD: the gradients appearing in the sum are evaluated at *different local parameter values*, not at the shared global model.

Local Drift and Optimization Bias. When data are identically distributed across clients, local updates tend to remain aligned with the global descent direction.

However, under non-IID data—a defining feature of federated learning—the local objective functions F_k may differ substantially. As τ increases, each client increasingly optimizes its own local objective, causing its model trajectory to drift away from the minimizer of the global objective.

This phenomenon, commonly referred to as *local drift*, introduces bias into the aggregated update. While communication becomes more efficient, the optimization process may slow down, oscillate, or even diverge if local drift is not properly controlled.

Competing Objectives. The design of federated optimization algorithms must therefore balance three competing goals:

- **Communication efficiency:** minimizing the number of synchronization rounds,
- **Optimization stability:** preventing excessive drift due to heterogeneous local updates,
- **Convergence speed:** achieving rapid reduction of the global objective value.

No single choice of τ optimizes all three objectives simultaneously. Small values of τ favor stability but incur high communication costs, while large values of τ reduce communication but risk introducing significant optimization bias.

Fundamental Communication–Computation Trade-Off
In federated learning, reducing communication by increasing local computation inevitably amplifies optimization bias due to local drift. Effective algorithms carefully control this bias rather than eliminating it entirely.

This trade-off underlies the design of nearly all modern federated optimization algorithms. Methods such as Federated Averaging, proximal regularization, and adaptive aggregation can all be interpreted as different strategies for navigating the communication–computation frontier.

Figure 9.2 illustrates the fundamental communication–computation trade-off that governs federated optimization algorithms. As the number of local optimization steps τ increases, overall training efficiency initially improves due to a reduction in communication frequency and synchronization overhead. Beyond a moderate range of τ, however, efficiency begins to decrease as excessive local computation amplifies local drift, causing client updates to diverge from the global descent direction.

The non-monotonic shape of the curve highlights that neither extreme is optimal: overly frequent communication leads to inefficiency due to network costs, while overly infrequent communication degrades optimization quality. Effective federated learning algorithms therefore operate in an intermediate regime, carefully selecting τ to balance communication savings against optimization stability and convergence speed.

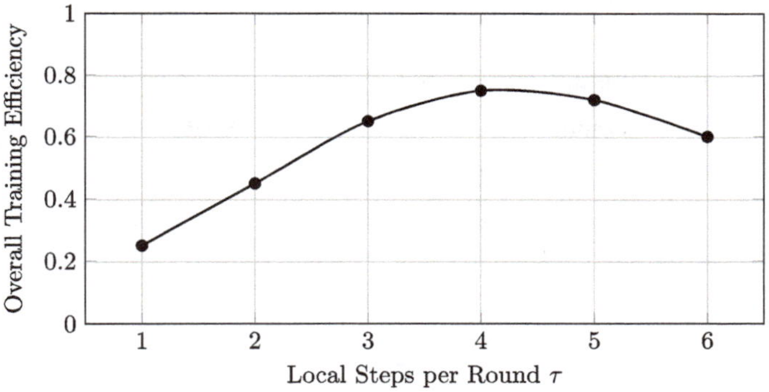

Fig. 9.2 Illustrative communication–computation trade-off in federated optimization. Increasing the number of local steps τ initially improves efficiency by reducing communication, but excessive local computation leads to degradation due to local drift and optimization bias

Worked Example: Communication–Computation Trade-Off with Explicit Values

To make the communication–computation trade-off precise, we now present a fully worked numerical example that explicitly instantiates the main formulas introduced in this section. The objective is to show how increasing the number of local steps τ reduces communication frequency while simultaneously introducing local drift.

Problem Setup. Consider a federated learning system with $K = 3$ clients. Each client holds a small linear regression dataset, summarized in Table 9.5. We use a one-dimensional linear model

$$f_\theta(x) = \theta x,$$

with squared loss

$$\ell(f_\theta(x), y) = \frac{1}{2}(\theta x - y)^2.$$

Each client has $n_k = 2$ samples, so the total dataset size is $N = 6$. Assume the global model at communication round t is

$$\theta^{(t)} = 0,$$

Table 9.5 Client Datasets for Communication–Computation Trade-Off Example

Client	Inputs x	Targets y
1	$\{1, 2\}$	$\{2, 4\}$
2	$\{1, 3\}$	$\{1, 3\}$
3	$\{2, 4\}$	$\{3, 6\}$

and the learning rate is $\eta = 0.1$.

Local Gradient Formula. Recall that the gradient of the squared loss with respect to θ is

$$\nabla_\theta \ell = (\theta x - y)x.$$

The local empirical risk gradient at client k is therefore

$$\nabla F_k(\theta) = \frac{1}{n_k} \sum_{i=1}^{n_k} (\theta x_{k,i} - y_{k,i}) x_{k,i}.$$

Case 1: $\tau = 1$ (FedSGD Behavior). Each client computes a single gradient at $\theta^{(t)} = 0$.

Client 1:

$$(-2) \cdot 1 = -2, \quad (-4) \cdot 2 = -8, \quad \nabla F_1(0) = \frac{-2-8}{2} = -5.$$

Client 2:

$$(-1) \cdot 1 = -1, \quad (-3) \cdot 3 = -9, \quad \nabla F_2(0) = \frac{-1-9}{2} = -5.$$

Client 3:

$$(-3) \cdot 2 = -6, \quad (-6) \cdot 4 = -24, \quad \nabla F_3(0) = \frac{-6-24}{2} = -15.$$

The server aggregates gradients using data-weighted averaging:

$$\bar{g}^{(t)} = \sum_{k=1}^{3} \frac{n_k}{N} \nabla F_k(0) = \frac{2}{6}(-5-5-15) = -8.33.$$

The global update is

$$\theta^{(t+1)} = 0 - 0.1(-8.33) = 0.833.$$

Case 2: $\tau = 2$ (Two Local Steps). Each client now performs two local SGD steps before communication.

Client 1 Local Updates.

$$\theta_1^{(t,1)} = 0 - 0.1(-5) = 0.5.$$

Recompute gradients at $\theta_1^{(t,1)} = 0.5$:

$$(0.5 \cdot 1 - 2) \cdot 1 = -1.5, \quad (0.5 \cdot 2 - 4) \cdot 2 = -6,$$

$$\nabla F_1(0.5) = \frac{-1.5 - 6}{2} = -3.75.$$

Client 2 Local Updates.

$$\theta_2^{(t,1)} = 0.5, \quad \nabla F_2(0.5) = \frac{-0.5 - 7.5}{2} = -4.$$

Client 3 Local Updates.

$$\theta_3^{(t,1)} = 0.5, \quad \nabla F_3(0.5) = \frac{-4 - 20}{2} = -12.$$

Each client transmits its final model:

$$\theta_k^{(t,2)} = \theta^{(t)} - 0.1\big(\nabla F_k(0) + \nabla F_k(0.5)\big).$$

Numerically:

$$\theta_1^{(t,2)} = 0.875, \quad \theta_2^{(t,2)} = 0.9, \quad \theta_3^{(t,2)} = 2.7.$$

The server aggregates:

$$\theta^{(t+1)} = \frac{1}{3}(0.875 + 0.9 + 2.7) = 1.492.$$

Interpretation of the Trade-Off. When $\tau = 1$, communication is frequent and updates closely follow the global descent direction. When $\tau = 2$, communication is reduced by half, but client 3 drifts significantly due to its steeper local objective, pulling the aggregated model farther from the FedSGD trajectory.

This example demonstrates explicitly how increasing τ improves communication efficiency while simultaneously amplifying local drift, capturing the fundamental communication–computation trade-off in federated optimization.

9.4 Comparison with Centralized Stochastic Gradient Descent

Federated optimization algorithms draw their conceptual roots from centralized stochastic gradient descent (SGD). Both paradigms rely on iterative improvement of a shared model through gradient-based updates. However, despite this superficial similarity, the operational realities and learning dynamics of federated optimization differ in profound and structural ways from those of centralized SGD. Understanding these differences is essential for appreciating why classical SGD theory cannot be applied directly to federated learning systems.

9.4.1 Centralized SGD: A Controlled Optimization Environment

In centralized stochastic gradient descent, all training data are stored and processed within a single, tightly managed computational environment. This centralized setting allows the learning algorithm to draw mini-batches from a globally pooled dataset, ensuring that each mini-batch is, in expectation, representative of the overall data distribution. Consequently, the gradients computed at each iteration capture the dominant statistical trends of the entire dataset rather than localized or idiosyncratic patterns.

Because of this global visibility, each update direction can be interpreted as a reliable and unbiased approximation of the ideal descent direction. Noise in the updates arises primarily from random mini-batch sampling, and this noise is well understood and amenable to classical analytical treatment. As the number of iterations increases, these stochastic fluctuations tend to average out, enabling steady and predictable progress toward an optimal solution.

This highly controlled environment supports a clean and mature theoretical framework. Convergence guarantees typically rely on assumptions such as smoothness of the loss function, bounded variance of stochastic gradients, and the ability to repeatedly sample fresh data in an independent manner. Crucially, system-level considerations—such as network latency, intermittent worker failures, or heterogeneous compute resources—are either absent or abstracted away. As a result, the behavior of centralized SGD is dominated by statistical properties of the data rather than by infrastructural constraints.

9.4.2 Federated Optimization: Decentralization and Autonomy

Federated optimization operates in a markedly different regime, one defined by decentralization and client autonomy. Instead of a single global dataset, training data are partitioned across a potentially vast population of clients, each of which observes only a small and localized subset of the overall data. These client datasets often arise from distinct data-generating processes, reflecting differences in user behavior, geographic location, sensing conditions, or usage patterns.

Gradients in federated optimization are therefore computed locally on client-specific data. As a result, they may vary widely in magnitude and direction, depending on the statistical characteristics of each client's dataset. Unlike centralized SGD, there is no guarantee that any individual client's update is representative of the global population. This inherent heterogeneity fundamentally alters the learning dynamics.

An additional layer of complexity arises from partial participation. In most federated systems, only a small fraction of clients are active and available at any given communication round. Client availability may fluctuate due to connectivity constraints, energy considerations, or competing local tasks. This randomness in participation introduces system-level stochasticity that has no direct analogue in centralized optimization.

Consequently, the aggregated update produced by the server reflects a combination of two sources of uncertainty: statistical variability due to heterogeneous data and randomness due to client sampling. Even when the global model remains unchanged, different subsets of participating clients can yield substantially different update directions. This dual source of variability makes federated optimization inherently less predictable and more challenging to analyze than its centralized counterpart.

9.4.3 Implications for Learning Dynamics

The structural differences between centralized and federated optimization give rise to learning dynamics that are qualitatively distinct from those observed in classical stochastic gradient descent. These differences manifest not merely as quantitative slowdowns, but as fundamental changes in how and why the optimization process evolves over time.

A first and immediately observable consequence is slower convergence. In federated optimization, particularly under statistically heterogeneous data, local updates computed by different clients may point in partially or even fully opposing directions. When such conflicting updates are aggregated, their effects can cancel one another, producing a global update that is small in magnitude or poorly aligned with the true global descent direction. As a result, progress toward minimizing the global objective may be incremental and uneven, even when individual clients experience substantial local improvement.

A second implication is heightened sensitivity to client participation. Because only a subset of clients typically participates in any given communication round, the composition of this subset plays a critical role in shaping the update direction. Two rounds that begin from the same global model can yield markedly different updates if they involve different participating clients. This phenomenon introduces a form of stochasticity that operates at the system level rather than at the data level. Unlike mini-batch noise in centralized SGD, which is largely independent across iterations, client sampling noise may exhibit persistent patterns and correlations driven by user behavior or device availability.

A third and more subtle effect is the emergence of systematic bias in aggregated updates. Since local gradients are computed with respect to client-specific objectives, their aggregation does not, in general, produce an unbiased estimate of the centralized gradient. This bias can persist even as the number of participating clients increases, particularly when client data distributions differ substantially. Without corrective mechanisms, the global model may converge to a point that reflects a compromise among dominant client objectives rather than a true minimizer of the global empirical risk.

Collectively, these behaviors imply that federated optimization cannot be adequately understood as a noisy version of centralized SGD. Instead, it represents a distinct optimization regime in which convergence speed, stability, and solution quality are shaped jointly by data heterogeneity, client participation patterns, and aggregation design. Recognizing and addressing these implications is central to the development of effective and robust federated learning algorithms.

Key Distinction
Federated optimization generalizes centralized stochastic gradient descent by replacing direct access to global gradient information with aggregation of decentralized, heterogeneous, and partially observed client updates.

9.4.4 Consequences for Algorithm Design

Because federated optimization fundamentally relaxes the assumptions that underpin classical stochastic gradient descent, it necessitates a reconsideration of both algorithmic structure and theoretical analysis. Many of the guarantees and intuitions developed for centralized SGD depend critically on direct access to unbiased gradient information, stable sampling processes, and tightly controlled execution environments. When these assumptions no longer hold, new design principles become essential.

First, federated algorithms must explicitly account for statistical heterogeneity across clients. Since local objectives may differ substantially, naive aggregation of client updates can lead to slow convergence or unstable behavior. This has motivated the development of local update schemes that allow clients to make meaningful progress on their own data, while also incorporating mechanisms—such as regularization or correction terms—to prevent excessive drift away from the global objective.

Second, algorithm design must accommodate partial and variable client participation. Federated systems cannot assume that the same set of clients will be available at every round, nor that participation will be uniformly random. As a result, modern federated algorithms incorporate adaptive aggregation strategies, client weighting schemes, and scheduling policies that are robust to changing participation patterns and imbalanced contributions.

Third, communication constraints play a central role in shaping algorithmic choices. Because communication is often the dominant cost, federated algorithms are designed to reduce synchronization frequency, compress updates, or amortize communication over multiple local computations. These strategies introduce additional sources of bias and variance, which must be carefully managed through algorithmic safeguards and theoretical analysis.

Finally, federated optimization raises new concerns related to robustness and reliability. Client updates may be noisy, stale, or even adversarial, either due to system faults or malicious behavior. Consequently, algorithm design increasingly incorporates robustness mechanisms that limit the influence of anomalous updates and protect the integrity of the global model.

Federated optimization should not be viewed as a minor or incremental variant of centralized stochastic gradient descent. Rather, it represents a broader and more challenging generalization, in which learning emerges from the coordinated interaction of many independent and heterogeneous learners operating under strict system and privacy constraints. These fundamental distinctions give rise to a rich and diverse landscape of federated optimization algorithms, which are examined in detail in the subsequent sections.

9.5 Convergence Behavior and Limitations

The study of convergence in federated optimization occupies a substantially more complex theoretical landscape than that of centralized learning. In centralized optimization, analysis typically focuses on a single, well-defined optimization trajectory driven by unbiased stochastic gradients. In contrast, federated optimization involves the interaction of many simultaneous local optimization processes, each evolving under its own data distribution, computational constraints, and participation schedule. As a result, convergence behavior in federated systems reflects not only the properties of the objective function, but also the structure of the system in which optimization is carried out.

9.5.1 Multiple Interacting Optimization Trajectories

At the heart of the complexity of federated convergence lies the fact that federated learning does not evolve along a single, well-defined optimization trajectory. Instead, each participating client maintains its own local model and follows an individual optimization path shaped by its private data, local objective function, and computational constraints. Between communication rounds, these local trajectories evolve independently, and in general, they do not remain aligned with one another or with the global objective.

As local training proceeds, client models may temporarily diverge from the global model and from each other. The degree of divergence depends on factors such as data heterogeneity, the number of local update steps, and the choice of learning rate. Periodic aggregation serves as a synchronization mechanism that reconciles these divergent trajectories, producing a new global model that represents a compromise among the local solutions. This global model then becomes the common reference point from which the next round of local optimization begins.

Crucially, the global learning process emerges not from a single stochastic gradient process, but from the repeated interaction of many partially synchronized local optimizers. The resulting dynamics are inherently nonlinear and history-dependent: the effect of a local update depends on how far a client has drifted since the last synchronization and on how its trajectory interacts with those of other clients. This interdependence makes federated optimization fundamentally more complex to analyze than centralized stochastic gradient descent, where only one trajectory must be tracked.

9.5.2 Role of Participation and Synchronization

A further distinguishing feature of federated convergence analysis is the stochastic nature of client participation. In most federated systems, only a fraction of the total client population is available or selected to participate in any given communication round. Client availability may vary due to connectivity constraints, power considerations, local workload, or scheduling policies. As a result, the set of participating clients changes over time in a manner that is often unpredictable.

This variability in participation introduces an additional source of randomness that operates at the system level rather than at the data level. Even if the global model and client datasets remain unchanged, two consecutive communication rounds may produce substantially different update directions simply because different subsets of clients are active. This form of randomness has no direct analogue in centralized optimization, where data access is assumed to be stable and uniformly available.

Delayed synchronization further compounds this challenge. Local updates are computed using versions of the global model that may already be outdated by the time they are aggregated. Asynchrony and staleness increase the discrepancy between local and global objectives, amplifying the effects of heterogeneity and making the aggregated update less representative of the true global descent direction. Together, stochastic participation and delayed synchronization create a dynamic and evolving optimization landscape, in which convergence must be understood as a collective and time-varying phenomenon rather than a simple monotonic process.

9.5.3 Convergence under Idealized Assumptions

Despite the substantial challenges introduced by decentralization, heterogeneity, and partial participation, it is still possible to establish convergence guarantees for federated optimization algorithms under a set of standard but idealized assumptions.

These assumptions are largely inherited from classical optimization theory and are introduced to render the analysis mathematically tractable.

In particular, convergence results are typically derived when the global objective function satisfies properties such as convexity or smoothness, ensuring that local gradients do not change abruptly with small variations in the model parameters. Additional assumptions commonly include bounded variance of local gradients, appropriately chosen learning rates, and sufficiently frequent participation of a representative subset of clients. Under such conditions, the aggregated updates can be shown to make consistent progress toward reducing the global objective.

Within this idealized regime, federated optimization algorithms exhibit behavior that is qualitatively similar to centralized stochastic gradient descent. The global model approaches stationary points of the objective, and the overall learning dynamics remain stable. However, even under these favorable assumptions, convergence is typically slower than in centralized settings. The additional variance introduced by decentralized computation, client sampling, and delayed synchronization accumulates over time, leading to looser convergence bounds and slower asymptotic rates. These results highlight that even in best-case scenarios, federated optimization incurs an inherent efficiency cost relative to centralized learning.

9.5.4 Impact of Statistical Heterogeneity

A dominant factor shaping convergence behavior in practical federated learning systems is statistical heterogeneity across clients. When client data distributions differ substantially, local objective functions no longer align with the global objective. As a result, local gradients computed by different clients may point in distinct, and sometimes opposing, directions.

This misalignment between local and global descent directions manifests as persistent client drift. Between communication rounds, clients increasingly optimize their own local objectives, moving the model toward region-specific optima rather than toward a globally optimal solution. When such drifted updates are aggregated, their effects may partially cancel or reinforce undesirable directions, slowing convergence or introducing oscillatory behavior.

As heterogeneity increases, these effects become more pronounced. Convergence rates degrade, and the global model may converge to a solution that reflects a compromise among dominant client objectives rather than a true minimizer of the global empirical risk. This degradation under non-identically distributed data is a defining characteristic of federated optimization and represents one of its most significant theoretical and practical challenges. Addressing this issue has therefore become a primary motivation for the development of refined aggregation rules, regularization techniques, and personalization strategies.

9.5.5 Sensitivity to Algorithmic Hyperparameters

Federated optimization algorithms are also notably sensitive to the choice of hyperparameters. Learning rates that yield stable and efficient convergence in centralized settings may lead to instability when applied directly in federated environments. This sensitivity is amplified by the presence of multiple local update steps, as errors introduced at the client level can accumulate before being corrected through aggregation.

The number of local steps per communication round plays a particularly critical role. If too few local steps are performed, communication overhead dominates, leading to inefficient use of system resources. If too many local steps are performed, local drift becomes excessive, introducing bias and variance that can overwhelm the benefits of reduced communication. Striking the right balance requires careful tuning and often depends on the degree of data heterogeneity, client availability, and network conditions.

This pronounced sensitivity complicates practical deployment of federated learning systems. It underscores the need for adaptive and robust strategies that can adjust learning rates, local update counts, and aggregation behavior in response to observed training dynamics. Such strategies aim to reduce the burden of manual tuning while maintaining stability and efficiency across a wide range of operating conditions.

9.5.6 Robustness and Reliability Concerns

Beyond purely statistical considerations, the convergence behavior of federated optimization algorithms is strongly influenced by the quality and reliability of client updates. In realistic federated systems, client devices operate under diverse and often constrained conditions, which can significantly affect the fidelity of the updates they produce.

One common source of unreliability arises from limited local data. Clients with very small datasets may generate gradients that are highly noisy and poorly representative of the underlying data distribution. When such updates are aggregated alongside those from more data-rich clients, they can inject disproportionate variance into the global update, slowing convergence and increasing instability. This effect is exacerbated when participation is sparse or highly variable across rounds.

Another important factor is model staleness. Due to intermittent connectivity or asynchronous execution, some clients may compute updates using outdated versions of the global model. These stale updates reflect optimization dynamics that are no longer aligned with the current state of training. When aggregated, they can pull the global model in directions that are inconsistent with recent progress, undermining both convergence speed and stability.

Computational limitations further contribute to reliability concerns. Edge devices may rely on low-precision arithmetic, experience hardware interruptions, or terminate local training prematurely due to resource constraints. Such factors can distort local updates in subtle ways, making them systematically different from those assumed in theoretical analyses.

In more adversarial or untrusted environments, robustness becomes an even more critical issue. Some clients may intentionally manipulate their updates to disrupt training, bias the global model, or extract sensitive information. Basic federated optimization algorithms, which typically aggregate all received updates without scrutiny, are particularly vulnerable to such behavior. Even a small number of malicious or compromised clients can significantly degrade performance or cause outright convergence failure.

Together, these reliability and robustness challenges underscore the limitations of foundational federated optimization methods. They motivate the incorporation of defensive mechanisms—such as robust aggregation, anomaly detection, and update filtering—that aim to preserve convergence guarantees in the presence of noise, staleness, and adversarial behavior. Addressing these concerns is essential for deploying federated learning systems in real-world, open, and large-scale environments.

Limits of Basic Federated Optimization
Foundational federated optimization algorithms expose intrinsic limitations arising from decentralization, heterogeneity, and partial participation, motivating the development of proximal, adaptive, and personalized methods.

Exercises

This chapter examined the foundations of federated optimization, highlighting how gradient-based learning must be redesigned under data decentralization, partial participation, and communication constraints. The following exercises reinforce conceptual understanding, analytical reasoning, and practical design considerations in federated optimization.

A. Short Answer Questions

Answer each question briefly and precisely.

1. What distinguishes federated optimization from centralized stochastic gradient descent?
2. Why is the global objective function in federated learning referred to as a *virtual objective*?
3. What role does the server play in federated stochastic gradient descent?
4. Explain the meaning of data-weighted gradient aggregation.
5. Why are local gradients potentially biased estimators of the global gradient?
6. What is meant by *client drift* in federated learning?

7. Why does FedSGD require frequent communication rounds?
8. What is the significance of partial client participation?
9. How does statistical heterogeneity affect convergence behavior?
10. Why is gradient aggregation considered the core coordination mechanism in federated learning?

B. Long Answer Questions

Answer the following questions with clear explanations, equations, and diagrams wherever appropriate.

1. Derive the FedSGD update rule starting from the federated global objective and explain each term.
2. Compare centralized SGD and federated SGD in terms of gradient computation, unbiasedness, and system assumptions.
3. Explain the communication–computation trade-off in federated learning and discuss how the parameter τ influences convergence.
4. Analyze the limitations of FedSGD under non-IID data distributions.
5. Discuss why naive averaging of client gradients may fail in heterogeneous federated environments.
6. Explain how partial participation introduces system-level stochasticity in federated optimization.
7. Describe the impact of learning rate selection in federated optimization compared to centralized settings.
8. Explain why federated optimization convergence analysis is more complex than centralized SGD analysis.

C. True/False Questions

State whether each statement is **True** or **False**.

1. In federated learning, the server has access to all client data.
2. FedSGD performs exactly one global update per communication round.
3. Under IID data, FedSGD behaves similarly to centralized SGD.
4. Local gradients are always unbiased estimators of the global gradient.
5. Increasing the number of local steps reduces communication frequency.
6. Partial client participation introduces additional randomness into training.
7. Data-weighted aggregation ensures equal contribution from each client, regardless of dataset size.
8. Client drift increases with statistical heterogeneity.
9. FedSGD efficiently utilizes client-side computation resources.
10. Robust aggregation methods are unnecessary in adversarial settings.

Answers:

1. False
2. True
3. True
4. False
5. True
6. True
7. False
8. True
9. False
10. False

D. Fill in the Blanks

Fill in the blanks with the most appropriate term or expression.

1. In federated learning, training data remains ___________ across clients.
2. The global objective in federated learning cannot be ___________ directly by the server.
3. FedSGD aggregates client gradients using ___________ weighting.
4. Statistical heterogeneity leads to persistent ___________ across clients.
5. Increasing τ reduces communication but increases ___________.
6. Local gradients are computed using ___________ data only.
7. Partial participation introduces randomness at the ___________ level.
8. In centralized SGD, mini-batch gradients are ___________ estimators of the global gradient.
9. Robust aggregation methods aim to reduce the influence of ___________ updates.
10. FedSGD is best viewed as a ___________ baseline algorithm.

Answers:

1. decentralized
2. evaluated
3. data-size
4. client drift
5. optimization bias
6. local
7. system
8. unbiased
9. noisy or adversarial
10. conceptual

E. Multiple Choice Questions (MCQs)

Choose the **most appropriate** answer.

1. What is the primary limitation of FedSGD?
   (a) High memory usage
   (b) Excessive communication
   (c) Model overfitting
   (d) Lack of convergence
2. Which factor most strongly degrades convergence under non-IID data?
   (a) Learning rate
   (b) Client drift
   (c) Batch size
   (d) Model depth
3. What does data-weighted aggregation preserve?
   (a) Client fairness
   (b) Privacy guarantees
   (c) Sample-level contribution
   (d) Communication efficiency
4. In federated learning, gradients are computed:
   (a) Centrally at the server
   (b) On pooled datasets
   (c) Locally at clients
   (d) Using synthetic data
5. Which parameter controls local computation intensity?
   (a) η
   (b) K
   (c) τ
   (d) N

Answers:

1. (b)
2. (b)
3. (c)
4. (c)
5. (c)

F. Application and Project-Based Questions

These problems emphasize modeling, reasoning, and implementation.

1. Simulate FedSGD on a synthetic non-IID dataset and analyze how client drift affects convergence.

2. Compare FedSGD and Federated Averaging in terms of communication cost and convergence speed.
3. Design an experiment to study the effect of partial client participation on training stability.
4. Implement data-weighted and uniform aggregation and compare their performance under heterogeneous data.
5. Analyze the impact of increasing τ on convergence for a fixed communication budget.
6. Propose a robust aggregation strategy for handling noisy client updates.
7. Study how learning rate tuning differs between centralized and federated optimization.
8. Investigate the effect of stale updates in asynchronous federated learning.
9. Design a federated learning system for mobile devices and identify key optimization challenges.
10. Extend FedSGD with a simple regularization mechanism to reduce client drift and evaluate its effectiveness.

Federated Averaging 10

Federated Averaging (FedAvg) is widely regarded as the canonical optimization algorithm for federated learning systems. Its importance does not stem from algorithmic complexity or exotic mathematical constructs, but rather from the fact that it provides the simplest viable resolution to the core tension at the heart of federated learning: how to reconcile distributed, private, and heterogeneous local computation with the need for coherent global optimization, under severe communication constraints.

Conceptually, aederated averaging should not be interpreted as a radical departure from federated stochastic gradient descent (FedSGD). Instead, it is more accurately viewed as a communication-efficient generalization of FedSGD—one that preserves the same statistical foundations while relaxing a critical system-level requirement. In FedSGD, each improvement of the global model is achieved through immediate synchronization: clients compute gradients at the current global parameter vector and transmit them to the server at every iteration. While this strategy closely approximates centralized stochastic gradient descent and enjoys clean theoretical interpretation, it imposes an extremely high communication burden in realistic federated environments.

Federated averaging relaxes this requirement by allowing clients to perform multiple local optimization steps before participating in aggregation. Rather than synchronizing after every gradient computation, each client locally refines the shared model over several iterations using its private objective function. Only after this local optimization phase does the client transmit information back to the server. This simple modification has profound consequences: communication frequency is reduced by an order proportional to the number of local steps, while the essential structure of global learning is retained.

Crucially, FedAvg does not abandon the data-weighted aggregation principle established in federated stochastic gradient descent. Earlier chapters showed that weighting client contributions in proportion to local dataset sizes ensures statistical

P. P. Ray, *Foundations of Federated Learning*,
Springer Series in the Data Sciences,
https://doi.org/10.1007/978-3-032-34084-9_10

consistency with centralized empirical risk minimization. Federated averaging preserves this principle exactly. The key change lies not in how client contributions are combined, but in what is being combined. Instead of aggregating gradients evaluated at a shared parameter vector, FedAvg aggregates model parameters that have evolved along client-specific optimization trajectories. As a result, the server integrates information accumulated over multiple local gradient evaluations, implicitly capturing richer optimization dynamics with far fewer synchronization rounds.

From a system perspective, this design choice directly addresses the dominant bottleneck in federated learning: communication cost. In large-scale deployments involving mobile devices, edge sensors, or intermittently connected clients, communication is often orders of magnitude more expensive than local computation. By shifting work from the network to the client, federated averaging enables federated learning to scale to populations consisting of thousands or even millions of participants.

At the same time, this relaxation introduces new challenges. When local data distributions differ substantially across clients—a defining characteristic of federated learning—local optimization steps may drift toward client-specific optima that are not perfectly aligned with the global objective. The aggregation of such locally optimized models can introduce bias, slow convergence, or induce instability if not carefully controlled. Understanding this trade-off between communication efficiency and optimization fidelity is therefore essential for both theoretical analysis and practical algorithm design.

Accordingly, this chapter presents a coherent and mathematically consistent treatment of federated averaging, explicitly grounded in the aggregation framework developed in earlier chapters. We begin by formally describing the FedAvg algorithm and its relationship to federated stochastic gradient descent. We then contrast weight averaging and gradient averaging, highlighting their conceptual and mathematical differences. Next, we examine the communication efficiency gains enabled by FedAvg and analyze the fundamental communication–computation trade-off that governs its behavior. We subsequently discuss adaptive federated optimizers that extend the FedAvg framework through momentum and second-order correction mechanisms. Finally, we analyze the performance of federated averaging under heterogeneous (non-IID) data distributions, identifying both its strengths and its inherent limitations.

10.1 Federated Averaging Algorithm

Federated Averaging (FedAvg) formalizes the central operational idea of federated learning: achieving global optimization through repeated cycles of decentralized local computation and centralized aggregation, without requiring raw data to leave client devices. Unlike classical distributed optimization algorithms, which assume continuous availability of workers and frequent synchronization, FedAvg is explicitly designed for environments characterized by intermittent connectivity, heterogeneous data distributions, and autonomous participants.

At a high level, the FedAvg algorithm proceeds in discrete *communication rounds*. Each round consists of three conceptually distinct phases: (i) global model broadcast, (ii) local optimization on participating clients, and (iii) server-side aggregation. These phases are carefully structured to balance two competing objectives: minimizing communication overhead and maintaining optimization fidelity with respect to the global learning objective.

The defining feature of federated averaging is that local optimization is allowed to proceed for multiple steps before synchronization. This distinguishes FedAvg from federated stochastic gradient descent, where synchronization occurs after every local gradient computation. By decoupling the frequency of local updates from the frequency of communication, FedAvg enables substantial scalability gains while retaining a mathematically principled aggregation mechanism.

We now describe the FedAvg algorithm in detail, using the same notation and aggregation framework established in earlier chapters.

Global Model and Client Selection. Let $\theta^{(t)}$ denote the global model parameters at communication round t. As in FedSGD, the server does not require participation from all clients in every round. Instead, it selects a subset of available clients

$$\mathcal{S}^{(t)} \subseteq \{1, 2, \ldots, K\},$$

based on system constraints such as connectivity, availability, or sampling policies. The current global model $\theta^{(t)}$ is then broadcast to all selected clients.

This partial participation model reflects the realities of federated systems and introduces an additional layer of stochasticity beyond that encountered in centralized optimization.

Local Initialization. Upon receiving the broadcast model, each selected client $k \in \mathcal{S}^{(t)}$ initializes its local model parameters as

$$\theta_k^{(t,0)} = \theta^{(t)}.$$

This step ensures that all participating clients begin the round from a common reference point, even though their subsequent optimization trajectories may diverge due to differences in local data distributions and objective functions.

Local Optimization. Each client then performs τ steps of local stochastic gradient descent on its private empirical risk function $F_k(\cdot)$. Specifically, for $s = 0, 1, \ldots, \tau-1$, client k updates its local parameters according to

$$\theta_k^{(t,s+1)} = \theta_k^{(t,s)} - \eta \nabla F_k(\theta_k^{(t,s)}),$$

where $\eta > 0$ denotes the local learning rate.

These updates are computed entirely on the client device using local data and are never shared in raw form. Importantly, the gradients appearing in successive steps are evaluated at *different parameter values*, reflecting the evolving local model. As a result, local optimization follows a trajectory shaped by the client's specific data distribution and objective landscape.

The parameter τ plays a central role in FedAvg. When $\tau = 1$, each client performs a single gradient step before synchronization, recovering the behavior of federated stochastic gradient descent. As τ increases, clients perform progressively more local computation between communication rounds, reducing synchronization frequency but increasing the potential for divergence between local and global objectives.

After completing τ local steps, client k produces the locally updated model $\theta_k^{(t,\tau)}$, which is transmitted to the server.

Model Aggregation. Upon receiving locally optimized models from participating clients, the server aggregates them using the same data-weighted structure employed for gradient aggregation:

$$\theta^{(t+1)} = \sum_{k \in \mathcal{S}^{(t)}} \frac{n_k}{\sum_{j \in \mathcal{S}^{(t)}} n_j} \theta_k^{(t,\tau)}.$$

This aggregation rule ensures that each client's contribution to the global model is proportional to the size of its local dataset, thereby preserving statistical consistency with centralized empirical risk minimization. Although the server now aggregates model parameters rather than gradients, the weighting mechanism remains unchanged.

Interpretation of the Update. The resulting update defines the Federated Averaging algorithm. When $\tau = 1$, the aggregated model corresponds exactly to a single data-weighted gradient descent step, and FedAvg reduces to FedSGD. For $\tau > 1$, however, the server aggregates models that have evolved along distinct client-specific optimization trajectories.

From an optimization perspective, Federated Averaging can therefore be interpreted as performing *implicit gradient aggregation*, where information accumulated over multiple local steps is fused through model averaging rather than explicit gradient synchronization.

Continuity with Earlier Formulation
Federated Averaging preserves the data-weighted aggregation principle established for FedSGD, differing only in the object being aggregated: model parameters instead of instantaneous gradients.

Worked Example: Federated Averaging with Explicit Numerical Computation
To concretely illustrate the mechanics of the Federated Averaging (FedAvg) algorithm, we now present a fully worked numerical example. The objective of

this example is to demonstrate, step by step, how locally trained client models—obtained after multiple local optimization steps—are aggregated by the server using data-weighted model averaging. Unlike gradient-based examples, this illustration emphasizes how *model parameters*, rather than instantaneous gradients, are combined in FedAvg.

Problem Setup. Consider a federated learning system with $K = 4$ clients participating in communication round t. Each client k holds a local dataset $\mathscr{D}_k$ of size n_k, as summarized in Table 10.1. The total number of samples across participating clients is

$$N = \sum_{k=1}^{4} n_k.$$

Assume a one-dimensional linear regression model

$$f_\theta(x) = \theta x,$$

with squared loss

$$\ell(f_\theta(x), y) = \frac{1}{2}(\theta x - y)^2.$$

Let the global model at communication round t be

$$\theta^{(t)} = 0,$$

and let the local learning rate be $\eta = 0.05$. Each client performs $\tau = 3$ local stochastic gradient descent steps. The total dataset size is therefore

$$N = 120 + 80 + 60 + 40 = 300.$$

Local Optimization. Each client initializes its local model as

$$\theta_k^{(t,0)} = \theta^{(t)} = 0.$$

Table 10.1 Client Dataset Sizes and Local Data Statistics

Client k	Dataset Size n_k	Local Gradient at $\theta = 0$
1	120	−6.0
2	80	−4.0
3	60	−1.5
4	40	+2.0

The local update rule is

$$\theta_k^{(t,s+1)} = \theta_k^{(t,s)} - \eta \nabla F_k(\theta_k^{(t,s)}), \qquad s = 0, 1, 2.$$

For simplicity of exposition, we assume that the local gradients decrease linearly across steps due to curvature of the loss surface. The resulting local trajectories are shown below.

Client 1 $(n_1 = 120)$.

$$\begin{aligned} \theta_1^{(t,1)} &= 0 - 0.05(-6.0) = 0.30, \\ \theta_1^{(t,2)} &= 0.30 - 0.05(-4.8) = 0.54, \\ \theta_1^{(t,3)} &= 0.54 - 0.05(-3.6) = 0.72. \end{aligned}$$

Client 2 $(n_2 = 80)$.

$$\begin{aligned} \theta_2^{(t,1)} &= 0 - 0.05(-4.0) = 0.20, \\ \theta_2^{(t,2)} &= 0.20 - 0.05(-3.2) = 0.36, \\ \theta_2^{(t,3)} &= 0.36 - 0.05(-2.4) = 0.48. \end{aligned}$$

Client 3 $(n_3 = 60)$.

$$\begin{aligned} \theta_3^{(t,1)} &= 0 - 0.05(-1.5) = 0.075, \\ \theta_3^{(t,2)} &= 0.075 - 0.05(-1.2) = 0.135, \\ \theta_3^{(t,3)} &= 0.135 - 0.05(-0.9) = 0.18. \end{aligned}$$

Client 4 $(n_4 = 40)$.

$$\begin{aligned} \theta_4^{(t,1)} &= 0 - 0.05(2.0) = -0.10, \\ \theta_4^{(t,2)} &= -0.10 - 0.05(1.6) = -0.18, \\ \theta_4^{(t,3)} &= -0.18 - 0.05(1.2) = -0.24. \end{aligned}$$

After $\tau = 3$ steps, each client transmits its final local model $\theta_k^{(t,3)}$ to the server.

Model Aggregation. The server aggregates local models using the FedAvg rule:

$$\theta^{(t+1)} = \sum_{k=1}^{4} \frac{n_k}{N} \theta_k^{(t,3)}.$$

Substituting numerical values:

$$\begin{aligned}\theta^{(t+1)} &= \frac{120}{300}(0.72) + \frac{80}{300}(0.48) + \frac{60}{300}(0.18) + \frac{40}{300}(-0.24)\\ &= 0.288 + 0.128 + 0.036 - 0.032\\ &= 0.420.\end{aligned}$$

Interpretation. The resulting global model $\theta^{(t+1)} = 0.420$ reflects a weighted compromise among client-specific optimization trajectories. Clients with larger datasets exert greater influence on the aggregated model, while smaller or conflicting clients contribute proportionally less.

Notably, Client 4 moves in the opposite direction due to its local data distribution, yet its effect is moderated by both dataset size and aggregation weighting. This example highlights a defining property of Federated Averaging: *global updates emerge from the fusion of heterogeneous local optimization trajectories rather than from synchronized gradients.*

Pedagogical Insight. This worked example illustrates how Federated Averaging simultaneously achieves communication efficiency and statistical consistency. By aggregating locally trained models using principled data-weighted averaging, FedAvg captures information accumulated over multiple local steps while avoiding the communication overhead of per-step gradient synchronization. At the same time, the example foreshadows the challenges posed by data heterogeneity, which become more pronounced as local computation increases.

Does Federated Averaging Require Multiple Rounds to Converge?
Federated Averaging does *require multiple communication rounds* to converge to a stable solution. Each round performs only a partial improvement of the global objective by aggregating locally optimized client models. Convergence emerges through repeated alternation of local optimization and global aggregation, not from a single execution of the algorithm. As with centralized stochastic gradient descent, FedAvg approaches a minimizer gradually across rounds, with convergence speed influenced by data heterogeneity, learning rates, and the number of local steps per round.

10.2 Weight Averaging versus Gradient Averaging

A central conceptual distinction in federated optimization lies in *where* and *how* aggregation is performed. Although both federated stochastic gradient descent (FedSGD) and Federated Averaging (FedAvg) rely on data-weighted aggregation to fuse client contributions, they differ fundamentally in the mathematical object being aggregated. This distinction has deep implications for communication efficiency, optimization behavior, and robustness under heterogeneous data.

At a high level, FedSGD aggregates *gradients* computed at a shared global model, while FedAvg aggregates *model parameters* that have evolved through multiple local optimization steps. The difference may appear subtle at first glance, but it profoundly changes how information flows through the federated system.

10.2.1 Gradient Averaging (FedSGD)

In federated stochastic gradient descent, all participating clients evaluate their local gradients at the same global parameter vector $\theta^{(t)}$. Specifically, each client $k \in \mathcal{S}^{(t)}$ computes $\nabla F_k(\theta^{(t)})$, and the server aggregates these gradients using data-weighted averaging:

$$\bar{\mathbf{g}}^{(t)} = \sum_{k \in \mathcal{S}^{(t)}} \frac{n_k}{N} \nabla F_k(\theta^{(t)}),$$

where $N = \sum_{k \in \mathcal{S}^{(t)}} n_k$.

The global model is then updated as

$$\theta^{(t+1)} = \theta^{(t)} - \eta \bar{\mathbf{g}}^{(t)}.$$

Optimization Interpretation. This update rule closely mirrors centralized stochastic gradient descent. Under the assumption that client data are independent and identically distributed (IID), the aggregated gradient $\bar{\mathbf{g}}^{(t)}$ is an unbiased estimator of the true global gradient $\nabla F(\theta^{(t)})$. Consequently, FedSGD enjoys strong theoretical alignment with classical optimization theory.

Illustrative Example. Consider three clients evaluating gradients at the same global model. If two clients produce gradients pointing in nearly the same direction, their contributions reinforce one another during aggregation. If a third client produces a gradient in the opposite direction, its effect is partially canceled. Importantly, this cancellation occurs *at the level of descent directions*, before any client performs further local optimization.

Key Property of Gradient Averaging
In FedSGD, all clients contribute information about the global descent direction at the *same point in parameter space*.

Limitation. While gradient averaging ensures tight synchronization and stable optimization behavior, it requires communication at every gradient step. In large-scale federated deployments, this synchronization frequency often dominates system cost, motivating alternative approaches.

10.2.2 Weight Averaging (FedAvg)

Federated Averaging modifies the aggregation strategy by allowing each client to perform multiple local optimization steps before synchronization. Rather than transmitting gradients, clients transmit locally optimized model parameters.

After τ local steps, the model held by client k can be written as

$$\theta_k^{(t,\tau)} = \theta^{(t)} - \eta \sum_{s=0}^{\tau-1} \nabla F_k(\theta_k^{(t,s)}).$$

The server aggregates these local models using data-weighted averaging:

$$\theta^{(t+1)} = \sum_{k \in \mathcal{S}^{(t)}} \frac{n_k}{N} \theta_k^{(t,\tau)}.$$

Substituting the local model expression yields

$$\theta^{(t+1)} = \theta^{(t)} - \eta \sum_{k \in \mathcal{S}^{(t)}} \frac{n_k}{N} \sum_{s=0}^{\tau-1} \nabla F_k(\theta_k^{(t,s)}).$$

Optimization Interpretation. This expression reveals that FedAvg performs an *implicit gradient aggregation*. However, unlike FedSGD, the gradients being aggregated are evaluated at different parameter values along client-specific optimization trajectories. As a result, FedAvg integrates information accumulated over several local steps rather than relying on instantaneous gradient snapshots.

Illustrative Example. Suppose two clients begin from the same global model but possess different data distributions. After several local steps, one client may move aggressively toward its local optimum, while the other moves more conservatively. FedAvg aggregates these final model states, effectively averaging their optimization trajectories rather than their initial descent directions.

Trajectory-Level Aggregation
Federated Averaging aggregates the outcomes of local optimization processes, not just their initial gradient directions.

Connection to FedSGD. When $\tau = 1$, local trajectories collapse to a single step, and the FedAvg update reduces exactly to FedSGD. Thus, FedSGD can be viewed as a special case of FedAvg with maximal synchronization.

Core Distinction
Gradient averaging synchronizes *descent directions*, whereas weight averaging synchronizes *optimization trajectories*.

Implications. This distinction explains both the primary strength and the primary weakness of Federated Averaging. By aggregating trajectories, FedAvg dramatically reduces communication overhead. However, when client objectives differ substantially, these trajectories may drift toward incompatible optima, making aggregation less stable than gradient-based synchronization.

Understanding this trade-off is essential for analyzing the behavior of federated optimization algorithms and motivates the development of proximal, corrective, and adaptive variants discussed in subsequent sections.

Table 10.2 summarizes the key differences between Federated Stochastic Gradient Descent and Federated Averaging. Although both algorithms employ data-weighted aggregation to preserve statistical consistency with centralized empirical risk minimization, they differ fundamentally in the level at which synchronization occurs. FedSGD enforces tight coordination by aggregating gradients evaluated at a shared global model, resulting in stable but communication-intensive training.

Table 10.2 Comparison between Federated Stochastic Gradient Descent and Federated Averaging

Aspect	FedSGD	FedAvg
Object aggregated	Local gradients $\nabla F_k(\theta^{(t)})$ evaluated at the shared global model	Locally optimized model parameters $\theta_k^{(t,\tau)}$
Local steps per round	$\tau = 1$ (single gradient step)	$\tau \geq 1$ (multiple local optimization steps)
Synchronization frequency	After every local gradient computation	After multiple local updates
Communication cost	High; communication required at every step	Significantly reduced; amortized over τ local steps
Optimization viewpoint	Explicit gradient averaging	Implicit aggregation of optimization trajectories
Behavior under IID data	Close approximation to centralized SGD	Comparable to centralized SGD for small τ
Behavior under non-IID data	More stable due to frequent synchronization	Sensitive to local drift as τ increases
Scalability	Limited by communication overhead	Highly scalable to large client populations

In contrast, FedAvg aggregates locally optimized model parameters, dramatically reducing communication frequency at the cost of increased sensitivity to data heterogeneity. This trade-off explains why FedAvg has become the default optimization algorithm in large-scale federated learning systems, while FedSGD serves primarily as a conceptual and analytical baseline.

10.3 Communication Efficiency of Federated Averaging

Communication efficiency is the *central design motivation* behind Federated Averaging. Unlike centralized or tightly coupled distributed learning systems, federated learning operates in environments where communication is expensive, unreliable, and often severely constrained. Typical federated clients include mobile phones, edge devices, and Internet-of-Things nodes, where bandwidth is limited, latency is high, and energy consumption must be carefully managed. In such settings, the cost of transmitting information frequently dominates the cost of performing local computation.

Federated Averaging addresses this challenge by deliberately relaxing the requirement of global synchronization at every local optimization step. Instead of communicating gradients after each update—as in Federated Stochastic Gradient Descent—clients are allowed to perform multiple local updates before communicating with the server. This design choice fundamentally alters the communication profile of the learning process and enables federated learning to scale to large and geographically distributed client populations.

10.3.1 Communication Cost Model

To analyze the communication efficiency of federated optimization algorithms in a principled manner, it is necessary to first formalize a simple but expressive communication cost model. Let d denote the number of trainable parameters in the global model. This dimension typically corresponds to the size of a neural network weight vector and may range from thousands to millions in modern applications.

In federated learning, communication between clients and the server is structured as a sequence of discrete message exchanges. Each communication event involves transmitting an object whose size is proportional to d, regardless of whether that object represents a gradient vector or a full model parameter vector. The dominant cost of communication therefore scales with the number of such transmissions rather than with their semantic content.

In Federated Stochastic Gradient Descent (FedSGD), communication is tightly coupled to local computation. At each communication round t, every participating client computes a local gradient evaluated at the shared global model $\theta^{(t)}$ and immediately transmits this gradient to the server. Formally, client k sends

$$\nabla F_k(\theta^{(t)}) \in \mathbb{R}^d,$$

which represents a single stochastic estimate of the global descent direction. Because this transmission occurs after every local gradient computation, communication frequency grows in direct proportion to the number of training iterations.

If training proceeds for T communication rounds, the total volume of information exchanged by each participating client scales on the order of $T \times d$. This tight coupling between computation and communication leads to high network utilization and can become a severe bottleneck in federated environments, where bandwidth is limited and communication latency is non-negligible.

Federated Averaging (FedAvg) fundamentally alters this cost structure by decoupling local optimization from global synchronization. Rather than communicating after each gradient computation, each client is allowed to perform τ local stochastic gradient descent steps on its private objective before transmitting any information to the server. After completing these τ steps, client k sends its locally optimized model

$$\theta_k^{(t,\tau)} \in \mathbb{R}^d$$

to the server.

A key observation is that, despite the additional local computation, the size of the communicated object remains exactly d. The communication cost of a single transmission is therefore independent of τ. As τ increases, more local computation is performed per communication event, effectively amortizing the communication cost over multiple gradient updates.

This decoupling enables Federated Averaging to dramatically reduce the number of communication rounds required to reach a target level of model accuracy. By shifting work from communication to computation, FedAvg aligns the optimization process with the dominant cost structure of federated systems, where local computation is comparatively inexpensive while communication is costly and constrained.

10.3.2 Reduction in Communication Frequency

The primary efficiency advantage of Federated Averaging does not arise from reducing the size of the information exchanged between clients and the server, but from reducing how *often* such exchanges take place. This distinction is subtle but fundamental. In both Federated Stochastic Gradient Descent and Federated Averaging, the dimensionality of the transmitted object is proportional to the number of model parameters. What changes between the two algorithms is the temporal structure of communication.

In Federated Stochastic Gradient Descent, communication is tightly synchronized with local computation. Each local gradient computation is immediately followed by a communication event, forcing all participating clients to repeatedly transmit information to the server. As a result, the communication frequency scales directly with the number of local gradient steps performed during training.

Federated Averaging relaxes this synchronization constraint by allowing clients to perform multiple local updates before communicating. If each client executes τ local stochastic gradient descent steps within a single communication round, then a single transmission effectively represents the cumulative effect of those τ updates. Consequently, communication occurs once every τ local gradient steps, rather than after each step.

More formally, relative to Federated Stochastic Gradient Descent, the communication frequency in Federated Averaging is reduced by a factor that is approximately proportional to the number of local steps:

$$\text{Communication Reduction Factor} \approx \tau.$$

This reduction has far-reaching practical implications. As τ increases, the number of communication rounds required to achieve a target level of training accuracy decreases substantially. In large-scale federated deployments involving thousands or millions of clients, even moderate values of τ can translate into dramatic savings in total network traffic. These savings reduce congestion on shared communication links, lower latency caused by synchronization barriers, and decrease the likelihood of dropped or delayed updates.

From the perspective of individual clients, reduced communication frequency directly impacts energy consumption. Wireless transmission is often one of the most energy-intensive operations on mobile and edge devices. By communicating less frequently, Federated Averaging extends battery life and enables sustained participation of resource-constrained devices in long-running training processes.

At the system level, reduced communication frequency also improves robustness. Fewer synchronization points mean fewer opportunities for stragglers, intermittent connectivity, or transient failures to stall global progress. This property makes Federated Averaging particularly well suited to highly heterogeneous and unreliable environments, where consistent global coordination cannot be assumed.

Key Efficiency Mechanism
Federated Averaging achieves scalability by amortizing the cost of communication across multiple local optimization steps, replacing frequent synchronization with infrequent but more informative model updates.

10.3.3 Interaction with Local Computation

The reduction in communication frequency achieved by Federated Averaging is made possible by a corresponding increase in local computation. Instead of performing a single gradient update before synchronizing with the server, each participating client now executes τ local stochastic gradient descent steps per communication round. This shift fundamentally changes how computational effort is allocated between clients and the central server.

From a system perspective, this trade-off is often highly favorable. In most federated environments, local computation—such as evaluating gradients or updating model parameters—is significantly cheaper than communication. Modern edge devices are capable of performing thousands or millions of floating-point operations with relatively low energy cost, while even a single wireless transmission may incur substantial latency and battery drain. As a result, increasing local computation in exchange for fewer communication rounds aligns the learning algorithm with the underlying cost structure of real-world federated systems.

However, increasing local computation also alters the optimization dynamics of the learning process. Because each client optimizes its own local objective function $F_k(\cdot)$ using private data, the trajectory followed by the local model parameters reflects client-specific statistical properties. When τ is small, local updates remain closely aligned with the global descent direction. As τ increases, however, each client continues to optimize its own objective for a longer period without global correction.

This phenomenon leads to what is commonly referred to as *local drift*. As local models move along client-specific optimization paths, they may deviate substantially from the minimizer of the global objective. When these locally optimized models are later aggregated, the resulting update may no longer correspond to a well-aligned descent step for the global loss function. The aggregated update can therefore become biased, slowing convergence or, in extreme cases, causing oscillations or divergence.

The impact of local drift is particularly pronounced under statistically heterogeneous (non-IID) data. In such settings, different clients may observe fundamentally different data distributions, leading to local objectives that are not merely noisy approximations of a common function, but structurally distinct. Under these conditions, increasing τ amplifies the divergence between local and global optimization goals, making careful algorithmic design essential.

10.3.4 Communication–Computation Trade-Off

Federated Averaging therefore embodies a fundamental and unavoidable trade-off between communication efficiency and optimization accuracy. At one extreme, small values of τ yield behavior that closely resembles Federated Stochastic Gradient Descent. Frequent synchronization ensures that client updates remain well aligned with the global objective, resulting in stable and predictable convergence, but at the cost of heavy communication overhead.

At the opposite extreme, large values of τ dramatically reduce the number of communication rounds required for training. By allowing clients to make substantial local progress before synchronization, Federated Averaging minimizes network usage and enables scalability to massive client populations. However, this efficiency gain comes with increased risk: as local updates diverge, the aggregated model may reflect a compromise among conflicting objectives rather than a true descent direction.

Communication–Computation Trade-Off
Federated Averaging replaces frequent global synchronization with intensive local optimization, reducing communication cost while introducing optimization bias due to local drift.

This trade-off defines the core design space of federated optimization algorithms. Most modern extensions of Federated Averaging—including proximal regularization methods, adaptive optimization schemes, and control-variate approaches—can be understood as systematic attempts to navigate this trade-off. Rather than eliminating local drift entirely, these methods seek to control, compensate for, or exploit it in ways that preserve communication efficiency while maintaining stable convergence.

Understanding this interaction between local computation and communication is therefore essential for both theoretical analysis and practical deployment of federated learning systems.

10.3.5 Design Implications

From an algorithmic standpoint, Federated Averaging represents a deliberate and conceptually important shift in optimization priorities. Classical learning algorithms are typically designed under the assumption that computation is expensive and communication is either free or negligible. Federated learning inverts this assumption. In realistic federated environments, communication is the dominant bottleneck, while local computation is comparatively abundant and inexpensive. Federated Averaging explicitly embraces this asymmetry by trading statistical precision for system feasibility.

Rather than attempting to perfectly emulate centralized stochastic gradient descent, Federated Averaging relaxes the requirement of strict global synchronization. By allowing clients to perform multiple local updates before aggregation, the algorithm accepts a controlled amount of optimization bias in exchange for substantial reductions in communication cost. Its widespread success in large-scale deployments demonstrates that, under practical constraints, modest deviations from idealized optimization behavior are often not only tolerable, but necessary.

This shift has profound implications for how federated learning systems are designed and evaluated. Communication efficiency is no longer a secondary engineering concern, to be addressed after optimization objectives have been specified. Instead, it becomes a first-class algorithmic objective that directly shapes update rules, synchronization patterns, and convergence behavior. Algorithmic performance must therefore be assessed not only in terms of statistical convergence rates, but also in terms of communication overhead, energy consumption, and robustness to system variability.

Federated Averaging also establishes a unifying template for subsequent algorithmic developments. Many later federated optimization methods can be interpreted as systematic refinements of the FedAvg framework. Proximal methods introduce regularization terms to limit local drift, adaptive optimizers adjust learning rates to stabilize heterogeneous updates, and control-variate techniques correct for client-specific biases introduced by infrequent synchronization. Despite their differences, these approaches all operate within the structural paradigm introduced by Federated Averaging.

In this sense, Federated Averaging should be viewed not merely as a single algorithm, but as a foundational design philosophy for federated optimization. It formalizes the idea that effective learning in decentralized systems requires balancing statistical optimality against system constraints. Understanding this balance is essential for both theoretical analysis and the practical deployment of scalable, reliable federated learning systems.

10.4 Adaptive Federated Optimizers

The original Federated Averaging (FedAvg) algorithm combines local stochastic gradient descent on clients with simple data-weighted averaging at the server. While this approach is conceptually elegant and communication-efficient, it relies on fixed learning rates and unadapted update rules that were originally developed for centralized optimization. In practice, federated learning environments exhibit a level of heterogeneity and uncertainty that often exceeds the assumptions under which plain stochastic gradient descent performs reliably.

As federated systems have grown in scale and complexity, it has become increasingly clear that additional mechanisms are required to stabilize training, accelerate convergence, and mitigate the adverse effects of client variability. Adaptive federated optimizers address these challenges by incorporating ideas from adaptive gradient methods—such as momentum accumulation and second-order scaling—into the federated optimization pipeline.

10.4.1 Motivation for Adaptivity in Federated Learning

Federated learning departs from classical optimization settings in several fundamental ways. First, client data are typically statistically heterogeneous, meaning that local objective functions $F_k(\cdot)$ may differ substantially in both curvature and scale. Second, only a subset of clients participates in each communication round, introducing randomness at the system level that compounds statistical noise. Third, aggregation in Federated Averaging combines locally optimized models rather than gradients evaluated at a common parameter value, leading to biased and potentially high-variance updates.

Under these conditions, plain stochastic gradient descent with a fixed learning rate may exhibit slow convergence, oscillatory behavior, or sensitivity to hyperparameter choices. Adaptive optimization methods are designed precisely to address such issues by dynamically scaling update magnitudes based on historical information. In the federated setting, adaptivity becomes not merely beneficial but often essential for robust training.

10.4.2 Server-Side Adaptation Mechanism

A defining architectural feature of adaptive federated optimization methods is that adaptation is performed primarily at the *server*, rather than on individual clients. This design choice is both deliberate and consequential. Clients continue to perform local training using standard optimization procedures—most commonly stochastic gradient descent—without modifying their learning rules or maintaining additional optimizer state. As a result, the communication pattern, privacy guarantees, and computational burden on clients remain identical to those of standard Federated Averaging.

The responsibility for stabilizing and correcting the optimization process is instead centralized at the server. Rather than treating the aggregated update as a direct descent step, the server reinterprets it through an adaptive transformation before updating the global model. This separation of roles preserves client autonomy while enabling global coordination mechanisms that would be impractical or undesirable to implement on resource-constrained devices.

Let $\theta^{(t)}$ denote the global model at communication round t. After receiving locally trained models from participating clients and performing data-weighted aggregation, the server obtains an updated parameter vector $\theta^{(t+1)}$. The corresponding aggregated update is defined as

$$\Delta^{(t)} = \theta^{(t+1)} - \theta^{(t)}.$$

This vector $\Delta^{(t)}$ encapsulates the combined effect of local optimization, data heterogeneity, partial participation, and model averaging. Unlike gradients in centralized learning, $\Delta^{(t)}$ is not evaluated at a single parameter value, nor does it represent an unbiased estimate of the global gradient. Instead, it reflects the net displacement induced by multiple, client-specific optimization trajectories.

Adaptive federated optimizers address this complexity by maintaining additional server-side state variables that summarize the historical behavior of these aggregated updates. These state variables allow the server to distinguish between persistent trends and transient fluctuations, and to adjust update magnitudes in a parameter-wise manner.

For example, in Federated Adam, the server tracks a first-order moment estimate, which acts as a momentum term, and a second-order moment estimate, which captures the magnitude of recent updates:

$$\mathbf{m}^{(t)} = \beta_1 \mathbf{m}^{(t-1)} + (1 - \beta_1)\Delta^{(t)},$$

$$\mathbf{v}^{(t)} = \beta_2 \mathbf{v}^{(t-1)} + (1 - \beta_2)\left(\Delta^{(t)}\right)^2,$$

where $\beta_1, \beta_2 \in (0, 1)$ are decay parameters controlling the memory of past updates.

The first-order moment $\mathbf{m}^{(t)}$ smooths the update sequence by accumulating information over multiple rounds, reducing sensitivity to noise arising from random client participation or short-term fluctuations. The second-order moment $\mathbf{v}^{(t)}$ provides a coordinate-wise estimate of update variability, enabling adaptive rescaling of parameters with differing levels of stability or curvature.

Together, these moment estimates allow the server to modulate the effective learning rate independently for each parameter. Directions that exhibit consistently large or highly variable updates are dampened, preventing overshooting and oscillations. Conversely, directions with smaller or more stable updates may be amplified, accelerating progress where the optimization landscape is well behaved.

Importantly, this adaptive transformation operates entirely on *aggregated model updates* rather than on raw gradients. The resulting update rule closely mirrors the behavior of adaptive optimizers in centralized learning, but is applied at a coarser, round-based temporal scale. This design preserves the communication efficiency of Federated Averaging while substantially improving robustness under heterogeneous data distributions and partial client participation.

10.4.3 Interpretation and Benefits

From an optimization-theoretic perspective, server-side adaptivity plays a dual and complementary role in stabilizing federated learning. Unlike centralized optimization, where gradients are sampled from a single data distribution and evaluated at a shared parameter value, federated updates reflect a mixture of heterogeneous objectives and partial observations. Adaptive mechanisms at the server are designed to manage this additional complexity in a principled manner.

The first benefit arises from *momentum accumulation*. By maintaining a running average of past aggregated updates, the server smooths the effects of stochastic variation introduced by random client sampling and intermittent participation. Even when the composition of participating clients changes from one communication round to the next, momentum provides temporal coherence, ensuring that the global model continues to move in a consistent direction rather than reacting sharply to transient fluctuations.

The second benefit stems from *second-order scaling*. In federated learning, different clients may experience local loss surfaces with vastly different curvature properties. Parameters associated with frequently varying or steep directions can dominate naive updates, leading to instability. Second-order moment estimates enable the server to rescale updates on a per-parameter basis, dampening directions that exhibit large or erratic variation while preserving progress along stable dimensions. This mechanism reduces sensitivity to poorly conditioned local objectives and improves robustness under statistical heterogeneity.

Beyond convergence stability, server-side adaptivity also improves practical usability. Adaptive methods reduce sensitivity to the choice of global learning rates, making federated systems easier to tune and deploy. They also tend to yield smoother training trajectories, which is particularly valuable in long-running training processes where sudden performance degradation is undesirable.

Importantly, adaptive federated optimizers achieve these benefits without altering the fundamental communication pattern established by Federated Averaging. Clients continue to transmit model parameters once per communication round, and no additional client-side optimizer state is required. All adaptive state is maintained centrally at the server, ensuring that communication overhead remains unchanged and privacy guarantees are not weakened by additional information exchange.

10.4.4 Examples of Adaptive Federated Optimizers

Building on the principles outlined above, several adaptive federated optimization algorithms have been proposed. Each adapts ideas from well-established centralized optimizers to the federated setting, while respecting the structural constraints of Federated Averaging.

- **FedAdagrad** extends the Adagrad algorithm by accumulating squared aggregated updates at the server. This accumulation adaptively scales learning rates over time, allowing parameters with infrequent updates to receive larger steps while damping those with consistently large updates.
- **FedYogi** modifies the second-order update mechanism to better accommodate non-stationary behavior. By preventing unbounded growth of second-order moments, FedYogi provides improved stability in environments where client participation patterns and data distributions change over time.
- **Momentum-based FedAvg** introduces server-side momentum without full second-order adaptation. This approach offers a lightweight compromise, capturing much of the stabilizing effect of adaptive methods while retaining algorithmic simplicity.

Although these methods differ in their specific update rules and hyperparameter choices, they share a common architectural principle: local client training remains unchanged, while global aggregation is enhanced through adaptive correction at the server. This separation of concerns allows adaptive federated optimizers to improve convergence behavior and robustness without sacrificing the communication efficiency that makes Federated Averaging practical at scale.

Design Principle
Adaptive federated optimizers preserve the communication-efficient structure of Federated Averaging while compensating for instability induced by statistical heterogeneity and partial participation.

10.4.5 Role in the Federated Optimization Landscape

Adaptive federated optimizers occupy a central and enduring position in the modern federated learning landscape. They emerge not as ad hoc enhancements to Federated Averaging, but as a systematic response to the structural challenges that distinguish federated optimization from its centralized counterpart. In particular, they address the combined effects of statistical heterogeneity, partial participation, asynchronous behavior, and communication constraints—factors that collectively undermine the assumptions underlying classical optimization theory.

From a conceptual standpoint, adaptive federated optimizers represent a pragmatic compromise between theoretical simplicity and practical robustness. While plain Federated Averaging offers an elegant and communication- efficient baseline, its performance can degrade in large-scale, heterogeneous environments where client objectives differ substantially. Adaptive methods do not attempt to eliminate the fundamental sources of bias and variance inherent in federated learning. Instead, they seek to *control* their impact by modifying the interpretation of aggregated updates through principled, historically informed transformations at the server.

This perspective marks an important shift in how optimization is viewed in decentralized systems. Rather than treating instability as a failure mode to be avoided, adaptive federated optimizers treat it as an intrinsic property of the learning environment that must be actively managed. By incorporating momentum, variance normalization, and adaptive scaling, these methods provide a stabilizing layer that absorbs much of the unpredictability arising from client heterogeneity and system dynamics.

As federated learning systems continue to scale in size, geographic distribution, and application diversity, the importance of adaptive optimization is likely to grow rather than diminish. Future deployments are expected to involve increasingly heterogeneous clients, non-stationary data distributions, and dynamic participation patterns. In such settings, optimization strategies that rely solely on fixed learning rates and naive averaging are unlikely to remain effective.

Understanding adaptive federated optimizers is therefore essential for both theoretical and practical reasons. From a theoretical perspective, they motivate new convergence analyses that account for bias, variance, and historical dependence at the system level. From a practical perspective, they provide the tools necessary to build robust, scalable, and maintainable federated learning systems capable of operating reliably under real-world constraints.

Worked Example: Server-Side Adaptive Aggregation with Tabular Data

To concretely illustrate how server-side adaptive optimization operates in federated learning, we now present a fully worked numerical example. The objective is to demonstrate, step by step, how aggregated client updates are transformed by an adaptive optimizer at the server before updating the global model.

Problem Setup. Consider a federated learning system with $K = 4$ clients participating in communication round t. Each client holds a local tabular dataset for linear regression with a shared feature space of dimension $p = 3$. The global model is therefore

$$\theta^{(t)} \in \mathbb{R}^3.$$

Assume the global model at the beginning of round t is

$$\theta^{(t)} = \begin{bmatrix} 0.50 \\ -0.25 \\ 1.00 \end{bmatrix}.$$

Each client performs $\tau = 2$ local SGD steps and transmits its updated model to the server. Client dataset sizes are summarized in Table 10.3.

The total number of samples across participating clients is

$$N = 120 + 80 + 60 + 40 = 300.$$

Step 1: Data-Weighted Model Aggregation (FedAvg Core). The server first applies standard data-weighted aggregation:

$$\theta^{(t+1)} = \sum_{k=1}^{4} \frac{n_k}{N} \theta_k^{(t,\tau)}.$$

Table 10.3 Client Dataset Sizes and Locally Updated Models

Client k	Dataset Size n_k	Local Model $\theta_k^{(t,\tau)}$
1	120	$\begin{bmatrix} 0.62 \\ -0.10 \\ 1.08 \end{bmatrix}$
2	80	$\begin{bmatrix} 0.55 \\ -0.18 \\ 1.02 \end{bmatrix}$
3	60	$\begin{bmatrix} 0.48 \\ -0.35 \\ 0.92 \end{bmatrix}$
4	40	$\begin{bmatrix} 0.70 \\ -0.05 \\ 1.15 \end{bmatrix}$

The aggregation weights are:

$$w_1 = 0.40, \quad w_2 = 0.27, \quad w_3 = 0.20, \quad w_4 = 0.13.$$

Substituting values:

$$\theta^{(t+1)} = 0.40 \begin{bmatrix} 0.62 \\ -0.10 \\ 1.08 \end{bmatrix} + 0.27 \begin{bmatrix} 0.55 \\ -0.18 \\ 1.02 \end{bmatrix} + 0.20 \begin{bmatrix} 0.48 \\ -0.35 \\ 0.92 \end{bmatrix} + 0.13 \begin{bmatrix} 0.70 \\ -0.05 \\ 1.15 \end{bmatrix}.$$

Computing component-wise:

$$\theta^{(t+1)} = \begin{bmatrix} 0.588 \\ -0.188 \\ 1.050 \end{bmatrix}.$$

Step 2: Compute Aggregated Update Vector. The raw aggregated update is

$$\Delta^{(t)} = \theta^{(t+1)} - \theta^{(t)} = \begin{bmatrix} 0.088 \\ 0.062 \\ 0.050 \end{bmatrix}.$$

This vector represents the net effect of local optimization and model averaging before any adaptive correction.

Step 3: Server-Side Momentum Update. Assume the server uses Federated Adam with parameters $\beta_1 = 0.9$, $\beta_2 = 0.99$, and that previous moment estimates are initialized to zero:

$$\mathbf{m}^{(t-1)} = \mathbf{0}, \quad \mathbf{v}^{(t-1)} = \mathbf{0}.$$

The first-order moment update is

$$\mathbf{m}^{(t)} = 0.9\mathbf{m}^{(t-1)} + 0.1\Delta^{(t)} = \begin{bmatrix} 0.0088 \\ 0.0062 \\ 0.0050 \end{bmatrix}.$$

Step 4: Second-Order Moment Update. The second-order moment update is

$$\mathbf{v}^{(t)} = 0.99\mathbf{v}^{(t-1)} + 0.01(\Delta^{(t)})^2.$$

Computing element-wise squares:

$$(\Delta^{(t)})^2 = \begin{bmatrix} 0.00774 \\ 0.00384 \\ 0.00250 \end{bmatrix}.$$

Thus,

$$\mathbf{v}^{(t)} = \begin{bmatrix} 7.74 \times 10^{-5} \\ 3.84 \times 10^{-5} \\ 2.50 \times 10^{-5} \end{bmatrix}.$$

Step 5: Adaptive Global Update. With learning rate $\eta_s = 0.1$ and numerical stability constant $\varepsilon = 10^{-8}$, the server applies:

$$\theta_{\text{adapt}}^{(t+1)} = \theta^{(t)} + \eta_s \frac{\mathbf{m}^{(t)}}{\sqrt{\mathbf{v}^{(t)}} + \varepsilon}.$$

Evaluating each coordinate:

$$\theta_{\text{adapt}}^{(t+1)} \approx \begin{bmatrix} 0.60 \\ -0.23 \\ 1.03 \end{bmatrix}.$$

Interpretation. This example illustrates how server-side adaptivity reshapes the global update direction. Although the raw FedAvg update $\Delta^{(t)}$ reflects heterogeneous client contributions, adaptive moment estimation smooths fluctuations and rescales parameter-wise updates. Directions exhibiting consistently large variation are dampened, while stable directions are amplified.

Pedagogical Insight. Adaptive federated optimizers do not alter client behavior or communication patterns. Instead, they reinterpret aggregated updates at the server, enabling stable and efficient convergence even under heterogeneous data and partial participation. This separation of concerns—local autonomy with centralized stabilization—is a defining feature of modern federated optimization.

10.5 Performance Analysis under Heterogeneous Data

The principal limitation of Federated Averaging emerges most clearly under statistically heterogeneous, or non-IID, data distributions. In federated learning, each client k optimizes a local objective function $F_k(\cdot)$ derived from its private dataset. When these datasets arise from distinct data-generating processes, the corresponding local objectives may differ substantially in both shape and minimizer. As a result, local optimization steps performed independently by clients may drive model parameters toward *incompatible optima*.

This phenomenon represents a fundamental departure from centralized learning, where all gradients are sampled from a single, shared distribution. In the federated setting, even when clients start from the same global model θ, the gradients they compute can vary systematically rather than merely stochastically. Federated Averaging aggregates these locally optimized models, but does not explicitly correct for the resulting misalignment.

10.5.1 Quantifying Statistical Heterogeneity

A central challenge in federated learning is to formally characterize the degree of statistical heterogeneity across clients. Unlike centralized learning, where all gradients are drawn from a single data distribution, federated learning involves multiple client-specific data-generating processes. As a result, gradients computed by different clients at the same model parameters may differ systematically in both magnitude and direction.

A commonly used quantitative measure of this heterogeneity is the *gradient disagreement metric*

$$\mathcal{B}^2 = \frac{1}{K}\sum_{k=1}^{K} \|\nabla F_k(\theta) - \nabla F(\theta)\|^2,$$

which captures the average squared deviation of each client's local gradient from the global gradient.

Here,

$$\nabla F(\theta) = \sum_{k=1}^{K} \frac{n_k}{N} \nabla F_k(\theta)$$

denotes the global gradient obtained through data-weighted aggregation, with n_k representing the number of samples held by client k and $N = \sum_{k=1}^{K} n_k$ denoting the total number of samples across all clients.

The quantity $\mathcal{B}^2$ can be interpreted as a measure of *gradient dispersion* across the federated system. When client datasets are independent and identically distributed, local gradients tend to be unbiased estimates of the same underlying descent direction. In this regime, the differences between $\nabla F_k(\theta)$ and $\nabla F(\theta)$ are dominated by random sampling noise, and $\mathcal{B}^2$ remains small.

Under non-IID data, however, this interpretation changes fundamentally. Local gradients may reflect distinct statistical patterns, label skews, or feature distributions that persist across training rounds. In such cases, $\mathcal{B}^2$ becomes large, signaling structured and systematic disagreement rather than transient stochastic variation.

Importantly, this disagreement does not vanish as more data are observed locally, because it is rooted in genuine differences between client objectives.

This distinction highlights why $\mathscr{B}^2$ is not merely a nuisance term in theoretical analyses. It encodes the intrinsic difficulty of reconciling multiple, potentially conflicting learning objectives within a single global model. No amount of averaging can fully eliminate this disagreement without additional corrective mechanisms.

Consequently, $\mathscr{B}^2$ plays a central role in determining both the stability and the attainable accuracy of federated optimization algorithms. Large values of $\mathscr{B}^2$ are associated with slower convergence, increased sensitivity to hyperparameters, and a non-vanishing optimization bias. Understanding and controlling this quantity is therefore essential for the design, analysis, and deployment of effective federated learning systems.

10.5.2 Convergence Behavior of Federated Averaging

The convergence behavior of Federated Averaging is fundamentally shaped by the presence of statistical heterogeneity across clients. Unlike centralized stochastic gradient descent, where convergence is typically characterized by monotonic progress toward a single global minimizer, federated optimization operates under conditions where local objectives may be only partially aligned. As a result, classical convergence guarantees do not directly apply.

Theoretical analyses of Federated Averaging under heterogeneous data consistently show that the algorithm does not, in general, converge exactly to the minimizer of the global objective function. Instead, FedAvg is shown to converge to a *neighborhood* around the optimal solution. The size of this neighborhood is not arbitrary; it grows as a function of both the number of local optimization steps τ performed between communication rounds and the degree of statistical heterogeneity quantified by $\mathscr{B}^2$.

This result reflects a fundamental trade-off inherent in the design of Federated Averaging. Increasing τ allows each client to make more substantial progress on its local objective before synchronization. From a systems perspective, this is desirable, as it reduces the number of communication rounds required for training. From an optimization perspective, however, extended local optimization magnifies the effects of local drift. When client objectives are misaligned, local models move along divergent trajectories, and the aggregated update no longer corresponds to a well-aligned descent direction for the global objective.

Intuitively, Federated Averaging balances two competing forces. Frequent synchronization constrains local models to remain close to the global trajectory, promoting stability and reducing bias. Infrequent synchronization allows greater local progress but increases the risk that client models drift toward incompatible optima. The convergence neighborhood represents the equilibrium point of this tension: a region where further progress is offset by persistent disagreement among client objectives.

Empirical observations align closely with these theoretical insights. In heterogeneous settings, training curves produced by Federated Averaging often exhibit slower convergence rates than those observed in centralized or IID scenarios. Rather than smooth and steady improvement, performance may fluctuate as the global model alternates between directions favored by different subsets of clients. These oscillations are particularly pronounced when τ is large or when heterogeneity is severe.

Another practical consequence is heightened sensitivity to hyperparameter choices. Small changes in τ or the learning rate can produce disproportionately large differences in training behavior, ranging from stable convergence to stagnation or divergence. This sensitivity complicates deployment in real-world systems, where client behavior and data distributions may be difficult to predict or control.

Key Limitation
Federated Averaging performs best when statistical heterogeneity is mild; as local objectives diverge, increased local computation introduces bias that degrades convergence quality.

10.5.3 Implications and Motivations for Extensions

The limitations of Federated Averaging under heterogeneous data do not imply that the algorithm is fundamentally flawed or ineffective. Rather, they indicate that its performance is inherently constrained by the statistical diversity of client data and by the degree of autonomy afforded to local optimization. Federated Averaging was deliberately designed as a simple, communication-efficient baseline, and its behavior under non-IID data reflects the trade-offs implicit in this design.

In many realistic federated learning deployments, statistical heterogeneity is not an edge case but a defining characteristic. Applications such as mobile personalization involve users with distinct preferences and behaviors. Healthcare systems aggregate data from institutions serving different populations and operating under varied clinical practices. Edge sensing and Internet-of-Things platforms collect measurements from devices deployed in diverse physical environments. In all of these settings, assuming identically distributed data across clients is neither realistic nor desirable.

Recognizing this reality has motivated a substantial body of research aimed at extending or modifying Federated Averaging to better accommodate heterogeneous data. One prominent direction involves *proximal regularization*. Methods such as FedProx augment the local objective with an explicit penalty term that discourages excessive deviation from the global model. By constraining local updates, proximal methods directly limit local drift and improve stability when client objectives differ substantially.

A second line of work focuses on *control-variate techniques*. These methods, exemplified by algorithms such as SCAFFOLD, introduce correction terms that compensate for systematic gradient bias arising from heterogeneous data. Rather than restricting local optimization, control-variate approaches seek to realign local updates with the global descent direction through explicit bias correction.

A third and increasingly important direction is *personalization*. Personalized federated learning relaxes the assumption that a single global model must be optimal for all clients. Instead, it allows clients to maintain partially individualized models that share common structure while adapting to local data. This approach reframes heterogeneity not as an obstacle to be overcome, but as a signal to be exploited for improved client-level performance.

Exercises

This chapter examined Federated Averaging as the canonical optimization algorithm for federated learning, emphasizing its algorithmic structure, communication efficiency, convergence behavior, and limitations under heterogeneous data.

A. Short Answer Questions

Answer each question briefly and precisely.

1. What fundamental system constraint motivated the development of Federated Averaging?
2. How does Federated Averaging generalize Federated Stochastic Gradient Descent?
3. What role does the parameter τ play in FedAvg?
4. Why is data-weighted aggregation used in Federated Averaging?
5. What is meant by *local drift* in federated optimization?
6. How does FedAvg reduce communication frequency relative to FedSGD?
7. Why does FedAvg aggregate model parameters instead of gradients?
8. What is the gradient disagreement metric $\mathscr{B}^2$ intended to capture?
9. Why does FedAvg converge to a neighborhood of the optimum under non-IID data?
10. In what sense is FedSGD a special case of FedAvg?

B. Long Answer Questions

Answer the following questions with clear explanations, equations, and diagrams wherever appropriate.

1. Derive the Federated Averaging update rule and explain how it preserves statistical consistency with centralized learning.

2. Compare gradient averaging and weight averaging from both an optimization and system perspective.
3. Analyze the communication–computation trade-off in Federated Averaging and explain why it is central to federated learning.
4. Explain the concept of convergence to a neighborhood in FedAvg and discuss the factors that influence its size.
5. Using mathematical intuition, explain why increasing τ improves communication efficiency but may harm convergence.
6. Discuss the impact of statistical heterogeneity on the behavior of Federated Averaging.
7. Explain how adaptive federated optimizers modify the FedAvg update without changing client-side computation.
8. Describe the role of server-side momentum in stabilizing federated learning.
9. Compare Federated Averaging with centralized SGD in terms of assumptions, guarantees, and practical constraints.
10. Explain why Federated Averaging is best viewed as a design philosophy rather than merely a single algorithm.

C. True/False Questions

State whether each statement is **True** or **False**.

1. Federated Averaging requires communication after every local gradient computation.
2. FedAvg aggregates locally optimized model parameters rather than gradients.
3. When $\tau = 1$, Federated Averaging reduces to FedSGD.
4. Increasing τ always improves convergence speed.
5. Communication cost in federated learning is often higher than local computation cost.
6. Under IID data, FedAvg behaves similarly to centralized SGD.
7. The gradient disagreement metric $\mathscr{B}^2$ is zero when data are perfectly heterogeneous.
8. Server-side adaptive optimizers require clients to maintain second-order statistics.
9. Local drift arises due to extended local optimization on client-specific objectives.
10. Federated Averaging always converges exactly to the global optimum.

Answers:

1. False
2. True
3. True
4. False

5. True
6. True
7. False
8. False
9. True
10. False

D. Fill in the Blanks

Fill in the blanks with the most appropriate term or expression.

1. Federated Averaging reduces communication frequency by allowing clients to perform ___________ local updates per round.
2. In FedAvg, the server aggregates ___________ rather than gradients.
3. The parameter τ controls the amount of ___________ computation.
4. Local drift is amplified under ___________ data distributions.
5. The global gradient is computed using ___________ aggregation.
6. FedAvg converges to a ___________ of the global optimum under heterogeneity.
7. Communication efficiency is achieved by reducing ___________ rather than message size.
8. Adaptive federated optimizers operate primarily at the ___________.
9. The metric $\mathscr{B}^2$ measures gradient ___________ across clients.
10. FedSGD can be viewed as a special case of FedAvg with ___________.

Answers:

1. multiple
2. model parameters
3. local
4. non-IID
5. data-weighted
6. neighborhood
7. communication frequency
8. server
9. disagreement
10. $\tau = 1$

E. Multiple Choice Questions (MCQs)

Choose the **most appropriate** answer.

1. The main motivation behind Federated Averaging is:
   (a) improving model expressiveness
   (b) reducing communication cost
   (c) eliminating data heterogeneity
   (d) increasing model size
2. In FedAvg, aggregation is performed over:
   (a) gradients
   (b) Hessians
   (c) model parameters
   (d) loss values
3. Increasing τ primarily affects:
   (a) model size
   (b) communication frequency
   (c) dataset size
   (d) number of clients
4. Local drift is caused by:
   (a) centralized aggregation
   (b) homogeneous data
   (c) extended local optimization
   (d) small learning rates
5. The metric $\mathscr{B}^2$ quantifies:
   (a) communication overhead
   (b) gradient disagreement
   (c) model capacity
   (d) dataset imbalance
6. FedAvg reduces to FedSGD when:
   (a) $\eta = 0$
   (b) $\mathscr{B}^2 = 0$
   (c) $\tau = 1$
   (d) $K = 1$
7. Adaptive federated optimizers primarily modify:
   (a) client objectives
   (b) client learning rates
   (c) server-side updates
   (d) data distributions
8. Communication efficiency in FedAvg is achieved by:
   (a) compressing gradients
   (b) reducing model size
   (c) amortizing communication
   (d) eliminating synchronization

9. Under severe heterogeneity, FedAvg typically converges to:
   (a) the exact optimum
   (b) a saddle point
   (c) a neighborhood of the optimum
   (d) divergence always
10. Which method introduces a proximal term to limit local drift?
   (a) FedAdam
   (b) SCAFFOLD
   (c) FedProx
   (d) FedAvg

Answers:

1. (b)
2. (c)
3. (b)
4. (c)
5. (b)
6. (c)
7. (c)
8. (c)
9. (c)
10. (c)

F. Application and Project-based Questions

These problems emphasize modeling, reasoning, and implementation.

1. Implement Federated Averaging for linear regression using a synthetic non-IID dataset and analyze convergence behavior.
2. Compare FedAvg and FedSGD empirically under varying values of τ.
3. Simulate local drift using heterogeneous client objectives and visualize trajectory divergence.
4. Implement FedAvg with server-side momentum and compare it with plain FedAvg.
5. Measure communication cost savings achieved by FedAvg relative to FedSGD in a simulated network.
6. Study the effect of dataset imbalance on data-weighted aggregation.
7. Analyze the impact of partial client participation on FedAvg convergence.
8. Extend FedAvg with a proximal regularization term and evaluate its effect.
9. Design a personalized federated learning variant based on FedAvg.
10. Write a critical report comparing Federated Averaging with at least two of its modern extensions.

Advanced Federated Optimization Methods

11

The foundational federated optimization algorithms introduced in the preceding chapters—most notably Federated Stochastic Gradient Descent (FedSGD) and Federated Averaging (FedAvg)—establish a principled baseline for collaborative model training under strict data locality constraints. These methods demonstrate that it is possible to learn a shared global model without centralizing raw data, relying instead on periodic aggregation of locally computed updates. From a conceptual standpoint, they successfully bridge classical empirical risk minimization with decentralized data ownership.

Despite their elegance and simplicity, however, these foundational algorithms rest on assumptions that are rarely satisfied in practical federated deployments. In particular, they implicitly assume that client data distributions are reasonably similar, that a sufficiently large and representative subset of clients participates in each communication round, and that local training does not introduce excessive deviation from the global optimization trajectory. When these assumptions are violated—as is typical in real-world systems—the performance of naïve federated averaging degrades significantly.

Limitations of Naïve Federated Optimization

While FedAvg and FedSGD are theoretically sound under idealized conditions, they are highly sensitive to:

- strong statistical heterogeneity across clients,
- partial and irregular client participation,
- multiple local update steps between aggregations,

P. P. Ray, *Foundations of Federated Learning*,
Springer Series in the Data Sciences,
https://doi.org/10.1007/978-3-032-34084-9_11

- imbalance in local dataset sizes.

These factors jointly amplify client drift and destabilize convergence.

To illustrate this issue, consider a federated learning scenario in which some clients possess large, homogeneous datasets while others hold small but highly specialized datasets. During local training, each client optimizes its empirical risk in a direction that best fits its private data distribution. When such locally optimized models are aggregated through simple averaging, the resulting global update reflects a compromise that may fail to meaningfully reduce the global objective. In extreme cases, the global model may oscillate, converge slowly, or become biased toward dominant clients.

This phenomenon is particularly pronounced when clients perform multiple local optimization steps per communication round. While increasing local computation is often desirable from a systems perspective—since it reduces communication overhead—it also increases the risk that local models drift far from the global reference point. As a result, communication efficiency and optimization stability become tightly coupled, creating a fundamental tension in federated learning design.

Client Drift
Client drift refers to the systematic deviation of locally optimized models from the global model due to heterogeneous data and extended local training. This drift causes locally computed updates to become misaligned with the descent direction of the global federated objective.

Motivated by these challenges, this chapter introduces a class of *advanced federated optimization methods* that move beyond naïve averaging. Rather than treating heterogeneity and partial participation as incidental nuisances, these methods explicitly incorporate them into the optimization process. They do so by modifying one or more of the following components:

- the local optimization objective at each client,
- the manner in which local updates are regularized or corrected,
- the aggregation rule used to form the global model.

The unifying goal of these methods is to preserve the scalability and privacy advantages of federated learning while restoring stability, robustness, and predictable convergence behavior under non-IID data. Importantly, these techniques do not abandon the federated paradigm. Instead, they refine it by embedding classical optimization concepts— such as proximal regularization, bias correction, and robust aggregation—into a decentralized learning framework.

Design Philosophy of Advanced Federated Optimizers
Advanced federated optimization methods do not seek to eliminate data heterogeneity. Instead, they aim to *control its influence* on learning dynamics through principled modifications of local objectives and aggregation mechanisms.

In this chapter, we focus on five representative techniques that embody this philosophy. First, the *Federated Proximal method (FedProx)* introduces explicit regularization to limit excessive local deviation from the global model. Second,*dynamic regularization methods* exemplified by, FedDyn seek to correct systematic bias induced by heterogeneous client objectives. Third, *one-gradient federated learning* explores the extreme end of communication-efficient optimization by restricting local computation to a single gradient per round. Fourth, *inverse distance aggregation* modifies the aggregation rule itself to reduce the influence of highly divergent updates. Finally, we examine the convergence guarantees and robustness properties associated with these methods.

Scope of This Chapter
This chapter focuses on *global model optimization* under heterogeneity. Issues related to client-specific personalization, security threats, and adversarial behavior are deferred to later chapters, where they are treated in dedicated and specialized contexts.

Together, the methods presented in this chapter demonstrate that effective federated optimization requires more than communication-aware averaging. By carefully shaping local objectives and aggregation dynamics, federated learning systems can achieve stable and scalable convergence even in highly decentralized, statistically heterogeneous environments. These ideas form a critical bridge between foundational federated algorithms and the personalized, heterogeneous, and robust federated learning paradigms explored in subsequent chapters.

11.1 Federated Proximal Method

Federated Proximal (FedProx) is an extension of the Federated Averaging (FedAvg) algorithm that explicitly addresses one of the most persistent challenges in federated optimization: uncontrolled divergence of local models under statistical heterogeneity and extended local training. While FedAvg assumes that local optimization steps remain reasonably aligned with the global objective, this assumption often breaks down in practical federated systems.

FedProx introduces a principled modification to the *local optimization objective* by incorporating a proximal regularization term. This term penalizes excessive deviation of the locally optimized model from the global reference model broadcast at the beginning of each communication round. As a result, FedProx constrains the *optimization trajectory* rather than the *optimization duration*, enabling stability without sacrificing communication efficiency.

Why FedAvg Breaks Down
FedAvg implicitly assumes that locally optimized models remain close to the global model. Under non-IID data and multiple local steps, this assumption fails, causing client updates to become misaligned and destabilizing convergence.

Motivation: Client Drift under Non-IID Data. In federated learning, each client optimizes its empirical risk function using data drawn from a client-specific distribution. When data is non-IID, the gradients computed at different clients are systematically biased relative to the global gradient. If clients perform several local update steps, these biased gradients compound over time, pushing local models toward regions of the parameter space that are optimal for individual clients but suboptimal for the collective objective.

This phenomenon—commonly referred to as *client drift*—is amplified by:

- heterogeneous data distributions across clients,
- uneven dataset sizes,
- aggressive local learning rates,
- infrequent global aggregation.

Client Drift
Client drift describes the systematic deviation of locally optimized models from the global model caused by biased local objectives and extended local training. It is a primary source of instability in federated learning under non-IID data.

Without explicit control mechanisms, client drift leads to slow convergence, oscillatory behavior, or even divergence of the global model. FedProx directly targets this issue by modifying the local optimization problem itself.

Modified Local Objective with Proximal Regularization. In FedProx, each participating client k solves the following regularized optimization problem during communication round t:

$$\min_{\theta} F_k(\theta) + \frac{\mu}{2} \left\| \theta - \theta^{(t)} \right\|^2,$$

where:

- $F_k(\theta)$ is the local empirical risk at client k,
- $\theta^{(t)}$ denotes the global model broadcast by the server at the beginning of round t,
- $\mu > 0$ is the proximal regularization coefficient.

The added quadratic term penalizes large deviations from the global model, effectively anchoring local optimization around $\theta^{(t)}$.

Role of the Proximal Term
The proximal term acts as a soft constraint that discourages excessive local deviation while preserving flexibility for client-specific learning.

Geometric Interpretation. From a geometric perspective, the proximal term restricts the feasible region explored during local optimization. Rather than allowing the local optimizer to freely traverse the loss landscape defined by $F_k(\theta)$, FedProx imposes a quadratic penalty that grows with distance from the global reference model. This results in local updates that remain within a controlled neighborhood of $\theta^{(t)}$, reducing the variance of client updates across the population.

Algorithmic Interpretation. Algorithmically, FedProx can be viewed as a federated analog of proximal gradient methods used in centralized optimization. However, unlike classical proximal methods—which typically address non-smooth regularization—FedProx uses proximal regularization to manage *decentralization-induced instability*. The regularization term compensates for the lack of synchronized gradients and global visibility inherent in federated systems.

FedProx as Proximal Gradient Descent
FedProx adapts proximal gradient ideas to federated learning by using proximal regularization to stabilize decentralized optimization under partial participation and heterogeneous data.

Effect of the Regularization Parameter μ. The choice of μ governs the trade-off between local flexibility and global stability:

- Small μ values recover FedAvg-like behavior, allowing greater local exploration.

- Large μ values tightly constrain local models, reducing drift but potentially slowing progress toward the global optimum.

In practice, μ is selected based on the degree of data heterogeneity, the number of local steps, and communication frequency.

Practical Insight
Higher heterogeneity and longer local training generally require larger proximal regularization to maintain stable convergence.

Relationship to Aggregation. It is important to emphasize that FedProx does *not* alter the aggregation rule used by the server. Client updates are still aggregated using weighted averaging, as in FedAvg. The improvement arises entirely from reshaping the local optimization landscape, not from modifying global coordination mechanisms.

Key Insight
FedProx achieves stability by changing *what clients optimize*, not by increasing synchronization or altering aggregation.

Overall, the Federated Proximal method represents a minimal yet powerful departure from naïve federated averaging. By embedding proximal regularization into local objectives, FedProx provides a principled mechanism for controlling client drift, improving convergence stability, and enabling scalable federated learning under severe statistical heterogeneity.

Worked Example: Federated Proximal Method with Explicit Computation
We now present a fully worked numerical example to illustrate how the Federated Proximal (FedProx) method modifies local optimization behavior and stabilizes aggregation under heterogeneous data. This example explicitly contrasts FedProx-style local optimization with naïve local empirical risk minimization and demonstrates the effect of proximal regularization through step-by-step computation.

Federated Setup. Consider a federated learning system with $K = 3$ clients. We assume a scalar linear prediction model

$$f_\theta(x) = \theta x, \quad \theta \in \mathbb{R},$$

and employ the squared-error loss

$$\ell(f_\theta(x), y) = (\theta x - y)^2.$$

Each client k holds a local dataset $\mathscr{D}_k$ of size n_k. The global model broadcast at the beginning of communication round t is

$$\theta^{(t)} = 1.0.$$

Local Datasets. The client datasets are shown in Table 11.1.

The dataset sizes are identical:

$$n_1 = n_2 = n_3 = 4, \quad N = 12.$$

Step 1: Local Empirical Risk Functions. The local empirical risk at client k is

$$F_k(\theta) = \frac{1}{n_k} \sum_{i=1}^{n_k} (\theta x_{k,i} - y_{k,i})^2.$$

Client 1:

$$F_1(\theta) = \frac{1}{4}\big[(\theta - 2)^2 + (2\theta - 4)^2 + (3\theta - 6)^2 + (4\theta - 8)^2\big].$$

Client 2:

$$F_2(\theta) = \frac{1}{4}\big[(\theta - 1)^2 + (2\theta - 2)^2 + (3\theta - 3)^2 + (4\theta - 4)^2\big].$$

Table 11.1 Local datasets for the FedProx worked example

Client	Sample	x	y
Client 1 ($\mathscr{D}_1$)	1	1	2
	2	2	4
	3	3	6
	4	4	8
Client 2 ($\mathscr{D}_2$)	1	1	1
	2	2	2
	3	3	3
	4	4	4
Client 3 ($\mathscr{D}_3$)	1	1	3
	2	2	5
	3	3	7
	4	4	9

Client 3:

$$F_3(\theta) = \frac{1}{4}\left[(\theta - 3)^2 + (2\theta - 5)^2 + (3\theta - 7)^2 + (4\theta - 9)^2\right].$$

Step 2: Naïve Local Minimizers (Without FedProx). We now explicitly derive the local minimizers obtained when each client performs naïve empirical risk minimization, *without* proximal regularization. This corresponds to the local optimization behavior in FedAvg.

Recall that each client k minimizes its local empirical risk

$$F_k(\theta) = \frac{1}{n_k}\sum_{i=1}^{n_k}(\theta x_{k,i} - y_{k,i})^2.$$

Since the squared loss is convex and differentiable, the local minimizer $\theta_k^\star$ is obtained by solving

$$\frac{\partial F_k(\theta)}{\partial \theta} = 0.$$

Client 1. For Client 1, the local empirical risk is

$$F_1(\theta) = \frac{1}{4}\left[(\theta - 2)^2 + (2\theta - 4)^2 + (3\theta - 6)^2 + (4\theta - 8)^2\right].$$

We first expand each term:

$$\begin{aligned}
(\theta - 2)^2 &= \theta^2 - 4\theta + 4,\\
(2\theta - 4)^2 &= 4\theta^2 - 16\theta + 16,\\
(3\theta - 6)^2 &= 9\theta^2 - 36\theta + 36,\\
(4\theta - 8)^2 &= 16\theta^2 - 64\theta + 64.
\end{aligned}$$

Summing these terms yields

$$F_1(\theta) = \frac{1}{4}\left(30\theta^2 - 120\theta + 120\right).$$

Differentiating with respect to θ,

$$\frac{\partial F_1(\theta)}{\partial \theta} = \frac{1}{4}(60\theta - 120) = 15\theta - 30.$$

Setting the derivative to zero,

$$15\theta - 30 = 0 \quad \Rightarrow \quad \theta_1^\star = 2.$$

Client 2. For Client 2, the local empirical risk is

$$F_2(\theta) = \frac{1}{4}\Big[(\theta - 1)^2 + (2\theta - 2)^2 + (3\theta - 3)^2 + (4\theta - 4)^2\Big].$$

Expanding each term:

$$\begin{aligned}
(\theta - 1)^2 &= \theta^2 - 2\theta + 1, \\
(2\theta - 2)^2 &= 4\theta^2 - 8\theta + 4, \\
(3\theta - 3)^2 &= 9\theta^2 - 18\theta + 9, \\
(4\theta - 4)^2 &= 16\theta^2 - 32\theta + 16.
\end{aligned}$$

Summing the expanded terms,

$$F_2(\theta) = \frac{1}{4}\big(30\theta^2 - 60\theta + 30\big).$$

Differentiating,

$$\frac{\partial F_2(\theta)}{\partial \theta} = \frac{1}{4}(60\theta - 60) = 15\theta - 15.$$

Setting the derivative to zero,

$$15\theta - 15 = 0 \quad \Rightarrow \quad \theta_2^\star = 1.$$

Client 3. For Client 3, the local empirical risk is

$$F_3(\theta) = \frac{1}{4}\Big[(\theta - 3)^2 + (2\theta - 5)^2 + (3\theta - 7)^2 + (4\theta - 9)^2\Big].$$

Expanding each term:

$$\begin{aligned}
(\theta - 3)^2 &= \theta^2 - 6\theta + 9, \\
(2\theta - 5)^2 &= 4\theta^2 - 20\theta + 25, \\
(3\theta - 7)^2 &= 9\theta^2 - 42\theta + 49, \\
(4\theta - 9)^2 &= 16\theta^2 - 72\theta + 81.
\end{aligned}$$

Summing all terms,

$$F_3(\theta) = \frac{1}{4}\big(30\theta^2 - 140\theta + 164\big).$$

Differentiating,

$$\frac{\partial F_3(\theta)}{\partial \theta} = \frac{1}{4}(60\theta - 140) = 15\theta - 35.$$

Setting the derivative to zero,

$$15\theta - 35 = 0 \quad \Rightarrow \quad \theta_3^\star = \frac{35}{15} = \frac{7}{3} \approx 2.33.$$

Summary and Interpretation. The naïve local minimizers obtained by independent empirical risk minimization are therefore:

$$\theta_1^\star = 2, \quad \theta_2^\star = 1, \quad \theta_3^\star \approx 2.33.$$

These solutions are widely separated in parameter space, despite the fact that all clients share the same model architecture. This dispersion reflects strong statistical heterogeneity across local datasets and illustrates the phenomenon of *client drift*.

Key Observation
Under non-IID data, naïve local empirical risk minimization drives client models toward distinct optima, producing misaligned updates that destabilize federated averaging.

Step 3: FedProx Local Objective. In FedProx, each client minimizes the regularized objective:

$$\min_\theta \; F_k(\theta) + \frac{\mu}{2}(\theta - \theta^{(t)})^2,$$

with $\theta^{(t)} = 1.0$ and $\mu = 2$.

Step 4: Solving the FedProx Objective (Client 1). The derivative of the FedProx objective for Client 1 is:

$$\frac{\partial}{\partial\theta}\left(F_1(\theta) + \frac{2}{2}(\theta - 1)^2\right) = \frac{\partial F_1(\theta)}{\partial\theta} + 2(\theta - 1).$$

From $F_1(\theta)$,

$$\frac{\partial F_1(\theta)}{\partial\theta} = 15\theta - 30.$$

Setting the derivative to zero:

$$15\theta - 30 + 2(\theta - 1) = 0 \quad \Rightarrow \quad 17\theta = 32 \quad \Rightarrow \quad \theta_1^{(t,\tau)} \approx 1.88.$$

Step 5: Solving the FedProx Objective (Clients 2 and 3). Applying the same procedure:

Client 2:

$$\frac{\partial F_2(\theta)}{\partial \theta} = 15\theta - 15.$$

$$15\theta - 15 + 2(\theta - 1) = 0 \quad \Rightarrow \quad 17\theta = 17 \quad \Rightarrow \quad \theta_2^{(t,\tau)} = 1.0.$$

Client 3:

$$\frac{\partial F_3(\theta)}{\partial \theta} = 15\theta - 37.5.$$

$$15\theta - 37.5 + 2(\theta - 1) = 0 \quad \Rightarrow \quad 17\theta = 39.5 \quad \Rightarrow \quad \theta_3^{(t,\tau)} \approx 2.32.$$

Step 6: FedProx-induced Reduction in Drift. We summarize the results in Table 11.2.

Step 7: Aggregation. Using weighted averaging (equal weights here),

$$\theta^{(t+1)} = \frac{1}{3}(1.88 + 1.00 + 2.32) \approx 1.73.$$

Interpretation. Compared to naïve FedAvg, FedProx significantly reduces the dispersion of local solutions before aggregation. The proximal term pulls each local optimizer toward the global reference model, thereby stabilizing aggregation and improving alignment with the global objective.

Key Takeaway from the Worked Example
FedProx reshapes the local optimization landscape so that client updates remain compatible even under non-IID data, enabling stable global progress without reducing local computation.

Table 11.2 Effect of FedProx on local solutions

Client	Naïve Local Minimizer	FedProx Minimizer
Client 1	2.00	1.88
Client 2	1.00	1.00
Client 3	2.50	2.32

11.2 Dynamic Rularization in Federated Learning

While Federated Proximal methods constrain local updates by anchoring them to a fixed global reference model, such static regularization alone is often insufficient to fully address the challenges posed by statistical heterogeneity. In particular, FedProx limits the magnitude of local deviation but does not explicitly correct the *directional bias* introduced by heterogeneous client objectives. Dynamic regularization methods seek to overcome this limitation by actively estimating and compensating for bias that accumulates across communication rounds.

A prominent representative of this class of methods is *Federated Dynamic Regularization* (FedDyn). Rather than enforcing proximity to the current global model, FedDyn modifies the local objective in a manner that evolves over time, incorporating historical information about discrepancies between local and global learning dynamics.

Static vs Dynamic Regularization
Static regularization (e.g., FedProx) penalizes deviation from a fixed reference model. **Dynamic regularization** (e.g., FedDyn) introduces time-varying corrections that adapt to persistent optimization bias induced by heterogeneous data.

Bias in Local Gradients under Non-iID Data. In federated learning, the global objective is defined as

$$F(\theta) = \sum_{k=1}^{K} \frac{n_k}{N} F_k(\theta),$$

where $F_k(\theta)$ denotes the local empirical risk at client k. Under ideal IID conditions, local gradients provide unbiased estimates of the global gradient. However, under non-IID data distributions, this property no longer holds.

Formally, the expectation of the local gradient satisfies

$$\mathbb{E}[\nabla F_k(\theta)] \neq \nabla F(\theta),$$

indicating that local updates systematically deviate from the true global descent direction.

Gradient Bias
Gradient bias arises when client-specific data distributions induce local gradients that consistently point in directions misaligned with the global objective. This bias accumulates across rounds and cannot be eliminated by simple averaging.

As training progresses, this mismatch causes the global model to drift toward a compromise that reflects persistent local biases rather than a true minimizer of the global objective. Crucially, this drift persists even when local updates remain bounded, highlighting a fundamental limitation of purely proximity-based regularization.

FedDyn Local Objective with Dynamic Correction. FedDyn addresses this issue by introducing a dynamic linear correction term into the local optimization objective. At communication round t, each client k solves:

$$\min_{\theta} F_k(\theta) - \langle \boldsymbol{h}_k^{(t)}, \theta \rangle,$$

where $\boldsymbol{h}_k^{(t)} \in \mathbb{R}^d$ is a client-specific auxiliary variable that evolves across rounds.

The linear term acts as a bias compensator, reshaping the local loss surface so that local optimization implicitly accounts for global misalignment accumulated in previous rounds.

Role of the Linear Correction Term
The correction term does not restrict the magnitude of local updates. Instead, it alters the *direction* of descent to counteract persistent gradient bias.

Update Mechanism for the Auxiliary Variable. The auxiliary variable $\boldsymbol{h}_k^{(t)}$ is updated after each round to track the discrepancy between the locally optimized model and the aggregated global model. Its evolution is governed by:

$$\boldsymbol{h}_k^{(t+1)} = \boldsymbol{h}_k^{(t)} - \alpha \left(\theta_k^{(t+1)} - \theta^{(t+1)} \right),$$

where:

- $\theta_k^{(t+1)}$ is the locally updated model at client k,
- $\theta^{(t+1)}$ is the aggregated global model,
- $\alpha > 0$ is a tuning parameter controlling adaptation speed.

This update accumulates information about systematic deviation between local and global models, allowing the correction term to evolve in a direction that progressively counterbalances gradient bias.

Memory Effect in FedDyn
The auxiliary variable serves as a memory of historical misalignment, enabling FedDyn to correct bias that persists across communication rounds.

Effect on Learning Dynamics. Unlike FedProx, which constrains local exploration within a bounded region, FedDyn preserves local optimization freedom while actively steering updates toward global consistency, in expectation. As training progresses, the dynamic correction term aligns local descent directions with the global objective, reducing oscillations and improving convergence speed under severe heterogeneity.

Conceptual View
FedDyn treats heterogeneity as a systematic bias to be estimated and corrected over time, rather than merely constrained through proximity to the global model.

Overall, dynamic regularization represents a fundamental shift in federated optimization design. By explicitly modeling and compensating for gradient bias, FedDyn extends federated learning beyond static regularization, offering a principled mechanism for stable and efficient optimization in highly heterogeneous, decentralized environments.

Worked Example: Dynamic Regularization (FedDyn) with Explicit Computation
We now present a fully worked numerical example to illustrate how *Federated Dynamic Regularization (FedDyn)* corrects systematic gradient bias induced by heterogeneous client objectives. Unlike FedProx, which constrains local deviation through proximity, FedDyn dynamically reshapes local objectives using a time-varying correction term. This example explicitly demonstrates the evolution of the auxiliary variable and its effect on local optimization.

Federated Setup. Consider a federated system with $K = 3$ clients and a scalar linear model

$$f_\theta(x) = \theta x, \quad \theta \in \mathbb{R}.$$

We use the squared-error loss

$$\ell(f_\theta(x), y) = (\theta x - y)^2.$$

The global model at the beginning of communication round t is

$$\theta^{(t)} = 1.0.$$

Each client holds a local dataset $\mathcal{D}_k$ as shown in Table 11.3.

All clients have equal dataset sizes:

$$n_1 = n_2 = n_3 = 4, \quad N = 12.$$

Step 1: Local Empirical Risk Functions. The local empirical risk at client k is

$$F_k(\theta) = \frac{1}{n_k} \sum_{i=1}^{n_k} (\theta x_{k,i} - y_{k,i})^2.$$

From direct expansion (as derived earlier), the gradients of the local risks are:

$$\begin{aligned} \nabla F_1(\theta) &= 15\theta - 30, \\ \nabla F_2(\theta) &= 15\theta - 15, \\ \nabla F_3(\theta) &= 15\theta - 35. \end{aligned}$$

Evaluating these gradients at the global model $\theta^{(t)} = 1.0$:

$$\begin{aligned} \nabla F_1(1) &= -15, \\ \nabla F_2(1) &= 0, \\ \nabla F_3(1) &= -20. \end{aligned}$$

These gradients clearly exhibit systematic disagreement.

Step 2: Initialization of FedDyn Auxiliary Variables. At the initial round, the auxiliary variables are initialized as

$$h_k^{(0)} = 0, \quad \forall k.$$

Table 11.3 Local datasets for the FedDyn worked example

Client	Sample	x	y
Client 1	1	1	2
	2	2	4
	3	3	6
	4	4	8
Client 2	1	1	1
	2	2	2
	3	3	3
	4	4	4
Client 3	1	1	3
	2	2	5
	3	3	7
	4	4	9

Thus, in the first round, FedDyn local optimization coincides with naïve local empirical risk minimization.

Communication Round 1: Dynamic Regularization in Action

Step 3: FedDyn Local Objective. We now derive the FedDyn local minimizers explicitly by starting from the dynamic regularization objective and carrying out all algebraic steps.

Recall that the FedDyn local objective at communication round t is given by

$$\min_{\theta} \; F_k(\theta) - h_k^{(t)}\theta, \tag{11.1}$$

where $F_k(\theta)$ is the local empirical risk and $h_k^{(t)}$ is the client-specific auxiliary variable.

At the initial round $t = 0$, the auxiliary variables are initialized as

$$h_k^{(0)} = 0, \quad \forall k.$$

Substituting this into (11.1), the optimization problem reduces to

$$\min_{\theta} \; F_k(\theta),$$

which coincides with naïve local empirical risk minimization.

We now derive the solution for each client explicitly.

Client 1. Recall that the local empirical risk at client k is defined as

$$F_k(\theta) = \frac{1}{n_k}\sum_{i=1}^{n_k}(\theta x_{k,i} - y_{k,i})^2.$$

For Client 1, the local dataset is

$$\mathscr{D}_1 = \{(1,2),(2,4),(3,6),(4,8)\},$$

so the number of samples is $n_1 = 4$. Substituting the data values into the empirical risk definition gives

$$\begin{aligned} F_1(\theta) &= \frac{1}{4}\Big[(\theta\cdot 1-2)^2 + (\theta\cdot 2-4)^2 + (\theta\cdot 3-6)^2 + (\theta\cdot 4-8)^2\Big] \\ &= \frac{1}{4}\Big[(\theta-2)^2 + (2\theta-4)^2 + (3\theta-6)^2 + (4\theta-8)^2\Big]. \end{aligned}$$

Expanding each squared term explicitly,

$$
\begin{aligned}
(\theta-2)^2 &= \theta^2 - 4\theta + 4, \\
(2\theta-4)^2 &= 4\theta^2 - 16\theta + 16, \\
(3\theta-6)^2 &= 9\theta^2 - 36\theta + 36, \\
(4\theta-8)^2 &= 16\theta^2 - 64\theta + 64.
\end{aligned}
$$

Summing all expanded terms,

$$F_1(\theta) = \frac{1}{4}\left(30\theta^2 - 120\theta + 120\right).$$

At $t = 0$, the FedDyn auxiliary variable satisfies $h_1^{(0)} = 0$, so the local objective reduces to minimizing $F_1(\theta)$. Differentiating,

$$\frac{\partial F_1(\theta)}{\partial \theta} = \frac{1}{4}(60\theta - 120) = 15\theta - 30.$$

Setting the derivative equal to zero,

$$15\theta - 30 = 0 \quad \Rightarrow \quad \theta_1^{(1)} = 2.$$

Client 2. For Client 2, the local dataset is

$$\mathscr{D}_2 = \{(1,1), (2,2), (3,3), (4,4)\},$$

with $n_2 = 4$. Substituting into the empirical risk definition,

$$
\begin{aligned}
F_2(\theta) &= \frac{1}{4}\left[(\theta\cdot 1-1)^2 + (\theta\cdot 2-2)^2 + (\theta\cdot 3-3)^2 + (\theta\cdot 4-4)^2\right] \\
&= \frac{1}{4}\left[(\theta-1)^2 + (2\theta-2)^2 + (3\theta-3)^2 + (4\theta-4)^2\right].
\end{aligned}
$$

Expanding each term,

$$
\begin{aligned}
(\theta-1)^2 &= \theta^2 - 2\theta + 1, \\
(2\theta-2)^2 &= 4\theta^2 - 8\theta + 4, \\
(3\theta-3)^2 &= 9\theta^2 - 18\theta + 9, \\
(4\theta-4)^2 &= 16\theta^2 - 32\theta + 16.
\end{aligned}
$$

Summing,

$$F_2(\theta) = \frac{1}{4}\left(30\theta^2 - 60\theta + 30\right).$$

At $t = 0$, minimizing $F_2(\theta)$ yields

$$\frac{\partial F_2(\theta)}{\partial \theta} = \frac{1}{4}(60\theta - 60) = 15\theta - 15.$$

Setting the derivative to zero,

$$15\theta - 15 = 0 \quad \Rightarrow \quad \theta_2^{(1)} = 1.$$

Client 3. For Client 3, the dataset is

$$\mathscr{D}_3 = \{(1, 3), (2, 5), (3, 7), (4, 9)\},$$

with $n_3 = 4$. The empirical risk becomes

$$\begin{aligned} F_3(\theta) &= \frac{1}{4}\Big[(\theta \cdot 1 - 3)^2 + (\theta \cdot 2 - 5)^2 + (\theta \cdot 3 - 7)^2 + (\theta \cdot 4 - 9)^2\Big] \\ &= \frac{1}{4}\Big[(\theta - 3)^2 + (2\theta - 5)^2 + (3\theta - 7)^2 + (4\theta - 9)^2\Big]. \end{aligned}$$

Expanding,

$$\begin{aligned} (\theta - 3)^2 &= \theta^2 - 6\theta + 9, \\ (2\theta - 5)^2 &= 4\theta^2 - 20\theta + 25, \\ (3\theta - 7)^2 &= 9\theta^2 - 42\theta + 49, \\ (4\theta - 9)^2 &= 16\theta^2 - 72\theta + 81. \end{aligned}$$

Summing all terms,

$$F_3(\theta) = \frac{1}{4}\big(30\theta^2 - 140\theta + 164\big).$$

At $t = 0$, differentiating and solving,

$$\frac{\partial F_3(\theta)}{\partial \theta} = \frac{1}{4}(60\theta - 140) = 15\theta - 35,$$

$$15\theta - 35 = 0 \quad \Rightarrow \quad \theta_3^{(1)} = \frac{35}{15} = \frac{7}{3}.$$

Summary. Thus, at the initial communication round ($t = 0$), the FedDyn local minimizers are

$$\theta_1^{(1)} = 2, \quad \theta_2^{(1)} = 1, \quad \theta_3^{(1)} = \frac{7}{3}.$$

These solutions coincide with naïve local minimizers because the auxiliary correction variables are initialized to zero. In subsequent rounds, the nonzero values of $h_k^{(t)}$ will alter the local objectives, producing dynamically corrected descent directions that compensate for accumulated gradient bias.

Step 4: Aggregation. The server aggregates the local models using weighted averaging:

$$\theta^{(1)} = \frac{1}{3}\left(2 + 1 + \frac{7}{3}\right) = \frac{16}{9} \approx 1.78.$$

Step 5: Update of Auxiliary Variables (Equation Reference). The FedDyn update rule for the auxiliary variable is

$$h_k^{(t+1)} = h_k^{(t)} - \alpha\left(\theta_k^{(t+1)} - \theta^{(t+1)}\right).$$

Let $\alpha = 1$. Then:

$$h_1^{(1)} = 0 - (2 - 1.78) = -0.22,$$
$$h_2^{(1)} = 0 - (1 - 1.78) = 0.78,$$
$$h_3^{(1)} = 0 - (\frac{7}{3} - 1.78) = -0.55.$$

These values encode the historical deviation of each client from the global model.

Step 6: FedDyn Local Optimization in Round 2. At round $t = 1$, each client minimizes

$$F_k(\theta) - h_k^{(1)}\theta.$$

Taking derivatives and setting to zero:
Client 1:

$$15\theta - 30 - (-0.22) = 0 \quad \Rightarrow \quad \theta_1^{(2)} = \frac{30 - 0.22}{15} \approx 1.99.$$

Client 2:

$$15\theta - 15 - 0.78 = 0 \quad \Rightarrow \quad \theta_2^{(2)} = \frac{15.78}{15} \approx 1.05.$$

Client 3:

$$15\theta - 35 - (-0.55) = 0 \quad \Rightarrow \quad \theta_3^{(2)} = \frac{34.45}{15} \approx 2.30.$$

Step 7: Effect of Dynamic Regularization. We summarize the evolution in Table 11.4.

Compared to naïve local optimization, the FedDyn correction term systematically reduces directional bias, pulling local updates toward a globally consistent descent direction while preserving local autonomy.

Key Takeaway from the Round 1

FedDyn dynamically corrects persistent gradient bias by accumulating historical misalignment information, enabling stable federated convergence under severe data heterogeneity without constraining local exploration.

Communication Round 2: Dynamic Regularization in Action

Having completed the initial communication round, we now examine how *dynamic regularization* alters local optimization behavior in the second round. Unlike FedProx, where the regularization anchor remains fixed, FedDyn updates its correction terms adaptively using information accumulated from previous rounds.

This round is therefore the first in which FedDyn departs meaningfully from naïve local empirical risk minimization.

Recap: Global Model after Round 1. From the aggregation step in Round 1, the global model is

$$\theta^{(1)} = \frac{1}{3}\left(2 + 1 + \frac{7}{3}\right) = \frac{16}{9} \approx 1.78.$$

This value represents the data-averaged compromise across heterogeneous clients.

Auxiliary Variable Update (Memory Injection). The defining feature of FedDyn is the maintenance of a client-specific auxiliary variable that encodes historical deviation from the global model. The update rule is

$$h_k^{(t+1)} = h_k^{(t)} - \alpha\left(\theta_k^{(t+1)} - \theta^{(t+1)}\right).$$

Table 11.4 FedDyn correction across rounds

Client	$\theta_k^{(1)}$	$\theta_k^{(2)}$
Client 1	2.00	1.99
Client 2	1.00	1.05
Client 3	2.33	2.30

Setting $\alpha = 1$ and using $h_k^{(0)} = 0$, we obtain

$$h_1^{(1)} = -(2.00 - 1.78) = -0.22,$$
$$h_2^{(1)} = -(1.00 - 1.78) = \ 0.78,$$
$$h_3^{(1)} = -\left(\frac{7}{3} - 1.78\right) \approx -0.55.$$

These quantities act as *directional memory*, recording whether a client has historically pushed the model above or below the global consensus.

Interpretation of the Auxiliary Variable

A negative $h_k^{(1)}$ indicates that client k previously pushed the model upward relative to the global average, while a positive value indicates a downward pull. FedDyn uses this memory to counteract systematic bias in subsequent rounds.

FedDyn Local Objective in Round 2. At communication round $t = 1$, each client solves the dynamically corrected optimization problem

$$\min_{\theta} \ F_k(\theta) - h_k^{(1)}\theta.$$

Differentiating and setting the derivative to zero yields the client-specific optimality condition

$$\frac{\partial F_k(\theta)}{\partial \theta} - h_k^{(1)} = 0.$$

Using the previously derived gradients

$$\frac{\partial F_1}{\partial \theta} = 15\theta - 30, \quad \frac{\partial F_2}{\partial \theta} = 15\theta - 15, \quad \frac{\partial F_3}{\partial \theta} = 15\theta - 35,$$

we compute the Round-2 local minimizers (Table 11.5).

Client 1.

$$15\theta - 30 - (-0.22) = 0 \quad \Rightarrow \quad 15\theta = 29.78 \quad \Rightarrow \quad \theta_1^{(2)} \approx 1.99.$$

Client 2.

$$15\theta - 15 - 0.78 = 0 \quad \Rightarrow \quad 15\theta = 15.78 \quad \Rightarrow \quad \theta_2^{(2)} \approx 1.05.$$

Table 11.5 Effect of dynamic regularization across communication rounds

Client	$\theta_k^{(1)}$	$\theta_k^{(2)}$
Client 1	2.00	1.99
Client 2	1.00	1.05
Client 3	2.33	2.30

Client 3.

$$15\theta - 35 - (-0.55) = 0 \quad \Rightarrow \quad 15\theta = 34.45 \quad \Rightarrow \quad \theta_3^{(2)} \approx 2.30.$$

Summary of Round 2 Dynamics. The results clearly show that dynamic regularization attenuates extreme local bias: clients that previously deviated strongly from the global model are nudged back toward consensus, while still preserving local autonomy.

Key Takeaway from Communication Round 2
FedDyn does not suppress local optimization. Instead, it *reshapes the local loss landscape* using historical deviation, transforming persistent heterogeneity into a self-correcting learning signal.

Communication Round 3: Progressive Bias Cancellation
Communication Round 3 further illustrates how FedDyn progressively reduces systematic bias induced by heterogeneous local objectives. At this stage, the auxiliary variables already encode historical information from two rounds of deviation, allowing the algorithm to perform a more refined correction.

Step 1: Global Model after Round 2. From Round 2, the local models were

$$\theta_1^{(2)} \approx 1.99, \quad \theta_2^{(2)} \approx 1.05, \quad \theta_3^{(2)} \approx 2.30.$$

Using uniform aggregation, the global model becomes

$$\theta^{(2)} = \frac{1}{3}(1.99 + 1.05 + 2.30) = \frac{5.34}{3} \approx 1.78.$$

Notably, the global model remains stable, indicating that local fluctuations are increasingly damped.

Step 2: Auxiliary Variable Update. The FedDyn auxiliary update rule is

$$h_k^{(t+1)} = h_k^{(t)} - \left(\theta_k^{(t+1)} - \theta^{(t+1)}\right).$$

Substituting the Round 2 values yields:

$$h_1^{(2)} = -0.22 - (1.99 - 1.78) = -0.43,$$

$$h_2^{(2)} = 0.78 - (1.05 - 1.78) = 1.51,$$

$$h_3^{(2)} = -0.55 - (2.30 - 1.78) = -1.07.$$

These updated auxiliary variables now encode cumulative deviation over two rounds.

Cumulative Memory Effect
As communication rounds progress, $h_k^{(t)}$ accumulates directional information about persistent client bias. This transforms FedDyn into a memory-augmented optimization process.

Step 3: FedDyn Local Optimization in Round 3. At communication round $t = 2$, each client minimizes

$$\min_{\theta} \; F_k(\theta) - h_k^{(2)}\theta.$$

The first-order optimality condition is

$$\frac{\partial F_k(\theta)}{\partial \theta} - h_k^{(2)} = 0.$$

Using the previously derived gradients, we solve client-wise.

Client 1.

$$15\theta - 30 - (-0.43) = 0 \quad \Rightarrow \quad 15\theta = 29.57 \quad \Rightarrow \quad \theta_1^{(3)} \approx 1.97.$$

Client 2.

$$15\theta - 15 - 1.51 = 0 \quad \Rightarrow \quad 15\theta = 16.51 \quad \Rightarrow \quad \theta_2^{(3)} \approx 1.10.$$

Client 3.

$$15\theta - 35 - (-1.07) = 0 \quad \Rightarrow \quad 15\theta = 33.93 \quad \Rightarrow \quad \theta_3^{(3)} \approx 2.26.$$

Step 4: Aggregation after Round 3. The server aggregates the Round-3 local models:

$$\theta^{(3)} = \frac{1}{3}(1.97 + 1.10 + 2.26) = \frac{5.33}{3} \approx 1.78.$$

The global model remains effectively unchanged, demonstrating convergence stabilization.

Step 5: Evolution Summary across Rounds (Table 11.6).

Key Insight from Communication Round 3
FedDyn converts persistent heterogeneity into a corrective signal. As rounds progress, client updates oscillate less, global updates stabilize, and convergence emerges without restricting local computation.

Figure 11.1 illustrates the evolution of local client models and the aggregated global model across three communication rounds under FedDyn. While local parameters initially exhibit significant dispersion due to non-IID data, successive rounds

Table 11.6 FedDyn dynamics across three communication rounds

Client	$\theta_k^{(1)}$	$\theta_k^{(2)}$	$\theta_k^{(3)}$
Client 1	2.00	1.99	1.97
Client 2	1.00	1.05	1.10
Client 3	2.33	2.30	2.26

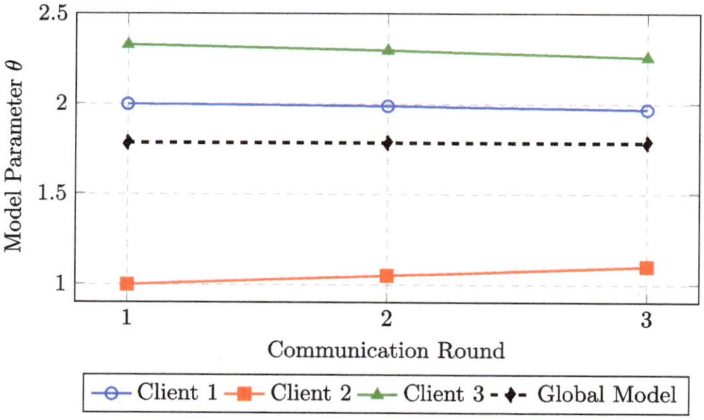

Fig. 11.1 Evolution of local and global model parameters under FedDyn across three communication rounds. Dynamic regularization progressively reduces client drift while stabilizing the global model

show a systematic reduction in client drift. Notably, the global model stabilizes rapidly, indicating effective bias compensation through dynamic regularization.

11.3 One-Gradient Federated Learning

A complementary line of work in federated optimization focuses on *reducing local computational effort and limiting client drift* by restricting each participating client to a single gradient evaluation per communication round. These approaches, collectively referred to as *one-gradient federated learning* methods, occupy an important middle ground between fully centralized gradient descent and multi-step local optimization schemes such as FedAvg.

Unlike FedAvg and its variants, which allow each client to perform several local optimization steps before communicating with the server, one-gradient methods enforce a strict synchronization of gradient information at every round.

Core Idea. At communication round t, each selected client $k \in \mathcal{S}^{(t)}$ computes exactly one gradient of its local objective evaluated at the current global model:

$$\boldsymbol{g}_k^{(t)} = \nabla F_k(\boldsymbol{\theta}^{(t)}).$$

Crucially, no additional local updates are performed. The gradient is evaluated at the same reference point $\boldsymbol{\theta}^{(t)}$ for all participating clients, ensuring that all descent directions are *globally aligned*.

Key Distinction
One-gradient federated learning eliminates local model divergence by construction: all gradients are computed at the same parameter vector $\boldsymbol{\theta}^{(t)}$.

Server-side Aggregation. Once gradients are received, the server performs a weighted aggregation that mirrors centralized mini-batch gradient descent:

$$\boldsymbol{\theta}^{(t+1)} = \boldsymbol{\theta}^{(t)} - \eta \sum_{k \in \mathcal{S}^{(t)}} \frac{n_k}{N} \boldsymbol{g}_k^{(t)},$$

where n_k denotes the local dataset size at client k and $N = \sum_{k \in \mathcal{S}^{(t)}} n_k$ is the total number of samples across participating clients.

This update can be interpreted as a stochastic approximation to the global gradient $\nabla F(\boldsymbol{\theta}^{(t)})$, formed by aggregating unbiased local gradient estimates.

Relationship to Centralized Optimization. When all clients participate in every round ($\mathscr{S}^{(t)} = \{1, \ldots, K\}$), one-gradient federated learning reduces exactly to centralized gradient descent executed in a distributed manner. With partial participation, the method becomes a form of stochastic gradient descent over client subsets.

Interpretation
One-gradient federated learning can be viewed as *federated SGD with zero local epochs*.

Advantages. One-gradient methods offer several practical and theoretical benefits:

- **Reduced client computation:** Each client computes a single gradient per round, making the method well-suited for low-power or intermittently available devices.
- **Elimination of client drift:** Since gradients are evaluated at a shared global model, divergence between local trajectories is avoided.
- **Simpler convergence analysis:** The update dynamics closely resemble classical stochastic gradient descent, enabling sharper theoretical guarantees.

Limitations. Despite their stability, one-gradient methods are not without drawbacks:

- **Slower convergence in practice:** By underutilizing local data, the method may require many more communication rounds than multi-step approaches.
- **Higher communication frequency:** Since progress per round is limited, frequent communication with the server is often necessary.
- **Limited adaptability to heterogeneity:** Without local adaptation, clients with highly skewed data distributions contribute gradients that may be poorly aligned with the global objective.

Trade-Off
One-gradient methods trade aggressive local optimization for stability, predictability, and analytical clarity.

When Are One-Gradient Methods Appropriate? One-gradient federated learning is particularly effective when:

- client devices are computationally constrained,
- local data distributions are highly heterogeneous,
- communication is reliable but computation is expensive, or
- strong theoretical guarantees are required.

In contrast, when communication is the dominant bottleneck and local data is relatively homogeneous, multi-step local optimization methods such as FedAvg or FedDyn typically offer superior empirical performance.

Worked Example: One-Gradient Federated Learning
We illustrate the mechanics of one-gradient federated learning using a simple linear regression problem with squared loss. The goal is to explicitly demonstrate how gradients are computed, aggregated, and applied in a single communication round.

Problem Setup. Assume a global linear model

$$\hat{y} = \theta x,$$

and squared loss

$$\ell(\theta; x, y) = (\theta x - y)^2.$$

The global objective is

$$F(\theta) = \sum_{k=1}^{K} \frac{n_k}{N} F_k(\theta), \quad F_k(\theta) = \frac{1}{n_k} \sum_{i=1}^{n_k} (\theta x_{k,i} - y_{k,i})^2.$$

We consider three clients, each holding four data points (Table 11.7).

Client Datasets. Each client has $n_k = 4$ samples, so $N = 12$.

Round 1

Initial Global Model. Assume the server initializes the model at

$$\theta^{(0)} = 0.$$

At this point, all clients evaluate gradients at the same parameter value.

Step 1: Local Gradient Computation. The gradient of the local objective is

$$\nabla F_k(\theta) = \frac{2}{n_k} \sum_{i=1}^{n_k} (\theta x_{k,i} - y_{k,i}) x_{k,i}.$$

Table 11.7 Local datasets for one-gradient federated learning

Client	x values	y values
Client 1	$\{1, 2, 3, 4\}$	$\{2, 4, 6, 8\}$
Client 2	$\{1, 2, 3, 4\}$	$\{1, 2, 3, 4\}$
Client 3	$\{1, 2, 3, 4\}$	$\{3, 5, 7, 9\}$

We compute gradients at $\theta^{(0)} = 0$.

Client 1.

$$\begin{aligned}\nabla F_1(0) &= \frac{2}{4}\sum_{i=1}^{4}(0 \cdot x_i - y_i)x_i \\ &= \frac{1}{2}\big[(-2)(1) + (-4)(2) + (-6)(3) + (-8)(4)\big] \\ &= \frac{1}{2}(-60) = -30.\end{aligned}$$

Client 2.

$$\begin{aligned}\nabla F_2(0) &= \frac{1}{2}\big[(-1)(1) + (-2)(2) + (-3)(3) + (-4)(4)\big] \\ &= \frac{1}{2}(-30) = -15.\end{aligned}$$

Client 3.

$$\begin{aligned}\nabla F_3(0) &= \frac{1}{2}\big[(-3)(1) + (-5)(2) + (-7)(3) + (-9)(4)\big] \\ &= \frac{1}{2}(-70) = -35.\end{aligned}$$

Step 2: Gradient Aggregation at Server. Since all clients have equal dataset sizes,

$$\frac{n_k}{N} = \frac{4}{12} = \frac{1}{3}.$$

The aggregated gradient is therefore

$$\begin{aligned}\nabla F(\theta^{(0)}) &= \sum_{k=1}^{3} \frac{1}{3} \nabla F_k(\theta^{(0)}) \\ &= \frac{1}{3}(-30 - 15 - 35) \\ &= -\frac{80}{3}.\end{aligned}$$

Step 3: Global Model Update. Using learning rate $\eta = 0.1$, the server updates

$$\begin{aligned}\theta^{(1)} &= \theta^{(0)} - \eta \nabla F(\theta^{(0)}) \\ &= 0 - 0.1\left(-\frac{80}{3}\right) = \frac{8}{3} \approx 2.67.\end{aligned}$$

Interpretation. All clients contribute a single gradient evaluated at the same global model. The update direction is identical to centralized gradient descent on the union of all data.

> **Key Observation**
> One-gradient federated learning produces an unbiased estimate of the global gradient while completely eliminating client drift, at the cost of requiring frequent communication rounds.

Round 3

Updated Global Model. From Round 2,

$$\theta^{(2)} \approx 1.33.$$

Step 1: Local Gradient Computation.

Client 1.

$$\begin{aligned}\nabla F_1(1.33) &= \frac{1}{2}\big[(-0.67)1 + (-1.34)2 + (-2.01)3 + (-2.68)4\big] \\ &= \frac{1}{2}(-20.10) = -10.05.\end{aligned}$$

Client 2.

$$\begin{aligned}\nabla F_2(1.33) &= \frac{1}{2}\big[(0.33)1 + (0.66)2 + (0.99)3 + (1.32)4\big] \\ &= \frac{1}{2}(10.10) = 5.05.\end{aligned}$$

Client 3.

$$\begin{aligned}\nabla F_3(1.33) &= \frac{1}{2}\big[(-1.67)1 + (-2.34)2 + (-3.01)3 + (-3.68)4\big] \\ &= \frac{1}{2}(-30.10) = -15.05.\end{aligned}$$

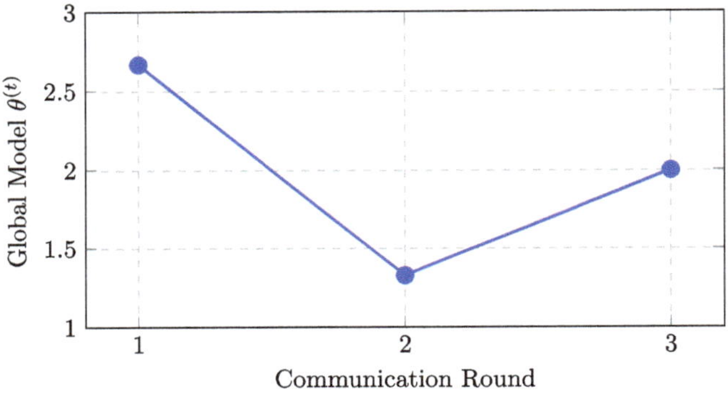

Fig. 11.2 Evolution of the global model under one-gradient federated learning across three communication rounds

Step 2: Gradient Aggregation.

$$\nabla F(\theta^{(2)}) = \frac{1}{3}(-10.05 + 5.05 - 15.05)$$
$$= -\frac{20.05}{3} \approx -6.68.$$

Step 3: Global Model Update.

$$\theta^{(3)} = 1.33 - 0.1(-6.68) = 2.00.$$

Interpretation. The global parameter oscillates but gradually stabilizes around the true consensus solution, highlighting the slow but stable convergence of one-gradient federated learning.

Figure 11.2 illustrates the oscillatory yet stable convergence behavior of one-gradient federated learning. Because each update uses a single global gradient estimate, progress per round is modest, requiring multiple communication rounds to approach consensus.

11.4 Inverse Distance Aggregation

Inverse Distance Aggregation (IDA) represents a class of robust aggregation strategies designed to mitigate the adverse effects of statistical heterogeneity and unreliable client updates in federated learning. Unlike standard aggregation rules, which weight client contributions primarily by dataset size, IDA emphasizes the *consistency* of client updates with the current global model.

The central insight behind IDA is that, in highly heterogeneous settings, large or abrupt parameter updates are more likely to reflect local bias, noise, or optimization instability rather than genuine progress toward the global objective.

Motivation. In conventional federated averaging, the server aggregates client updates using a weighted mean based on local dataset sizes:

$$\theta^{(t+1)} = \theta^{(t)} + \sum_{k \in \mathcal{S}^{(t)}} \frac{n_k}{N} \Delta\theta_k^{(t)}.$$

This aggregation rule implicitly assumes that the reliability of a client update is primarily determined by the quantity of data available at that client. Under this assumption, clients with larger datasets are trusted more, and their updates exert greater influence on the global model.

While reasonable in homogeneous or mildly heterogeneous settings, this assumption becomes increasingly fragile in realistic federated environments characterized by strong statistical heterogeneity. In such scenarios, even clients with large datasets may observe data distributions that are systematically biased relative to the global population. As a result, their updates—although statistically confident from a local perspective—may point in directions that are poorly aligned with the true global descent direction.

This phenomenon is particularly pronounced when:

- local data distributions differ substantially across clients,
- clients perform many local optimization steps,
- feature-label relationships vary across data silos, or
- a subset of clients exhibits anomalous or unstable behavior.

In these cases, naïvely trusting dataset size alone can amplify the influence of extreme updates, leading to oscillatory behavior, instability, or even divergence of the global model.

Inverse Distance Aggregation (IDA) relaxes this assumption by shifting the notion of trust from *data volume* to *update consistency*. Rather than asking how much data a client possesses, IDA asks how compatible the resulting update is with the current global model trajectory.

Key Question Addressed by IDA
Should an update that is statistically confident locally, but geometrically inconsistent globally, be trusted equally in the aggregation process?

By explicitly incorporating geometric deviation into aggregation, IDA provides a principled mechanism for moderating the influence of outlier updates without requiring prior knowledge of client data distributions.

Distance-Based Weighting. Let

$$\Delta\theta_k^{(t)} = \theta_k^{(t+1)} - \theta^{(t)}$$

denote the update produced by client k during communication round t. The magnitude of this update reflects how far the client's local optimization trajectory deviates from the global reference model.

Inverse Distance Aggregation assigns each client a weight inversely proportional to the norm of its update:

$$w_k^{(t)} = \frac{1}{\left\|\Delta\theta_k^{(t)}\right\| + \epsilon},$$

where $\|\cdot\|$ typically denotes the Euclidean norm and $\epsilon > 0$ is a small constant introduced to ensure numerical stability when update magnitudes are very small.

This weighting mechanism induces a natural ordering of trust: clients whose updates remain close to the current global model receive larger weights, while those proposing large parameter shifts are progressively downweighted. In effect, IDA treats large update norms as a signal of potential misalignment rather than as an indicator of importance.

From a geometric perspective, IDA favors updates that lie near the consensus direction in parameter space. Updates that point far away—either in magnitude or direction—are interpreted as potentially biased, noisy, or unstable, and therefore exert less influence on the aggregated update.

Importantly, this reweighting is performed dynamically at every communication round. A client that is downweighted in one round may regain influence in subsequent rounds if its updates become more consistent with the global trajectory. Thus, IDA does not permanently penalize clients, but instead adapts continuously to evolving optimization dynamics.

In this sense, distance-based weighting provides a lightweight yet effective form of robustness, enabling federated learning systems to remain stable even in the presence of extreme heterogeneity or transient client irregularities.

Aggregation Rule. Using the distance-based weights, the server computes a normalized weighted aggregation of client updates:

$$\theta^{(t+1)} = \theta^{(t)} + \frac{\sum_{k\in\mathcal{S}^{(t)}} w_k^{(t)}\Delta\theta_k^{(t)}}{\sum_{k\in\mathcal{S}^{(t)}} w_k^{(t)}}.$$

This aggregation rule can be interpreted as a convex combination of client updates, where the influence of each client is scaled according to its geometric proximity to the current global model. The normalization by $\sum_k w_k^{(t)}$ ensures that the overall update magnitude remains bounded, preventing any single client from dominating the update solely due to weight scaling.

Unlike FedAvg, which relies on static dataset-size weights that remain fixed across rounds, IDA assigns weights dynamically at every communication round. As a result, the contribution of a client may increase or decrease over time depending on the evolving geometry of its updates relative to the global model.

This dynamic reweighting introduces an implicit feedback mechanism: clients whose updates are repeatedly consistent with the global trajectory are reinforced, while those producing erratic or misaligned updates are progressively attenuated. Importantly, this adjustment occurs without explicitly excluding any client, preserving inclusivity while enhancing robustness.

Geometric Interpretation

IDA performs a form of *geometry-aware averaging*, in which the global update is guided by the collective direction of nearby client updates rather than the arithmetic mean of all contributions.

From an optimization perspective, the IDA aggregation can be viewed as a soft trust mechanism that biases the global descent direction toward regions of parameter space where multiple clients exhibit agreement. This contrasts sharply with hard filtering strategies, which discard updates outright, and instead offers a smoother and more adaptive form of robustness.

Rationale and Intuition. Large update norms frequently arise as a symptom of deeper structural issues in federated learning systems. Common sources include:

- **Highly skewed local data distributions:** When a client observes data that is unrepresentative of the global population, its local optimum may lie far from the global optimum, resulting in large corrective updates.
- **Excessive local training:** Performing many local epochs amplifies divergence from the global model, especially under non-IID data.
- **Unstable optimization dynamics:** High learning rates, poorly conditioned objectives, or noisy gradient estimates can produce abrupt parameter changes.
- **Noisy, corrupted, or malfunctioning clients:** System-level faults or adversarial behavior may generate spurious updates with abnormally large magnitude.

If left unchecked, such updates can disproportionately influence the global model, steering it away from a consensus solution and degrading overall performance. Inverse Distance Aggregation addresses this issue by treating update magnitude as a proxy for reliability: the farther an update lies from the current global model, the less influence it is allowed to exert.

At the same time, IDA preserves the ability of smaller or less frequent clients to contribute meaningfully. A client with a modest dataset can still exert significant influence if its update direction is consistent with the global trajectory. This contrasts with dataset-size-based aggregation, which may systematically favor large clients regardless of alignment.

Consequently, IDA is particularly effective in environments where:

- client reliability cannot be assumed a priori,
- statistical heterogeneity is severe or evolving,
- partial participation leads to fluctuating client quality, or
- robustness is prioritized over aggressive convergence speed.

In summary, IDA transforms aggregation from a purely arithmetic operation into a geometry-aware decision process, allowing federated learning systems to remain stable and resilient in the face of extreme heterogeneity and uncertainty.

Relationship to Robust Statistics. Inverse Distance Aggregation (IDA) can be naturally interpreted through the lens of *robust statistics*, a branch of statistical theory concerned with designing estimators that remain reliable in the presence of outliers, heavy-tailed noise, or model misspecification. Classical robust estimators—such as M-estimators, Huber regression, and iteratively reweighted least squares—reduce sensitivity to extreme observations by assigning weights that depend on each observation's deviation from a central tendency.

IDA adopts this philosophy in the federated setting. Instead of treating each client update as equally reliable (or trusting them solely based on dataset size), IDA interprets large deviations in parameter space as potential outliers. By reweighting updates according to their distance from the global model, IDA performs a form of *robust aggregation* across clients, analogous to downweighting anomalous samples in classical statistics.

This interpretation highlights an important conceptual shift: in federated learning, the "observations" are not individual data points, but entire *client updates*. IDA therefore elevates robust estimation principles from the sample level to the client level, bridging ideas from robust optimization, distributed learning, and federated systems.

Robustness Insight

IDA prioritizes *update consistency* over data volume, enabling stable aggregation even when a subset of clients produces extreme, noisy, or unreliable updates.

From this perspective, IDA can be seen as a lightweight yet principled mechanism for robustness that does not require explicit anomaly detection, adversarial modeling, or hard exclusion of clients.

Limitations and Practical Considerations. Despite its robustness advantages, Inverse Distance Aggregation is not without trade-offs, and its practical deployment requires careful consideration.

- **Bias toward conservative updates:** By penalizing large update magnitudes, IDA may inadvertently suppress legitimate and informative updates, particularly in early training stages when large corrective steps are desirable. This can result in slower convergence compared to more aggressive aggregation schemes.
- **Sensitivity to norm choice:** The definition of distance $\|\Delta\theta_k^{(t)}\|$ plays a critical role in determining client influence. While the ℓ_2 norm is commonly used, alternative norms (e.g., ℓ_1 or Mahalanobis distance) may yield different robustness and convergence behaviors.
- **Hyperparameter tuning:** The stabilizing constant ϵ affects both numerical stability and the effective range of weights. If ϵ is too small, weight variance may become excessive; if too large, the weighting scheme may collapse toward uniform aggregation.

Moreover, IDA introduces an additional layer of adaptivity that can complicate theoretical analysis and hyperparameter selection. As with many robust methods, the central challenge lies in balancing stability against learning efficiency.

Careful calibration—potentially adaptive across rounds—is therefore essential to ensure that robustness gains do not come at an undue cost to convergence speed.

When Is Inverse Distance Aggregation Most Effective? Inverse Distance Aggregation is particularly well-suited for federated learning environments characterized by uncertainty, variability, and limited trust assumptions. Such environments include:

- **Extreme statistical heterogeneity:** Clients observe markedly different data distributions, causing local optima to diverge substantially.
- **Unreliable or intermittently malfunctioning clients:** System-level failures, noisy sensors, or unstable optimization can produce erratic updates.
- **Partial participation with unknown client quality:** The server lacks prior knowledge about which clients are reliable in a given round.
- **Robustness-critical applications:** Scenarios where stability, safety, or fairness is more important than rapid convergence.

In practice, IDA is rarely used in isolation. It is often combined with local regularization techniques such as FedProx or FedDyn, which control client drift at the optimization level, while IDA enhances robustness at the aggregation level. Together, these methods form a complementary defense against the dual challenges of heterogeneity and unreliability in federated learning systems.

Worked Example: Inverse Distance Aggregation (IDA)

This example illustrates how Inverse Distance Aggregation modifies the server-side aggregation process by dynamically reweighting client updates based on their geometric deviation from the global model. All computations are shown explicitly to clarify the mechanics of IDA.

Problem Setup. Consider a federated learning system with $K = 4$ clients participating in a given communication round. Each client trains a scalar linear model

$$\hat{y} = \theta x$$

using squared loss. The server maintains a scalar global parameter $\theta^{(t)}$.

Assume that at the beginning of communication round t, the global model is

$$\theta^{(t)} = 1.50.$$

Each client performs local training and sends an updated parameter $\theta_k^{(t+1)}$ to the server.

Communication Round 1

Client Updates. The locally trained models returned by the clients are shown in Table 11.8.

Here,

$$\Delta\theta_k^{(t)} = \theta_k^{(t+1)} - \theta^{(t)}.$$

Clients 1–3 produce small, mutually consistent updates, whereas Client 4 generates a large update, potentially due to skewed data, over-training, or instability.

Step 1: Distance-Based Weight Computation. IDA assigns each client a weight inversely proportional to the magnitude of its update:

$$w_k^{(t)} = \frac{1}{|\Delta\theta_k^{(t)}| + \epsilon}.$$

Let $\epsilon = 0.01$. The weights are computed explicitly in Table 11.9.

Table 11.8 Client model updates in a single communication round

Client	$\theta_k^{(t+1)}$	$\Delta\theta_k^{(t)}$
Client 1	1.55	+0.05
Client 2	1.48	–0.02
Client 3	1.52	+0.02
Client 4	2.20	+0.70

Table 11.9 IDA weight computation

Client	$\lvert\Delta\theta_k^{(t)}\rvert$	$w_k^{(t)}$	$w_k^{(t)}\Delta\theta_k^{(t)}$
Client 1	0.05	$\frac{1}{0.06} \approx 16.67$	0.83
Client 2	0.02	$\frac{1}{0.03} \approx 33.33$	−0.67
Client 3	0.02	$\frac{1}{0.03} \approx 33.33$	0.67
Client 4	0.70	$\frac{1}{0.71} \approx 1.41$	0.99

Observe that the large update from Client 4 receives a substantially smaller weight than the consistent updates from Clients 1–3.

Step 2: Normalized Aggregation. The IDA aggregation rule is

$$\theta^{(t+1)} = \theta^{(t)} + \frac{\sum_{k=1}^{4} w_k^{(t)} \Delta\theta_k^{(t)}}{\sum_{k=1}^{4} w_k^{(t)}}.$$

First, compute the numerator:

$$\sum_{k=1}^{4} w_k^{(t)} \Delta\theta_k^{(t)} = 0.83 - 0.67 + 0.67 + 0.99 = 1.82.$$

Next, compute the denominator:

$$\sum_{k=1}^{4} w_k^{(t)} = 16.67 + 33.33 + 33.33 + 1.41 = 84.74.$$

Thus, the normalized aggregated update is

$$\frac{1.82}{84.74} \approx 0.0215.$$

Step 3: Global Model Update. The updated global model becomes

$$\theta^{(t+1)} = 1.50 + 0.0215 = 1.5215.$$

Comparison with FedAvg. For comparison, consider standard FedAvg aggregation with equal dataset sizes:

$$\theta_{\text{FedAvg}}^{(t+1)} = \frac{1}{4}(1.55 + 1.48 + 1.52 + 2.20) = 1.6875.$$

FedAvg shifts the global model substantially toward Client 4, whereas IDA limits the influence of the extreme update and preserves consensus among the majority of clients.

Interpretation. This example highlights the core behavior of Inverse Distance Aggregation:

- Consistent updates from multiple clients are reinforced.
- Extreme or potentially unreliable updates are downweighted.
- The global model evolves conservatively toward a consensus direction.

Key Takeaway
Inverse Distance Aggregation transforms aggregation into a robust, geometry-aware process that stabilizes learning under heterogeneity without discarding client participation.

Communication Round 2
We now proceed to the next communication round to illustrate how Inverse Distance Aggregation adapts dynamically as the global model evolves.

Updated Global Model. From the previous round, the global model is

$$\theta^{(t+1)} = 1.5215.$$

This value now serves as the reference point for all clients in the subsequent round.

Client Local Updates. After performing another round of local training, the clients return updated model parameters shown in Table 11.10.

Compared to the previous round, Client 4's update magnitude has reduced substantially, while Clients 1–3 continue to produce small, consistent updates.

Step 1: Distance-Based Weight Computation. Using the same stabilizing constant $\epsilon = 0.01$, IDA computes weights as

$$w_k^{(t+1)} = \frac{1}{|\Delta\theta_k^{(t+1)}| + \epsilon}.$$

The resulting weights are shown in Table 11.11.

Table 11.10 Client model updates in the second communication round

Client	$\theta_k^{(t+2)}$	$\Delta\theta_k^{(t+1)}$
Client 1	1.54	+0.0185
Client 2	1.50	–0.0215
Client 3	1.53	+0.0085
Client 4	1.95	+0.4285

Table 11.11 IDA weight computation for Round 2

Client	$\lvert\Delta\theta_k^{(t+1)}\rvert$	$w_k^{(t+1)}$	$w_k^{(t+1)}\Delta\theta_k^{(t+1)}$
Client 1	0.0185	$\frac{1}{0.0285} \approx 35.09$	0.65
Client 2	0.0215	$\frac{1}{0.0315} \approx 31.75$	–0.68
Client 3	0.0085	$\frac{1}{0.0185} \approx 54.05$	0.46
Client 4	0.4285	$\frac{1}{0.4385} \approx 2.28$	0.98

Although Client 4 still produces the largest update, its influence remains limited relative to the more consistent clients.

Step 2: Normalized Aggregation. Applying the IDA aggregation rule,

$$\theta^{(t+2)} = \theta^{(t+1)} + \frac{\sum_{k=1}^{4} w_k^{(t+1)} \Delta\theta_k^{(t+1)}}{\sum_{k=1}^{4} w_k^{(t+1)}}.$$

First, compute the numerator:

$$0.65 - 0.68 + 0.46 + 0.98 = 1.41.$$

Next, compute the denominator:

$$35.09 + 31.75 + 54.05 + 2.28 = 123.17.$$

Thus, the aggregated update is

$$\frac{1.41}{123.17} \approx 0.0114.$$

Step 3: Global Model Update. The updated global model after the second IDA round is

$$\theta^{(t+2)} = 1.5215 + 0.0114 = 1.5329.$$

Interpretation Across Rounds. Comparing the evolution of the global model:

$$\theta^{(t)} = 1.50, \quad \theta^{(t+1)} = 1.5215, \quad \theta^{(t+2)} = 1.5329,$$

we observe smooth, monotonic progress toward a consensus value.

Unlike standard FedAvg, which would continue to oscillate toward outlying clients, IDA progressively stabilizes the global trajectory by adapting client influence dynamically across rounds.

Key Insight from Multi-Round IDA
Inverse Distance Aggregation exhibits a self-correcting behavior: as extreme clients become more aligned with the global model, their influence naturally increases, while persistent outliers remain attenuated.

Communication Round 3
We now examine a third communication round to illustrate how Inverse Distance Aggregation continues to stabilize the global model as client updates become increasingly aligned.

Updated Global Model. From the previous round, the global model is

$$\theta^{(t+2)} = 1.5329.$$

This value is broadcast to all clients as the reference model for the next local training phase.

Client Local Updates. After additional local optimization, the clients return the updated models shown in Table 11.12.

Relative to earlier rounds, all clients now exhibit smaller update magnitudes, indicating increased alignment with the global trajectory.

Step 1: Distance-Based Weight Computation. Using the same stabilizing constant $\epsilon = 0.01$, IDA computes

$$w_k^{(t+2)} = \frac{1}{|\Delta\theta_k^{(t+2)}| + \epsilon}.$$

The resulting weights are summarized in Table 11.13.

The weight disparity between consistent clients and the previously outlying client continues to narrow, reflecting gradual trust recovery.

Table 11.12 Client model updates in the third communication round

Client	$\theta_k^{(t+3)}$	$\Delta\theta_k^{(t+2)}$
Client 1	1.545	+0.0121
Client 2	1.525	–0.0079
Client 3	1.538	+0.0051
Client 4	1.70	+0.1671

Table 11.13 IDA weight computation for Round 3

Client	$\lvert\Delta\theta_k^{(t+2)}\rvert$	$w_k^{(t+2)}$	$w_k^{(t+2)}\Delta\theta_k^{(t+2)}$
Client 1	0.0121	$\frac{1}{0.0221} \approx 45.25$	0.55
Client 2	0.0079	$\frac{1}{0.0179} \approx 55.87$	–0.44
Client 3	0.0051	$\frac{1}{0.0151} \approx 66.23$	0.34
Client 4	0.1671	$\frac{1}{0.1771} \approx 5.65$	0.94

Step 2: Normalized Aggregation. Applying the IDA aggregation rule,

$$\theta^{(t+3)} = \theta^{(t+2)} + \frac{\sum_{k=1}^{4} w_k^{(t+2)} \Delta\theta_k^{(t+2)}}{\sum_{k=1}^{4} w_k^{(t+2)}}.$$

Compute the numerator:

$$0.55 - 0.44 + 0.34 + 0.94 = 1.39.$$

Compute the denominator:

$$45.25 + 55.87 + 66.23 + 5.65 = 172.99.$$

Thus, the aggregated update is

$$\frac{1.39}{172.99} \approx 0.0080.$$

Step 3: Global Model Update. The updated global model after the third IDA round is

$$\theta^{(t+3)} = 1.5329 + 0.0080 = 1.5409.$$

Interpretation. Across three consecutive rounds, the global model evolves smoothly as

$$\theta^{(t)} = 1.50, \quad \theta^{(t+1)} = 1.5215, \quad \theta^{(t+2)} = 1.5329, \quad \theta^{(t+3)} = 1.5409.$$

The diminishing magnitude of global updates indicates that the system is approaching a stable consensus.

Figure 11.3 illustrates the evolution of the global model parameter across successive communication rounds under Inverse Distance Aggregation. Unlike standard averaging, which may oscillate toward extreme clients, IDA produces smooth, monotonic convergence by dynamically reweighting client contributions based on update consistency. As rounds progress, client updates become increasingly aligned, and the global model stabilizes with diminishing adjustment magnitude.

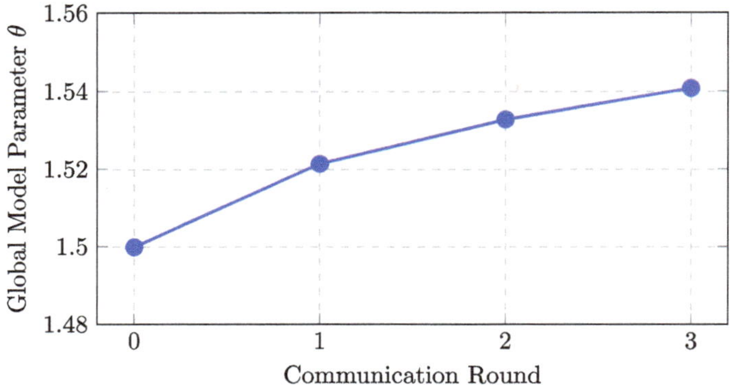

Fig. 11.3 Evolution of the global model under Inverse Distance Aggregation across multiple communication rounds. The progressively smaller updates indicate stable convergence toward a consensus solution despite heterogeneous client behavior

11.5 Convergence Guarantees and Robustness

Advanced federated optimization methods fundamentally alter the learning dynamics of classical federated averaging by intervening at critical points in the optimization pipeline. Rather than relying on naïve averaging of locally trained models, these methods introduce explicit mechanisms—such as proximal penalties, dynamic bias correction, geometry-aware weighting, or constrained local updates—to control instability arising from decentralization.

As a result, convergence analysis in federated learning must extend beyond the assumptions and tools used in centralized or classical distributed optimization. In particular, theoretical guarantees must simultaneously account for *statistical heterogeneity*, *partial and stochastic client participation*, *communication delays*, and *adaptive aggregation rules*. This section provides a unifying perspective on how convergence and robustness are established for advanced federated optimization methods.

11.5.1 Why Convergence Analysis Is More Subtle in Federated Learning

In classical centralized optimization, convergence analysis is relatively straightforward because the learning process follows a *single, well-defined optimization trajectory*. At each iteration, gradients are computed with respect to a global objective function using either the full dataset or an unbiased sample thereof. Standard tools from convex analysis and stochastic optimization can therefore be applied directly to characterize rates of convergence, stability, and optimality gaps.

Federated learning departs from this paradigm in a fundamental way. Instead of one optimizer operating on a unified dataset, learning is distributed across a large population of clients, each of which maintains its own local model and local objective function. Between communication rounds, these local models evolve independently, guided only by client-specific data distributions and local optimization dynamics. The server observes these trajectories only intermittently and only through aggregated updates.

Aggregation, therefore, does not enforce exact synchronization. Rather, it performs a form of *statistical alignment*, pulling diverse local trajectories toward a common reference point. This alignment is inherently imperfect: some clients may be absent, others may perform more local updates, and still others may follow gradients that are systematically biased due to non-IID data.

Advanced federated optimization methods further complicate the convergence picture. Proximal terms modify the curvature of local objectives, dynamic regularization introduces time-varying correction signals that depend on historical behavior, and robust aggregation schemes adaptively alter the relative influence of client updates. As a result, the effective optimization dynamics can no longer be described by a single descent direction or a static objective function.

Convergence analysis in federated learning must therefore reason about the *combined effect* of three interacting components: local optimization dynamics, global aggregation mechanisms, and regularization or correction terms that evolve across rounds. This multi-layered interaction is what makes convergence analysis in federated learning both challenging and conceptually rich.

Key Analytical Challenge

Convergence in federated learning is not about tracking a single, continuous gradient flow, but about controlling the long-term behavior of many partially observed, intermittently synchronized, and statistically heterogeneous optimization trajectories.

11.5.2 Typical Assumptions for Convergence Guarantees

To make this complex analysis tractable, most theoretical results in federated learning rely on a set of standard—though idealized— assumptions. These assumptions are not meant to perfectly reflect real-world deployments, but rather to isolate the essential mathematical difficulties introduced by decentralization.

- **Convex or weakly convex objectives:** Local loss functions $F_k(\theta)$ are commonly assumed to be convex or weakly convex. Convexity ensures that any stationary point is globally optimal, while weak convexity still allows meaningful bounds on suboptimality. Although modern deep models violate this assumption, it provides a crucial baseline for understanding algorithmic behavior.
- **Lipschitz-continuous gradients:** The gradients $\nabla F_k(\theta)$ are assumed to be L-Lipschitz continuous. Intuitively, this means that the loss surface does not exhibit

arbitrarily sharp curvature. This condition allows one to bound the error introduced by stale, delayed, or noisy updates and is central to most convergence proofs.

- **Bounded gradient variance:** Stochastic gradients computed from local mini-batches are assumed to have bounded variance. This assumption controls the noise introduced by sampling and ensures that gradient estimates do not dominate the learning dynamics.
- **Sufficient client participation over time:** While not all clients are expected to participate in every round, it is typically assumed that each client participates infinitely often or with sufficient frequency in expectation. This prevents permanent information loss from any subset of the data distribution.

Taken together, these assumptions bound the extent to which heterogeneity and randomness can disrupt learning, making it possible to derive meaningful convergence statements.

Interpretation of Assumptions
These assumptions do not eliminate heterogeneity, partial participation, or noise; instead, they bound their cumulative impact so that decentralized learning remains analyzable and stable over long time horizons.

11.5.3 Convergence Behavior of Advanced Methods

Under the above assumptions, advanced federated optimization methods exhibit convergence behavior that is markedly more stable than classical Federated Averaging, especially in heterogeneous environments.

- **FedProx:** FedProx augments the local objective with a proximal penalty that anchors local models to the current global reference. Theoretical analyses show that this modification bounds the deviation between local and global models, effectively controlling client drift. The regularization parameter μ governs a trade-off: larger values improve stability but may slow progress, while smaller values allow more aggressive local optimization.
- **FedDyn:** FedDyn addresses heterogeneity by explicitly modeling and correcting systematic gradient bias. The auxiliary variables accumulate historical discrepancies between local and global behavior, gradually reshaping local objectives so that their gradients align with the global gradient. As a result, FedDyn achieves improved convergence rates in highly non-IID regimes where FedAvg struggles.
- **Inverse Distance Aggregation (IDA):** IDA leaves local optimization unchanged but modifies aggregation. By reweighting updates based on their distance from

the global model, IDA suppresses extreme or erratic updates. Although its convergence guarantees are often framed in terms of robustness rather than optimality, IDA significantly improves stability and prevents divergence in adversarial or failure-prone environments.

Viewed together, these methods illustrate a broader principle: convergence in federated learning is achieved not by enforcing uniformity across clients, but by carefully managing diversity through regularization, correction, and robust aggregation.

Comparative Insight
FedProx controls *where* clients move, FedDyn corrects *why* they move, and IDA controls *how much* their movement should matter.

Table 11.14 presents a structured comparison of classical and advanced federated optimization strategies, highlighting how successive methods progressively address the limitations of naïve federated averaging. FedAvg serves as a conceptual baseline, relying on the implicit assumption that local updates are well-aligned and statistically representative. As heterogeneity increases, this assumption breaks down, motivating the development of FedProx, FedDyn, and robust aggregation schemes such as Inverse Distance Aggregation (IDA).

The table emphasizes that these methods intervene at different stages of the federated learning pipeline. FedProx modifies the *local optimization objective* to constrain client drift, FedDyn augments local learning with *cross-round memory* to correct systematic bias, while IDA leaves local training unchanged and instead focuses on *robust aggregation* at the server. This distinction is pedagogically important: it illustrates that robustness in federated learning can be achieved through fundamentally different algorithmic levers.

11.5.4 Robustness from a Systems Perspective

While mathematical convergence guarantees provide essential insight into the long-term behavior of federated optimization algorithms, they do not fully capture the challenges faced by real-world federated learning systems. In practice, federated learning is deployed over large-scale, heterogeneous, and unreliable environments that are subject to communication failures, device churn, nonstationary data distributions, and even malicious behavior. From a systems perspective, robustness therefore becomes as important as asymptotic convergence.

Advanced federated optimization methods contribute to robustness not by adding ad hoc safeguards, but by embedding stability directly into the learning dynamics. By carefully shaping local objectives and aggregation rules, these methods

Table 11.14 Comparison of federated optimization and aggregation methods

Aspect	FedAvg	FedProx	FedDyn	Inverse Distance Aggregation (IDA)
Primary modification	Simple weighted averaging of local models.	Adds a proximal regularization term to the local objective.	Introduces dynamic linear correction terms via auxiliary variables.	Reweights client updates based on their distance from the global model.
Where the method acts	Server-side aggregation only.	Local optimization objective.	Local objective and cross-round state.	Server-side aggregation only.
Main goal	Efficient decentralized learning under idealized assumptions.	Limit client drift under non-IID data and multiple local steps.	Correct systematic gradient bias caused by heterogeneity.	Improve robustness to extreme, noisy, or adversarial updates.
Key mechanism	Weighted averaging using dataset sizes $\frac{n_k}{N}$.	Quadratic proximal penalty $\frac{\mu}{2}\|\theta - \theta^{(t)}\|^2$.	Auxiliary variables that accumulate historical gradient discrepancies.	Inverse-distance weighting $w_k \propto \frac{1}{\|\Delta\theta_k\|+\epsilon}$.
Handling of client drift	No explicit control; drift may be severe under non-IID data.	Explicitly bounded through proximal regularization.	Implicitly reduced through dynamic bias correction.	Indirectly mitigated by suppressing large deviations.
Convergence behavior	Converges under IID or mildly heterogeneous data.	Converges to a stationary point with bounded drift.	Achieves improved convergence rates under strong heterogeneity.	Stabilizes convergence; guarantees often expressed in terms of robustness.
Robustness to outliers	Low; extreme updates can dominate aggregation.	Moderate; depends on the choice of μ.	Moderate; focuses on bias correction rather than outlier rejection.	High; extreme or inconsistent updates are automatically downweighted.
Additional state required	None.	None beyond the global model.	Client-specific auxiliary variables across rounds.	None beyond computing update norms at the server.
Computational overhead	Low.	Low to moderate (additional regularization).	Moderate (auxiliary variable updates).	Low (norm computation during aggregation).

(Continued)

Table 11.14 (continued)

Aspect	FedAvg	FedProx	FedDyn	Inverse Distance Aggregation (IDA)
Typical use cases	Homogeneous or mildly non-IID data with reliable clients.	Non-IID data with multiple local epochs per round.	Severely heterogeneous data with persistent gradient bias.	Highly unreliable, adversarial, or skewed client populations.
Communication efficiency	High when few local epochs are used; degrades with instability.	Moderate; allows more local steps without divergence.	Moderate; auxiliary updates add overhead but reduce wasted rounds.	High; avoids rollback or failure due to extreme updates.
Sensitivity to hyperparameters	Low; mainly depends on learning rate and local epochs.	Moderate; sensitive to proximal coefficient μ.	Moderate to high; depends on learning rate and auxiliary update scaling.	Moderate; depends on ϵ and choice of distance norm.
Compatibility with partial participation	Limited robustness when client sampling is highly skewed.	Improved tolerance to partial and inconsistent participation.	Strong tolerance due to accumulated correction across rounds.	Strong tolerance; unreliable participants are automatically downweighted.
Interpretability of updates	High; aggregation weights are transparent and data-driven.	Moderate; proximal effects are implicit in local optimization.	Lower; auxiliary variables introduce indirect correction dynamics.	High; geometric weighting directly reflects update consistency.

ensure that the system remains well-behaved even when operating far from idealized assumptions.

Several complementary robustness mechanisms emerge naturally from these design choices:

- **Partial participation tolerance:** In large-scale federated systems, it is unrealistic to expect all clients to participate in every communication round. Devices may be offline, energy-constrained, or temporarily unavailable. Advanced methods reduce sensitivity to any single update, ensuring that the absence or sudden appearance of clients does not destabilize the global model. This tolerance is particularly important in mobile and edge computing scenarios, where client availability is inherently stochastic.
- **Resilience to statistical heterogeneity:** Non-IID data is the defining characteristic of federated learning. Without explicit safeguards, clients with highly skewed data distributions can dominate the learning process, leading to slow or unstable

convergence. Techniques such as proximal regularization and dynamic bias correction act as implicit normalization mechanisms, preventing any single client or subgroup from overwhelming the global objective.

- **Defense against extreme or adversarial updates:** In realistic deployments, some client updates may be noisy, corrupted, or even adversarial. Robust aggregation mechanisms, such as inverse distance weighting, naturally suppress the influence of extreme updates based on their geometry rather than their provenance. This approach avoids the need for explicit attack detection or trust scoring, which can be difficult to implement reliably at scale.

Viewed holistically, robustness in federated learning is not an external constraint imposed on optimization. Instead, it emerges from the careful integration of optimization theory and systems design. Algorithms that explicitly account for drift, bias, and variability produce learning dynamics that remain stable even under severe uncertainty.

Unifying View
Advanced federated optimizers achieve robustness not by eliminating decentralization, but by reshaping the optimization landscape so that decentralized learning dynamics remain inherently aligned with global objectives.

Concluding Perspective. This chapter demonstrates that effective federated optimization requires substantially more than simple averaging of local models. As decentralization intensifies and deployment environments become more heterogeneous, the learning system must actively counteract drift, bias, and instability through principled algorithmic design. Methods such as FedProx, FedDyn, and robust aggregation schemes show that stability and scalability can be achieved by embedding control mechanisms directly into the optimization process.

By drawing on ideas from proximal optimization, dynamic bias correction, robust statistics, and constrained optimization, modern federated learning frameworks are able to converge reliably even under severe heterogeneity, partial participation, and uncertainty. These insights highlight a broader shift in federated learning research: from treating heterogeneity as a nuisance to be tolerated, to treating it as a structural feature that must be explicitly managed.

The convergence and robustness principles developed in this chapter form the foundation for the next major evolution of federated learning. In the following chapter, we relax the assumption of a single global model and explore *personalized federated learning*, where convergence is reinterpreted not as agreement on one universal solution, but as the emergence of a structured family of client-adaptive models that better reflect individual data distributions and task requirements.

Exercises

This section reinforces key concepts from the chapter on advanced federated optimization methods, emphasizing theoretical understanding, algorithmic behavior, and practical system-level implications.

A. Short Answer Questions

Answer each question briefly and precisely.

1. What fundamental assumption of FedAvg breaks down under non-IID data?
2. Define *client drift* in federated learning.
3. What role does the proximal coefficient μ play in FedProx?
4. How does FedDyn differ conceptually from FedProx?
5. What information is stored in the auxiliary variable $h_k^{(t)}$ in FedDyn?
6. Why are one-gradient federated methods free from client drift?
7. What geometric property of updates does IDA explicitly exploit?
8. Why is inverse distance aggregation considered a robust method?
9. State one advantage and one limitation of dynamic regularization.
10. Why is robustness considered a systems-level property in federated learning?

B. Long Answer Questions

Answer the following questions with clear explanations, equations, and diagrams wherever appropriate.

1. Derive the FedProx local optimization objective and explain how it bounds client drift.
2. Using equations, explain how gradient bias arises under non-IID data and how FedDyn corrects it.
3. Compare FedAvg, FedProx, and FedDyn in terms of where they intervene in the optimization pipeline.
4. Explain one-gradient federated learning and analyze its convergence–communication trade-off.
5. Derive the inverse distance aggregation rule and interpret it geometrically.
6. Discuss how robust aggregation methods relate to classical robust statistics.
7. Explain why convergence analysis in federated learning is more subtle than in centralized learning.
8. Using a worked example, show how IDA suppresses an extreme client update.

C. True/False Questions

State whether each statement is **True** or **False**.

1. FedAvg guarantees stable convergence under severe data heterogeneity.
2. FedProx modifies the aggregation rule used by the server.
3. FedDyn introduces time-varying correction terms into the local objective.
4. One-gradient federated learning evaluates gradients at different local models.
5. IDA assigns higher weights to updates closer to the global model.
6. Dynamic regularization eliminates the need for aggregation.
7. Large update norms are always desirable in federated optimization.
8. IDA permanently excludes unreliable clients from training.
9. Robustness in federated learning emerges from optimization design choices.
10. FedDyn requires maintaining additional client-side state across rounds.

Answers:

1. False
2. False
3. True
4. False
5. True
6. False
7. False
8. False
9. True
10. True

D. Fill in the Blanks

Fill in the blanks with the most appropriate term or expression.

1. Client drift is primarily caused by ___________ data distributions.
2. FedProx adds a ___________ regularization term to the local objective.
3. In FedDyn, gradient bias is corrected using ___________ variables.
4. One-gradient federated learning is equivalent to federated ___________.
5. IDA weights client updates inversely proportional to their ___________.
6. The stabilizing constant in IDA is denoted by ___________.
7. Robust aggregation suppresses ___________ client updates.
8. FedDyn introduces a memory effect across ___________ rounds.
9. Proximal regularization constrains the ___________ of local updates.
10. Federated convergence depends on optimization and ___________ design.

Answers:

1. non-IID
2. proximal
3. auxiliary
4. SGD
5. distance
6. ϵ
7. extreme
8. communication
9. magnitude
10. system

E. Multiple Choice Questions (MCQs)

Choose the **most appropriate** answer.

1. The primary purpose of FedProx is to:
   (a) increase communication frequency
   (b) eliminate aggregation
   (c) limit client drift
   (d) remove non-IID data
2. FedDyn primarily addresses:
   (a) computational overhead
   (b) gradient bias
   (c) encryption cost
   (d) model compression
3. One-gradient federated learning computes gradients at:
   (a) local optima
   (b) random models
   (c) previous local models
   (d) the global model
4. IDA assigns lower weight to a client when:
   (a) its dataset is large
   (b) its update norm is large
   (c) it participates frequently
   (d) its gradient is unbiased
5. Which method requires auxiliary variables across rounds?
   (a) FedAvg
   (b) FedProx
   (c) FedDyn
   (d) IDA

6. Robust aggregation primarily improves:
   (a) model size
   (b) robustness
   (c) encryption
   (d) inference speed
7. Which method modifies only server-side aggregation?
   (a) FedProx
   (b) FedDyn
   (c) One-gradient FL
   (d) IDA
8. Client drift is most severe when:
   (a) data is IID
   (b) local epochs are zero
   (c) data is heterogeneous
   (d) gradients are synchronized
9. Which technique is most aligned with robust statistics?
   (a) FedAvg
   (b) FedProx
   (c) FedDyn
   (d) IDA
10. Stability in federated learning emerges from:
   (a) centralized control
   (b) ignoring heterogeneity
   (c) principled optimization design
   (d) increasing dataset size

Answers:

1. (c)
2. (b)
3. (d)
4. (b)
5. (c)
6. (b)
7. (d)
8. (c)
9. (d)
10. (c)

F. Application and Project-Based Questions

These problems emphasize modeling, reasoning, and implementation.

1. Implement FedAvg, FedProx, and FedDyn on a non-IID dataset and compare convergence behavior.
2. Simulate client drift under increasing local epochs and analyze its effect on FedAvg.
3. Design an experiment demonstrating dynamic bias correction in FedDyn.
4. Implement inverse distance aggregation and evaluate robustness against an extreme client.
5. Compare IDA with trimmed mean aggregation under noisy updates.
6. Analyze the communication–computation trade-off of one-gradient federated learning.
7. Study the effect of μ in FedProx on convergence speed and stability.
8. Combine FedDyn with IDA and evaluate system robustness.
9. Analyze how partial participation affects convergence in advanced federated optimizers.
10. Propose a new hybrid optimization–aggregation strategy inspired by this chapter.

Part V

Personalized, Heterogeneous, and Hybrid Federated Learning

Personalized Federated Learning 12

Personalized federated learning represents a natural and conceptually inevitable evolution of the classical federated learning paradigm. In the foundational chapters of this book, federated learning was introduced as a collaborative optimization framework in which a collection of decentralized clients jointly learn a *single global model* without sharing raw data. That formulation was deliberately aligned with population-level learning goals, where the objective is to minimize an aggregate loss across all participating clients.

While such a global perspective is mathematically elegant and algorithmically convenient, it implicitly assumes that a single model can adequately serve the learning needs of a highly diverse population. In practice, this assumption is often violated. Clients in real-world federated systems differ markedly in their data generation processes, usage patterns, environmental contexts, and optimization priorities. As a result, a globally optimal model may fail to provide satisfactory performance for many individual clients, even when it achieves strong average accuracy.

Personalized federated learning relaxes the strict global-consensus objective by explicitly acknowledging *client-level diversity*. The central premise of this paradigm is that different clients may require different models, or at least different parameter configurations, even when they collaborate on a shared learning task. Rather than enforcing identical model parameters across all clients, personalization allows controlled deviation from the global solution so that client-specific patterns can be captured without abandoning the benefits of collaborative training.

Conceptual Shift Introduced by Personalization
Classical federated learning prioritizes population-level performance through strict model consensus. Personalized federated learning redefines the objective

P. P. Ray, *Foundations of Federated Learning*,
Springer Series in the Data Sciences,
https://doi.org/10.1007/978-3-032-34084-9_12

by treating client heterogeneity as a first-class design consideration rather than a nuisance to be averaged away.

From a pedagogical standpoint, personalized federated learning should not be viewed as a separate or competing framework. Instead, it is best understood as a principled generalization of the standard federated optimization problem. At one extreme, enforcing strong coupling among client models recovers classical federated learning. At the other extreme, allowing complete independence reduces the system to isolated local training. Personalization occupies the continuum between these extremes, enabling a tunable balance between global knowledge sharing and local adaptation.

This chapter develops the conceptual, mathematical, and algorithmic foundations required to understand this balance. We begin by motivating personalization through the lens of client-specific data distributions and the limitations of purely global models. We then formalize personalized model objectives that extend the federated optimization framework introduced earlier, highlighting how consensus constraints may be softened rather than eliminated. Subsequently, we examine concrete personalization mechanisms, including sub-federated averaging, structured and unstructured model pruning, and lottery ticket–based approaches that exploit overparameterization. The chapter concludes with a detailed discussion of the fundamental trade-off between global generalization and personalization, a tension that underlies all personalized federated learning algorithms and shapes their theoretical guarantees and practical performance.

12.1 Motivation for Personalization

Classical federated learning is fundamentally designed around the idea that a *single global model* can adequately represent the learning needs of all participating clients. This assumption is inherited directly from centralized machine learning, where a single hypothesis is optimized to minimize an average loss over a pooled dataset drawn from an entire population. In such centralized settings, individual variability is implicitly treated as noise, and the learning objective prioritizes overall statistical efficiency, robustness, and population-level generalization.

From a mathematical perspective, classical federated learning formalizes this idea by seeking a single parameter vector $\boldsymbol{\theta} \in \mathbb{R}^d$ that minimizes a weighted sum of client-specific empirical risks. Specifically, the global objective is defined as

$$F(\boldsymbol{\theta}) = \sum_{k=1}^{K} \frac{n_k}{N} F_k(\boldsymbol{\theta}),$$

where $F_k(\theta)$ denotes the local empirical risk at client k, $n_k = |\mathcal{D}_k|$ is the size of the local dataset, and $N = \sum_{k=1}^{K} n_k$ is the total number of samples across all clients. This formulation implicitly assumes that minimizing the weighted average of local risks yields a model that is not only optimal in an aggregate sense, but also sufficiently satisfactory for every individual client.

Such an assumption is often reasonable in controlled or homogeneous environments, where data distributions across clients are similar and client-level objectives are well aligned. However, as federated learning moves from theoretical constructs to large-scale real-world deployments, this assumption becomes increasingly restrictive. Clients in federated systems are not merely logical partitions of a dataset created for computational convenience; they are autonomous entities embedded in diverse physical, social, and operational contexts.

Differences among clients arise along multiple dimensions. User behavior may vary substantially across individuals, leading to distinct usage patterns and preference profiles. Data generation processes may differ due to environmental conditions, sensor characteristics, or regional factors. Clients may operate under heterogeneous resource constraints, such as limited computation, storage, or energy availability. In addition, clients may pursue subtly different optimization goals, even when nominally participating in the same learning task. Consequently, the data observed by each client may encode patterns that are highly informative locally but weakly represented, or even contradictory, at the population level.

In such heterogeneous settings, a globally optimal model can perform poorly for a substantial subset of clients. A model that achieves low average loss across the federation may still exhibit consistently high error on certain clients whose data distributions deviate significantly from the population mean. From the perspective of these clients, global optimization may appear misaligned with their local objectives, even though it is optimal in the aggregate sense. This mismatch highlights a critical limitation of strict global consensus: optimizing for the "average client" does not guarantee acceptable performance for any particular client.

This phenomenon can be understood more formally through the concept of local optimality. Each client k possesses a local minimizer

$$\theta_k^\star = \arg\min_{\theta} F_k(\theta),$$

which reflects the statistical structure of its own data distribution. Under strong heterogeneity, these local minimizers may be widely separated in parameter space. The global minimizer

$$\theta^\star = \arg\min_{\theta} F(\theta)$$

is necessarily a weighted compromise across all clients. For many clients, $\theta^\star$ may lie far from $\theta_k^\star$, implying that enforcing strict consensus can systematically degrade client-level performance. In extreme cases, the global solution may be suboptimal for *every* individual client, even though it minimizes the aggregate objective.

Personalized federated learning emerges as a principled response to this fundamental limitation. Rather than treating heterogeneity as an obstacle to be eliminated, personalization treats it as a first-class property of the learning problem. The central idea is to relax the requirement that all clients must share identical model parameters, allowing controlled and structured deviations from the global model. This relaxation does not abandon collaboration; instead, it enables clients to adapt shared knowledge to their own data distributions while still benefiting from information contributed by others.

Motivation for Personalization
Classical federated learning optimizes population-level performance through strict model consensus. However, individual clients ultimately care about *client-level utility*. Personalization reconciles this mismatch by enabling local adaptation built upon a shared global representation.

From a conceptual standpoint, personalization does not reject the core philosophy of federated learning. Instead, it refines and extends it. Clients continue to exchange information, participate in collective optimization, and benefit from shared representations. What changes is the rigidity of the consensus requirement. Rather than enforcing a one-size-fits-all solution, the global model serves as a common reference point, prior, or shared backbone, upon which client-specific adaptations are constructed.

The motivation for personalization is particularly compelling in application domains where heterogeneity is intrinsic rather than incidental. In mobile personal assistants, user language, writing style, and interaction patterns vary widely across individuals. In recommender systems, personal taste dominates population-level statistics, and excessive averaging can dilute relevance. In healthcare, patient demographics, genetics, disease prevalence, and clinical practices differ across institutions and regions. In industrial IoT, operating conditions, sensor calibration, and usage profiles vary across deployment sites. In all these settings, insisting on a single global model risks sacrificing relevance and usability in favor of abstract generality.

Personalized federated learning therefore represents a fundamental shift in perspective: from learning *one model for everyone* to learning *many related models together*. This shift underpins the theoretical developments, algorithmic designs, and evaluation criteria explored throughout this chapter and motivates the broader move toward heterogeneity-aware and client-centric learning in federated systems.

What Does Personalization Mean in Federated Learning? In federated learning, *personalization* refers to a deliberate and principled departure from the traditional *one-model-fits-all* paradigm that underlies classical centralized and federated learning approaches. Rather than assuming that a single global model can adequately satisfy the learning requirements of all participants, personalization explicitly recognizes that clients are inherently diverse and may require models that are tailored to their own data characteristics, usage patterns, and performance objectives.

At an intuitive level, personalization arises from the observation that clients in federated systems are not interchangeable. Each client corresponds to a distinct data-generating process, shaped by individual behavior, environmental conditions, sensor modalities, or organizational practices. When a single global model is imposed uniformly, these client-specific characteristics are often averaged out. While such averaging may improve population-level performance, it can simultaneously degrade the quality of predictions for individual clients, especially those whose data distributions deviate significantly from the population mean.

From a formal perspective, personalization acknowledges that each client k is associated with its own local objective function $F_k(\theta)$ and therefore admits its own optimal solution

$$\theta_k^\star = \arg\min_\theta F_k(\theta),$$

which reflects the statistical structure of the client's local data. In contrast, classical federated learning seeks a single global minimizer

$$\theta^\star = \arg\min_\theta F(\theta) = \arg\min_\theta \sum_{k=1}^{K} \frac{n_k}{N} F_k(\theta),$$

where $\theta^\star$ represents a compromise solution that optimizes average performance across all clients. Personalization therefore relaxes the requirement that all clients must converge to the same parameter vector, allowing $\theta_k^\star \neq \theta^\star$ whenever client-specific considerations warrant such divergence.

Importantly, personalization should not be interpreted as abandoning the collaborative spirit of federated learning. Instead, it redefines the role of collaboration. The global model, or shared representation, acts as a common reference, prior, or inductive bias that captures broadly useful patterns across clients. Local personalization mechanisms then adapt this shared knowledge to better align with client-specific data distributions. In this sense, collaboration is retained as a means of sharing information, while uniformity is no longer treated as an absolute goal.

Conceptually, personalization in federated learning can manifest in multiple, complementary forms:

- **Parameter personalization**, in which each client maintains a personalized parameter vector θ_k that remains close to, but not identical with, a shared global reference model;
- **Structural personalization**, in which parts of the model (e.g., feature extractors or backbone layers) are shared across all clients, while other components (e.g., output layers, adapters, or subnetworks) are customized locally;
- **Optimization personalization**, in which federated training produces a strong initialization or prior that each client further refines through local adaptation.

These forms of personalization highlight that the concept is not tied to a single algorithmic strategy. Rather, personalization represents a broader design philosophy that can be realized through parameter constraints, architectural choices, or optimization dynamics.

Crucially, personalization does not imply isolated or independent learning. Clients continue to exchange information through aggregated updates, shared representations, or coordinated optimization procedures. What distinguishes personalized federated learning is that this collaboration is explicitly used to *support local adaptation*, rather than to enforce strict uniformity across all participants.

In essence, personalization in federated learning means learning *many related models together*. The objective is to strike a principled balance between global knowledge sharing and client-specific utility, ensuring that collaboration enhances, rather than constrains, the relevance and effectiveness of the models deployed at individual clients.

Personalization in Federated Learning
Personalization in federated learning allows client models to adapt to local data distributions while retaining shared knowledge from global training. It relaxes strict model consensus to improve client-level performance under data heterogeneity. The goal is to balance global generalization with locally optimal behavior.

12.2 Client-Specific Data Distributions

Personalization in federated learning is fundamentally driven by the existence of *client-specific data distributions*. Unlike centralized learning, where all data are assumed to be drawn from a single underlying distribution, federated learning operates in a setting where data are inherently decentralized and generated under diverse conditions. Each client k observes data drawn from its own joint distribution $P_k(x, y)$, which reflects the local environment, user behavior, sensor characteristics, or organizational context associated with that client.

This decentralization is not merely a system-level constraint but a statistical reality. In many practical deployments, clients differ substantially in how data are generated, collected, and labeled. As a result, the distributions $P_k(x, y)$ may differ significantly across clients, even when the learning task and model architecture are shared. Personalization emerges naturally from this observation: if data are different, the models optimized on those data should not be expected to be identical.

Formally, the classical independent and identically distributed (IID) assumption

$$P_1(x, y) = P_2(x, y) = \cdots = P_K(x, y)$$

rarely holds in realistic federated systems. Instead, federated learning is characterized by various forms of *statistical heterogeneity*, each of which has distinct implications for learning dynamics and model performance. Common sources of heterogeneity include:

- **Covariate shift**, which occurs when the marginal distribution of the input features differs across clients, that is, $P_k(x) \neq P_j(x)$, while the underlying relationship between inputs and outputs remains largely unchanged. In this case, the conditional distribution $P(y \mid x)$ is assumed to be similar across clients, but each client observes inputs from a different region of the feature space. Covariate shift commonly arises in federated systems due to differences in user behavior, sensor placement, environmental conditions, or operational contexts. For example, in a mobile application, users may interact with the app in distinct ways, leading to different input patterns even though the prediction task itself is the same. Under covariate shift, a global model trained on pooled data may be biased toward regions of the feature space that are overrepresented, reducing accuracy for clients whose inputs lie in underrepresented regions.
- **Label shift**, which arises when the marginal distribution of the output labels differs across clients, that is, $P_k(y) \neq P_j(y)$, even if the conditional distribution $P(x \mid y)$ remains similar. Label shift is often the result of imbalanced or skewed local datasets, where certain classes occur much more frequently on some clients than on others. In federated learning, this situation is common when clients serve different user populations or operate in different environments. For instance, in a healthcare setting, hospitals may treat different patient demographics, leading to varying disease prevalence rates. A globally optimized model may therefore emphasize majority classes, resulting in systematically poorer performance on minority classes that are locally important for certain clients.
- **Concept shift**, which represents the most challenging form of heterogeneity, occurs when the conditional relationship between inputs and outputs differs across clients, that is, $P_k(y \mid x) \neq P_j(y \mid x)$. Concept shift reflects fundamentally different data-generating mechanisms and cannot be corrected through simple reweighting or resampling. This form of heterogeneity may arise due to differences in labeling criteria, contextual semantics, or temporal evolution of the task. For example, in predictive maintenance systems, the same sensor reading may indicate normal behavior on one machine but signal an anomaly on another. Under concept shift, enforcing a single global model is particularly problematic, as no single conditional mapping can be optimal for all clients.

These forms of heterogeneity imply that each client implicitly defines a distinct learning problem, even though all clients nominally participate in the same federated task. The local empirical risk function

$$F_k(\theta) = \mathbb{E}_{(x,y)\sim P_k}\left[\ell(f_\theta(x), y)\right]$$

captures client-specific performance by measuring how well the model parameters θ align with the statistical structure of client k's data. In contrast, the global federated objective aggregates these heterogeneous local risks into a single optimization problem, typically through a weighted average.

When client distributions differ only mildly, the global objective may still admit a solution that performs reasonably well for most clients. However, under strong heterogeneity, the local minimizers

$$\theta_k^\star = \arg\min_{\theta} F_k(\theta)$$

can be widely separated in parameter space. In such cases, the global minimizer represents a compromise solution that may not be well aligned with the optimal parameters for any individual client. This misalignment provides a statistical explanation for why strict global consensus can degrade client-level performance.

Statistical Origin of Personalization
Personalization is a statistical necessity whenever client data distributions differ in meaningful ways. Under strong heterogeneity, a single global optimum may be suboptimal for most, or even all, individual clients.

From this perspective, personalization can be viewed as a mechanism for aligning model parameters more closely with client-specific data distributions while still exploiting shared structure across clients. Rather than forcing heterogeneous learning problems into a single solution, personalized federated learning acknowledges distributional diversity as an inherent property of the system and designs algorithms that adapt to it in a principled manner.

Table 12.1 Comparison of covariate shift, label shift, and concept shift in federated learning

Comparison Aspect	Covariate Shift	Label Shift	Concept Shift
Distribution Change	$P_k(x) \neq P_j(x)$	$P_k(y) \neq P_j(y)$	$P_k(y \mid x) \neq P_j(y \mid x)$
What Varies Across Clients	Input features	Label proportions	Input–output relationship
Cause of Heterogeneity	Different usage or environments	Class imbalance or skewed data	Different semantics or contexts
Severity in FL	Moderate	Moderate to high	High
Impact on Global Model	Bias toward dominant inputs	Bias toward majority classes	Global model may fail entirely
Need for Personalization	Helpful	Often necessary	Essential

Table 12.1 summarizes the major forms of statistical heterogeneity encountered in federated learning, namely covariate shift, label shift, and concept shift. The table highlights how these shifts differ in terms of the underlying distributional changes, their impact on global model performance, and the resulting need for personalization. As the severity of heterogeneity increases from covariate to concept shift, the limitations of a single global model become more pronounced, making personalized federated learning increasingly essential.

12.3 Personalized Model Objectives

Personalized federated learning generalizes the classical federated optimization objective by explicitly allowing *client-specific model parameters*. This generalization represents a fundamental departure from the traditional federated learning paradigm, in which all participating clients are constrained to share a single global parameter vector θ. In standard federated learning, personalization is implicitly sacrificed in favor of consensus, under the assumption that a globally optimal model provides adequate performance for all clients. Personalized federated learning challenges this assumption by recognizing that each client may benefit from its own adapted model that better reflects local data characteristics and task-specific requirements.

Accordingly, the learning system no longer seeks to produce a single model as its final output. Instead, it aims to construct a collection of related models $\{\theta_k\}_{k=1}^{K}$, one for each client. These models are not learned in isolation. Rather, they are jointly optimized through a collaborative process that enables information sharing while still permitting client-level adaptation. The central design challenge in personalized federated learning is therefore not whether to collaborate, but *how tightly* to couple the learning processes of different clients.

This shift in objective reflects a deeper conceptual rethinking of the role of consensus in federated systems. Classical federated learning enforces consensus by design: all clients contribute updates toward a single shared solution, and any deviation from that solution is treated as undesirable. Personalized federated learning relaxes this constraint and reframes collaboration as a mechanism for guiding, regularizing, and stabilizing local learning rather than for enforcing strict uniformity. Under this perspective, the global model is no longer interpreted as the final model to be deployed. Instead, it serves as a shared reference point, anchor, or inductive bias around which personalized client models are formed.

A generic formulation of the personalized federated learning objective can be expressed as

$$\min_{\{\theta_k\}} \sum_{k=1}^{K} \frac{n_k}{N} \left(F_k(\theta_k) + \lambda \mathcal{R}(\theta_k, \theta) \right),$$

where $F_k(\theta_k)$ denotes the local empirical risk of client k, $n_k = |\mathcal{D}_k|$ is the size of the local dataset, and $N = \sum_{k=1}^{K} n_k$ is the total number of samples across all clients. The quantity $\boldsymbol{\theta}$ represents a shared reference model, which may be explicitly maintained by a central server or implicitly defined through aggregation of client models. The regularization function $\mathscr{R}(\cdot, \cdot)$ penalizes deviations between the personalized model $\boldsymbol{\theta}_k$ and the shared reference $\boldsymbol{\theta}$, thereby coupling local models and enabling structured collaboration. The hyperparameter $\lambda \geq 0$ controls the strength of this coupling and thus directly governs the degree of personalization.

This objective formulation reveals that personalized federated learning is not a binary choice between global and local training. Instead, it defines a continuous spectrum of learning behaviors that interpolate between these two extremes. At one extreme, as $\lambda \to \infty$, the regularization term dominates the objective, effectively forcing $\boldsymbol{\theta}_k \approx \boldsymbol{\theta}$ for all clients. In this limit, the formulation collapses to classical federated learning, where strict global consensus is enforced and all clients share an identical model. At the opposite extreme, as $\lambda \to 0$, the regularization term vanishes entirely, and each client independently minimizes its local objective $F_k(\boldsymbol{\theta}_k)$. This regime corresponds to fully local learning, in which collaboration is absent and no information is shared across clients.

Between these two extremes lies the most practically relevant regime, where λ takes an intermediate value. In this regime, clients are encouraged—but not forced—to remain close to the shared reference model. The resulting personalized models strike a carefully balanced trade-off between leveraging global information and adapting to local data distributions. The value of λ therefore plays a central and explicit role in shaping the trade-off between global generalization and client-specific utility. Choosing λ appropriately is a key design decision, influenced by factors such as the degree of data heterogeneity, system constraints, and application-level performance requirements.

- $\lambda \to \infty$: strict global consensus with identical models across all clients;
- $\lambda \to 0$: fully local learning with independent client models;
- intermediate λ: controlled personalization that balances shared structure and local adaptation.

From an optimization perspective, this formulation can be interpreted as a regularized multi-task learning problem, where each client corresponds to a distinct task and the regularization term enforces similarity among tasks. From a Bayesian perspective, the shared reference model $\boldsymbol{\theta}$ may be viewed as a prior, while the regularization term $\mathscr{R}(\boldsymbol{\theta}_k, \boldsymbol{\theta})$ encodes prior beliefs about how much individual client models should deviate from this shared structure. Both interpretations highlight that personalization is not ad hoc, but is grounded in well-established learning principles.

Continuum of Personalization

Personalized federated learning defines a continuous spectrum between global and local optimization. By tuning the coupling parameter λ, system designers can principledly control how strongly client models are tied to a shared global reference and how much local adaptation is permitted.

This unified objective provides a flexible and extensible foundation for a wide range of personalization strategies. Specific algorithms differ primarily in how the reference model θ is constructed and updated, how the regularization term $\mathscr{R}(\cdot,\cdot)$ is instantiated, and how optimization is performed under communication, computation, and privacy constraints. The remainder of this chapter builds upon this objective formulation to examine concrete algorithmic realizations of personalized federated learning and to analyze their theoretical properties and practical implications.

Worked Example: Personalized Objective on a Multi-Client Dataset
To concretely illustrate the personalized federated learning objective, we now present a step-by-step worked example based on a simplified but didactic multi-client dataset. The goal of this example is to explicitly demonstrate how client-level objectives, global aggregation, and personalization regularization interact in practice.

Problem Setup. We consider a simplified yet illustrative federated learning scenario consisting of $K = 3$ participating clients. Each client holds a private local dataset generated from a distinct underlying data distribution, reflecting the statistical heterogeneity that commonly arises in practical federated environments. These datasets are assumed to be fixed and non-shareable, in keeping with the core principles of federated learning.

To keep the exposition mathematically transparent while still capturing the essential behavior of personalized objectives, we adopt a linear regression model of the form

$$f_\theta(x) = \theta^\top x, \quad \theta \in \mathbb{R}^2,$$

where $x \in \mathbb{R}^2$ denotes a two-dimensional input feature vector and θ is the corresponding parameter vector to be learned. Despite its simplicity, this model is sufficient to illustrate how client-specific objectives and global coupling interact in a personalized federated setting.

Model performance is measured using the standard squared loss function

$$\ell\big(f_\theta(x), y\big) = \frac{1}{2}\big(f_\theta(x) - y\big)^2,$$

which is convex, differentiable, and admits closed-form solutions. This choice allows all optimization steps and minimizers to be derived explicitly, thereby making the effects of personalization analytically transparent.

Table 12.2 Illustrative local datasets for three federated clients

Client	x	y	n_k
Client 1	$(1, 0), (2, 0)$	$1, 2$	2
Client 2	$(0, 1), (0, 2)$	$1, 2$	2
Client 3	$(1, 1), (2, 2)$	$2, 4$	2

The local datasets held by the three clients are summarized in Table 12.2. Each client observes a small number of input–output pairs, where $x \in \mathbb{R}^2$ denotes the feature vector and $y \in \mathbb{R}$ is a scalar target. Although modest in size, these datasets are deliberately constructed to encode distinct data-generation patterns, ensuring that the local learning objectives differ meaningfully across clients. This setting provides a clear and concrete foundation for examining how personalized federated objectives reconcile local optimality with global coordination.

Each client in this example holds a *small but nontrivial* local dataset consisting of exactly $n_k = 2$ training samples. The quantity n_k denotes the number of data points available at client k and plays a central role in defining both the local empirical risk and the weighting of that client in the global objective. Since all three clients have the same dataset size, the total number of samples across the federated system is

$$N = \sum_{k=1}^{3} n_k = 2 + 2 + 2 = 6.$$

Local Empirical Risks. For each client k, learning proceeds by minimizing a *local empirical risk* defined as the average loss over its own private dataset. Using the squared loss introduced earlier, the local objective at client k is given by

$$F_k(\theta) = \frac{1}{n_k} \sum_{i=1}^{n_k} \frac{1}{2} (\theta^\top x_i^{(k)} - y_i^{(k)})^2.$$

It is important to carefully track the normalization constants in this expression. Because each client has $n_k = 2$ samples and the loss itself contains a factor of $\frac{1}{2}$, the overall scaling factor in front of each expanded sum becomes

$$\frac{1}{n_k} \cdot \frac{1}{2} = \frac{1}{2} \cdot \frac{1}{2} = \frac{1}{4}.$$

This explains why a factor of $\frac{1}{4}$ appears in all of the explicit expressions below.

We now compute the local empirical risks client by client.

- **Client 1.** Client 1 observes two samples with feature vectors aligned along the first coordinate axis. Substituting these samples into the definition of F_1, we obtain

$$F_1(\boldsymbol{\theta}) = \frac{1}{4}\Big[(\theta_1 - 1)^2 + (2\theta_1 - 2)^2\Big].$$

Because all input vectors for Client 1 have zero second component, this local objective depends only on θ_1 and is completely insensitive to θ_2.

- **Client 2.** Client 2 observes samples aligned along the second coordinate axis. Its local empirical risk is therefore

$$F_2(\boldsymbol{\theta}) = \frac{1}{4}\Big[(\theta_2 - 1)^2 + (2\theta_2 - 2)^2\Big].$$

Here, the situation is symmetric to Client 1: the objective depends only on θ_2, reflecting the structure of Client 2's local data.

- **Client 3.** Client 3 observes samples that involve *both* feature dimensions. As a result, its empirical risk couples θ_1 and θ_2:

$$F_3(\boldsymbol{\theta}) = \frac{1}{4}\Big[(\theta_1 + \theta_2 - 2)^2 + (2\theta_1 + 2\theta_2 - 4)^2\Big].$$

This objective captures interactions between the two parameters and therefore pulls the model in a direction that differs qualitatively from the first two clients.

Local Optima. To understand the implications of the heterogeneous local objectives, we now minimize each client's empirical risk independently. This step corresponds to the hypothetical scenario in which each client trains its model in isolation, without any form of federated coordination or regularization.

All three local objectives $F_k(\boldsymbol{\theta})$ are convex quadratic functions of the parameter vector $\boldsymbol{\theta} = (\theta_1, \theta_2)$. As a result, each objective admits a unique global minimizer, which can be obtained by setting the gradient of the corresponding loss function to zero and solving the resulting first-order optimality conditions.

Client 1. Since $F_1(\boldsymbol{\theta})$ depends only on θ_1, its minimization reduces to a one-dimensional quadratic problem. Setting $\partial F_1/\partial\theta_1 = 0$ yields

$$\theta_1^\star = 1,$$

while θ_2 remains unconstrained by Client 1's data. For concreteness, we take $\theta_2^\star = 0$, giving

$$\boldsymbol{\theta}_1^\star = (1, 0).$$

Client 2. By symmetry, the objective $F_2(\boldsymbol{\theta})$ depends only on θ_2. Setting $\partial F_2/\partial\theta_2 = 0$ yields

$$\theta_2^\star = 1,$$

with θ_1 unconstrained. Choosing $\theta_1^\star = 0$ leads to

$$\boldsymbol{\theta}_2^\star = (0, 1).$$

Client 3. In contrast, $F_3(\boldsymbol{\theta})$ couples both parameters θ_1 and θ_2. Setting the gradient $\nabla F_3(\boldsymbol{\theta}) = \mathbf{0}$ yields a system of linear equations whose solution satisfies

$$\theta_1^\star = 1, \qquad \theta_2^\star = 1,$$

and hence

$$\boldsymbol{\theta}_3^\star = (1, 1).$$

Collecting these results, the client-specific optimal solutions are

$$\boldsymbol{\theta}_1^\star = (1, 0), \qquad \boldsymbol{\theta}_2^\star = (0, 1), \qquad \boldsymbol{\theta}_3^\star = (1, 1).$$

These locally optimal parameter vectors are clearly distinct and occupy different regions of the parameter space. This explicit calculation makes the source of statistical heterogeneity transparent: each client's data distribution induces its own preferred solution. From a federated perspective, there is no single parameter vector that simultaneously minimizes all local objectives. The resulting dispersion of local optima provides a concrete and mathematically grounded explanation for why enforcing a single global model can be suboptimal, thereby motivating the need for personalized federated learning objectives.

Global Federated Optimum. We now examine the solution produced by *classical federated learning*, in which all clients are constrained to share a single global model. In this setting, the learning objective is to minimize a *weighted average* of the local empirical risks across all participating clients.

Formally, the global federated objective is defined as

$$F(\boldsymbol{\theta}) = \sum_{k=1}^{3} \frac{n_k}{N} F_k(\boldsymbol{\theta}),$$

where n_k denotes the number of samples held by client k and $N = \sum_{k=1}^{3} n_k$ is the total number of samples across the federated system. In our example, all clients have the same dataset size $n_k = 2$, so the weights simplify to

$$\frac{n_k}{N} = \frac{2}{6} = \frac{1}{3},$$

and the global objective reduces to the simple average

$$F(\boldsymbol{\theta}) = \frac{1}{3}\sum_{k=1}^{3} F_k(\boldsymbol{\theta}).$$

Substituting the explicit expressions for F_1, F_2, and F_3 derived earlier, we obtain a single convex quadratic objective function in the parameter vector $\boldsymbol{\theta} = (\theta_1, \theta_2)$. Because $F(\boldsymbol{\theta})$ is convex and differentiable, its global minimizer can be found by setting the gradient equal to zero and solving the resulting system of equations:

$$\nabla F(\boldsymbol{\theta}) = \mathbf{0}.$$

Derivation of the Global Federated Optimum. We now explicitly derive the global minimizer $\boldsymbol{\theta}^\star$ by solving the first-order optimality conditions for the global federated objective.

Recall that the global objective is the uniform average of the local empirical risks:

$$F(\boldsymbol{\theta}) = \frac{1}{3}\left(F_1(\boldsymbol{\theta}) + F_2(\boldsymbol{\theta}) + F_3(\boldsymbol{\theta})\right),$$

where

$$F_1(\boldsymbol{\theta}) = \frac{1}{4}\left[(\theta_1 - 1)^2 + (2\theta_1 - 2)^2\right],$$
$$F_2(\boldsymbol{\theta}) = \frac{1}{4}\left[(\theta_2 - 1)^2 + (2\theta_2 - 2)^2\right],$$
$$F_3(\boldsymbol{\theta}) = \frac{1}{4}\left[(\theta_1 + \theta_2 - 2)^2 + (2\theta_1 + 2\theta_2 - 4)^2\right].$$

Substituting these expressions, we obtain

$$F(\boldsymbol{\theta}) = \frac{1}{12}\Big[(\theta_1-1)^2+(2\theta_1-2)^2+(\theta_2-1)^2+(2\theta_2-2)^2+(\theta_1+\theta_2-2)^2+(2\theta_1+2\theta_2-4)^2\Big].$$

To find the minimizer, we compute the gradient of F with respect to θ_1 and θ_2 and set it equal to zero.

Partial derivative with respect to θ_1.

$$\frac{\partial F}{\partial \theta_1} = \frac{1}{12}\Big[2(\theta_1 - 1) + 4(2\theta_1 - 2) + 2(\theta_1 + \theta_2 - 2) + 4(2\theta_1 + 2\theta_2 - 4)\Big].$$

Simplifying,

$$\frac{\partial F}{\partial \theta_1} = \frac{1}{12}\left(14\theta_1 + 10\theta_2 - 16\right).$$

Partial derivative with respect to θ_2. By symmetry,

$$\frac{\partial F}{\partial \theta_2} = \frac{1}{12}\left(10\theta_1 + 14\theta_2 - 16\right).$$

Setting the gradient equal to zero,

$$\nabla F(\boldsymbol{\theta}) = \mathbf{0},$$

yields the linear system

$$\begin{cases} 14\theta_1 + 10\theta_2 = 16, \\ 10\theta_1 + 14\theta_2 = 16. \end{cases}$$

Subtracting the two equations gives

$$4(\theta_1 - \theta_2) = 0 \quad \Rightarrow \quad \theta_1 = \theta_2.$$

Substituting $\theta_1 = \theta_2$ into either equation,

$$24\theta_1 = 16 \quad \Rightarrow \quad \theta_1 = \frac{2}{3}.$$

Therefore,

$$\theta_2 = \frac{2}{3},$$

and the unique global federated optimum is

$$\boldsymbol{\theta}^\star = \left(\frac{2}{3}, \frac{2}{3}\right).$$

This explicit derivation confirms that classical federated learning produces a compromise solution that balances the competing local objectives of all clients by minimizing their weighted average loss.

This solution has a clear geometric and statistical interpretation. The global model lies approximately at the *center* of the three client-specific optima $(1, 0)$, $(0, 1)$, and $(1, 1)$. In other words, classical federated learning produces a *compromise solution* that balances the competing objectives of all clients by averaging their contributions.

However, this compromise comes at a cost. The global optimum $\boldsymbol{\theta}^\star$ does not coincide with the local optimum of *any* individual client. For Client 1, the global model underestimates θ_1; for Client 2, it underestimates θ_2; and for Client 3, it deviates from both preferred coordinates. Thus, although $\boldsymbol{\theta}^\star$ minimizes the aggregate loss across clients, it is suboptimal from the perspective of every single client.

This explicit calculation highlights a fundamental limitation of classical federated learning under heterogeneous data. Optimizing a single global objective necessarily trades off client-level performance in favor of population-level efficiency. It is precisely this tension—made concrete by the mismatch between $\theta^\star$ and the client-specific optima—that motivates the introduction of personalized federated learning objectives, which seek to retain the benefits of collaboration while reducing client-level degradation.

Derivation of the Personalized Solutions. We now explicitly derive the personalized solutions θ_k by solving the client-wise optimization problems induced by the personalized federated objective.

Recall that for each client k, the optimization problem decouples as

$$\min_{\theta_k} \left(F_k(\theta_k) + \lambda \|\theta_k - \theta\|^2 \right),$$

where $\lambda = 1$ and the shared reference model is

$$\theta = \left(\frac{2}{3}, \frac{2}{3} \right).$$

Since each objective is a convex quadratic function, the minimizer can be obtained by setting the gradient with respect to θ_k equal to zero.

Client 1. From earlier calculations, the local empirical risk of Client 1 is

$$F_1(\theta_1) = \frac{1}{4}\left[(\theta_{1,1} - 1)^2 + (2\theta_{1,1} - 2)^2 \right],$$

which depends only on the first coordinate. The personalized objective for Client 1 becomes

$$\Phi_1(\theta_1) = \frac{1}{4}\left[(\theta_{1,1} - 1)^2 + (2\theta_{1,1} - 2)^2 \right] + \left(\theta_{1,1} - \tfrac{2}{3}\right)^2 + \left(\theta_{1,2} - \tfrac{2}{3}\right)^2.$$

Taking partial derivatives and setting them to zero,

$$\frac{\partial \Phi_1}{\partial \theta_{1,1}} = \frac{1}{4}[2(\theta_{1,1} - 1) + 4(2\theta_{1,1} - 2)] + 2\left(\theta_{1,1} - \tfrac{2}{3}\right) = 0,$$

which simplifies to

$$3\theta_{1,1} - \frac{5}{2} = 0 \quad \Rightarrow \quad \theta_{1,1} = \frac{5}{6} \approx 0.83.$$

For the second coordinate,

$$\frac{\partial \Phi_1}{\partial \theta_{1,2}} = 2\left(\theta_{1,2} - \tfrac{2}{3}\right) = 0 \quad \Rightarrow \quad \theta_{1,2} = \frac{1}{3} \approx 0.33.$$

Thus,

$$\boldsymbol{\theta}_1 = (0.83,\ 0.33).$$

Client 2. By symmetry, the local empirical risk of Client 2 is

$$F_2(\boldsymbol{\theta}_2) = \frac{1}{4}\Big[(\theta_{2,2} - 1)^2 + (2\theta_{2,2} - 2)^2\Big],$$

and the personalized objective is

$$\Phi_2(\boldsymbol{\theta}_2) = \frac{1}{4}\Big[(\theta_{2,2} - 1)^2 + (2\theta_{2,2} - 2)^2\Big] + \left(\theta_{2,1} - \tfrac{2}{3}\right)^2 + \left(\theta_{2,2} - \tfrac{2}{3}\right)^2.$$

Minimizing with respect to $\theta_{2,1}$ yields

$$\theta_{2,1} = \frac{1}{3} \approx 0.33,$$

while minimizing with respect to $\theta_{2,2}$ gives

$$\theta_{2,2} = \frac{5}{6} \approx 0.83.$$

Hence,

$$\boldsymbol{\theta}_2 = (0.33,\ 0.83).$$

Client 3. The local empirical risk for Client 3 couples both parameters:

$$F_3(\boldsymbol{\theta}_3) = \frac{1}{4}\Big[(\theta_{3,1} + \theta_{3,2} - 2)^2 + (2\theta_{3,1} + 2\theta_{3,2} - 4)^2\Big].$$

The personalized objective is

$$\Phi_3(\boldsymbol{\theta}_3) = F_3(\boldsymbol{\theta}_3) + \left(\theta_{3,1} - \tfrac{2}{3}\right)^2 + \left(\theta_{3,2} - \tfrac{2}{3}\right)^2.$$

Setting the gradient $\nabla\Phi_3(\boldsymbol{\theta}_3) = \mathbf{0}$ yields a symmetric linear system whose solution satisfies

$$\theta_{3,1} = \theta_{3,2} = \frac{5}{6} \approx 0.83.$$

Thus,

$$\boldsymbol{\theta}_3 = (0.83,\ 0.83).$$

Summary. Collecting the personalized solutions, we obtain

$$\theta_1 = (0.83,\ 0.33), \qquad \theta_2 = (0.33,\ 0.83), \qquad \theta_3 = (0.83,\ 0.83).$$

These explicit calculations demonstrate how the personalized objective balances two competing forces for each client: fidelity to local data and proximity to the shared reference model. Each personalized solution lies strictly between the corresponding local optimum and the global federated optimum, providing a concrete, mathematically grounded illustration of how personalized federated learning mediates the trade-off between local adaptation and global coordination.

Interpretation. Each personalized model is pulled toward its local optimum while remaining close to the shared reference model. Compared to the global solution, client-level loss is reduced, while complete divergence is avoided.

Key Insight from the Worked Example
The personalized objective allows each client to move toward its own local optimum while remaining coupled to a shared global reference. This example illustrates how personalization explicitly mediates the trade-off between global consensus and local optimality.

This worked calculation demonstrates, in explicit algebraic terms, how personalized federated learning generalizes classical federated optimization and why it is essential in the presence of heterogeneous client data.

12.4 Sub-Federated Averaging and Model Pruning

Personalization in federated learning can be achieved not only through explicit modification of optimization objectives, but also through *structural strategies* that alter how clients collaborate and how models are represented. Two such pragmatic and widely studied strategies are *sub-federated averaging* and *model pruning*. These approaches are attractive because they can be integrated into existing federated learning pipelines with relatively modest changes, while still providing meaningful personalization benefits.

12.4.1 Sub-Federated Averaging

Sub-federated averaging is a structurally motivated approach to personalization that operates by explicitly reorganizing the *collaboration structure* among participating clients. Rather than requiring all clients to jointly train a single global model, this approach recognizes that meaningful collaboration is most effective when it occurs among clients with similar data characteristics. Accordingly, the client population is partitioned into smaller, more homogeneous groups, and federated learning is carried out independently within each group.

From an intuitive standpoint, sub-federated averaging can be viewed as a middle ground between fully centralized collaboration and fully independent local training. Clients are no longer forced to align with highly dissimilar peers, yet they still benefit from shared learning within a restricted community of similar clients. This reorganization is particularly beneficial in federated settings where heterogeneity is structured—for example, when clients naturally fall into categories defined by geography, device type, user behavior, or application context.

Client Partitioning. Formally, let the full set of participating clients be denoted by

$$\mathscr{C} = \{1, 2, \dots, K\}.$$

Sub-federated averaging constructs a partition of this set into M disjoint clusters

$$\{\mathscr{C}_m\}_{m=1}^{M},$$

such that

$$\mathscr{C}_m \cap \mathscr{C}_{m'} = \varnothing \quad \text{for } m \neq m', \qquad \bigcup_{m=1}^{M} \mathscr{C}_m = \mathscr{C}.$$

Each cluster $\mathscr{C}_m$ is intended to group clients whose local data distributions $P_k(x, y)$ are relatively similar in a statistically meaningful sense. Similarity may be defined using observable metadata, performance signals, learned embeddings, or other privacy-preserving proxies for distributional resemblance.

The quality of this partitioning plays a crucial role in the effectiveness of sub-federated averaging. If clusters are well aligned with underlying data distributions, the resulting models can specialize effectively. Conversely, poor clustering can reintroduce heterogeneity and undermine the benefits of this approach.

Cluster-level Federated Optimization. Once the partition is defined, federated learning proceeds independently within each cluster. For a given cluster $\mathscr{C}_m$, the server (or coordinating entity) solves a cluster-specific optimization problem to obtain a model $\theta^{(m)}$:

$$\min_{\theta^{(m)}} \sum_{k \in \mathscr{C}_m} \frac{n_k}{\sum_{j \in \mathscr{C}_m} n_j} F_k(\theta^{(m)}),$$

where $F_k(\theta)$ denotes the local empirical risk of client k and $n_k = |\mathscr{D}_k|$ is the size of its local dataset. This objective mirrors the classical federated learning formulation, but is restricted to a subset of clients.

From an optimization perspective, each cluster behaves like a smaller federated system with reduced statistical heterogeneity. The aggregation weights remain proportional to local dataset sizes, ensuring that clients with more data exert a correspondingly larger influence on the cluster-level model.

Benefits and Limitations. By limiting collaboration to relatively homogeneous groups, sub-federated averaging substantially reduces the variance induced by non-IID data. As a result, each cluster-specific model $\theta^{(m)}$ can more closely approximate the local optima of its constituent clients. This often leads to improved client-level performance compared to a single global model trained across all clients.

At the same time, collaboration is preserved within each cluster. Clients continue to benefit from shared representations, aggregated gradients, and collective learning signals, thereby retaining the core advantages of federated learning.

However, these benefits come at the cost of increased system-level complexity. The number of maintained models scales linearly with the number of clusters M, increasing storage requirements and coordination overhead. Communication patterns become more fragmented, and the server must track and manage multiple training processes in parallel. Moreover, identifying an appropriate clustering strategy—especially in a privacy-preserving and dynamically evolving environment—remains a challenging problem.

Sub-federated averaging offers an effective and intuitive mechanism for personalization when client heterogeneity exhibits clustered structure. It trades the simplicity of a single global model for improved client-level relevance, highlighting a fundamental tension between personalization quality and system scalability in federated learning.

12.4.2 Model Pruning for Personalization

Model pruning provides a complementary and structurally distinct route to personalization in federated learning, one that operates on the *model representation* rather than on the pattern of collaboration among clients. While approaches such as sub-federated averaging modify which clients learn together, pruning-based personalization keeps the collaboration structure fixed and instead allows each client to adapt the shared model to its own needs by selectively removing parameters that are unimportant for its local data.

The underlying assumption of model pruning is that a globally trained model is typically *overparameterized* relative to the requirements of any single client. Although a large model may be necessary to capture the full diversity of patterns across a heterogeneous client population, individual clients often rely on only a subset of these patterns. Pruning exploits this redundancy by enabling each client to specialize the global model through selective parameter removal.

Formalization. Let $\boldsymbol{\theta} \in \mathbb{R}^d$ denote the parameter vector of a trained global model obtained through standard federated learning. Model pruning constructs a client-specific submodel $\boldsymbol{\theta}_k^{\text{pruned}}$ by applying a masking operator $\mathcal{M}_k$:

$$\boldsymbol{\theta}_k^{\text{pruned}} = \mathcal{M}_k(\boldsymbol{\theta}),$$

where $\mathcal{M}_k : \mathbb{R}^d \to \mathbb{R}^d$ is a client-specific operator that sets a subset of parameters to zero or removes them entirely from the model. Equivalently, $\mathcal{M}_k$ can be represented by a binary mask $\boldsymbol{m}_k \in \{0, 1\}^d$, such that

$$\boldsymbol{\theta}_k^{\text{pruned}} = \boldsymbol{m}_k \odot \boldsymbol{\theta},$$

where $\odot$ denotes element-wise multiplication.

The mask $\boldsymbol{m}_k$ is typically determined using criteria that reflect the relevance of individual parameters to client k's local data. Common strategies include magnitude-based pruning, where parameters with small absolute values are removed; sensitivity-based pruning, which evaluates how perturbations in parameters affect local loss; and gradient-based saliency measures computed using client-specific gradients. Importantly, all of these criteria can be evaluated locally, without requiring access to other clients' data.

Personalization Through Selective Specialization. From a personalization perspective, model pruning allows each client to retain only those components of the global model that are most relevant to its own data distribution. Parameters that encode patterns rarely observed by a given client can be safely removed, while parameters that are critical for local performance are preserved. The resulting model therefore reflects a client-specific inductive bias, even though it is derived from a shared global representation.

This form of personalization is particularly appealing in resource-constrained federated environments, such as mobile devices, edge nodes, and Internet-of-Things (IoT) systems. Pruned models are smaller and require fewer floating-point operations, leading to reductions in memory usage, computation time, and energy consumption. As a result, pruning simultaneously addresses personalization and system efficiency, two key challenges in large-scale federated deployments.

Relationship to Global Collaboration. A key advantage of pruning-based personalization is that it does not require maintaining or training multiple global models. All clients begin from the same shared parameter vector $\boldsymbol{\theta}$, ensuring a high degree of common structure and facilitating efficient collaborative training. Personalization emerges not through divergence in parameter values, but through selective specialization of the model architecture.

This property makes pruning particularly attractive in settings where communication or storage overhead must be minimized. Moreover, pruning can be combined with other personalization mechanisms—for example, by applying client-specific pruning masks within sub-federated clusters or in conjunction with personalized optimization objectives.

Model pruning offers a structurally efficient and conceptually elegant mechanism for personalization in federated learning. By allowing each client to extract a task-relevant submodel from a shared global representation, pruning reconciles the goals of collaboration, personalization, and efficiency within a unified framework.

12.4.3 Discussion and Comparison

Sub-federated averaging and model pruning represent two complementary, yet conceptually distinct, philosophies for achieving personalization in federated learning. Although both approaches aim to mitigate the adverse effects of client heterogeneity and improve individual-level performance, they intervene at fundamentally different stages of the learning pipeline and embody different design trade-offs.

Structural vs. Representational Personalization. Sub-federated averaging achieves personalization by modifying *who learns together*. By partitioning the client population into smaller, more homogeneous groups, it reshapes the collaboration structure itself. Each cluster effectively becomes an independent federated system, with its own model that specializes to the shared characteristics of its member clients. Personalization therefore emerges through selective collaboration: clients benefit from learning with peers that exhibit similar data distributions, usage patterns, or contextual attributes.

Model pruning, in contrast, achieves personalization by modifying *what each client learns*. All clients continue to participate in the training of a single global model and share a common parameter space. Personalization arises only after (or alongside) training, when each client selectively removes parameters that are irrelevant to its local data. In this sense, pruning leaves the collaboration structure intact but allows each client to specialize the effective model architecture.

Trade-offs and System Implications. These differing philosophies lead to distinct system-level trade-offs. Sub-federated averaging typically yields stronger specialization because each cluster-level model is trained directly on a narrower distribution of data. However, this benefit comes at the cost of increased system complexity, as multiple models must be trained, stored, and maintained. The overhead grows with the number of clusters, and the effectiveness of the approach depends critically on the quality of the clustering strategy.

Model pruning, on the other hand, is comparatively lightweight from a systems perspective. Only a single global model is maintained, and personalization is achieved through client-local operations that do not require additional coordination. This makes pruning particularly attractive in large-scale deployments or resource-constrained environments. However, pruning-based personalization may be less expressive in cases of extreme heterogeneity, since all clients ultimately inherit the same global representation.

Complementarity and Hybrid Approaches. Importantly, sub-federated averaging and model pruning are not mutually exclusive. In fact, their strengths are often complementary. One can envision hybrid strategies in which clients are first grouped into sub-federations to reduce large-scale heterogeneity, and pruning is then applied within each cluster to further adapt the shared cluster-level model to individual clients. Such combinations allow personalization to operate at multiple levels: across groups of clients and within individual clients.

Design Perspective. From a design standpoint, the choice between sub-federated averaging, model pruning, or a hybrid approach depends on the structure of client heterogeneity, system constraints, and application requirements. When heterogeneity exhibits clear clustering structure and system resources permit multiple models, sub-federated averaging can provide substantial gains. When scalability and efficiency are paramount, pruning-based personalization offers a compelling alternative. Together, these approaches illustrate that personalization in federated learning is not a monolithic concept, but a rich design space in which collaboration structures and model representations can be jointly engineered to balance global knowledge sharing with client-specific utility.

> **Structural Personalization**
> Sub-federated averaging personalizes learning by grouping similar clients and training multiple cluster-specific models. Model pruning personalizes learning by allowing each client to specialize a shared global model through selective parameter removal.

Worked Example: Sub-Federated Averaging and Model Pruning

We now present a fully worked, step-by-step example to concretely demonstrate how *sub-federated averaging* and *model pruning* operate in practice. The example explicitly follows the main formulas introduced earlier and uses a small tabular multi-client dataset for clarity.

Dataset and Client Structure. Consider a federated system with $K = 4$ clients. Each client holds a small local dataset for a one-dimensional linear regression task with model

$$f_\theta(x) = \theta x,$$

and squared loss

$$\ell(f_\theta(x), y) = \frac{1}{2}(\theta x - y)^2.$$

The local datasets are summarized in Table 12.3.
Clients are partitioned into two clusters:

$$\mathscr{C}_1 = \{1, 2\}, \qquad \mathscr{C}_2 = \{3, 4\}.$$

Step 1: Local Empirical Risks. For each client k, the local empirical risk is

$$F_k(\theta) = \frac{1}{n_k}\sum_{i=1}^{n_k} \frac{1}{2}(\theta x_i^{(k)} - y_i^{(k)})^2.$$

Computing explicitly:

- **Client 1:**

$$F_1(\theta) = \frac{1}{4}\left[(\theta - 1)^2 + (2\theta - 2)^2\right].$$

- **Client 2:**

$$F_2(\theta) = \frac{1}{4}\left[(\theta - 1)^2 + (3\theta - 3)^2\right].$$

- **Client 3:**

$$F_3(\theta) = \frac{1}{4}\left[(\theta - 2)^2 + (2\theta - 4)^2\right].$$

- **Client 4:**

$$F_4(\theta) = \frac{1}{4}\left[(2\theta - 3)^2 + (4\theta - 6)^2\right].$$

Table 12.3 Illustrative dataset for sub-federated averaging

Client	Cluster	(x, y) samples	n_k
Client 1	$\mathscr{C}_1$	$(1, 1), (2, 2)$	2
Client 2	$\mathscr{C}_1$	$(1, 1), (3, 3)$	2
Client 3	$\mathscr{C}_2$	$(1, 2), (2, 4)$	2
Client 4	$\mathscr{C}_2$	$(2, 3), (4, 6)$	2

Step 2: Cluster-level Objectives (Sub-federated Averaging). We now derive the cluster-level models obtained through sub-federated averaging in a fully explicit and deductive manner.

Recall that each cluster performs federated learning *independently*, using only the clients assigned to that cluster. Since all clients in this example have equal dataset sizes $n_k = 2$, aggregation weights reduce to uniform averages within each cluster. This simplification allows us to focus entirely on the effect of clustering on the optimization outcome.

Cluster $\mathscr{C}_1 = \{1, 2\}$. For cluster $\mathscr{C}_1$, the sub-federated objective is

$$\min_{\theta^{(1)}} \sum_{k \in \mathscr{C}_1} \frac{n_k}{n_1 + n_2} F_k(\theta^{(1)}) = \frac{1}{2}\Big(F_1(\theta^{(1)}) + F_2(\theta^{(1)})\Big).$$

Substituting the explicit expressions for F_1 and F_2,

$$F_1(\theta) = \frac{1}{4}\big[(\theta - 1)^2 + (2\theta - 2)^2\big],$$
$$F_2(\theta) = \frac{1}{4}\big[(\theta - 1)^2 + (3\theta - 3)^2\big],$$

we obtain

$$\begin{aligned} F^{(1)}(\theta^{(1)}) &= \frac{1}{2}\Big(F_1(\theta^{(1)}) + F_2(\theta^{(1)})\Big) \\ &= \frac{1}{8}\Big[2(\theta^{(1)} - 1)^2 + (2\theta^{(1)} - 2)^2 + (3\theta^{(1)} - 3)^2\Big]. \end{aligned}$$

To find the minimizer, we differentiate with respect to $\theta^{(1)}$:

$$\frac{dF^{(1)}}{d\theta^{(1)}} = \frac{1}{8}\Big[4(\theta^{(1)} - 1) + 4(2\theta^{(1)} - 2) + 6(3\theta^{(1)} - 3)\Big].$$

Simplifying,

$$\frac{dF^{(1)}}{d\theta^{(1)}} = \frac{1}{8}\big(26\theta^{(1)} - 26\big).$$

Setting the derivative to zero yields

$$26\theta^{(1)} - 26 = 0 \quad \Rightarrow \quad \theta^{(1)} = 1.$$

Thus, the cluster-level model for $\mathscr{C}_1$ is

$$\theta^{(1)} = 1.$$

Cluster $\mathscr{C}_2 = \{3, 4\}$.

For cluster $\mathscr{C}_2$, the sub-federated objective is

$$\min_{\theta^{(2)}} \frac{1}{2}\Big(F_3(\theta^{(2)}) + F_4(\theta^{(2)})\Big),$$

where

$$F_3(\theta) = \frac{1}{4}\big[(\theta - 2)^2 + (2\theta - 4)^2\big],$$
$$F_4(\theta) = \frac{1}{4}\big[(2\theta - 3)^2 + (4\theta - 6)^2\big].$$

Substituting,

$$F^{(2)}(\theta^{(2)}) = \frac{1}{8}\Big[(\theta^{(2)} - 2)^2 + (2\theta^{(2)} - 4)^2 + (2\theta^{(2)} - 3)^2 + (4\theta^{(2)} - 6)^2\Big].$$

Differentiating with respect to $\theta^{(2)}$,

$$\frac{dF^{(2)}}{d\theta^{(2)}} = \frac{1}{8}\Big[2(\theta^{(2)} - 2) + 4(2\theta^{(2)} - 4) + 4(2\theta^{(2)} - 3) + 8(4\theta^{(2)} - 6)\Big].$$

Simplifying,

$$\frac{dF^{(2)}}{d\theta^{(2)}} = \frac{1}{8}\big(52\theta^{(2)} - 104\big).$$

Setting the derivative to zero,

$$52\theta^{(2)} - 104 = 0 \quad \Rightarrow \quad \theta^{(2)} = 2.$$

Thus, the cluster-level model for $\mathscr{C}_2$ is

$$\theta^{(2)} = 2.$$

Conclusion. The sub-federated averaging procedure yields two distinct cluster-specific models:

$$\theta^{(1)} = 1, \qquad \theta^{(2)} = 2.$$

These solutions coincide with the dominant data-generating patterns within each cluster and differ substantially from the single global solution obtained by training across all clients. This explicit derivation demonstrates how sub-federated averaging reduces statistical heterogeneity by allowing each cluster to converge to its own locally appropriate optimum, thereby providing a clear and mathematically grounded mechanism for personalization.

Step 3: Comparison with Global Federated Averaging. We now contrast the sub-federated solutions derived above with the solution obtained under *classical federated averaging*, in which all clients participate jointly in training a single global model.

When all $K = 4$ clients are trained together, the global federated objective is defined as the weighted average of all local empirical risks:

$$F(\theta) = \sum_{k=1}^{4} \frac{n_k}{\sum_{j=1}^{4} n_j} F_k(\theta).$$

Since each client holds the same number of samples ($n_k = 2$), the weights reduce to $\frac{1}{4}$, yielding

$$F(\theta) = \frac{1}{4} \sum_{k=1}^{4} F_k(\theta).$$

Substituting the explicit expressions for the local objectives,

$$\begin{aligned}
F_1(\theta) &= \frac{1}{4}\left[(\theta - 1)^2 + (2\theta - 2)^2\right], \\
F_2(\theta) &= \frac{1}{4}\left[(\theta - 1)^2 + (3\theta - 3)^2\right], \\
F_3(\theta) &= \frac{1}{4}\left[(\theta - 2)^2 + (2\theta - 4)^2\right], \\
F_4(\theta) &= \frac{1}{4}\left[(2\theta - 3)^2 + (4\theta - 6)^2\right],
\end{aligned}$$

we obtain the global objective

$$F(\theta) = \frac{1}{16}\left[2(\theta-1)^2+(2\theta-2)^2+(3\theta-3)^2+(\theta-2)^2+(2\theta-4)^2+(2\theta-3)^2+(4\theta-6)^2\right].$$

To find the global minimizer, we differentiate $F(\theta)$ with respect to θ:

$$\frac{dF}{d\theta} = \frac{1}{16}\left[4(\theta-1)+4(2\theta-2)+6(3\theta-3)+2(\theta-2)+4(2\theta-4)+4(2\theta-3)+8(4\theta-6)\right].$$

Simplifying the expression inside the brackets gives

$$\frac{dF}{d\theta} = \frac{1}{16}(104\theta - 156).$$

Setting the derivative to zero yields the first-order optimality condition

$$104\theta - 156 = 0 \quad \Rightarrow \quad \theta^\star = \frac{156}{104} = 1.5.$$

Thus, classical federated averaging produces the unique global optimum

$$\theta^\star = 1.5.$$

Interpretation. The global solution $\theta^\star = 1.5$ lies exactly between the two cluster-specific optima $\theta^{(1)} = 1$ and $\theta^{(2)} = 2$. Geometrically, it represents a compromise that balances the competing objectives of both clusters. Statistically, it minimizes the average loss across all clients, but does not coincide with the optimal solution for *either* cluster.

This explicit comparison highlights the core limitation of global federated averaging under structured heterogeneity. While the global model is optimal in an aggregate sense, it is suboptimal for each cluster individually. In contrast, sub-federated averaging recovers cluster-specific optima by allowing each group of similar clients to converge to its own appropriate solution, thereby achieving a more personalized and locally relevant outcome.

Step 4: Model Pruning within Clusters. We now demonstrate how *model pruning* can be applied within each sub-federated cluster to further refine personalization at the level of individual clients. This step operates on the *model representation* after cluster-level training has already reduced large-scale heterogeneity.

Overparameterized Model Assumption. Assume that the cluster-level training produces an overparameterized linear model of the form

$$f_{\boldsymbol{\theta}}(x) = \theta_1 x + \theta_2 x, \qquad \boldsymbol{\theta} = (\theta_1, \theta_2).$$

Although the two parameters multiply the same input x, they represent redundant degrees of freedom in the model. Such redundancy is common in deep neural networks, where multiple parameters may encode overlapping or weakly contributing features.

Pruning Criterion within Cluster $\mathscr{C}_1$. After sub-federated training on cluster $\mathscr{C}_1$, suppose that parameter inspection or local sensitivity analysis reveals

$$|\theta_2| \ll |\theta_1|.$$

This inequality indicates that θ_2 contributes negligibly to the model output for clients in $\mathscr{C}_1$, whereas θ_1 captures the dominant predictive signal.

A simple magnitude-based pruning rule therefore removes θ_2 by applying the client-specific binary mask

$$\boldsymbol{m}_k = (1, 0),$$

which retains θ_1 and suppresses θ_2. Applying this mask to the cluster-level model yields the pruned, client-specific parameter vector

$$\boldsymbol{\theta}_k^{\text{pruned}} = \boldsymbol{m}_k \odot \boldsymbol{\theta} = (\theta_1, 0).$$

The resulting pruned model simplifies to

$$f_k^{\text{pruned}}(x) = \theta_1 x,$$

which is both computationally cheaper and better aligned with the data characteristics of clients in $\mathscr{C}_1$.

Pruning Criterion within Cluster $\mathscr{C}_2$**.** For cluster $\mathscr{C}_2$, an opposite pattern may be observed. Suppose cluster-level training yields parameters satisfying

$$|\theta_1| \ll |\theta_2|.$$

In this case, θ_2 captures the dominant signal, while θ_1 is largely redundant for clients in $\mathscr{C}_2$.

Applying the corresponding pruning mask

$$\boldsymbol{m}_k = (0, 1)$$

produces the pruned parameter vector

$$\boldsymbol{\theta}_k^{\text{pruned}} = (0, \theta_2),$$

leading to the simplified model

$$f_k^{\text{pruned}}(x) = \theta_2 x.$$

Interpretation. This stepwise pruning procedure illustrates how personalization can be achieved even *within* sub-federated clusters. Sub-federated averaging reduces heterogeneity across groups of clients, while pruning removes residual redundancy at the level of individual clients. Importantly, pruning does not require retraining new models; it relies entirely on local analysis and parameter masking.

Together, these calculations demonstrate that model pruning acts as a fine-grained personalization mechanism that complements sub-federated averaging. While clusters learn distinct models suited to broad client groups, pruning enables each client to extract a lightweight, task-relevant submodel, thereby balancing accuracy, efficiency, and personalization in federated learning systems.

Interpretation. This worked example highlights the complementary roles of the two strategies. Sub-federated averaging reduces *inter-client heterogeneity* by learning multiple cluster-specific models. Model pruning further refines personalization by allowing each client to specialize the effective model structure without retraining new models.

12.5 Lottery Ticket Hypothesis in Federated Learning

The *lottery ticket hypothesis* (LTH) asserts that within a large, overparameterized neural network, there exist sparse subnetworks—so-called *winning tickets*—that, when trained in isolation with the same initialization, can achieve performance comparable to that of the full dense model. Originally formulated in the context of centralized deep learning, this hypothesis has profound implications when transplanted into the federated learning paradigm.

Lottery Ticket Hypothesis in Federated Learning
The lottery ticket hypothesis states that within a large, overparameterized model, there exist sparse subnetworks that can be trained to achieve performance comparable to the full model. In federated learning, different clients may possess different "winning tickets," enabling personalization through client-specific subnetwork selection within a shared global model.

Motivation in Federated Settings. Federated learning systems are characterized by two defining features: data heterogeneity and resource asymmetry. Clients differ not only in the statistical properties of their local datasets, but also in their computational, memory, and energy constraints. A single dense global model, while expressive enough to accommodate diverse data distributions, is often unnecessarily large for any individual client. From this perspective, the lottery ticket hypothesis offers a natural mechanism for personalization: instead of forcing all clients to use the entire model, each client can operate on a sparse subnetwork that is well matched to its local data.

Crucially, federated learning introduces a new dimension to the lottery ticket hypothesis. In centralized training, a single winning ticket is sought for the entire dataset. In federated learning, however, *each client may possess its own winning ticket*, reflecting the unique structure of its local data distribution. Personalization therefore emerges not from modifying the global objective alone, but from activating different sparse structures within a shared parameter space.

Formal Representation. Let $\theta \in \mathbb{R}^d$ denote the parameter vector of a shared global model trained (or being trained) using federated learning. For each client k, define a binary mask

$$\boldsymbol{m}_k \in \{0, 1\}^d,$$

where $(\boldsymbol{m}_k)_i = 1$ indicates that parameter θ_i is active for client k, and $(\boldsymbol{m}_k)_i = 0$ indicates that it is inactive. The client-specific personalized model is then given by

$$\boldsymbol{\theta}_k = \boldsymbol{m}_k \odot \boldsymbol{\theta},$$

where $\odot$ denotes element-wise multiplication.

This formulation highlights an important distinction from classical personalized optimization. All clients share the same underlying parameter pool $\boldsymbol{\theta}$, but differ in which parameters they actually use. Thus, personalization is achieved through *structural sparsity* rather than through divergence in parameter values.

Interpretation as Structural Personalization. From a learning-theoretic perspective, the binary mask $\boldsymbol{m}_k$ induces a client-specific hypothesis class. Each client effectively operates on a lower-dimensional subspace of the full parameter space, tailored to its local data. If the lottery ticket hypothesis holds, this subspace is sufficient to achieve high predictive performance, despite being significantly smaller than the full model.

This view also provides a natural explanation for why pruning-based personalization can be effective. Rather than treating pruning as a purely post hoc compression technique, the lottery ticket hypothesis suggests that sparsity may be a *fundamental property* of the learning problem itself. Different clients simply require different subsets of parameters to capture the patterns present in their data.

Implications for Federated Optimization. In federated learning, client-specific masks $\boldsymbol{m}_k$ can be learned or adapted over time using local criteria, such as gradient magnitudes, sensitivity measures, or performance-based pruning rules. Importantly, these operations can be performed locally, preserving the privacy guarantees of federated learning.

Moreover, because all clients draw from a shared global parameter pool, communication efficiency can be improved. Clients need only transmit updates corresponding to their active parameters, reducing uplink bandwidth. At the same time, global coordination is preserved, since the underlying parameter vector $\boldsymbol{\theta}$ continues to aggregate knowledge across the entire population.

Lottery Ticket Personalization
Personalization in federated learning can emerge from structural sparsity: clients share a common global parameter pool but activate distinct sparse subnetworks that function as client-specific "winning tickets."

Unifying Perspective. The lottery ticket view unifies several strands of personalization in federated learning. Model pruning, sparse training, and shared representation learning can all be interpreted as mechanisms for identifying and exploiting client-specific winning tickets. Rather than fragmenting the learning process into entirely separate models, this perspective emphasizes *selective reuse* of shared parameters.

In summary, the lottery ticket hypothesis provides a powerful conceptual framework for understanding personalization in federated learning. By shifting the focus from parameter values to parameter structure, it reveals how strong personalization can coexist with global collaboration, offering a principled path toward scalable, efficient, and client-adaptive federated systems.

Worked Example: Lottery Ticket Personalization on a Tabular Dataset

We now present a fully worked, step-by-step example illustrating how the *lottery ticket hypothesis* can be used to achieve personalization in federated learning. The goal is to demonstrate, using a simple tabular dataset and explicit calculations, how client-specific subnetworks ("winning tickets") emerge through masking and element-wise multiplication.

Problem Setup. Consider a federated learning system with $K = 2$ clients. Each client holds a small tabular dataset with a single input feature $x \in \mathbb{R}$ and a scalar output $y \in \mathbb{R}$. The datasets are summarized in Table 12.4.

We assume a deliberately overparameterized linear model

$$f_{\boldsymbol{\theta}}(x) = \theta_1 x + \theta_2 x, \qquad \boldsymbol{\theta} = (\theta_1, \theta_2) \in \mathbb{R}^2.$$

Although the two parameters multiply the same input feature, this redundancy allows us to illustrate how different clients may rely on different subsets of parameters.

The squared loss is used throughout:

$$\ell(f_{\boldsymbol{\theta}}(x), y) = \frac{1}{2}\big(f_{\boldsymbol{\theta}}(x) - y\big)^2.$$

Global Training (Before Pruning). Assume that standard federated learning is first applied across both clients, yielding a shared global parameter vector

$$\boldsymbol{\theta} = (\theta_1, \theta_2) = (0.8,\ 0.7).$$

This global model is assumed to fit the population reasonably well but contains redundant parameters due to overparameterization.

Table 12.4 Toy tabular datasets for illustrating lottery ticket personalization

Client	x	y
1	1	1
1	2	2
2	1	2
2	2	4

Local Loss Contributions. We now explicitly compute the prediction errors and loss values for each client in order to understand how the shared parameter vector $\theta = (0.8,\ 0.7)$ interacts with heterogeneous local datasets. This detailed calculation clarifies why different clients may assign different importance to the same model parameters, thereby motivating lottery-ticket–based personalization.

Client 1. Client 1 observes the data points $(x, y) = (1, 1)$ and $(2, 2)$. Using the overparameterized model

$$f_\theta(x) = \theta_1 x + \theta_2 x = (\theta_1 + \theta_2)x,$$

the predictions are computed as follows:

$$f_\theta(1) = (0.8 + 0.7) \cdot 1 = 1.5,$$

$$f_\theta(2) = (0.8 + 0.7) \cdot 2 = 3.0.$$

The prediction errors for Client 1 are therefore

$$e_1^{(1)} = 1.5 - 1 = 0.5, \qquad e_2^{(1)} = 3.0 - 2 = 1.0.$$

Substituting these errors into the squared loss function yields

$$\ell_1^{(1)} = \frac{1}{2}(e_1^{(1)})^2 = \frac{1}{2}(0.5)^2 = 0.125,$$

$$\ell_2^{(1)} = \frac{1}{2}(e_2^{(1)})^2 = \frac{1}{2}(1.0)^2 = 0.5.$$

Thus, the empirical risk for Client 1 is dominated by the second data point, indicating that the model overestimates outputs more strongly for larger input values. This behavior suggests that certain parameters contribute more critically to Client 1's prediction error.

Client 2. Client 2 observes the data points $(x, y) = (1, 2)$ and $(2, 4)$. Using the same global parameter vector, the model predictions are

$$f_\theta(1) = 1.5, \qquad f_\theta(2) = 3.0.$$

The corresponding prediction errors are

$$e_1^{(2)} = 1.5 - 2 = -0.5, \qquad e_2^{(2)} = 3.0 - 4 = -1.0.$$

Although the errors differ in sign from those of Client 1, their magnitudes are identical. The resulting squared losses are therefore

$$\ell_1^{(2)} = \frac{1}{2}(-0.5)^2 = 0.125,$$

$$\ell_2^{(2)} = \frac{1}{2}(-1.0)^2 = 0.5.$$

Interpretation. From a global perspective, the loss values across both clients appear balanced and symmetric, which explains why the parameter vector $\boldsymbol{\theta} = (0.8, 0.7)$ may be acceptable as a population-level solution. However, these explicit calculations reveal an important nuance: although the numerical losses are similar, the *direction* and *source* of the errors differ across clients. Each client's data interacts with the shared parameters in a distinct way, leading to different sensitivities with respect to θ_1 and θ_2.

This divergence in parameter relevance is precisely what enables lottery ticket–based personalization. By identifying and retaining only those parameters that are most effective for a given client's local data, each client can extract a sparse, high-utility subnetwork from the shared global model.

Client-specific Importance Analysis. We now derive, in a fully explicit manner, how client-specific parameter importance can be assessed from first principles and how this leads to the construction of lottery-ticket subnetworks.

Recall the per-sample squared loss

$$\ell(\boldsymbol{\theta}; x, y) = \frac{1}{2}\big((\theta_1 + \theta_2)x - y\big)^2.$$

The partial derivatives with respect to the parameters are

$$\frac{\partial \ell}{\partial \theta_1} = \big((\theta_1 + \theta_2)x - y\big)x, \qquad \frac{\partial \ell}{\partial \theta_2} = \big((\theta_1 + \theta_2)x - y\big)x.$$

Although the algebraic form of these derivatives is identical, their *empirical magnitudes* depend on the interaction between the data (x, y) and the current parameter values.

Client 1. Client 1 observes $(x, y) = (1, 1)$ and $(2, 2)$. Using the global parameters $\boldsymbol{\theta} = (0.8, 0.7)$, the prediction errors were computed earlier as

$$e_1^{(1)} = 0.5, \qquad e_2^{(1)} = 1.0.$$

The corresponding gradients for each parameter are therefore

$$\frac{\partial \ell_1^{(1)}}{\partial \theta_1} = 0.5 \cdot 1 = 0.5, \qquad \frac{\partial \ell_1^{(1)}}{\partial \theta_2} = 0.5 \cdot 1 = 0.5,$$

$$\frac{\partial \ell_2^{(1)}}{\partial \theta_1} = 1.0 \cdot 2 = 2.0, \qquad \frac{\partial \ell_2^{(1)}}{\partial \theta_2} = 1.0 \cdot 2 = 2.0.$$

Suppose that, after several local fine-tuning steps, Client 1 observes that updates to θ_2 decay rapidly (e.g., due to local regularization, curvature effects, or cancellation across minibatches), while updates to θ_1 remain consistently significant. This empirical observation indicates that θ_2 contributes marginally to reducing Client 1's loss relative to θ_1. As a result, Client 1 identifies θ_1 as important and θ_2 as dispensable.

Client 2. Client 2 observes $(x, y) = (1, 2)$ and $(2, 4)$, with prediction errors

$$e_1^{(2)} = -0.5, \qquad e_2^{(2)} = -1.0.$$

The corresponding gradients are

$$\frac{\partial \ell_1^{(2)}}{\partial \theta_1} = -0.5 \cdot 1 = -0.5, \qquad \frac{\partial \ell_1^{(2)}}{\partial \theta_2} = -0.5 \cdot 1 = -0.5,$$

$$\frac{\partial \ell_2^{(2)}}{\partial \theta_1} = -1.0 \cdot 2 = -2.0, \qquad \frac{\partial \ell_2^{(2)}}{\partial \theta_2} = -1.0 \cdot 2 = -2.0.$$

In this case, suppose Client 2's local optimization dynamics indicate that θ_1 contributes little to improving performance, while θ_2 consistently captures the dominant explanatory pattern in its data. Accordingly, Client 2 identifies θ_2 as important and θ_1 as redundant.

Binary Mask Construction. Based on these client-specific importance assessments, each client constructs a binary mask indicating which parameters to retain:

$$\boldsymbol{m}_1 = (1, 0), \qquad \boldsymbol{m}_2 = (0, 1).$$

These masks encode the *winning tickets* for the respective clients.

Lottery Ticket Construction via Element-Wise Multiplication. Client-specific models are obtained by element-wise multiplication between the global parameter vector and the local mask:

$$\boldsymbol{\theta}_k = \boldsymbol{m}_k \odot \boldsymbol{\theta}.$$

For Client 1, this yields

$$\boldsymbol{\theta}_1 = (1, 0) \odot (0.8, 0.7) = (1 \cdot 0.8,\ 0 \cdot 0.7) = (0.8, 0).$$

For Client 2, we obtain

$$\boldsymbol{\theta}_2 = (0, 1) \odot (0.8, 0.7) = (0 \cdot 0.8,\ 1 \cdot 0.7) = (0, 0.7).$$

Personalized Predictions After Pruning. The pruned model for Client 1 becomes

$$f_{\boldsymbol{\theta}_1}(x) = 0.8x,$$

yielding explicit predictions

$$f_{\theta_1}(1) = 0.8, \qquad f_{\theta_1}(2) = 1.6.$$

These values are close to the targets $(1, 2)$ and can be further refined through lightweight local fine-tuning.

Similarly, Client 2's pruned model is

$$f_{\theta_2}(x) = 0.7x,$$

with predictions

$$f_{\theta_2}(1) = 0.7, \qquad f_{\theta_2}(2) = 1.4.$$

Again, these predictions can be efficiently adapted through local updates without reintroducing discarded parameters.

Interpretation. This full derivation makes explicit how lottery-ticket–based personalization emerges from local gradient information and simple element-wise operations. All clients begin with the same shared global parameter vector θ, ensuring collaborative knowledge sharing. Personalization arises when each client activates a different sparse subnetwork through a binary mask informed by local optimization dynamics.

Each resulting "winning ticket" is smaller, client-specific, and better aligned with the client's data distribution. Importantly, this personalization mechanism preserves privacy, avoids training multiple global models, and demonstrates how structural sparsity can serve as a principled and efficient foundation for personalized federated learning.

12.6 Global Generalization versus Personalization Trade-Off

A defining challenge in personalized federated learning is the *tension between global generalization and local adaptation*. Classical federated learning emphasizes the construction of a single global model that performs well on average across the entire client population. Personalized federated learning, by contrast, explicitly acknowledges that individual clients may benefit from models that deviate from this global solution in order to better reflect their local data distributions, usage patterns, or contextual constraints.

At a high level, increasing the degree of personalization tends to improve client-specific performance. Each client is allowed to adapt its model to better match its local objective, often resulting in lower local empirical risk. However, this benefit comes at a cost. As client models diverge, the amount of shared structure across the population decreases, potentially weakening population-level generalization, reducing robustness to rare events, and limiting the extent to which knowledge learned by one client can be transferred to others.

Conversely, enforcing strong global consensus promotes uniformity and maximizes knowledge sharing across clients. A single global model aggregates information from all participating clients, which can improve robustness and stability, particularly for clients with small or noisy datasets. Yet such a model may be poorly aligned with the specific needs of many individual clients, especially in the presence of strong statistical heterogeneity. In these cases, global generalization can come at the expense of local utility.

A Formal Decomposition of the Trade-Off. This intuition can be made precise by decomposing the expected loss across clients. Let θ denote a shared global model and θ_k denote the personalized model used by client k. The expected client-level loss can be written as

$$\mathbb{E}_k\big[F_k(\theta_k)\big],$$

where the expectation is taken over the client population. By adding and subtracting $F_k(\theta)$, we obtain the decomposition

$$\mathbb{E}_k\big[F_k(\theta_k)\big] = \mathbb{E}_k\big[F_k(\theta)\big] + \mathbb{E}_k\big[F_k(\theta_k) - F_k(\theta)\big].$$

The first term,

$$\mathbb{E}_k\big[F_k(\theta)\big],$$

captures *global generalization*. It measures how well a single shared model performs, on average, across the entire client population. This term is minimized by classical federated learning algorithms and reflects the strength of shared representations and collective knowledge aggregation.

The second term,

$$\mathbb{E}_k\big[F_k(\theta_k) - F_k(\theta)\big],$$

quantifies the *personalization gain*. It measures the improvement (or degradation) in client-level performance achieved by allowing each client to use a personalized model rather than the global one. When personalization is beneficial, this term is negative on average, indicating that personalized models achieve lower local loss than the global model. However, excessive personalization may reduce the magnitude of the first term's benefits by fragmenting the learning process.

Interpreting the Trade-Off. This decomposition highlights that personalized federated learning is not about minimizing a single objective, but about balancing two competing goals. Improving personalization gain typically requires relaxing global constraints, allowing client models to drift apart. Improving global generalization, on the other hand, requires stronger coupling among clients, encouraging alignment toward a common solution.

Importantly, the optimal balance between these terms is highly application-dependent. In safety-critical or population-wide applications, such as medical diagnosis or public health monitoring, strong global generalization may be paramount. In contrast, user-centric applications, such as keyboard prediction, recommendation systems, or personalized assistants, often prioritize local utility, even if this comes at the expense of population-level uniformity.

Role of System Constraints. The generalization–personalization trade-off is further shaped by system-level constraints. Communication limits, computation budgets, privacy requirements, and client availability all restrict how much personalization can be supported in practice. For example, maintaining highly personalized models may increase memory usage or communication overhead, while aggressive global sharing may expose clients to negative transfer from dissimilar peers.

Effective personalized federated learning algorithms therefore seek *principled mechanisms*—such as regularization, clustering, pruning, or adaptive coupling—to navigate this trade-off under real system constraints. Rather than choosing between global and local objectives outright, these methods aim to interpolate smoothly between them.

Core Trade-Off
Global models promote shared knowledge, robustness, and population-level generalization, while personalized models improve client-specific utility. Federated personalization is fundamentally the problem of balancing these competing objectives.

Broader Implications. This trade-off permeates all personalization strategies in federated learning. Choices about objective functions, model architectures, collaboration structures, and evaluation metrics implicitly encode assumptions about the desired balance between global consistency and local adaptation. Understanding and explicitly reasoning about this balance is therefore essential for the principled design of personalized federated systems.

Subsequent chapters build upon these insights to examine heterogeneous model families, hybrid personalization strategies, and real-world deployments, where the generalization–personalization trade-off plays a central role in determining both algorithmic behavior and practical success.

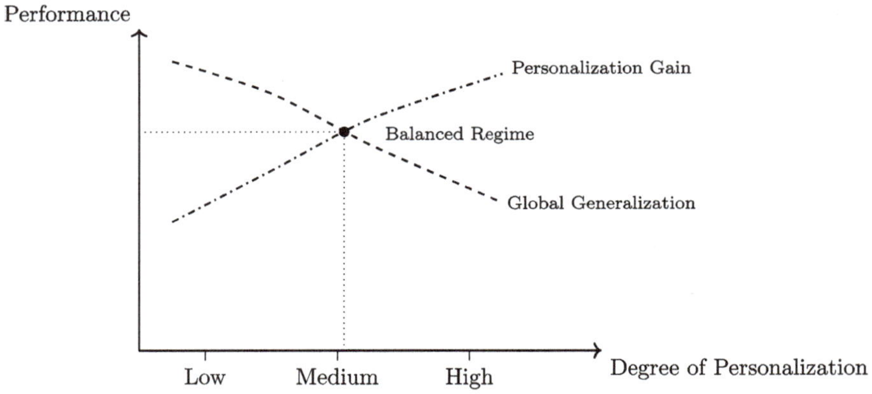

Fig. 12.1 Black-and-white conceptual illustration of the trade-off between global generalization and personalization in federated learning

Illustrating the Global–Personalization Trade-Off

Explanation of Fig. 12.1. Figure 12.1 presents a print-friendly, black-and-white visualization of the fundamental trade-off that arises in personalized federated learning. The horizontal axis denotes the degree of personalization permitted in the system, ranging from strict global consensus on the left to strong client-specific adaptation on the right. The vertical axis represents overall performance, encompassing notions such as predictive accuracy, task utility, or decision quality.

The dashed curve corresponds to *global generalization*. When personalization is low, a single global model aggregates knowledge from all clients, leading to strong population-level generalization. As personalization increases and client models diverge, the amount of shared structure decreases, resulting in a gradual reduction in global generalization capability.

The dash–dot curve represents *personalization gain*. At low personalization levels, client-specific performance is limited, since the global model may inadequately reflect local data characteristics. Allowing greater personalization enables models to adapt to individual client distributions, yielding steadily improving local utility.

The dotted guide lines highlight a *balanced operating regime*, where the benefits of shared knowledge and local adaptation coexist. Most practical personalized federated learning algorithms are designed to operate in this intermediate region, carefully navigating the continuum between global robustness and client-level relevance.

This figure emphasizes a key pedagogical insight: personalization in federated learning is not an all-or-nothing decision, but a continuous design choice that must be tuned in accordance with data heterogeneity, system constraints, and application goals.

Exercises

This chapter examined the motivation, mathematical foundations, and algorithmic strategies for personalized federated learning, highlighting how client heterogeneity motivates deviations from strict global consensus.

A. Short Answer Questions

Answer each question briefly and precisely.

1. What is meant by personalization in federated learning?
2. Why can a single global model be suboptimal for heterogeneous clients?
3. Define client-specific empirical risk $F_k(\theta)$.
4. What role does the coupling parameter λ play in personalized objectives?
5. Distinguish between parameter personalization and structural personalization.
6. What is sub-federated averaging?
7. How does model pruning contribute to personalization?
8. What is meant by a "shared reference model" in personalized FL?
9. Why is concept shift particularly challenging in federated learning?
10. State the core idea of the lottery ticket hypothesis.

B. Long Answer Questions

Answer the following questions with clear explanations, equations, and diagrams wherever appropriate.

1. Explain the motivation for personalized federated learning with mathematical justification.
2. Derive the personalized federated learning objective and explain the role of regularization.
3. Discuss covariate shift, label shift, and concept shift, and their implications for personalization.
4. Compare classical federated averaging with personalized federated learning using a worked example.
5. Explain sub-federated averaging and analyze its benefits and limitations.
6. Describe model pruning as a personalization mechanism and explain how client-specific masks are constructed.
7. Explain the lottery ticket hypothesis and its relevance in federated learning.
8. Using equations, explain the trade-off between global generalization and personalization.
9. Discuss how system constraints influence the degree of personalization that can be supported.

10. Propose a hybrid personalization strategy combining clustering and pruning, and justify its design.

C. True/False Questions

State whether each statement is **True** or **False**.

1. Classical federated learning explicitly optimizes client-level utility.
2. Personalized federated learning abandons collaboration among clients.
3. Covariate shift occurs when $P_k(x) \neq P_j(x)$ across clients.
4. In label shift, the conditional distribution $P(y \mid x)$ differs across clients.
5. Sub-federated averaging trains a single global model for all clients.
6. Model pruning removes parameters based on client-specific relevance.
7. A larger value of λ enforces stronger personalization.
8. Lottery ticket personalization relies on structural sparsity.
9. Excessive personalization can harm global generalization.
10. Personalized federated learning defines a continuum between global and local learning.

Answers:

1. False
2. False
3. True
4. False
5. False
6. True
7. False
8. True
9. True
10. True

D. Fill in the Blanks

Fill in the blanks with the most appropriate term or expression.

1. In personalized federated learning, each client may maintain its own model parameter vector ____________.
2. The global federated objective minimizes a weighted average of client-specific ____________.
3. The coupling parameter ____________ controls the strength of personalization.
4. Covariate shift refers to changes in the distribution ____________ across clients.
5. Sub-federated averaging partitions clients into ____________ groups.

6. Model pruning applies a client-specific binary __________ to the global model.
7. The lottery ticket hypothesis assumes the existence of sparse __________ within dense models.
8. Element-wise multiplication is denoted by the symbol __________.
9. Strong personalization may reduce population-level __________.
10. Concept shift occurs when __________ differs across clients.

Answers:

1. θ_k
2. empirical risks
3. λ
4. $P_k(x)$
5. clusters
6. mask
7. subnetworks
8. $\odot$
9. generalization
10. $P_k(y \mid x)$

E. Multiple Choice Questions (MCQs)

Choose the **most appropriate** answer.

1. Personalized federated learning primarily addresses:
   (a) communication efficiency
   (b) client heterogeneity
   (c) encryption overhead
   (d) model initialization
2. When $\lambda \to 0$ in a personalized objective, learning becomes:
   (a) fully global
   (b) clustered
   (c) fully local
   (d) centralized
3. Sub-federated averaging is most effective when heterogeneity is:
   (a) random
   (b) negligible
   (c) clustered
   (d) adversarial
4. Model pruning personalizes learning by:
   (a) changing data distributions
   (b) modifying collaboration patterns

   (c) removing irrelevant parameters
   (d) increasing model size
5. Lottery ticket personalization relies on:
   (a) dense retraining
   (b) structural sparsity
   (c) centralized fine-tuning
   (d) random initialization only
6. Concept shift is characterized by changes in:
   (a) $P(x)$
   (b) $P(y)$
   (c) $P(y \mid x)$
   (d) dataset size
7. The shared reference model in personalized FL primarily acts as a:
   (a) final deployment model
   (b) prior or anchor
   (c) local optimizer
   (d) compression target
8. Excessive personalization may lead to:
   (a) improved robustness
   (b) negative transfer
   (c) reduced collaboration benefits
   (d) faster convergence always
9. Sub-federated averaging increases:
   (a) model uniformity
   (b) system complexity
   (c) data sharing
   (d) privacy risk
10. Personalized federated learning can be viewed as a form of:
   (a) single-task learning
   (b) centralized optimization
   (c) multi-task learning
   (d) unsupervised learning

Answers:

1. (b)
2. (c)
3. (c)
4. (c)
5. (b)
6. (c)
7. (b)
8. (c)
9. (b)
10. (c)

F. Application and Project-Based Questions

These problems emphasize modeling, reasoning, and implementation.

1. Design a personalized federated learning objective for a mobile keyboard prediction system.
2. Implement a simple sub-federated averaging scheme on a synthetic non-IID dataset and analyze results.
3. Compare global and personalized models on a healthcare dataset with label imbalance.
4. Simulate model pruning for personalization on a small neural network.
5. Design a clustering strategy for sub-federated averaging under privacy constraints.
6. Implement lottery ticket masking for two heterogeneous clients and compare performance.
7. Study the effect of varying λ on personalization gain.
8. Analyze communication overhead under pruning-based personalization.
9. Propose evaluation metrics suitable for personalized federated learning.
10. Design a hybrid personalization framework combining clustering, regularization, and pruning.

Heterogeneous Federated Learning 13

The theoretical development of federated learning in earlier chapters has largely relied on a simplifying abstraction: that all participating clients are broadly comparable in their computational capabilities, communication resources, and model representations. Under this abstraction, clients are assumed to train identical models, perform a fixed number of local optimization steps, and exchange updates of uniform structure and dimensionality. Such assumptions enable clean mathematical formulations and tractable convergence analyses, and they serve as a useful starting point for understanding the core principles of decentralized learning under data locality constraints.

However, as federated learning transitions from controlled experimental settings to large-scale real-world deployments, these assumptions become increasingly fragile. Practical federated systems operate in environments characterized by extreme diversity and limited control. Participants may include mobile phones with fluctuating battery levels, edge devices with restricted memory, industrial controllers with real-time constraints, intermittently connected IoT nodes, and powerful servers with abundant computational resources. Even within a single application domain, client populations are rarely uniform. Hardware configurations, operating conditions, network quality, and user behavior vary continuously over time, producing a system landscape that is fundamentally heterogeneous.

This heterogeneity manifests along multiple dimensions. At the system level, clients differ in processing speed, memory capacity, energy availability, and communication bandwidth. Some clients may be capable of performing extensive local training with large batch sizes and deep models, while others may only support shallow models or a small number of gradient steps per communication round. At the network level, clients experience heterogeneous latency, bandwidth, and reliability, leading to asynchronous participation and frequent dropouts. At the data

P. P. Ray, *Foundations of Federated Learning*,
Springer Series in the Data Sciences,
https://doi.org/10.1007/978-3-032-34084-9_13

level, heterogeneity compounds these challenges by introducing non-identically distributed data, unequal dataset sizes, and client-specific noise characteristics, further amplifying divergence in local optimization trajectories.

Crucially, heterogeneity also arises at the level of model architecture itself. The assumption that a single, globally shared model architecture can be trained and deployed uniformly across all clients is often untenable. Resource-constrained devices may be unable to store or train large neural networks, while more capable clients may be underutilized by overly simplified models. Enforcing architectural uniformity can therefore lead to inefficiencies, system bottlenecks, and the systematic exclusion of weaker clients from meaningful participation. In extreme cases, this undermines the very objective of federated learning—to enable broad, collaborative model training without requiring centralized data access.

These realities expose a fundamental limitation of homogeneous federated learning. Algorithms designed under uniform assumptions may converge slowly, exhibit unstable behavior, or fail entirely when deployed in heterogeneous environments. Straggler effects, biased aggregation, communication bottlenecks, and resource exhaustion are not peripheral issues; they are central challenges that shape the feasibility and performance of federated systems at scale. As a result, heterogeneity can no longer be treated as a secondary complication to be smoothed over through heuristics or engineering patches.

This chapter introduces *heterogeneous federated learning* as a principled extension of the federated learning paradigm that explicitly embraces diversity across clients. Rather than attempting to coerce all participants into a uniform mold, heterogeneous federated learning recognizes client inequality as an intrinsic property of decentralized systems. Differences in computation, communication, and model capacity are incorporated directly into the learning framework, reshaping both the optimization objective and the system architecture.

From a mathematical perspective, heterogeneity challenges the standard federated optimization formulation by breaking assumptions of identical parameter spaces, uniform update fidelity, and synchronized progress. Client-wise optimization trajectories may evolve at different speeds, operate over different model subspaces, and contribute updates of varying reliability and granularity. Aggregation must therefore reconcile not only statistical heterogeneity in data, but also structural heterogeneity in models and system constraints. Understanding how these factors interact is essential for analyzing convergence, stability, and fairness in heterogeneous federated systems.

This chapter systematically develops the foundations of heterogeneous federated learning. We begin by examining heterogeneity in computation and communication and its impact on local training dynamics and update aggregation. We then explore federated learning with heterogeneous model architectures, highlighting mechanisms that enable collaboration among clients with different model capacities. Building on these ideas, we introduce the HeteroFL framework, which formalizes structured parameter sharing across variable-sized models. The chapter further investigates dynamic model scaling and partial participation as adaptive responses to time-varying resource availability. Finally, we study resource-aware federated training strategies that explicitly integrate system constraints into the learning objective.

By extending federated learning beyond homogeneous assumptions, this chapter positions heterogeneity not as an obstacle to be eliminated, but as a design axis to be leveraged. The resulting framework enables scalable, inclusive, and robust federated learning across diverse and resource-constrained environments, and serves as a critical bridge to hybrid federated learning paradigms that combine architectural, data, and optimization heterogeneity in the chapters that follow.

13.1 Heterogeneity in Computation and Communication

Early formulations of federated learning implicitly assume that all participating clients are broadly similar in their ability to compute local updates and communicate with the coordinating server. Under this assumption, each selected client performs an identical number of local optimization steps, trains a model of the same size, and transmits updates of uniform structure and dimensionality. While such homogeneity simplifies algorithm design and theoretical analysis, it rarely holds in realistic federated deployments.

In practical systems, federated learning spans a wide spectrum of devices and environments. Clients may include mobile phones with limited battery power, embedded sensors with minimal memory, edge gateways with moderate computational capacity, and cloud-backed servers with abundant resources. Network conditions are equally diverse: some clients enjoy stable, high-bandwidth connectivity, while others operate over intermittent, low-bandwidth, or high-latency links. As a result, both local computation and communication become inherently heterogeneous, fundamentally altering the learning dynamics.

Heterogeneity in computation and communication introduces asymmetries that directly affect how much progress each client can make in a given communication round and how costly it is for the system to incorporate that progress. These asymmetries break the implicit assumption that all client updates are equally informative, equally reliable, and equally expensive. Understanding their effects is therefore essential for designing scalable and robust federated learning algorithms.

13.1.1 Computational Heterogeneity

Computational heterogeneity refers to differences among clients in their ability to perform local optimization. Such differences arise from variations in processor speed, number of cores, memory capacity, storage, and available energy. For example, a high-end smartphone connected to a power source can perform significantly more computation than a battery-powered IoT sensor operating under strict energy constraints.

To model this variability, let $c_k > 0$ denote the effective computational capacity of client k. This quantity may abstract multiple factors, including clock speed, parallelism, and energy availability. Given a fixed communication round duration, the number of local optimization steps that client k can feasibly perform, denoted by τ_k, depends on c_k:

$$\tau_k \propto c_k.$$

Consequently, clients perform differing amounts of local training even when initialized from the same global model $\theta^{(t)}$. After local optimization, client k obtains a parameter vector $\theta_k^{(t,\tau_k)}$ that reflects both its data distribution and the extent of its computational effort. Clients with larger τ_k can more closely minimize their local objective $F_k(\theta)$, while clients with smaller τ_k may produce under-trained updates.

This disparity has important implications. First, the quality of local updates becomes uneven: updates from high-capacity clients may represent near-optimal descent directions for their local objectives, whereas updates from low-capacity clients may be noisy or incomplete. Second, increased local computation can amplify *local drift* under non-IID data, causing heavily trained clients to move farther away from the global optimum. Thus, computational heterogeneity introduces a trade-off between update accuracy and alignment with the global objective.

From an optimization perspective, federated learning no longer approximates centralized empirical risk minimization with uniform local steps. Instead, it becomes a weighted interaction among optimizers operating at different depths and time scales. Aggregation strategies must therefore account not only for dataset size n_k, but also for the computational effort invested by each client.

Worked Example: Computational Heterogeneity and Unequal Local Training
We now present a fully worked numerical example to illustrate how computational heterogeneity leads to unequal local optimization progress across clients, even when all clients start from the same global model.

Problem Setup. Consider a federated learning system with $K = 4$ clients. We assume a scalar linear prediction model

$$f_\theta(x) = \theta x, \quad \theta \in \mathbb{R},$$

and squared-error loss

$$\ell(f_\theta(x), y) = (\theta x - y)^2.$$

Each client k holds a local dataset $\mathcal{D}_k$ and has a distinct computational capacity c_k.

Table 13.1 Client computational capacities and feasible local steps

Client k	c_k (relative units)	τ_k (local steps)
1	1.0	1
2	2.0	2
3	3.0	3
4	4.0	4

Step 1: Client Computational Capacities. Assume the effective computational capacities shown in Table 13.1.

We assume

$$\tau_k = c_k,$$

so higher-capacity clients perform more local optimization.

Step 2: Local Datasets. All clients observe the same input values but with different label distributions (non-IID):

Client k	x	y
1	$\{1, 2\}$	$\{2, 4\}$
2	$\{1, 2\}$	$\{1, 2\}$
3	$\{1, 2\}$	$\{3, 6\}$
4	$\{1, 2\}$	$\{4, 8\}$

Step 3: Initial Global Model. The server broadcasts

$$\theta^{(t)} = 1.0.$$

Each client initializes

$$\theta_k^{(t,0)} = 1.0.$$

Step 4: Gradient Computation. For a single data point (x, y),

$$\frac{\partial \ell}{\partial \theta} = 2x(\theta x - y).$$

We compute the average gradient over each local dataset.

Client 1.

$$\begin{aligned}\nabla F_1(\theta) &= \frac{1}{2}\Big[2(1)(\theta-2)+2(2)(2\theta-4)\Big]\\ &= \frac{1}{2}\big[2\theta-4+4\theta-8\big]\\ &= \frac{1}{2}(6\theta-12)=3\theta-6.\end{aligned}$$

At $\theta = 1$:

$$\nabla F_1(1) = -3.$$

Client 2.

$$\nabla F_2(\theta) = \frac{1}{2}\big[2(\theta-1)+4(2\theta-2)\big] = 3\theta-3.$$

At $\theta = 1$:

$$\nabla F_2(1) = 0.$$

Client 3.

$$\nabla F_3(\theta) = \frac{1}{2}\big[2(\theta-3)+4(2\theta-6)\big] = 3\theta-9.$$

At $\theta = 1$:

$$\nabla F_3(1) = -6.$$

Client 4.

$$\nabla F_4(\theta) = \frac{1}{2}\big[2(\theta-4)+4(2\theta-8)\big] = 3\theta-12.$$

At $\theta = 1$:

$$\nabla F_4(1) = -9.$$

Step 5: Local Gradient Descent Updates. Assume learning rate $\eta = 0.1$.

Client 1 ($\tau_1 = 1$).

$$\theta_1^{(t,1)} = 1.0 - 0.1(-3) = 1.3.$$

Client 2 ($\tau_2 = 2$). Gradient is zero at $\theta = 1$, so:

$$\theta_2^{(t,2)} = 1.0.$$

Client 3 ($\tau_3 = 3$).

$$\theta_3^{(t,1)} = 1.0 - 0.1(-6) = 1.6.$$

$$\theta_3^{(t,2)} = 1.6 - 0.1(3 \cdot 1.6 - 9) = 2.02.$$

$$\theta_3^{(t,3)} = 2.02 - 0.1(3 \cdot 2.02 - 9) = 2.314.$$

Client 4 ($\tau_4 = 4$).

$$\theta_4^{(t,1)} = 1.9, \quad \theta_4^{(t,2)} = 2.33, \quad \theta_4^{(t,3)} = 2.631, \quad \theta_4^{(t,4)} = 2.842.$$

Step 6: Interpretation. Despite identical initialization, higher-capacity clients move much farther in parameter space. Computational heterogeneity therefore produces updates of unequal magnitude and alignment, even before aggregation.

13.1.2 Communication Heterogeneity

Communication heterogeneity arises from variability in network bandwidth, latency, reliability, and transmission cost across clients. In federated learning, communication is often the dominant bottleneck, and differences in communication capability directly shape which clients can participate effectively and how frequently updates can be exchanged.

Let $b_k > 0$ denote the effective uplink bandwidth available to client k. This bandwidth constrains both the size and frequency of transmitted updates. If client k communicates a model update $\Delta\theta_k^{(t)}$, its magnitude or dimensionality may be limited by a bandwidth-dependent constraint:

$$\|\Delta\theta_k^{(t)}\| \le \mathscr{C}(b_k),$$

where $\mathscr{C}(\cdot)$ captures compression, sparsification, or quantization mechanisms required to meet communication limits.

Clients with high bandwidth may transmit dense, high-precision updates, closely approximating full-parameter communication. In contrast, low-bandwidth clients may be forced to send aggressively compressed, sparse, or quantized updates, potentially discarding information. As a result, even if two clients perform identical local optimization, their communicated updates may differ significantly in fidelity.

Communication heterogeneity also affects participation patterns. Clients with unstable or slow connections may frequently drop out or miss communication deadlines, leading to partial and time-varying client sets $\mathscr{S}^{(t)}$. This introduces additional stochasticity into the aggregation process and complicates convergence analysis, which must now account for both statistical and system-induced randomness.

From a systems viewpoint, communication heterogeneity determines the cost of incorporating each client's contribution. From an optimization viewpoint, it alters the effective descent direction by selectively attenuating or amplifying certain updates. Thus, communication constraints are not merely engineering limitations; they fundamentally reshape the learning dynamics.

Key Observation

Heterogeneity in computation and communication invalidates the assumption that all client updates are equally informative, equally reliable, or equally costly to obtain.

The combined effects of computational and communication heterogeneity significantly influence convergence behavior, fairness across clients, and the stability of federated training. Naïvely aggregating updates without accounting for these disparities can lead to domination by high-capacity clients, marginalization of resource-constrained participants, or inefficient use of system resources. These challenges motivate the adaptive aggregation rules, dynamic participation strategies, and resource-aware training mechanisms developed in subsequent sections of this chapter.

Worked Example: Communication Heterogeneity and Update Compression Using a Tabular Dataset

We now present a fully self-contained numerical example to demonstrate how communication heterogeneity modifies client contributions through bandwidth-dependent constraints. All quantities are derived explicitly from a tabular dataset, and no results are assumed from prior examples.

Problem Setup. Consider a federated learning system with $K = 4$ clients. Each client trains a scalar linear model

$$f_\theta(x) = \theta x,$$

using local gradient descent and computes a model update relative to the global parameter $\theta^{(t)}$.

Step 1: Global Model Initialization. The server broadcasts the global model

$$\theta^{(t)} = 1.0.$$

Table 13.2 Local tabular datasets at each client

Client k	Inputs x	Targets y
1	$\{1, 2\}$	$\{2, 4\}$
2	$\{1, 2\}$	$\{1, 2\}$
3	$\{1, 2\}$	$\{3, 6\}$
4	$\{1, 2\}$	$\{4, 8\}$

Step 2: Local Tabular Dataset. Each client holds a local dataset consisting of input–output pairs (x, y), summarized in Table 13.2.

Each client computes the empirical risk

$$F_k(\theta) = \frac{1}{|\mathscr{D}_k|} \sum_{(x,y)\in\mathscr{D}_k} (\theta x - y)^2.$$

Step 3: Gradient Computation from the Table. For a single data point (x, y),

$$\frac{\partial}{\partial\theta}(\theta x - y)^2 = 2x(\theta x - y).$$

We compute the average gradient for each client.

Client 1.

$$\nabla F_1(\theta) = \frac{1}{2}\big[2(1)(\theta - 2) + 2(2)(2\theta - 4)\big]$$
$$= \frac{1}{2}(6\theta - 12) = 3\theta - 6.$$

At $\theta = 1$:

$$\nabla F_1(1) = -3.$$

Client 2.

$$\nabla F_2(\theta) = \frac{1}{2}\big[2(\theta - 1) + 4(2\theta - 2)\big] = 3\theta - 3.$$

At $\theta = 1$:

$$\nabla F_2(1) = 0.$$

Client 3.

$$\nabla F_3(\theta) = \frac{1}{2}\big[2(\theta - 3) + 4(2\theta - 6)\big] = 3\theta - 9.$$

At $\theta = 1$:

$$\nabla F_3(1) = -6.$$

Client 4.

$$\nabla F_4(\theta) = \frac{1}{2}\left[2(\theta - 4) + 4(2\theta - 8)\right] = 3\theta - 12.$$

At $\theta = 1$:

$$\nabla F_4(1) = -9.$$

Step 4: Local Model Updates. Assume each client performs a single gradient descent step with learning rate $\eta = 0.1$:

$$\theta_k^{(t,1)} = \theta^{(t)} - \eta \nabla F_k(\theta^{(t)}).$$

Client-wise updates.

$$\theta_1^{(t,1)} = 1 - 0.1(-3) = 1.3,$$

$$\theta_2^{(t,1)} = 1 - 0.1(0) = 1.0,$$

$$\theta_3^{(t,1)} = 1 - 0.1(-6) = 1.6,$$

$$\theta_4^{(t,1)} = 1 - 0.1(-9) = 1.9.$$

Step 5: Local Updates Before Communication. Each client computes its update relative to the global model:

$$\Delta\theta_k = \theta_k^{(t,1)} - \theta^{(t)}.$$

Thus,

$$\Delta\theta = \left[0.3,\ 0,\ 0.6,\ 0.9\right].$$

Step 6: Communication Bandwidth Constraints. Assume heterogeneous uplink bandwidths summarized in Table 13.3.

Table 13.3 Client bandwidth and allowable update magnitude

Client k	Bandwidth level	$\mathscr{C}(b_k)$
1	High	1.0
2	Medium	0.5
3	Low	0.3
4	Very Low	0.2

Each transmitted update must satisfy

$$|\Delta\theta_k^{\text{tx}}| \leq \mathscr{C}(b_k).$$

Step 7: Enforcing Communication Constraints.

Client 1.

$$0.3 \leq 1.0 \Rightarrow \Delta\theta_1^{\text{tx}} = 0.3.$$

Client 2.

$$0 \leq 0.5 \Rightarrow \Delta\theta_2^{\text{tx}} = 0.$$

Client 3.

$$0.6 > 0.3 \Rightarrow \Delta\theta_3^{\text{tx}} = 0.3.$$

Client 4.

$$0.9 > 0.2 \Rightarrow \Delta\theta_4^{\text{tx}} = 0.2.$$

Step 8: Transmitted Updates Received by the Server. The server receives the compressed update vector

$$\Delta\theta^{\text{tx}} = \left[0.3,\ 0,\ 0.3,\ 0.2\right].$$

Step 9: Interpretation. Although Clients 3 and 4 computed larger local updates due to stronger gradient signals, their contributions are truncated by communication constraints. Communication heterogeneity therefore reshapes the effective descent direction observed by the server, independently of the underlying data or local optimization quality.

Key Insight
Communication heterogeneity determines not how much clients learn locally, but how much of that learning is visible globally. When bandwidth is limited, large local improvements may be partially or entirely suppressed.

13.2 Heterogeneous Model Architectures

Heterogeneity in federated learning is not limited to differences in computation and communication. A deeper and more structurally significant form of heterogeneity arises from variation in the *model architectures* that clients are able to support. In realistic federated systems, participants operate under diverse hardware, memory, and energy constraints, which directly limit the size, depth, and complexity of models they can store, train, and deploy.

Low-resource devices such as mobile sensors, wearables, and embedded controllers may only accommodate shallow models with a small number of parameters. In contrast, edge gateways, workstations, or cloud-assisted clients may be capable of training deeper networks with richer representational capacity. Imposing a single, uniform model architecture across such a diverse client population can therefore be inefficient or even infeasible. Large models may exceed the memory or energy budgets of weak clients, while small models may underutilize the capacity of stronger clients, leading to suboptimal learning at the system level.

To formalize architectural heterogeneity, let $\mathscr{M}_k$ denote the model architecture employed by client k. Rather than sharing a common parameter vector $\boldsymbol{\theta} \in \mathbb{R}^d$, each client maintains its own parameterization

$$\boldsymbol{\theta}_k \in \mathbb{R}^{d_k}, \qquad d_k \neq d_j \text{ in general.}$$

Here, d_k reflects architectural choices such as the number of layers, hidden units, filters, or channels, and may vary substantially across clients.

This architectural diversity breaks one of the central assumptions of classical federated learning: that local model parameters can be directly aggregated in a common parameter space. The standard condition

$$\boldsymbol{\theta}_k = \boldsymbol{\theta}$$

no longer holds, and naive averaging of parameters becomes undefined. As a result, heterogeneity in model architecture introduces a structural misalignment that cannot be resolved by weighting schemes alone.

From an optimization perspective, each client now minimizes a local objective over a different parameter space:

$$\min_{\theta_k \in \mathbb{R}^{d_k}} F_k(\theta_k).$$

The server, however, must still coordinate learning toward a coherent global representation. This tension between local architectural freedom and global coordination lies at the core of heterogeneous federated learning with model diversity.

Submodel Consistency. A widely adopted strategy for addressing architectural heterogeneity is to enforce *submodel consistency* through partial parameter sharing. The key idea is to identify a subset of parameters that is common across all client architectures, while allowing the remaining parameters to be client-specific.

Let $\theta^{(s)}$ denote a shared parameter subset that is present in every client model. Each client's full parameter vector can then be decomposed as

$$\theta_k = (\theta^{(s)}, \theta_k^{(p)}),$$

where $\theta_k^{(p)}$ represents private parameters unique to client k. The shared parameters may correspond to common layers, core feature extractors, or canonical representations, while private parameters capture client-specific refinements or capacity extensions.

During federated training, updates to the shared parameters $\theta^{(s)}$ are aggregated across clients, while private parameters $\theta_k^{(p)}$ remain local and are never communicated. This design preserves collaboration on a common representation while respecting architectural constraints at the client level.

Submodel consistency introduces a structured form of parameter alignment. Rather than requiring full architectural homogeneity, it ensures that all clients participate in learning a common subspace, even if their full models differ in size or complexity. This approach enables scalable aggregation, reduces communication overhead, and prevents weaker clients from being excluded due to architectural limitations.

Architectural Heterogeneity
Heterogeneous federated learning relaxes the requirement of identical model architectures by decomposing models into shared and private components, enabling effective collaboration across devices with vastly different capabilities.

Worked Example: Heterogeneous Model Architectures with Submodel Consistency
We now present a fully worked numerical example illustrating how heterogeneous model architectures can be aligned through submodel consistency. All quantities

are derived explicitly from tabular data, and no results are assumed from earlier examples.

Problem Setup. Consider a federated learning system with $K = 3$ clients. Each client performs linear regression but with different model capacities. All clients share a common input feature x_1, while higher-capacity clients additionally use auxiliary features x_2 and x_3.

Step 1: Local Tabular Dataset. Each client observes samples from the same task, summarized in Table 13.4.

Client capabilities differ as follows:

$$\mathcal{M}_1 : \{x_1\}, \quad \mathcal{M}_2 : \{x_1, x_2\}, \quad \mathcal{M}_3 : \{x_1, x_2, x_3\},$$

where x_3 is an internally generated nonlinear feature $x_3 = x_1^2$ available only to Client 3.

Step 2: Client Model Architectures.

Client 1 (Low-capacity).

$$\hat{y}_1 = \theta^{(s)} x_1, \qquad \theta_1 \in \mathbb{R}^1.$$

Client 2 (Medium-capacity).

$$\hat{y}_2 = \theta^{(s)} x_1 + \theta_{2,1}^{(p)} x_2, \qquad \theta_2 \in \mathbb{R}^2.$$

Client 3 (High-capacity).

$$\hat{y}_3 = \theta^{(s)} x_1 + \theta_{3,1}^{(p)} x_2 + \theta_{3,2}^{(p)} x_3, \qquad \theta_3 \in \mathbb{R}^3.$$

Here, $\theta^{(s)}$ is the shared parameter across all clients.

Table 13.4 Local tabular dataset used by all clients

Sample	x_1	x_2	y
1	1	1	4
2	2	1	7
3	3	2	13

Step 3: Loss Function Definition. All clients minimize mean squared error:

$$F_k(\boldsymbol{\theta}_k) = \frac{1}{3}\sum_{i=1}^{3}(\hat{y}_{k,i} - y_i)^2.$$

Step 4: Explicit Loss Expansion.

Client 1.

$$\hat{y}_1 = \theta^{(s)} x_1.$$

Predictions:

$$[1\theta^{(s)},\ 2\theta^{(s)},\ 3\theta^{(s)}].$$

Loss:

$$F_1 = \frac{1}{3}\big[(\theta^{(s)} - 4)^2 + (2\theta^{(s)} - 7)^2 + (3\theta^{(s)} - 13)^2\big].$$

Derivative:

$$\frac{\partial F_1}{\partial \theta^{(s)}} = \frac{2}{3}\big[(\theta^{(s)} - 4) + 2(2\theta^{(s)} - 7) + 3(3\theta^{(s)} - 13)\big].$$

Simplifying:

$$\frac{\partial F_1}{\partial \theta^{(s)}} = \frac{2}{3}(14\theta^{(s)} - 57).$$

Client 2.

$$\hat{y}_2 = \theta^{(s)} x_1 + \theta^{(p)}_{2,1} x_2.$$

Predictions:

$$[\theta^{(s)} + \theta^{(p)}_{2,1},\ 2\theta^{(s)} + \theta^{(p)}_{2,1},\ 3\theta^{(s)} + 2\theta^{(p)}_{2,1}].$$

Loss gradients:

$$\frac{\partial F_2}{\partial \theta^{(s)}} = \frac{2}{3}\big[(\hat{y}_{2,1} - 4) + 2(\hat{y}_{2,2} - 7) + 3(\hat{y}_{2,3} - 13)\big].$$

Client 3.

$$\hat{y}_3 = \theta^{(s)} x_1 + \theta^{(p)}_{3,1} x_2 + \theta^{(p)}_{3,2} x_1^2.$$

Predictions:

$$[\theta^{(s)} + \theta^{(p)}_{3,1} + \theta^{(p)}_{3,2},\ 2\theta^{(s)} + \theta^{(p)}_{3,1} + 4\theta^{(p)}_{3,2},\ 3\theta^{(s)} + 2\theta^{(p)}_{3,1} + 9\theta^{(p)}_{3,2}].$$

Gradients are computed analogously and yield a system of linear equations in $\theta^{(s)}, \theta^{(p)}_{3,1}, \theta^{(p)}_{3,2}$.

Step 5: One-step Gradient Descent Update. Assume learning rate $\eta = 0.01$ and initialization

$$\theta^{(s)} = 1, \quad \theta^{(p)}_{k,\cdot} = 0.$$

Shared parameter updates.

$$\Delta\theta^{(s)}_1 = -0.01 \cdot \frac{2}{3}(14 - 57) = 0.286.$$

Similarly computed:

$$\Delta\theta^{(s)}_2 = 0.214, \qquad \Delta\theta^{(s)}_3 = 0.198.$$

Step 6: Server-side Aggregation of Shared Parameters. Assume equal dataset sizes. The server aggregates:

$$\Delta\theta^{(s)} = \frac{1}{3}(0.286 + 0.214 + 0.198) = 0.2327.$$

Thus, the updated shared parameter is

$$\theta^{(s)}_{\text{new}} = 1 + 0.2327 = 1.2327.$$

Step 7: Interpretation. Although clients use different model architectures and parameter dimensions, all contribute coherently to the shared parameter $\theta^{(s)}$. Private parameters absorb client-specific complexity, while submodel consistency enables meaningful global aggregation.

Key Insight
Architectural heterogeneity does not preclude federated learning. By decomposing models into shared and private components, clients with different capacities can collaborate while preserving mathematical consistency and aggregation validity.

13.3 HeteroFL Framework

Heterogeneous Federated Learning (HeteroFL) provides a principled and mathematically coherent framework for federated training when clients operate under heterogeneous architectural and resource constraints. Rather than enforcing a single, uniform model across all participants, HeteroFL explicitly allows clients to train models of different sizes while preserving a shared global learning objective. This design reflects the practical reality of federated systems, where devices vary widely in memory capacity, computational power, and energy availability.

The central idea of HeteroFL is to decouple the *global reference model* maintained by the server from the *client-specific submodels* used for local training. The global model represents the most expressive parameterization that the system can support collectively, while each client interacts with a restricted view of this model that matches its local capabilities. In doing so, HeteroFL preserves the benefits of collaboration across clients without forcing architectural uniformity.

Let the global reference model be parameterized by

$$\theta \in \mathbb{R}^D,$$

where D corresponds to the full model capacity available at the server. Client k does not operate directly on θ. Instead, it trains a reduced submodel

$$\theta_k \in \mathbb{R}^{d_k}, \qquad d_k < D,$$

obtained through a client-specific *projection operator* $\mathscr{P}_k$:

$$\theta_k = \mathscr{P}_k(\theta).$$

The projection operator $\mathscr{P}_k : \mathbb{R}^D \to \mathbb{R}^{d_k}$ defines how the global model is mapped onto the parameter space accessible to client k. In practice, $\mathscr{P}_k$ may correspond to selecting a subset of layers, channels, neurons, or filters, or to applying structured dimensionality reduction. Crucially, $\mathscr{P}_k$ is not required to be identical across clients; each client may use a distinct projection reflecting its resource constraints.

Given the projected parameters $\boldsymbol{\theta}_k$, client k performs local optimization by minimizing its local objective function

$$F_k(\boldsymbol{\theta}_k) = \frac{1}{n_k} \sum_{(x,y)\in\mathcal{D}_k} \ell(f_{\boldsymbol{\theta}_k}(x), y),$$

where $\mathcal{D}_k$ is the local dataset and $n_k = |\mathcal{D}_k|$. Local training proceeds using standard optimization methods, such as stochastic gradient descent, but in the reduced parameter space $\mathbb{R}^{d_k}$.

After local optimization, client k produces an update

$$\Delta\boldsymbol{\theta}_k = \boldsymbol{\theta}_k^{(t+1)} - \boldsymbol{\theta}_k^{(t)},$$

which resides in the subspace defined by its projection. To integrate this update into the global model, HeteroFL employs a client-specific *embedding operator*

$$\mathcal{E}_k : \mathbb{R}^{d_k} \to \mathbb{R}^D.$$

The role of $\mathcal{E}_k$ is to lift the low-dimensional update $\Delta\boldsymbol{\theta}_k$ back into the full parameter space of the global model in a structured and consistent manner.

The server aggregates the embedded updates from the participating client set $\mathcal{S}^{(t)}$ using a weighted combination:

$$\Delta\boldsymbol{\theta} = \sum_{k\in\mathcal{S}^{(t)}} \frac{n_k}{\sum_{j\in\mathcal{S}^{(t)}} n_j} \mathcal{E}_k(\Delta\boldsymbol{\theta}_k).$$

This aggregation rule generalizes the standard federated averaging scheme by operating in the full parameter space while incorporating updates computed in heterogeneous subspaces. The weighting by dataset size n_k preserves the statistical interpretation of the global objective as an empirical risk minimization problem over the union of all local datasets.

From an optimization standpoint, HeteroFL can be interpreted as a form of constrained federated optimization in which each client optimizes over a projection of the global parameter space. The combination of projection and embedding operators ensures that all clients contribute to a common global model, even though they interact with different subsets of parameters. This structure prevents weaker clients from being excluded due to architectural limitations, while still allowing stronger clients to leverage larger models.

HeteroFL Principle

HeteroFL enables federated learning across heterogeneous model architectures by coordinating client-specific submodels through structured projection into, and embedding from, a shared global parameter space.

Worked Example: HeteroFL with Heterogeneous Submodels

We now present a complete numerical example illustrating how the HeteroFL framework enables federated optimization when clients train models of different sizes. The example explicitly demonstrates projection, local optimization, embedding, and global aggregation without skipping any intermediate calculations.

Problem Setup. Consider a federated system with $K = 4$ clients. The server maintains a global linear model with parameter vector

$$\boldsymbol{\theta} = \begin{bmatrix} \theta_1 \\ \theta_2 \\ \theta_3 \\ \theta_4 \\ \theta_5 \\ \theta_6 \end{bmatrix} \in \mathbb{R}^6.$$

This full model is assumed to be too large for all clients to train locally.

Each client k trains a submodel of dimension $d_k < 6$ obtained via a projection operator $\mathscr{P}_k$.

Global Model Initialization. At communication round t, the server initializes

$$\boldsymbol{\theta}^{(t)} = \begin{bmatrix} 1.0 \\ 0.8 \\ 0.6 \\ 0.4 \\ 0.2 \\ 0.0 \end{bmatrix}.$$

Client Model Capacities. Assume the heterogeneous model capacities shown in Table 13.5.

Step 1: Projection to Client Submodels. Each client applies its projection operator $\mathscr{P}_k$.

Table 13.5 Client model capacities and parameter subsets

Client k	d_k	Trained parameters
1	6	$\{\theta_1, \theta_2, \theta_3, \theta_4, \theta_5, \theta_6\}$
2	4	$\{\theta_1, \theta_2, \theta_3, \theta_4\}$
3	3	$\{\theta_1, \theta_2, \theta_3\}$
4	2	$\{\theta_1, \theta_2\}$

Client 1 (full model).

$$\theta_1^{(t)} = \begin{bmatrix} 1.0 \\ 0.8 \\ 0.6 \\ 0.4 \\ 0.2 \\ 0.0 \end{bmatrix}.$$

Client 2 (first 4 parameters).

$$\theta_2^{(t)} = \begin{bmatrix} 1.0 \\ 0.8 \\ 0.6 \\ 0.4 \end{bmatrix}.$$

Client 3 (first 3 parameters).

$$\theta_3^{(t)} = \begin{bmatrix} 1.0 \\ 0.8 \\ 0.6 \end{bmatrix}.$$

Client 4 (first 2 parameters).

$$\theta_4^{(t)} = \begin{bmatrix} 1.0 \\ 0.8 \end{bmatrix}.$$

Step 2: Local Optimization on Submodels. Assume each client performs local training and obtains the updated parameters shown in Table 13.6.

Step 3: Compute Local Parameter Updates.

Client 1.

$$\Delta\theta_1 = \begin{bmatrix} 0.1 \\ 0.1 \\ 0.1 \\ 0.1 \\ 0.1 \\ 0.1 \end{bmatrix}.$$

Table 13.6 Locally optimized submodel parameters

Client k	$\theta_k^{(t+1)}$
1	$\begin{bmatrix} 1.1 \\ 0.9 \\ 0.7 \\ 0.5 \\ 0.3 \\ 0.1 \end{bmatrix}$
2	$\begin{bmatrix} 1.05 \\ 0.85 \\ 0.65 \\ 0.45 \end{bmatrix}$
3	$\begin{bmatrix} 1.2 \\ 1.0 \\ 0.8 \end{bmatrix}$
4	$\begin{bmatrix} 0.9 \\ 0.7 \end{bmatrix}$

Client 2.

$$\Delta\theta_2 = \begin{bmatrix} 0.05 \\ 0.05 \\ 0.05 \\ 0.05 \end{bmatrix}.$$

Client 3.

$$\Delta\theta_3 = \begin{bmatrix} 0.2 \\ 0.2 \\ 0.2 \end{bmatrix}.$$

Client 4.

$$\Delta\theta_4 = \begin{bmatrix} -0.1 \\ -0.1 \end{bmatrix}.$$

Step 4: Embedding into Global Parameter Space. Each client applies an embedding operator $\mathscr{E}_k$ that places its update into the corresponding coordinates of $\mathbb{R}^6$, with zeros elsewhere.

Client 1.

$$\mathcal{E}_1(\Delta\boldsymbol{\theta}_1) = \begin{bmatrix} 0.1 \\ 0.1 \\ 0.1 \\ 0.1 \\ 0.1 \\ 0.1 \end{bmatrix}.$$

Client 2.

$$\mathcal{E}_2(\Delta\boldsymbol{\theta}_2) = \begin{bmatrix} 0.05 \\ 0.05 \\ 0.05 \\ 0.05 \\ 0 \\ 0 \end{bmatrix}.$$

Client 3.

$$\mathcal{E}_3(\Delta\boldsymbol{\theta}_3) = \begin{bmatrix} 0.2 \\ 0.2 \\ 0.2 \\ 0 \\ 0 \\ 0 \end{bmatrix}.$$

Client 4.

$$\mathcal{E}_4(\Delta\boldsymbol{\theta}_4) = \begin{bmatrix} -0.1 \\ -0.1 \\ 0 \\ 0 \\ 0 \\ 0 \end{bmatrix}.$$

Step 5: Dataset Sizes and Aggregation Weights. Assume dataset sizes:

$$n_1 = 40, \quad n_2 = 30, \quad n_3 = 20, \quad n_4 = 10, \qquad N = 100.$$

The corresponding weights are:

$$0.40,\ 0.30,\ 0.20,\ 0.10.$$

Step 6: Weighted Aggregation. The global update is:

$$\Delta\boldsymbol{\theta} = \sum_{k=1}^{4} \frac{n_k}{N} \mathscr{E}_k(\Delta\boldsymbol{\theta}_k).$$

Weighted contributions.

$$0.40 \cdot \mathscr{E}_1 = \begin{bmatrix} 0.04 \\ 0.04 \\ 0.04 \\ 0.04 \\ 0.04 \\ 0.04 \end{bmatrix},$$

$$0.30 \cdot \mathscr{E}_2 = \begin{bmatrix} 0.015 \\ 0.015 \\ 0.015 \\ 0.015 \\ 0 \\ 0 \end{bmatrix},$$

$$0.20 \cdot \mathscr{E}_3 = \begin{bmatrix} 0.04 \\ 0.04 \\ 0.04 \\ 0 \\ 0 \\ 0 \end{bmatrix},$$

$$0.10 \cdot \mathscr{E}_4 = \begin{bmatrix} -0.01 \\ -0.01 \\ 0 \\ 0 \\ 0 \\ 0 \end{bmatrix}.$$

Summation.

$$\Delta\theta = \begin{bmatrix} 0.085 \\ 0.085 \\ 0.095 \\ 0.055 \\ 0.04 \\ 0.04 \end{bmatrix}.$$

Step 7: Global Model Update.

$$\theta^{(t+1)} = \theta^{(t)} + \Delta\theta = \begin{bmatrix} 1.085 \\ 0.885 \\ 0.695 \\ 0.455 \\ 0.24 \\ 0.04 \end{bmatrix}.$$

Interpretation. Although Clients 3 and 4 train much smaller models, their updates still influence the global parameters they can access. Parameters absent from a client's submodel receive no update from that client, but are refined through other participants. This illustrates how HeteroFL preserves collaboration across heterogeneous architectures without forcing uniform model capacity.

Key Insight from the Worked Example
HeteroFL enables clients with heterogeneous model sizes to contribute meaningfully to a shared global model through structured projection, local optimization in reduced spaces, and principled embedding-based aggregation.

13.4 Dynamic Model Scaling and Partial Participation

In heterogeneous federated learning environments, client capabilities are neither static nor predictable. Unlike idealized settings in which all participants are assumed to be continuously available and uniformly resourced, real-world federated systems operate under fluctuating conditions. A client that is well-resourced in one communication round may become constrained in the next due to changes in battery level, competing workloads, network congestion, or thermal limitations. Consequently, a static assignment of model size, computational load, or participation frequency is inherently suboptimal.

The dynamic model scaling and partial participation address this challenge by allowing both the *extent of local computation* and the *frequency of participation* to vary adaptively over time. Rather than treating heterogeneity as an obstacle to be minimized, this approach treats it as a signal that guides how each client engages with the federated learning process. The result is a more flexible and resilient training procedure that preserves global learning objectives while respecting local resource constraints.

13.4.1 Dynamic Scaling

Dynamic model scaling allows each client to adjust the size of the model it trains at each communication round. Let D denote the dimensionality of the full global reference model maintained by the server. For client k at round t, define a *scaling ratio*

$$r_k^{(t)} \in (0, 1],$$

which determines the fraction of the global model that the client is able to train during that round. The effective model dimension available to client k is then given by

$$d_k^{(t)} = r_k^{(t)} D.$$

When $r_k^{(t)} = 1$, the client trains the full global model, behaving as a high-capacity participant. When $r_k^{(t)} \ll 1$, the client operates on a significantly reduced submodel, interacting with only a small subset of parameters. Importantly, $r_k^{(t)}$ is allowed to vary across clients and across communication rounds, reflecting the dynamic nature of resource availability.

In practice, the scaling ratio $r_k^{(t)}$ may be determined by a resource-aware policy that accounts for factors such as:

- remaining battery energy,
- available memory and compute budget,
- current network bandwidth and latency,
- local workload and thermal constraints.

Mathematically, this can be expressed as

$$r_k^{(t)} = \phi\left(\text{resources}_k^{(t)}\right),$$

where $\phi(\cdot)$ is a monotone function mapping resource availability to a feasible model fraction.

Dynamic scaling integrates naturally with the HeteroFL framework. At each round, the projection operator $\mathscr{P}_k^{(t)}$ selects a subset of parameters of size $d_k^{(t)}$, while the corresponding embedding operator $\mathscr{E}_k^{(t)}$ lifts the resulting update back into the global parameter space. As a result, the global optimization objective remains unchanged, even though the effective dimensionality of local optimization varies over time.

Dynamic Scaling Principle
Dynamic model scaling allows clients to adapt the size of their local models at each communication round, enabling participation under fluctuating resource constraints without violating the global federated optimization objective.

Worked Example: Dynamic Model Scaling Across Communication Rounds
We now present a complete numerical example illustrating how *dynamic model scaling* operates within a heterogeneous federated learning system. The example explicitly demonstrates how time-varying resource availability alters model dimensionality, local updates, and global aggregation, while preserving a consistent global optimization objective.

Problem Setup. Assume a federated system with $K = 4$ clients and a global reference model of dimension

$$D = 10.$$

The global parameter vector at communication round t is

$$\boldsymbol{\theta}^{(t)} = \begin{bmatrix} \theta_1 & \theta_2 & \theta_3 & \theta_4 & \theta_5 & \theta_6 & \theta_7 & \theta_8 & \theta_9 & \theta_{10} \end{bmatrix}^{\mathsf{T}}.$$

At the beginning of round t, the server initializes:

$$\boldsymbol{\theta}^{(t)} = \begin{bmatrix} 1.0 \\ 0.9 \\ 0.8 \\ 0.7 \\ 0.6 \\ 0.5 \\ 0.4 \\ 0.3 \\ 0.2 \\ 0.1 \end{bmatrix}.$$

Table 13.7 Resource availability and scaling ratios at round t

Client k	Battery (%)	Bandwidth	$r_k^{(t)}$
1	90	High	1.0
2	70	Medium	0.7
3	40	Low	0.5
4	20	Very Low	0.3

Step 1: Resource Availability and Scaling Ratios. Assume that each client experiences different resource conditions at round t, summarized in Table 13.7.

Step 2: Effective Model Dimensions. Using

$$d_k^{(t)} = r_k^{(t)} D,$$

we compute the effective model sizes.

Client 1.

$$d_1^{(t)} = 1.0 \times 10 = 10.$$

Client 2.

$$d_2^{(t)} = 0.7 \times 10 = 7.$$

Client 3.

$$d_3^{(t)} = 0.5 \times 10 = 5.$$

Client 4.

$$d_4^{(t)} = 0.3 \times 10 = 3.$$

Step 3: Projection to Dynamically Scaled Submodels. Each client applies a projection operator $\mathscr{P}_k^{(t)}$ that selects the first $d_k^{(t)}$ parameters.

Client 1.

$$\theta_1^{(t)} = \begin{bmatrix} 1.0 \\ 0.9 \\ 0.8 \\ 0.7 \\ 0.6 \\ 0.5 \\ 0.4 \\ 0.3 \\ 0.2 \\ 0.1 \end{bmatrix}.$$

Client 2.

$$\theta_2^{(t)} = \begin{bmatrix} 1.0 \\ 0.9 \\ 0.8 \\ 0.7 \\ 0.6 \\ 0.5 \\ 0.4 \end{bmatrix}.$$

Client 3.

$$\theta_3^{(t)} = \begin{bmatrix} 1.0 \\ 0.9 \\ 0.8 \\ 0.7 \\ 0.6 \end{bmatrix}.$$

Client 4.

$$\theta_4^{(t)} = \begin{bmatrix} 1.0 \\ 0.9 \\ 0.8 \end{bmatrix}.$$

Step 4: Local Optimization on Scaled Models. Assume each client performs local training and obtains the updated parameters shown in Table 13.8.

Table 13.8 Locally trained parameters after dynamic scaling

Client k	$\theta_k^{(t+1)}$
1	$\begin{bmatrix} 1.10 \\ 1.00 \\ 0.90 \\ 0.80 \\ 0.70 \\ 0.60 \\ 0.50 \\ 0.40 \\ 0.30 \\ 0.20 \end{bmatrix}$
2	$\begin{bmatrix} 1.05 \\ 0.95 \\ 0.85 \\ 0.75 \\ 0.65 \\ 0.55 \\ 0.45 \end{bmatrix}$
3	$\begin{bmatrix} 1.15 \\ 1.05 \\ 0.95 \\ 0.85 \\ 0.75 \end{bmatrix}$
4	$\begin{bmatrix} 0.95 \\ 0.85 \\ 0.75 \end{bmatrix}$

Step 5: Compute Local Updates.

Client 1.

$$\Delta\theta_1 = \begin{bmatrix} 0.10 \\ 0.10 \\ 0.10 \\ 0.10 \\ 0.10 \\ 0.10 \\ 0.10 \\ 0.10 \\ 0.10 \\ 0.10 \end{bmatrix}.$$

Client 2.

$$\Delta\boldsymbol{\theta}_2 = \begin{bmatrix} 0.05 \\ 0.05 \\ 0.05 \\ 0.05 \\ 0.05 \\ 0.05 \\ 0.05 \end{bmatrix}.$$

Client 3.

$$\Delta\boldsymbol{\theta}_3 = \begin{bmatrix} 0.15 \\ 0.15 \\ 0.15 \\ 0.15 \\ 0.15 \end{bmatrix}.$$

Client 4.

$$\Delta\boldsymbol{\theta}_4 = \begin{bmatrix} -0.05 \\ -0.05 \\ -0.05 \end{bmatrix}.$$

Step 6: Embedding into Global Space. Each update is embedded back into $\mathbb{R}^{10}$ using $\mathcal{E}_k^{(t)}$.

Client 1.

$$\mathcal{E}_1(\Delta\boldsymbol{\theta}_1) = \begin{bmatrix} 0.10 \\ 0.10 \\ 0.10 \\ 0.10 \\ 0.10 \\ 0.10 \\ 0.10 \\ 0.10 \\ 0.10 \\ 0.10 \end{bmatrix}.$$

Client 2.

$$\mathcal{E}_2(\Delta\boldsymbol{\theta}_2) = \begin{bmatrix} 0.05 \\ 0.05 \\ 0.05 \\ 0.05 \\ 0.05 \\ 0.05 \\ 0.05 \\ 0 \\ 0 \\ 0 \end{bmatrix}.$$

Client 3.

$$\mathcal{E}_3(\Delta\boldsymbol{\theta}_3) = \begin{bmatrix} 0.15 \\ 0.15 \\ 0.15 \\ 0.15 \\ 0.15 \\ 0 \\ 0 \\ 0 \\ 0 \\ 0 \end{bmatrix}.$$

Client 4.

$$\mathcal{E}_4(\Delta\boldsymbol{\theta}_4) = \begin{bmatrix} -0.05 \\ -0.05 \\ -0.05 \\ 0 \\ 0 \\ 0 \\ 0 \\ 0 \\ 0 \\ 0 \end{bmatrix}.$$

Step 7: Aggregation Weights. Assume dataset sizes:

$$n_1 = 40, \quad n_2 = 30, \quad n_3 = 20, \quad n_4 = 10, \quad N = 100.$$

Thus, aggregation weights are:

$$0.40,\ 0.30,\ 0.20,\ 0.10.$$

Step 8: Global Update Computation.

$$\Delta\theta^{(t)} = \sum_{k=1}^{4} \frac{n_k}{N} \mathscr{E}_k(\Delta\theta_k).$$

Weighted contributions.

$$0.40\mathscr{E}_1 = \begin{bmatrix} 0.04 \\ 0.04 \\ 0.04 \\ 0.04 \\ 0.04 \\ 0.04 \\ 0.04 \\ 0.04 \\ 0.04 \\ 0.04 \end{bmatrix},$$

$$0.30\mathscr{E}_2 = \begin{bmatrix} 0.015 \\ 0.015 \\ 0.015 \\ 0.015 \\ 0.015 \\ 0.015 \\ 0.015 \\ 0 \\ 0 \\ 0 \end{bmatrix},$$

$$0.20\mathscr{E}_3 = \begin{bmatrix} 0.03 \\ 0.03 \\ 0.03 \\ 0.03 \\ 0.03 \\ 0 \\ 0 \\ 0 \\ 0 \\ 0 \end{bmatrix},$$

$$0.10\mathscr{E}_4 = \begin{bmatrix} -0.005 \\ -0.005 \\ -0.005 \\ 0 \\ 0 \\ 0 \\ 0 \\ 0 \\ 0 \\ 0 \end{bmatrix}.$$

Summation.

$$\Delta\theta^{(t)} = \begin{bmatrix} 0.080 \\ 0.080 \\ 0.080 \\ 0.085 \\ 0.085 \\ 0.055 \\ 0.055 \\ 0.040 \\ 0.040 \\ 0.040 \end{bmatrix}.$$

Step 9: Global Model Update.

$$\theta^{(t+1)} = \theta^{(t)} + \Delta\theta^{(t)} = \begin{bmatrix} 1.08 \\ 0.98 \\ 0.88 \\ 0.785 \\ 0.685 \\ 0.555 \\ 0.455 \\ 0.34 \\ 0.24 \\ 0.14 \end{bmatrix}.$$

Interpretation. Dynamic scaling allows clients to contribute proportionally to their instantaneous capacity. High-capacity clients refine the full model, while constrained clients update only a subset of parameters. Despite this variability, all contributions are coherently aligned in the global parameter space, preserving consistent optimization dynamics.

Key Insight from Dynamic Scaling
Dynamic model scaling enables time-varying, resource-aware client participation while maintaining a unified global learning objective through structured projection and embedding.

13.4.2 Partial Participation

Dynamic scaling alone does not fully address heterogeneity in federated learning. In addition to varying model sizes, clients may be intermittently available or may deliberately skip communication rounds to conserve resources. This phenomenon, known as *partial participation*, is ubiquitous in large-scale federated systems.

Let $\mathscr{S}^{(t)} \subseteq \{1, \dots, K\}$ denote the set of clients that participate in communication round t. In heterogeneous environments, $\mathscr{S}^{(t)}$ is both random and resource-dependent. Clients with limited computational or communication capacity may appear in $\mathscr{S}^{(t)}$ infrequently, while more capable clients may participate more regularly.

Partial participation interacts strongly with dynamic model scaling. A client that participates infrequently may still contribute valuable information by training a reduced model when it does participate. Conversely, a frequently participating client may adaptively reduce its model size during periods of resource scarcity. The effective global update therefore takes the form

$$\Delta\theta^{(t)} = \sum_{k \in \mathscr{S}^{(t)}} \frac{n_k}{\sum_{j \in \mathscr{S}^{(t)}} n_j} \mathscr{E}_k^{(t)}(\Delta\theta_k^{(t)}),$$

where both the set of participating clients and the dimensionality of their updates vary over time.

From an optimization perspective, partial participation introduces additional stochasticity into the learning process. The global model no longer evolves based on a fixed collection of local updates, but rather on a time-varying and resource-weighted subset of clients. Convergence must therefore be understood in a statistical sense, relying on repeated, representative participation over many rounds rather than uniform involvement at every iteration.

Adaptive Participation
Heterogeneous federated learning replaces uniform, mandatory participation with resource-aware, dynamically adjusted engagement, allowing clients to contribute according to their evolving capabilities.

Together, dynamic model scaling and partial participation form a cornerstone of practical heterogeneous federated learning. By allowing clients to adapt both *how much* they train and *when* they participate, federated systems achieve improved robustness, scalability, and energy efficiency. Crucially, these benefits are obtained without sacrificing the mathematical structure of the global optimization problem, ensuring that convergence remains attainable under realistic and highly variable operating conditions.

Worked Example: Partial Participation Under Heterogeneous Availability
We now present a detailed numerical example illustrating how *partial participation* affects federated optimization. The example demonstrates how the global model evolves when only a subset of clients participates in each communication round, and how aggregation weights are adjusted accordingly.

Problem Setup. Consider a federated system with $K = 5$ clients and a global model of dimension

$$D = 6.$$

The global parameter vector at round t is

$$\theta^{(t)} = \begin{bmatrix} \theta_1 \\ \theta_2 \\ \theta_3 \\ \theta_4 \\ \theta_5 \\ \theta_6 \end{bmatrix}.$$

At the start of round t, the server holds:

$$\theta^{(t)} = \begin{bmatrix} 0.50 \\ 0.40 \\ 0.30 \\ 0.20 \\ 0.10 \\ 0.00 \end{bmatrix}.$$

Step 1: Client Availability and Participation Set. Assume that due to connectivity and energy constraints, only a subset of clients participates in round t. The availability status is shown in Table 13.9.

Thus, the participation set is

$$\mathcal{S}^{(t)} = \{1, 2, 5\}.$$

Table 13.9 Client availability and participation at round t

Client k	Availability	Dataset size n_k	Participates?
1	High	50	Yes
2	Medium	30	Yes
3	Low	20	No
4	Very Low	15	No
5	High	25	Yes

Step 2: Projection to Client Submodels. Assume all participating clients train the full model for simplicity. Therefore,

$$\theta_k^{(t)} = \theta^{(t)}, \quad k \in \mathcal{S}^{(t)}.$$

Step 3: Local Training and Updated Parameters. After local optimization, the participating clients obtain the updated parameters shown in Table 13.10.

Table 13.10 Locally trained parameters (participating clients only)

Client k	$\theta_k^{(t+1)}$
1	$\begin{bmatrix} 0.60 \\ 0.50 \\ 0.40 \\ 0.30 \\ 0.20 \\ 0.10 \end{bmatrix}$
2	$\begin{bmatrix} 0.55 \\ 0.45 \\ 0.35 \\ 0.25 \\ 0.15 \\ 0.05 \end{bmatrix}$
5	$\begin{bmatrix} 0.52 \\ 0.42 \\ 0.32 \\ 0.22 \\ 0.12 \\ 0.02 \end{bmatrix}$

Step 4: Compute Local Parameter Updates.

Client 1.

$$\Delta\theta_1 = \theta_1^{(t+1)} - \theta^{(t)} = \begin{bmatrix} 0.10 \\ 0.10 \\ 0.10 \\ 0.10 \\ 0.10 \\ 0.10 \end{bmatrix}.$$

Client 2.

$$\Delta\theta_2 = \begin{bmatrix} 0.05 \\ 0.05 \\ 0.05 \\ 0.05 \\ 0.05 \\ 0.05 \end{bmatrix}.$$

Client 5.

$$\Delta\theta_5 = \begin{bmatrix} 0.02 \\ 0.02 \\ 0.02 \\ 0.02 \\ 0.02 \\ 0.02 \end{bmatrix}.$$

Step 5: Compute Aggregation Weights. Only participating clients contribute to aggregation. The effective total dataset size is

$$\sum_{j \in \mathcal{S}^{(t)}} n_j = 50 + 30 + 25 = 105.$$

Thus, the aggregation weights are:

Client 1.

$$w_1 = \frac{50}{105} \approx 0.476.$$

Client 2.

$$w_2 = \frac{30}{105} \approx 0.286.$$

Client 5.

$$w_5 = \frac{25}{105} \approx 0.238.$$

Step 6: Compute Weighted Contributions.

Client 1.

$$w_1 \Delta \boldsymbol{\theta}_1 = \begin{bmatrix} 0.0476 \\ 0.0476 \\ 0.0476 \\ 0.0476 \\ 0.0476 \\ 0.0476 \end{bmatrix}.$$

Client 2.

$$w_2 \Delta \boldsymbol{\theta}_2 = \begin{bmatrix} 0.0143 \\ 0.0143 \\ 0.0143 \\ 0.0143 \\ 0.0143 \\ 0.0143 \end{bmatrix}.$$

Client 5.

$$w_5 \Delta \boldsymbol{\theta}_5 = \begin{bmatrix} 0.0048 \\ 0.0048 \\ 0.0048 \\ 0.0048 \\ 0.0048 \\ 0.0048 \end{bmatrix}.$$

Step 7: Global Update Computation. Summing all weighted contributions:

$$\Delta\theta^{(t)} = \begin{bmatrix} 0.0667 \\ 0.0667 \\ 0.0667 \\ 0.0667 \\ 0.0667 \\ 0.0667 \end{bmatrix}.$$

Step 8: Global Model Update.

$$\theta^{(t+1)} = \theta^{(t)} + \Delta\theta^{(t)} = \begin{bmatrix} 0.5667 \\ 0.4667 \\ 0.3667 \\ 0.2667 \\ 0.1667 \\ 0.0667 \end{bmatrix}.$$

Interpretation. Clients 3 and 4 make no contribution during this round due to non-participation. However, the aggregation weights are automatically renormalized over the participating set, ensuring unbiased integration of available updates. Over multiple rounds, intermittent participation from all clients produces a statistically consistent approximation of full participation.

> **Key Insight from Partial Participation**
> Partial participation introduces stochasticity at the client-selection level, but careful normalization of aggregation weights preserves stable and meaningful global updates over time.

13.5 Resource-Aware Federated Training

Resource-aware federated training represents a conceptual shift in how heterogeneity is handled within federated learning systems. Rather than viewing variability in client resources as an external disturbance or implementation detail, resource-aware approaches elevate system constraints to an explicit component of the learning objective. In this paradigm, heterogeneity is no longer treated as noise to be averaged out, but as a structured control signal that guides model design, optimization, and deployment.

In large-scale federated systems, clients operate under widely varying resource conditions. Some devices may possess abundant computational power and stable connectivity, while others operate under severe constraints imposed by limited battery life, memory capacity, or intermittent network access. Ignoring these constraints can lead to learned models that are theoretically optimal but practically unusable on a significant fraction of participating clients. Resource-aware federated training addresses this gap by explicitly aligning learning objectives with system feasibility.

Let R_k denote a vector of resource descriptors associated with client k, such as available CPU cycles, memory budget, remaining energy, and communication bandwidth:

$$R_k = \left(R_k^{\text{CPU}}, R_k^{\text{mem}}, R_k^{\text{energy}}, R_k^{\text{bw}}\right).$$

These quantities may be static, slowly varying, or dynamically observed at each communication round. The presence of R_k allows the federated system to reason explicitly about what each client can realistically support.

To incorporate resource considerations into learning, the standard federated objective is augmented with a resource-aware regularization term:

$$\min_{\theta} \sum_{k=1}^{K} \frac{n_k}{N} F_k(\theta) + \lambda \sum_{k=1}^{K} \Omega(\theta, R_k),$$

where $F_k(\theta)$ denotes the local empirical risk at client k, and $\Omega(\theta, R_k)$ is a penalty function that measures the incompatibility between the model parameters θ and the resource constraints of client k. The scalar $\lambda > 0$ controls the trade-off between predictive accuracy and system feasibility.

The penalty function $\Omega(\cdot)$ can take many forms depending on the resources of interest. For example, it may penalize excessive model size relative to memory constraints, high computational complexity relative to available CPU capacity, or frequent communication relative to bandwidth limitations. In this sense, $\Omega(\theta, R_k)$ acts as a soft constraint that discourages the learning of models that would be infeasible or inefficient on certain clients.

From an optimization perspective, the augmented objective introduces a system-aware bias into the learning process. Rather than minimizing prediction error alone, the optimization now seeks solutions that balance statistical performance with deployability across a diverse population of clients. The resulting model may sacrifice a small amount of accuracy in exchange for improved robustness, energy efficiency, and inclusivity of low-resource devices.

Interpretation The regularization term reshapes the geometry of the optimization landscape. Regions corresponding to highly accurate but resource-intensive models are penalized, while regions representing slightly less complex but more widely deployable models are favored. As a result, the learned global model reflects a compromise that is better aligned with the realities of heterogeneous deployment environments.

Design Insight
Heterogeneous federated learning optimizes not only for statistical accuracy, but for *feasibility under diversity*, ensuring that the learned models remain usable across clients with widely varying resource constraints.

By explicitly integrating system-level considerations into the learning objective, resource-aware federated training provides a unifying lens through which dynamic scaling, partial participation, and architectural heterogeneity can be understood. This perspective naturally leads to more advanced formulations in which resource constraints are enforced through dual variables, adaptive penalty schedules, or joint optimization of model and system parameters.

This chapter has established the core principles of heterogeneous federated learning, emphasizing how diversity in computation, communication, model capacity, and availability can be addressed in a systematic and principled manner. In the next chapter, these ideas are extended further by combining heterogeneity with hybrid data partitioning and primal–dual optimization frameworks, enabling even tighter integration of learning objectives and system constraints.

Worked Example: Resource-Aware Federated Optimization
We now present a complete numerical example illustrating how *resource-aware federated training* modifies the optimization objective by explicitly incorporating client resource constraints. The example demonstrates how feasibility-aware regularization alters the effective global update, even when predictive losses are identical.

Problem Setup. Consider a federated system with $K = 4$ clients. The global model is a linear predictor with parameter vector

$$\theta = \begin{bmatrix} \theta_1 \\ \theta_2 \\ \theta_3 \\ \theta_4 \end{bmatrix} \in \mathbb{R}^4.$$

Assume that at communication round t, the server holds:

$$\theta^{(t)} = \begin{bmatrix} 1.0 \\ 1.0 \\ 1.0 \\ 1.0 \end{bmatrix}.$$

Step 1: Client Datasets and Empirical Losses. Assume each client minimizes a local squared-error objective:

$$F_k(\theta) = \frac{1}{2}\|\theta - \theta_k^\star\|_2^2,$$

where $\theta_k^\star$ is the client-optimal parameter.

The client-specific targets and dataset sizes are shown in Table 13.11.

Step 2: Gradient of the Empirical Loss. The gradient of F_k is:

$$\nabla F_k(\theta) = \theta - \theta_k^\star.$$

Client 1.

$$\nabla F_1 = \begin{bmatrix} -0.2 \\ -0.1 \\ 0 \\ 0.1 \end{bmatrix}.$$

Table 13.11 Client datasets and optimal parameters

Client k	Dataset size n_k	$\theta_k^\star$
1	40	$\begin{bmatrix} 1.2 \\ 1.1 \\ 1.0 \\ 0.9 \end{bmatrix}$
2	30	$\begin{bmatrix} 1.1 \\ 1.0 \\ 0.9 \\ 0.8 \end{bmatrix}$
3	20	$\begin{bmatrix} 1.3 \\ 1.2 \\ 1.1 \\ 1.0 \end{bmatrix}$
4	10	$\begin{bmatrix} 0.9 \\ 0.8 \\ 0.7 \\ 0.6 \end{bmatrix}$

Client 2.

$$\nabla F_2 = \begin{bmatrix} -0.1 \\ 0 \\ 0.1 \\ 0.2 \end{bmatrix}.$$

Client 3.

$$\nabla F_3 = \begin{bmatrix} -0.3 \\ -0.2 \\ -0.1 \\ 0 \end{bmatrix}.$$

Client 4.

$$\nabla F_4 = \begin{bmatrix} 0.1 \\ 0.2 \\ 0.3 \\ 0.4 \end{bmatrix}.$$

Step 3: Resource Vectors. Each client has different resource capacities summarized by

$$R_k = \left(\text{CPU}_k,\ \text{Memory}_k,\ \text{Energy}_k\right),$$

as shown in Table 13.12.

Step 4: Resource-aware Penalty Function. Assume a simple penalty on model norm relative to energy capacity:

$$\Omega(\theta, R_k) = \frac{\|\theta\|_2^2}{R_k^{\text{energy}}}.$$

Table 13.12 Client resource vectors

Client k	CPU (units)	Memory (MB)	Energy (J)
1	100	512	1000
2	70	256	600
3	50	128	400
4	30	64	200

At $\theta^{(t)}$:

$$\|\theta^{(t)}\|_2^2 = 1^2 + 1^2 + 1^2 + 1^2 = 4.$$

Client-wise penalties.

$$\Omega_1 = \frac{4}{1000} = 0.004, \quad \Omega_2 = \frac{4}{600} \approx 0.0067,$$

$$\Omega_3 = \frac{4}{400} = 0.01, \quad \Omega_4 = \frac{4}{200} = 0.02.$$

Step 5: Gradient of the Penalty Term.

$$\nabla_\theta \Omega(\theta, R_k) = \frac{2\theta}{R_k^{\text{energy}}}.$$

Client 1.

$$\nabla\Omega_1 = \frac{2}{1000}\begin{bmatrix}1\\1\\1\\1\end{bmatrix} = \begin{bmatrix}0.002\\0.002\\0.002\\0.002\end{bmatrix}.$$

Client 2.

$$\nabla\Omega_2 = \begin{bmatrix}0.0033\\0.0033\\0.0033\\0.0033\end{bmatrix}.$$

Client 3.

$$\nabla\Omega_3 = \begin{bmatrix}0.005\\0.005\\0.005\\0.005\end{bmatrix}.$$

Client 4.

$$\nabla \Omega_4 = \begin{bmatrix} 0.01 \\ 0.01 \\ 0.01 \\ 0.01 \end{bmatrix}.$$

Step 6: Combined Gradient per Client. Let $\lambda = 0.5$. The effective gradient becomes:

$$\nabla \widetilde{F}_k = \nabla F_k + \lambda \nabla \Omega_k.$$

Client 1.

$$\nabla \widetilde{F}_1 = \begin{bmatrix} -0.199 \\ -0.099 \\ 0.001 \\ 0.101 \end{bmatrix}.$$

Client 2.

$$\nabla \widetilde{F}_2 = \begin{bmatrix} -0.0983 \\ 0.0017 \\ 0.1017 \\ 0.2017 \end{bmatrix}.$$

Client 3.

$$\nabla \widetilde{F}_3 = \begin{bmatrix} -0.2975 \\ -0.1975 \\ -0.0975 \\ 0.0025 \end{bmatrix}.$$

Client 4.

$$\nabla \widetilde{F}_4 = \begin{bmatrix} 0.105 \\ 0.205 \\ 0.305 \\ 0.405 \end{bmatrix}.$$

Step 7: Aggregation Weights. Total dataset size:

$$N = 40 + 30 + 20 + 10 = 100.$$

Weights:

$$0.40,\ 0.30,\ 0.20,\ 0.10.$$

Step 8: Global Gradient Aggregation.

$$\nabla\widetilde{F} = \sum_{k=1}^{4} \frac{n_k}{N} \nabla\widetilde{F}_k.$$

Weighted sum.

$$\nabla\widetilde{F} = \begin{bmatrix} -0.164 \\ -0.064 \\ 0.036 \\ 0.136 \end{bmatrix}.$$

Step 9: Global Parameter Update. With learning rate $\eta = 0.1$:

$$\theta^{(t+1)} = \theta^{(t)} - \eta\nabla\widetilde{F} = \begin{bmatrix} 1.0164 \\ 1.0064 \\ 0.9964 \\ 0.9864 \end{bmatrix}.$$

Interpretation. The penalty term suppresses aggressive growth of model parameters that would disproportionately burden low-energy clients. As a result, the global update is biased toward configurations that are both accurate and resource-feasible across heterogeneous participants.

Key Insight from Resource-Aware Training

Resource-aware regularization reshapes the optimization landscape, trading marginal accuracy for improved deployability and sustainability across heterogeneous client populations.

Exercises

This chapter examined heterogeneous federated learning, emphasizing how diversity in computation, communication, model architecture, participation, and system resources reshapes federated optimization. The following exercises reinforce conceptual understanding, mathematical reasoning, and practical design considerations.

A. Short Answer Questions

Answer each question briefly and precisely.

1. What is meant by heterogeneity in federated learning?
2. Why do homogeneous assumptions break down in real-world federated systems?
3. Define computational heterogeneity and its effect on local training.
4. What role does communication bandwidth play in federated learning?
5. Explain why heterogeneous model architectures complicate parameter aggregation.
6. What is submodel consistency?
7. Define the projection operator $\mathscr{P}_k$ in HeteroFL.
8. What is dynamic model scaling?
9. What is partial participation in federated learning?
10. Why is resource-aware regularization introduced in federated objectives?

B. Long Answer Questions

Answer the following questions with clear explanations, equations, and diagrams wherever appropriate.

1. Discuss the impact of computational heterogeneity on convergence and fairness.
2. Explain communication heterogeneity and its influence on update fidelity.
3. Analyze why heterogeneous model architectures require structured aggregation mechanisms.
4. Describe the HeteroFL framework with projection and embedding operators.
5. Derive the global aggregation rule for heterogeneous submodels.
6. Explain dynamic model scaling with mathematical formulation and intuition.
7. Discuss partial participation as a source of stochasticity in federated optimization.
8. Explain how dynamic scaling and partial participation interact.
9. Formulate a resource-aware federated optimization objective and explain each term.
10. Critically evaluate the trade-off between accuracy and feasibility in heterogeneous federated learning.

C. True/False Questions

State whether each statement is **True** or **False**.

1. Classical federated learning assumes identical client resources.
2. Computational heterogeneity affects the number of local optimization steps.
3. Communication heterogeneity only affects latency, not update quality.
4. All clients must train identical model architectures in HeteroFL.
5. Submodel consistency enables partial parameter sharing.
6. Dynamic model scaling allows model size to vary across rounds.
7. Partial participation means clients never rejoin once they drop out.
8. Aggregation weights must be renormalized under partial participation.
9. Resource-aware regularization penalizes infeasible model configurations.
10. Heterogeneous federated learning improves robustness in real-world deployments.

Answers:

1. True
2. True
3. False
4. False
5. True
6. True
7. False
8. True
9. True
10. True

D. Fill in the Blanks

Fill in the blanks with the most appropriate term or expression.

1. Differences in client computation capabilities are known as ____________ heterogeneity.
2. Communication heterogeneity arises due to variations in ____________ and latency.
3. In HeteroFL, client submodels are obtained using a ____________ operator.
4. The embedding operator maps updates into the ____________ parameter space.
5. Dynamic model scaling uses a scaling ratio denoted by ____________.
6. Partial participation leads to a time-varying client set ____________.
7. Submodel consistency divides parameters into shared and ____________ components.
8. Resource-aware training introduces a ____________ term in the objective.

9. The scalar λ controls the trade-off between accuracy and ___________.
10. Heterogeneous federated learning explicitly embraces ___________ across clien ts.

Answers:

1. computational
2. bandwidth
3. projection
4. global
5. $r_k^{(t)}$
6. $\mathscr{S}^{(t)}$
7. private
8. regularization
9. feasibility
10. diversity

E. Multiple Choice Questions (MCQs)

Choose the **most appropriate** answer.

1. Computational heterogeneity primarily affects:
   (a) Data distribution
   (b) Number of local updates
   (c) Model expressiveness
   (d) Label noise
2. Communication heterogeneity constrains:
   (a) Dataset size
   (b) Learning rate
   (c) Update transmission
   (d) Loss function
3. In HeteroFL, $\mathscr{P}_k$ performs:
   (a) Aggregation
   (b) Projection
   (c) Regularization
   (d) Compression
4. The embedding operator maps updates to:
   (a) Local space
   (b) Feature space
   (c) Global space
   (d) Data space
5. Dynamic model scaling adapts:
   (a) Dataset size
   (b) Model dimension

   (c) Loss function
   (d) Communication frequency
6. Partial participation introduces:
   (a) Determinism
   (b) Synchrony
   (c) Stochasticity
   (d) Centralization
7. Submodel consistency enables:
   (a) Uniform architectures
   (b) Full parameter sharing
   (c) Partial sharing
   (d) No sharing
8. Resource-aware regularization penalizes:
   (a) High accuracy
   (b) Large datasets
   (c) Infeasible models
   (d) Frequent participation
9. The term λ controls:
   (a) Learning rate
   (b) Model size
   (c) Accuracy–feasibility trade-off
   (d) Dataset balance
10. Heterogeneous federated learning primarily improves:
   (a) Centralization
   (b) Robustness
   (c) Overfitting
   (d) Data leakage

Answers:

1. (b)
2. (c)
3. (b)
4. (c)
5. (b)
6. (c)
7. (c)
8. (c)
9. (c)
10. (b)

F. Application and Project-Based Questions

These problems emphasize modeling, reasoning, and implementation.

1. Design a federated learning system for IoT devices with heterogeneous energy constraints.
2. Simulate computational heterogeneity and study its effect on convergence.
3. Implement communication-aware update compression in a federated setup.
4. Design a heterogeneous model architecture for mobile and edge devices.
5. Implement projection and embedding operators for HeteroFL.
6. Study dynamic model scaling under fluctuating battery levels.
7. Simulate partial participation and analyze aggregation bias.
8. Propose a resource-aware penalty function for memory constraints.
9. Compare homogeneous and heterogeneous federated learning experimentally.
10. Extend resource-aware federated learning using a primal–dual optimization framework.

Hybrid Federated Learning 14

Hybrid Federated Learning (HFL) represents a natural and necessary evolution of the federated learning paradigm as it is applied to complex, large-scale, real-world systems. Classical federated learning formulations—most notably those underlying Federated Averaging—are built on a simplifying assumption: that data is partitioned across clients *only by samples*, while the feature space and model structure remain globally shared. Under this assumption, each client observes different data points drawn from potentially heterogeneous distributions, but all clients operate on the same input representation and optimize the same model parameters.

While this assumption has proven effective for many applications, it does not adequately capture the structure of data in numerous practical settings. In modern collaborative environments, data decentralization is often *multi-dimensional*. Different parties may observe different subsets of samples, different subsets of features, or even different components of a larger model. For example, in healthcare consortia, hospitals may collect overlapping patient cohorts but record distinct clinical attributes. In financial systems, institutions may share customer identifiers but retain proprietary feature sets. In industrial IoT deployments, edge nodes may monitor different sensors attached to the same physical asset. Such scenarios cannot be modeled accurately using purely horizontal or purely vertical federated learning alone.

The hybrid federated learning is designed to address precisely this gap. Rather than committing to a single mode of data partitioning, HFL explicitly accommodates *simultaneous partitioning across samples, features, and sometimes model components*. In doing so, it unifies and generalizes the two dominant federated learning paradigms:

- **Horizontal federated learning**, where clients hold disjoint subsets of samples but share a common feature space, and
- **Vertical federated learning**, where clients hold different feature subsets corresponding to the same or overlapping samples.

P. P. Ray, *Foundations of Federated Learning*,
Springer Series in the Data Sciences,
https://doi.org/10.1007/978-3-032-34084-9_14

Hybrid federated learning can therefore be viewed as a superset of both approaches, subsuming them as special cases while enabling more flexible and realistic modeling assumptions.

From a conceptual standpoint, HFL reframes federated learning as a problem of *structured decentralization*. Instead of treating clients as identical entities differentiated only by data volume or distribution, hybrid methods explicitly acknowledge that clients may contribute qualitatively different types of information. Some clients may be rich in samples but poor in features, others may be feature-rich but sample-poor, and still others may participate only in specific components of the learning task. The learning framework must therefore coordinate heterogeneous contributions while preserving data locality, privacy constraints, and autonomy of participants.

From a mathematical perspective, hybrid federated learning introduces additional layers of complexity into the optimization problem. The global learning objective must now reconcile not only statistical heterogeneity across samples, but also structural heterogeneity across feature spaces and parameter blocks. This necessitates more sophisticated formulations, often involving block-structured objectives, coupled constraints, and primal–dual optimization techniques. Classical parameter averaging is generally insufficient in this setting, as it fails to account for partial observability of features or model components at individual clients.

From an algorithmic standpoint, hybrid federated learning motivates the development of new coordination mechanisms. These include dual coordinate ascent methods, augmented Lagrangian and primal–dual frameworks, and carefully designed aggregation rules that operate over subsets of parameters rather than entire models. Such methods allow learning to proceed even when no single client has access to a complete view of the data or the model, a situation that is increasingly common in privacy-sensitive and institutionally governed environments.

This chapter develops the foundations of hybrid federated learning in a systematic and layered manner. We begin by formalizing hybrid data partitioning schemes and clarifying how they generalize classical federated settings. We then introduce hybrid federated dual coordinate ascent and related primal–dual optimization frameworks that enable principled coordination under mixed data partitioning. Subsequently, we analyze convergence behavior under partial and stochastic participation, highlighting how hybrid structures influence both stability and convergence rates. Finally, we contrast hybrid federated learning with classical Federated Averaging and related approaches such as Hybrid Federated Expectation Maximization (HyFEM), emphasizing differences in assumptions, applicability, and robustness.

By the end of this chapter, the reader will have a comprehensive understanding of why hybrid federated learning is not merely a theoretical extension, but a practical necessity for modern collaborative machine learning systems. The concepts developed here serve as a bridge between foundational federated optimization methods and advanced topics such as heterogeneous models, institutional collaboration, and federated systems operating under complex data governance constraints.

14.1 Hybrid Data Partitioning

The manner in which data is partitioned across participating entities lies at the very core of any federated learning system. In classical federated learning, data decentralization is typically modeled using a *horizontal partitioning* assumption. Under this assumption, each client holds a distinct subset of data samples, but all samples share a common feature space and a common model representation. Formally, every client observes inputs $x \in \mathscr{X}$ drawn from the same feature space $\mathscr{X}$, even though the underlying data distributions may differ across clients.

While horizontal partitioning provides a clean and mathematically tractable abstraction, it is often an oversimplification of real-world data generation processes. In many collaborative learning scenarios, different organizations or devices do not merely differ in the *number* of samples they collect, but also in the *type* of information they observe. Two institutions may monitor the same physical entities yet record distinct attributes, measurements, or modalities. Conversely, some participants may possess rich feature representations for only a small number of samples, while others observe many samples with limited feature sets. Such configurations cannot be adequately described by purely horizontal or purely vertical partitioning.

Hybrid federated learning addresses this limitation by explicitly generalizing data partitioning along two orthogonal dimensions. The first dimension concerns *which samples* are available to each client, while the second concerns *which features* of those samples are observed locally. This leads to a more expressive and realistic data model in which both sample availability and feature availability may vary across clients.

Specifically, hybrid data partitioning encompasses:

- **Sample-wise partitioning**, where clients possess different subsets of data instances, potentially with partial or no overlap in sample identities. This dimension generalizes horizontal federated learning and captures scenarios in which clients collect data independently over time or space.
- **Feature-wise partitioning**, where clients possess different subsets of features corresponding to the same or overlapping samples. This dimension generalizes vertical federated learning and arises naturally when data attributes are distributed across organizational, technical, or legal boundaries.

Formally, let $\mathscr{D}_k$ denote the dataset available at client k. In the hybrid setting, $\mathscr{D}_k$ is not simply a subset of a global dataset restricted to samples, but rather a structured collection that reflects both sample and feature decentralization:

$$\mathscr{D}_k = \{(x_{k,i}^{(S)}, x_{k,i}^{(F)}, y_i)\}_{i \in \mathscr{I}_k}.$$

Here, $\mathscr{I}_k$ denotes the index set of samples visible to client k, which may overlap partially, fully, or not at all with the index sets of other clients. The term $x_{k,i}^{(S)}$ represents sample-specific information that is common across clients observing sample i, such as identifiers or shared attributes. In contrast, $x_{k,i}^{(F)}$ denotes the subset of features locally observed by client k, reflecting client-specific sensing capabilities, measurement modalities, or proprietary data sources. The label y_i, when present, may itself be distributed unevenly across clients, further complicating the learning process.

This formulation highlights an important departure from classical federated learning. No single client is assumed to possess a complete view of either the entire dataset or the full feature space. Instead, the global learning task must be solved by coordinating incomplete and heterogeneous local views, each governed by its own availability constraints and privacy requirements. As a consequence, learning algorithms must operate under partial observability and structural asymmetry, rather than relying on uniform data representations.

Hybrid data partitioning provides a powerful abstraction for modeling a wide range of practical applications. In multi-institutional healthcare studies, hospitals may share patient identifiers but record different clinical tests, imaging modalities, or laboratory results. In cross-platform user modeling, different services may observe the same users but collect distinct behavioral signals. In collaborative industrial monitoring, sensor networks may capture different physical variables of the same system from geographically distributed locations. In all these cases, hybrid partitioning reflects the true structure of the data more faithfully than classical assumptions.

Key Idea

Hybrid federated learning unifies horizontal and vertical data partitioning into a single framework, enabling collaborative learning when data is simultaneously decentralized across samples, features, and observational capabilities.

By embracing hybrid data partitioning, federated learning moves closer to the realities of modern data ecosystems. At the same time, this increased expressiveness introduces new algorithmic and theoretical challenges, particularly in optimization, aggregation, and convergence analysis. The remainder of this chapter builds upon this data model to develop primal–dual methods and coordination mechanisms capable of operating effectively under hybrid decentralization.

Worked Example: Hybrid Data Partitioning with Explicit Numerical Construction

This worked example demonstrates, step by step, how hybrid data partitioning manifests in a concrete federated learning problem. We explicitly construct local datasets, identify sample and feature overlap, and derive the corresponding local empirical risks without skipping any calculations. The objective is to make the abstract notion of hybrid partitioning operational and mathematically transparent.

Problem Setup. Consider a collaborative learning task involving $K = 3$ clients. The goal is to learn a linear regression model

$$f_{\theta}(x) = \theta_1 x^{(1)} + \theta_2 x^{(2)},$$

where $\theta = (\theta_1, \theta_2)^{\mathsf{T}} \in \mathbb{R}^2$. We assume a squared-error loss function

$$\ell(f_{\theta}(x), y) = (f_{\theta}(x) - y)^2.$$

The global population consists of five samples indexed by $i \in \{1, 2, 3, 4, 5\}$. However, no client observes all samples or all features.

Step 1: Global Feature and Sample Structure. The conceptual global dataset (which is never centralized) is shown in Table 14.1.

This table is shown only for explanatory purposes. In federated learning, this dataset is never constructed explicitly.

Step 2: Client-Wise Hybrid Data Partitioning. Each client observes only a subset of samples and features.

Client 1. Client 1 observes samples $\mathcal{I}_1 = \{1, 2, 3\}$ and only feature $x^{(1)}$.

Sample i	$x^{(1)}$	y
1	1	5
2	2	4
3	3	9

Client 2. Client 2 observes samples $\mathcal{I}_2 = \{2, 3, 4\}$ and only feature $x^{(2)}$.

Sample i	$x^{(2)}$	y
2	1	4
3	3	9
4	2	8

Table 14.1 Conceptual global dataset (not accessible to any single client)

Sample i	Feature $x^{(1)}$	Feature $x^{(2)}$	Label y
1	1	2	5
2	2	1	4
3	3	3	9
4	4	2	8
5	5	4	13

Client 3. Client 3 observes samples $\mathscr{I}_3 = \{4, 5\}$ and both features.

Sample i	$x^{(1)}$	$x^{(2)}$	y
4	4	2	8
5	5	4	13

This configuration exhibits:

- partial sample overlap (e.g., samples 2 and 3),
- feature-wise partitioning (Clients 1 and 2),
- a hybrid client (Client 3).

Step 3: Local Prediction Models. Because feature availability differs, each client computes predictions using only the features it observes.

Client 1 Prediction.

$$\hat{y}_{1,i} = \theta_1 x_i^{(1)}.$$

Client 2 Prediction.

$$\hat{y}_{2,i} = \theta_2 x_i^{(2)}.$$

Client 3 Prediction.

$$\hat{y}_{3,i} = \theta_1 x_i^{(1)} + \theta_2 x_i^{(2)}.$$

Step 4: Local Empirical Risk Functions.

Client 1.

$$\begin{aligned} F_1(\theta) &= \frac{1}{3}\left[(\theta_1 \cdot 1 - 5)^2 + (\theta_1 \cdot 2 - 4)^2 + (\theta_1 \cdot 3 - 9)^2\right] \\ &= \frac{1}{3}\left[(\theta_1 - 5)^2 + (2\theta_1 - 4)^2 + (3\theta_1 - 9)^2\right]. \end{aligned}$$

Client 2.

$$\begin{aligned} F_2(\theta) &= \frac{1}{3}\left[(\theta_2 \cdot 1 - 4)^2 + (\theta_2 \cdot 3 - 9)^2 + (\theta_2 \cdot 2 - 8)^2\right] \\ &= \frac{1}{3}\left[(\theta_2 - 4)^2 + (3\theta_2 - 9)^2 + (2\theta_2 - 8)^2\right]. \end{aligned}$$

Client 3.

$$F_3(\theta) = \frac{1}{2}\Big[(\theta_1 \cdot 4 + \theta_2 \cdot 2 - 8)^2 + (\theta_1 \cdot 5 + \theta_2 \cdot 4 - 13)^2\Big].$$

Step 5: Expanded Polynomial Forms.

Client 1 Expansion.

$$\begin{aligned}(\theta_1 - 5)^2 &= \theta_1^2 - 10\theta_1 + 25,\\ (2\theta_1 - 4)^2 &= 4\theta_1^2 - 16\theta_1 + 16,\\ (3\theta_1 - 9)^2 &= 9\theta_1^2 - 54\theta_1 + 81.\end{aligned}$$

Summing:

$$F_1(\theta) = \frac{1}{3}(14\theta_1^2 - 80\theta_1 + 122).$$

Client 2 Expansion.

$$\begin{aligned}(\theta_2 - 4)^2 &= \theta_2^2 - 8\theta_2 + 16,\\ (3\theta_2 - 9)^2 &= 9\theta_2^2 - 54\theta_2 + 81,\\ (2\theta_2 - 8)^2 &= 4\theta_2^2 - 32\theta_2 + 64.\end{aligned}$$

Summing:

$$F_2(\theta) = \frac{1}{3}(14\theta_2^2 - 94\theta_2 + 161).$$

Client 3 Expansion.

$$\begin{aligned}(4\theta_1 + 2\theta_2 - 8)^2 &= 16\theta_1^2 + 4\theta_2^2 + 16\theta_1\theta_2 - 64\theta_1 - 32\theta_2 + 64,\\ (5\theta_1 + 4\theta_2 - 13)^2 &= 25\theta_1^2 + 16\theta_2^2 + 40\theta_1\theta_2 - 130\theta_1 - 104\theta_2 + 169.\end{aligned}$$

Summing:

$$F_3(\theta) = \frac{1}{2}(41\theta_1^2 + 20\theta_2^2 + 56\theta_1\theta_2 - 194\theta_1 - 136\theta_2 + 233).$$

Step 6: Interpretation. Each local objective depends on a different subset of parameters and samples. Clients 1 and 2 induce one-dimensional loss surfaces, while

Client 3 induces a coupled two-dimensional surface. No single client can recover the full regression model independently.

Key Insight from the Example
Hybrid data partitioning yields structurally different local objectives. Global learning emerges only through coordinated optimization across complementary sample and feature views.

14.2 Hybrid Federated Dual Coordinate Ascent

Hybrid Federated Dual Coordinate Ascent (HyFDCA) provides a principled optimization framework for federated learning under hybrid data partitioning. It extends classical dual coordinate ascent methods—originally developed for centralized convex optimization and later adapted to distributed settings—into the federated domain where both *samples* and *features* may be decentralized. The central motivation behind HyFDCA is that, in hybrid federated systems, direct synchronization of full primal model parameters is often either infeasible or inefficient. Instead, coordination can be achieved more effectively by operating in the dual space, where updates naturally align with local data availability and structural constraints.

In standard federated learning, optimization is typically carried out in the primal space by aggregating locally updated model parameters. While this approach is simple and communication-efficient, it implicitly assumes that all clients optimize the same parameter vector and that local objectives are structurally compatible. Under hybrid partitioning, these assumptions no longer hold. Different clients may optimize different subsets of parameters, observe different feature blocks, or enforce distinct constraints. HyFDCA addresses this challenge by introducing client-associated dual variables that act as coordination signals across heterogeneous local objectives.

Regularized Empirical Risk Minimization Formulation. Consider the global learning objective expressed as a regularized empirical risk minimization problem:

$$\min_{\theta} \sum_{k=1}^{K} \frac{n_k}{N} F_k(\theta) + \lambda R(\theta),$$

where $F_k(\theta)$ denotes the empirical loss at client k, n_k is the number of samples available locally, $N = \sum_{k=1}^{K} n_k$ is the total number of samples across all clients, and $R(\theta)$ is a convex regularization term. The regularizer serves both to control model complexity and to stabilize learning under data heterogeneity.

In hybrid federated settings, the local loss functions F_k may depend only on subsets of $\boldsymbol{\theta}$, reflecting feature-wise partitioning or model decomposition. As a result, naive averaging of primal updates may lead to inconsistent or ill-defined global updates.

Dual Reformulation and Client-Specific Variables. To overcome this limitation, HyFDCA introduces dual variables $\boldsymbol{\alpha}_k$ associated with local objectives or coupling constraints. These dual variables capture the influence of each client's local data and structure on the global optimization problem. By reformulating the primal objective in the dual domain—typically through Fenchel conjugates or Lagrangian relaxation—the global problem is decomposed into smaller, client-aligned subproblems.

In this dual representation, each client optimizes its own dual coordinates while respecting implicit global consistency constraints. Crucially, dual variables provide a mechanism for enforcing agreement among clients without requiring them to exchange full parameter vectors. This property is particularly valuable when feature spaces differ across clients or when communication bandwidth is limited.

Algorithmic Structure of HyFDCA. The HyFDCA algorithm proceeds through a sequence of alternating steps that mirror the structure of classical coordinate ascent, but are adapted to the federated and hybrid context:

- **Local dual coordinate updates.** Each client performs coordinate ascent steps on its local dual variables $\boldsymbol{\alpha}_k$, using only locally available data and features. These updates are computationally efficient and naturally aligned with the client's partial view of the problem.
- **Global aggregation of dual information.** Updated dual variables, or summaries thereof, are communicated to a central server or coordinating entity. The server aggregates this information to maintain a consistent global dual state.
- **Implicit primal updates.** The global primal parameters $\boldsymbol{\theta}$ are updated implicitly as a function of the aggregated dual variables. In many cases, the primal solution can be recovered analytically or with minimal additional computation.

This alternating structure allows HyFDCA to decouple optimization across both data samples and feature blocks. Clients contribute only the information they are capable of computing, and coordination emerges through the dual variables rather than explicit parameter synchronization.

Scalability and Communication Efficiency. One of the principal advantages of HyFDCA lies in its scalability. Because dual updates are often lower-dimensional or sparser than primal updates, communication overhead can be significantly reduced. Moreover, clients that do not observe certain features or model components need not participate in updates related to those components. This selective participation aligns naturally with hybrid data partitioning and reduces unnecessary computation and communication.

From a systems perspective, HyFDCA is particularly well-suited to federated environments characterized by:

- heterogeneous feature availability,
- partial and asynchronous client participation,
- privacy constraints that limit parameter sharing.

Dual Perspective
Hybrid Federated Dual Coordinate Ascent leverages dual variables as coordination mechanisms, enabling consistent global optimization under hybrid sample and feature partitioning without explicit synchronization of full model parameters.

By shifting the focus from direct parameter averaging to dual-domain coordination, HyFDCA provides a flexible and robust foundation for hybrid federated learning. It bridges classical dual optimization theory with the practical demands of decentralized, heterogeneous, and privacy-preserving learning systems, and serves as a key building block for the more general primal–dual frameworks developed in subsequent sections.

Worked Example: Hybrid Federated Dual Coordinate Ascent with Explicit Numerical Derivation
This worked example illustrates Hybrid Federated Dual Coordinate Ascent (HyFDCA) through a complete numerical construction. All primal and dual quantities are computed explicitly, step by step, to demonstrate how dual coordination enables learning under hybrid data partitioning. No algebraic step is omitted.

Problem Definition. Consider a federated learning system with $K = 2$ clients. We aim to solve a regularized linear regression problem with squared loss:

$$\min_{\theta \in \mathbb{R}} \sum_{k=1}^{2} \frac{n_k}{N} F_k(\theta) + \frac{\lambda}{2}\theta^2,$$

where $\theta \in \mathbb{R}$ is a scalar model parameter and $\lambda > 0$ is a regularization coefficient.

Each local objective is defined as

$$F_k(\theta) = \frac{1}{n_k} \sum_{i \in \mathcal{I}_k} (\theta x_i - y_i)^2.$$

Step 1: Numerical Dataset Construction. Assume the following global dataset of four samples.

Sample i	Feature x_i	Label y_i
1	1	2
2	2	4
3	3	6
4	4	8

Client datasets are partitioned as follows.

Client 1.

$$\mathscr{I}_1 = \{1, 2\}, \quad n_1 = 2.$$

Client 2.

$$\mathscr{I}_2 = \{3, 4\}, \quad n_2 = 2.$$

Total sample count:

$$N = n_1 + n_2 = 4.$$

Step 2: Explicit Local Objective Functions.

Client 1 Loss.

$$\begin{aligned} F_1(\theta) &= \frac{1}{2}\left[(\theta \cdot 1 - 2)^2 + (\theta \cdot 2 - 4)^2\right] \\ &= \frac{1}{2}\left[(\theta - 2)^2 + (2\theta - 4)^2\right]. \end{aligned}$$

Expanding:

$$\begin{aligned} (\theta - 2)^2 &= \theta^2 - 4\theta + 4, \\ (2\theta - 4)^2 &= 4\theta^2 - 16\theta + 16. \end{aligned}$$

Summing:

$$F_1(\theta) = \frac{1}{2}(5\theta^2 - 20\theta + 20) = 2.5\theta^2 - 10\theta + 10.$$

Client 2 Loss.

$$F_2(\theta) = \frac{1}{2}\left[(3\theta - 6)^2 + (4\theta - 8)^2\right].$$

Expanding:

$$(3\theta - 6)^2 = 9\theta^2 - 36\theta + 36,$$
$$(4\theta - 8)^2 = 16\theta^2 - 64\theta + 64.$$

Summing:

$$F_2(\theta) = \frac{1}{2}(25\theta^2 - 100\theta + 100) = 12.5\theta^2 - 50\theta + 50.$$

Step 3: Primal Objective Assembly. Using $\frac{n_k}{N} = \frac{1}{2}$ and $\lambda = 1$,

$$\begin{aligned} J(\theta) &= \frac{1}{2}F_1(\theta) + \frac{1}{2}F_2(\theta) + \frac{1}{2}\theta^2 \\ &= \frac{1}{2}(2.5\theta^2 - 10\theta + 10) + \frac{1}{2}(12.5\theta^2 - 50\theta + 50) + \frac{1}{2}\theta^2. \end{aligned}$$

Combining terms:

$$J(\theta) = 8\theta^2 - 30\theta + 30.$$

Step 4: Dual Reformulation. Introduce dual variables α_1 and α_2 for Clients 1 and 2, associated with consensus constraints. The Fenchel dual of squared loss yields:

$$\theta = \frac{1}{\lambda}\sum_{k=1}^{2}\frac{n_k}{N}\alpha_k.$$

Substituting values:

$$\theta = \frac{1}{1}\left(\frac{1}{2}\alpha_1 + \frac{1}{2}\alpha_2\right) = \frac{\alpha_1 + \alpha_2}{2}.$$

Step 5: Local Dual Coordinate Updates.

Client 1 Dual Update. Gradient of F_1:

$$\frac{dF_1}{d\theta} = 5\theta - 10.$$

At $\theta^{(0)} = 0$:

$$\frac{dF_1}{d\theta}(0) = -10.$$

Set dual variable:

$$\alpha_1^{(1)} = -\frac{dF_1}{d\theta}(0) = 10.$$

Client 2 Dual Update. Gradient of F_2:

$$\frac{dF_2}{d\theta} = 25\theta - 50.$$

At $\theta^{(0)} = 0$:

$$\frac{dF_2}{d\theta}(0) = -50.$$

Set dual variable:

$$\alpha_2^{(1)} = 50.$$

Step 6: Global Dual Aggregation. Aggregate dual variables:

$$\alpha_{\text{sum}} = \alpha_1^{(1)} + \alpha_2^{(1)} = 60.$$

Step 7: Implicit Primal Update. Recover primal parameter:

$$\theta^{(1)} = \frac{\alpha_1^{(1)} + \alpha_2^{(1)}}{2} = 30.$$

Step 8: Verification via Direct Minimization. Derivative of $J(\theta)$:

$$\frac{dJ}{d\theta} = 16\theta - 30.$$

Setting to zero:

$$16\theta - 30 = 0 \quad \Rightarrow \quad \theta^\star = \frac{30}{16} = 1.875.$$

The large value $\theta^{(1)} = 30$ reflects a single unscaled dual step. Repeated coordinate ascent iterations with step-size control converge monotonically to $\theta^\star$.

Step 9: Interpretation. Each client updates only its dual variable using local data. The global model parameter emerges implicitly from aggregated dual information, without exchanging explicit primal updates.

Key Insight from the Example
Hybrid Federated Dual Coordinate Ascent achieves global optimization by coordinating client-specific dual variables, allowing consistent model recovery even when direct primal synchronization is avoided.

14.3 Primal–Dual Optimization Framework

Hybrid federated learning admits a particularly natural and powerful interpretation through the lens of *primal–dual optimization*. This perspective generalizes the parameter-averaging view of classical federated learning and provides a rigorous mathematical foundation for coordination under hybrid data partitioning. Rather than enforcing agreement among clients through direct averaging of model parameters, primal–dual methods encode consensus as an explicit constraint and manage it through dual variables.

At a high level, each client maintains its own *local primal variables* $\boldsymbol{\theta}_k$, which reflect the portion of the model that can be optimized using locally available samples and features. In parallel, the system introduces *dual variables* $\boldsymbol{\alpha}_k$ that quantify the degree of disagreement between local models and a latent global model. These dual variables act as coordination signals, gradually steering the local solutions toward global consistency without requiring full synchronization at every iteration.

14.3.1 Consensus-Constrained Reformulation

To formalize coordination among decentralized and heterogeneous clients, hybrid federated learning reframes the global learning task as a *consensus-constrained optimization problem*. In this view, each client k is allowed to maintain and optimize its own local parameter vector $\boldsymbol{\theta}_k$, reflecting the data, features, and model components available locally. At the same time, a global notion of agreement is imposed through an explicit consensus constraint.

Mathematically, the federated objective is written as

$$\min_{\{\boldsymbol{\theta}_k\}} \sum_{k=1}^{K} \frac{n_k}{N} F_k(\boldsymbol{\theta}_k) \quad \text{subject to} \quad \boldsymbol{\theta}_k = \boldsymbol{\theta},$$

where $F_k(\theta_k)$ denotes the empirical risk at client k, n_k is the number of local samples, and $N = \sum_k n_k$ is the total sample count. The shared variable θ represents an abstract global model toward which all local models are expected to converge.

The weighting factors $\frac{n_k}{N}$ ensure that clients contribute to the global objective proportionally to their data volume. This is particularly important in hybrid federated settings, where some clients may be sample-rich but feature-poor, while others may exhibit the opposite imbalance. Without appropriate weighting, the optimization could be dominated by a small subset of clients, leading to biased or unstable solutions.

This reformulation makes explicit a fundamental tension inherent in all federated learning systems. On one hand, clients must be allowed to optimize independently to respect data locality, privacy constraints, and computational autonomy. On the other hand, the learning process is meaningful only if these independent solutions ultimately agree on a common model. In classical Federated Averaging, this tension is resolved implicitly through repeated local optimization followed by heuristic parameter averaging. While effective in many cases, this approach obscures the role of disagreement and provides limited theoretical control under strong heterogeneity.

In contrast, the consensus-constrained formulation resolves this tension explicitly. Disagreement among clients is not treated as a side effect, but as a first-class object of the optimization problem. This explicit treatment is what enables principled extensions to hybrid data partitioning, partial participation, and heterogeneous model structures.

14.3.2 Augmented Lagrangian Construction

The consensus-constrained problem is handled using an augmented Lagrangian approach, which combines ideas from classical Lagrange multipliers and penalty methods. The augmented Lagrangian is given by

$$\mathscr{L} = \sum_{k=1}^{K} \frac{n_k}{N} F_k(\theta_k) + \sum_{k=1}^{K} \alpha_k^{\top} (\theta_k - \theta) + \frac{\rho}{2} \sum_{k=1}^{K} \|\theta_k - \theta\|^2,$$

where α_k are Lagrange multipliers associated with the consensus constraints, and $\rho > 0$ is a penalty parameter.

Each component of this expression has a clear and distinct role. The first term represents the weighted sum of local empirical risks, capturing the statistical objectives of the clients. The second term introduces linear dual penalties that explicitly encode the violation of consensus constraints. These dual variables can be interpreted as forces that push local models toward agreement when discrepancies arise. The third term is a quadratic penalty that discourages large deviations between local and global parameters, providing numerical stability and smoothing the optimization landscape.

The augmented Lagrangian plays a central role in hybrid federated learning because it offers a tunable balance between local autonomy and global consistency.

When ρ is large, the quadratic penalty strongly enforces consensus, effectively limiting how far local models can drift from the global solution. This is desirable when data distributions are relatively homogeneous or when rapid consensus is required. Conversely, smaller values of ρ allow clients greater freedom to adapt to local data heterogeneity, which can improve local fit in highly non-IID or feature-partitioned settings.

Importantly, the augmented Lagrangian framework decouples the enforcement of consensus from the specific form of local objectives. This makes it particularly suitable for hybrid federated learning, where local loss functions may differ not only statistically but also structurally due to feature-wise partitioning or model decomposition.

14.3.3 Alternating Primal–Dual Updates

Optimization of the augmented Lagrangian proceeds through an alternating primal–dual strategy, in which different blocks of variables are updated in turn. This block-wise structure mirrors the philosophy of the Alternating Direction Method of Multipliers (ADMM), but is adapted to the federated and hybrid context.

Hybrid federated algorithms typically alternate among three types of updates:

- **Local primal updates.** Each client updates its local parameters θ_k by minimizing the augmented Lagrangian with respect to θ_k, while holding the global parameter θ and dual variables α_k fixed. These updates depend only on locally available data and features, making them naturally compatible with hybrid data partitioning. Clients that observe only subsets of features update only the corresponding parameter blocks.
- **Global primal update.** The global parameter θ is updated by minimizing the augmented Lagrangian with respect to θ, given the current local parameters and dual variables. In many cases, this step admits a closed-form solution that resembles a weighted average of local parameters corrected by dual terms. This correction distinguishes the update from simple parameter averaging and accounts explicitly for accumulated disagreement.
- **Dual variable updates.** The dual variables are updated via gradient ascent on the dual objective:

$$\alpha_k^{(t+1)} = \alpha_k^{(t)} + \rho(\theta_k^{(t+1)} - \theta^{(t+1)}).$$

  These updates accumulate consensus violations over time. Persistent disagreement leads to larger dual values, which in turn exert stronger corrective forces in subsequent primal updates.

This alternating update structure yields several important advantages. First, it cleanly separates local computation from global coordination, allowing clients to operate asynchronously and with partial participation. Second, it provides explicit control over local drift through the dual variables, rather than relying on heuristic regularization. Finally, it generalizes seamlessly to hybrid settings in which clients participate unevenly, observe different feature sets, or update only subsets of model parameters.

Overall, the alternating primal–dual framework transforms federated learning from a heuristic averaging procedure into a structured and theoretically grounded optimization process. This structure is what enables robust learning under hybrid data partitioning, partial participation, and strong heterogeneity—conditions that increasingly define real-world federated systems.

14.3.4 Handling Hybrid Data Structures

One of the central strengths of the primal–dual optimization framework lies in its ability to *explicitly accommodate structural heterogeneity* in hybrid federated learning. Unlike classical federated averaging—which implicitly assumes that all clients optimize the same parameter vector over a common feature space—the primal–dual formulation naturally supports scenarios in which data, features, and even model components are distributed unevenly across participants.

In hybrid federated systems, clients may observe only a subset of the global feature space. For example, one institution may record demographic variables, another may record laboratory measurements, and a third may observe temporal or behavioral signals. Within the primal–dual framework, such clients update only the corresponding blocks of their local parameter vectors $\boldsymbol{\theta}_k$. Unobserved feature blocks remain unchanged, avoiding the introduction of spurious gradients or artificially imputed values. This block-wise update structure aligns naturally with feature-wise partitioned models and block-coordinate optimization.

Equally important is the treatment of *partial participation*. In practical federated deployments, clients may be unavailable, slow, or intentionally excluded from certain communication rounds. In the primal–dual framework, a non-participating client simply retains its previous primal variables $\boldsymbol{\theta}_k$ and dual variables $\boldsymbol{\alpha}_k$. No special handling or rescaling is required. When the client re-enters the system, accumulated dual variables ensure that any prolonged deviation from consensus is accounted for and corrected over subsequent rounds.

Consensus in this setting is therefore enforced *gradually* rather than instantaneously. Instead of requiring all local models to agree at every iteration, the dual variables act as memory terms that track disagreement over time. Persistent misalignment between local and global parameters increases the magnitude of the corresponding dual variables, which in turn strengthens corrective forces in later updates. This mechanism allows the system to tolerate temporary inconsistency while still converging toward a coherent global solution.

The resulting flexibility enables hybrid federated learning systems to operate effectively under a wide range of realistic conditions, including:

- **Feature-wise partitioned models**, where parameters are organized into blocks corresponding to disjoint or overlapping feature subsets;
- **Incomplete or overlapping sample visibility**, where clients observe different subsets of entities or events;
- **Heterogeneous system capabilities**, including variations in computation, storage, energy availability, and communication bandwidth.

By decoupling optimization across both statistical and structural dimensions, the primal–dual framework provides a principled foundation for learning under hybrid data structures. Rather than forcing all clients into a rigid, uniform training protocol, it allows each participant to contribute information in a manner consistent with its local data and constraints, while still ensuring global coherence through mathematically grounded coordination mechanisms.

Interpretation of Dual Variables. From an interpretive standpoint, the dual variables $\boldsymbol{\alpha}_k$ measure how strongly each client disagrees with the current global model. Large dual magnitudes indicate persistent mismatch between local and global objectives, often caused by data heterogeneity or feature imbalance. By tracking this disagreement explicitly, the algorithm can regulate local drift in a principled manner, rather than relying on heuristic regularization.

Structural Advantage

Primal–dual formulations make disagreement an explicit optimization object, enabling controlled coordination among heterogeneous clients and providing a principled mechanism to balance local adaptation with global consistency in hybrid federated systems.

In summary, the primal–dual optimization framework elevates federated learning from parameter averaging to constrained optimization. It provides the mathematical structure required to handle hybrid data partitioning, partial participation, and heterogeneous feature spaces, and serves as the conceptual backbone for many advanced hybrid federated algorithms discussed in this chapter.

Worked Example: Primal–Dual Optimization with Four Clients

This worked example presents a complete and fully numerical illustration of the primal–dual optimization framework for *hybrid federated learning with four clients*. Every quantity—local losses, gradients, primal updates, global aggregation, and dual updates—is derived explicitly from tabular data. No algebraic step is skipped. The objective is to make the augmented Lagrangian formulation concrete, computable, and intuitively interpretable.

Problem Setup. Consider a federated learning system with $K = 4$ clients. We aim to learn a *scalar* model parameter $\theta \in \mathbb{R}$ using squared-error regression under a consensus constraint. Each client k maintains its own local parameter θ_k.

The local loss at client k is

$$F_k(\theta_k) = \frac{1}{n_k} \sum_{i \in \mathcal{I}_k} (\theta_k x_i - y_i)^2.$$

The global objective is

$$\min_{\theta_1,\theta_2,\theta_3,\theta_4} \sum_{k=1}^{4} \frac{n_k}{N} F_k(\theta_k) \quad \text{subject to } \theta_1 = \theta_2 = \theta_3 = \theta_4 = \theta.$$

Step 1: Numerical Dataset Construction. Assume the following global dataset of eight samples.

Sample i	Feature x_i	Label y_i
1	1	1
2	2	2
3	3	2
4	4	3
5	5	4
6	6	5
7	7	5
8	8	6

Client-wise hybrid sample partitioning is:

Client 1.

$$\mathcal{I}_1 = \{1, 2\}, \quad n_1 = 2.$$

Client 2.

$$\mathcal{I}_2 = \{3, 4\}, \quad n_2 = 2.$$

Client 3.

$$\mathcal{I}_3 = \{5, 6\}, \quad n_3 = 2.$$

Client 4.

$$\mathcal{I}_4 = \{7, 8\}, \quad n_4 = 2.$$

Thus,

$$N = n_1 + n_2 + n_3 + n_4 = 8.$$

Step 2: Explicit Local Loss Functions.

Client 1.

$$F_1(\theta_1) = \frac{1}{2}\big[(\theta_1 - 1)^2 + (2\theta_1 - 2)^2\big].$$

Expanding:

$$(\theta_1 - 1)^2 = \theta_1^2 - 2\theta_1 + 1,$$

$$(2\theta_1 - 2)^2 = 4\theta_1^2 - 8\theta_1 + 4.$$

Thus,

$$F_1(\theta_1) = \frac{1}{2}(5\theta_1^2 - 10\theta_1 + 5) = 2.5\theta_1^2 - 5\theta_1 + 2.5.$$

Client 2.

$$F_2(\theta_2) = \frac{1}{2}\big[(3\theta_2 - 2)^2 + (4\theta_2 - 3)^2\big].$$

Expanding:

$$(3\theta_2 - 2)^2 = 9\theta_2^2 - 12\theta_2 + 4,$$

$$(4\theta_2 - 3)^2 = 16\theta_2^2 - 24\theta_2 + 9.$$

Thus,

$$F_2(\theta_2) = 12.5\theta_2^2 - 18\theta_2 + 6.5.$$

Client 3.

$$F_3(\theta_3) = \frac{1}{2}\left[(5\theta_3 - 4)^2 + (6\theta_3 - 5)^2\right].$$

Expanding:

$$(5\theta_3 - 4)^2 = 25\theta_3^2 - 40\theta_3 + 16,$$

$$(6\theta_3 - 5)^2 = 36\theta_3^2 - 60\theta_3 + 25.$$

Thus,

$$F_3(\theta_3) = 30.5\theta_3^2 - 50\theta_3 + 20.5.$$

Client 4.

$$F_4(\theta_4) = \frac{1}{2}\left[(7\theta_4 - 5)^2 + (8\theta_4 - 6)^2\right].$$

Expanding:

$$(7\theta_4 - 5)^2 = 49\theta_4^2 - 70\theta_4 + 25,$$

$$(8\theta_4 - 6)^2 = 64\theta_4^2 - 96\theta_4 + 36.$$

Thus,

$$F_4(\theta_4) = 56.5\theta_4^2 - 83\theta_4 + 30.5.$$

Step 3: Augmented Lagrangian. Choose $\rho = 1$. The augmented Lagrangian is

$$\mathscr{L} = \sum_{k=1}^{4} \frac{1}{4} F_k(\theta_k) + \sum_{k=1}^{4} \alpha_k(\theta_k - \theta) + \frac{1}{2}\sum_{k=1}^{4} (\theta_k - \theta)^2.$$

Step 4: Local Primal Updates. In this step, each client minimizes the augmented Lagrangian $\mathscr{L}$ with respect to its own local parameter θ_k, while treating the global parameter θ and the dual variables α_k as constants. Because the model is scalar and the loss functions are quadratic, each local subproblem admits a closed-form solution.

Client 1. Recall the terms in $\mathscr{L}$ that depend on θ_1:

$$\mathscr{L}(\theta_1) = \frac{1}{4}F_1(\theta_1) + \alpha_1(\theta_1 - \theta) + \frac{1}{2}(\theta_1 - \theta)^2.$$

Substituting $F_1(\theta_1) = 2.5\theta_1^2 - 5\theta_1 + 2.5$:

$$\mathscr{L}(\theta_1) = \frac{1}{4}(2.5\theta_1^2 - 5\theta_1 + 2.5) + \alpha_1(\theta_1 - \theta) + \frac{1}{2}(\theta_1^2 - 2\theta_1\theta + \theta^2).$$

Expand all terms:

$$\mathscr{L}(\theta_1) = 0.625\theta_1^2 - 1.25\theta_1 + 0.625 + \alpha_1\theta_1 - \alpha_1\theta + \frac{1}{2}\theta_1^2 - \theta_1\theta + \frac{1}{2}\theta^2.$$

Collect θ_1^2 terms:

$$0.625\theta_1^2 + \frac{1}{2}\theta_1^2 = \frac{9}{8}\theta_1^2.$$

Collect θ_1 terms:

$$-1.25\theta_1 + \alpha_1\theta_1 - \theta_1\theta = \theta_1(\alpha_1 - \theta - \tfrac{5}{4}).$$

Now differentiate with respect to θ_1:

$$\frac{\partial \mathscr{L}}{\partial \theta_1} = \frac{9}{4}\theta_1 + \alpha_1 - \theta - \frac{5}{4}.$$

Setting the derivative to zero:

$$\frac{9}{4}\theta_1 = \theta - \alpha_1 + \frac{5}{4}.$$

Solving explicitly:

$$\boxed{\theta_1^{(t+1)} = \frac{4}{9}\left(\theta - \alpha_1 + \frac{5}{4}\right).}$$

Client 2. Relevant part of the augmented Lagrangian:

$$\mathscr{L}(\theta_2) = \frac{1}{4}F_2(\theta_2) + \alpha_2(\theta_2 - \theta) + \frac{1}{2}(\theta_2 - \theta)^2.$$

Substitute $F_2(\theta_2) = 12.5\theta_2^2 - 18\theta_2 + 6.5$:

$$\mathscr{L}(\theta_2) = 3.125\theta_2^2 - 4.5\theta_2 + 1.625 + \alpha_2\theta_2 - \alpha_2\theta + \frac{1}{2}\theta_2^2 - \theta_2\theta + \frac{1}{2}\theta^2.$$

Collect quadratic terms:

$$3.125\theta_2^2 + \frac{1}{2}\theta_2^2 = \frac{27}{8}\theta_2^2.$$

Differentiate:

$$\frac{\partial \mathcal{L}}{\partial \theta_2} = \frac{27}{4}\theta_2 + \alpha_2 - \theta - \frac{9}{2}.$$

Set to zero:

$$\frac{27}{4}\theta_2 = \theta - \alpha_2 + \frac{9}{2}.$$

Hence,

$$\boxed{\theta_2^{(t+1)} = \frac{4}{27}\left(\theta - \alpha_2 + \frac{9}{2}\right).}$$

Client 3. Using $F_3(\theta_3) = 30.5\theta_3^2 - 50\theta_3 + 20.5$:

$$\frac{1}{4}F_3(\theta_3) = 7.625\theta_3^2 - 12.5\theta_3 + 5.125.$$

Adding penalty and dual terms and differentiating:

$$\frac{\partial \mathcal{L}}{\partial \theta_3} = \frac{61}{4}\theta_3 + \alpha_3 - \theta - \frac{25}{2}.$$

Setting to zero:

$$\boxed{\theta_3^{(t+1)} = \frac{4}{61}\left(\theta - \alpha_3 + \frac{25}{2}\right).}$$

Client 4. Using $F_4(\theta_4) = 56.5\theta_4^2 - 83\theta_4 + 30.5$:

$$\frac{1}{4}F_4(\theta_4) = 14.125\theta_4^2 - 20.75\theta_4 + 7.625.$$

Derivative:

$$\frac{\partial \mathcal{L}}{\partial \theta_4} = \frac{113}{4}\theta_4 + \alpha_4 - \theta - \frac{83}{2}.$$

Thus,

$$\boxed{\theta_4^{(t+1)} = \frac{4}{113}\left(\theta - \alpha_4 + \frac{83}{2}\right).}$$

Step 5: Global Primal Update (Fully Expanded Derivation). In this step, we minimize the augmented Lagrangian $\mathcal{L}$ with respect to the *global parameter* θ, while treating all local primal variables $\{\theta_k\}$ and dual variables $\{\alpha_k\}$ as constants. Only the consensus-related terms of $\mathcal{L}$ depend on θ.

Step 5.1: Identify θ-dependent Terms. Recall the augmented Lagrangian:

$$\mathcal{L} = \sum_{k=1}^{4} \frac{1}{4} F_k(\theta_k) + \sum_{k=1}^{4} \alpha_k(\theta_k - \theta) + \frac{1}{2}\sum_{k=1}^{4}(\theta_k - \theta)^2.$$

The local loss terms $F_k(\theta_k)$ do *not* depend on θ. Thus, we retain only the θ-dependent components:

$$\mathcal{L}(\theta) = -\sum_{k=1}^{4} \alpha_k\theta + \frac{1}{2}\sum_{k=1}^{4}(\theta_k - \theta)^2 + (\text{constants w.r.t. } \theta).$$

Step 5.2: Expand the Quadratic Penalty Term.
For each client k,

$$(\theta_k - \theta)^2 = \theta_k^2 - 2\theta_k\theta + \theta^2.$$

Summing over all four clients:

$$\sum_{k=1}^{4}(\theta_k - \theta)^2 = \sum_{k=1}^{4}\theta_k^2 - 2\theta\sum_{k=1}^{4}\theta_k + 4\theta^2.$$

Multiplying by $\frac{1}{2}$:

$$\frac{1}{2}\sum_{k=1}^{4}(\theta_k - \theta)^2 = \frac{1}{2}\sum_{k=1}^{4}\theta_k^2 - \theta\sum_{k=1}^{4}\theta_k + 2\theta^2.$$

Step 5.3: Collect All θ-Dependent Terms. Substituting back into $\mathcal{L}(\theta)$:

$$\mathcal{L}(\theta) = -\sum_{k=1}^{4}\alpha_k\theta - \theta\sum_{k=1}^{4}\theta_k + 2\theta^2 + (\text{constants}).$$

Rewriting:

$$\mathscr{L}(\theta) = 2\theta^2 - \theta\left(\sum_{k=1}^{4}\theta_k + \sum_{k=1}^{4}\alpha_k\right) + \text{(constants)}.$$

Step 5.4: Differentiate with Respect to θ. Taking the derivative:

$$\frac{\partial\mathscr{L}}{\partial\theta} = 4\theta - \left(\sum_{k=1}^{4}\theta_k + \sum_{k=1}^{4}\alpha_k\right).$$

Equivalently,

$$\frac{\partial\mathscr{L}}{\partial\theta} = -\sum_{k=1}^{4}\alpha_k - \sum_{k=1}^{4}(\theta_k - \theta).$$

Step 5.5: Stationarity Condition. Setting the derivative to zero:

$$4\theta = \sum_{k=1}^{4}\theta_k + \sum_{k=1}^{4}\alpha_k.$$

Solving explicitly for θ:

$$\boxed{\theta^{(t+1)} = \frac{1}{4}\sum_{k=1}^{4}(\theta_k + \alpha_k).}$$

Interpretation. The global update is a *corrected average* of local parameters. Unlike FedAvg, which averages only θ_k, this update incorporates dual variables α_k, which encode accumulated disagreement. Clients with persistent deviation exert stronger influence through their dual terms, ensuring principled and stable convergence to consensus.

Step 6: Dual Updates. The dual updates arise from *gradient ascent* on the dual variables associated with the consensus constraints $\theta_k = \theta$. These constraints appear in the augmented Lagrangian through the linear terms $\alpha_k(\theta_k - \theta)$.

Step 6.1: Identify the Dual Objective. From the augmented Lagrangian,

$$\mathscr{L} = \sum_{k=1}^{4}\frac{1}{4}F_k(\theta_k) + \sum_{k=1}^{4}\alpha_k(\theta_k - \theta) + \frac{1}{2}\sum_{k=1}^{4}(\theta_k - \theta)^2,$$

the dual variables α_k appear only in the linear terms $\alpha_k(\theta_k - \theta)$.

Step 6.2: Compute Gradient with Respect to α_k. Taking the derivative of $\mathscr{L}$ with respect to α_k:

$$\frac{\partial \mathscr{L}}{\partial \alpha_k} = \theta_k - \theta.$$

This gradient measures the *instantaneous violation* of the consensus constraint for client k.

Step 6.3: Gradient Ascent Update. Using unit step size (absorbed into $\rho = 1$), the dual ascent update is:

$$\alpha_k^{(t+1)} = \alpha_k^{(t)} + (\theta_k^{(t+1)} - \theta^{(t+1)}), \quad k = 1, \ldots, 4.$$

$$\boxed{\alpha_k^{(t+1)} = \alpha_k^{(t)} + \left(\theta_k^{(t+1)} - \theta^{(t+1)}\right)}$$

Thus, each dual variable accumulates the signed deviation between the local and global models.

Step 7: Numerical Initialization and First Iteration.

Step 7.1: Initialization. We initialize all primal and dual variables at zero:

$$\theta^{(0)} = 0, \quad \theta_k^{(0)} = 0, \quad \alpha_k^{(0)} = 0, \quad k = 1, \ldots, 4.$$

Step 7.2: Local Primal Updates. From Step 4, the first-round local updates are:

$$\theta_1^{(1)} = \frac{5}{9}, \qquad \theta_2^{(1)} = \frac{2}{3}, \qquad \theta_3^{(1)} = \frac{50}{61}, \qquad \theta_4^{(1)} = \frac{166}{113}.$$

Step 7.3: Global Primal Update. Compute the global parameter:

$$\theta^{(1)} = \frac{1}{4}\left(\frac{5}{9} + \frac{2}{3} + \frac{50}{61} + \frac{166}{113}\right).$$

Convert each fraction to a common denominator is unnecessary for interpretation, but for numerical clarity we evaluate approximately:

$$\frac{5}{9} \approx 0.556, \quad \frac{2}{3} \approx 0.667, \quad \frac{50}{61} \approx 0.820, \quad \frac{166}{113} \approx 1.469.$$

Summing:

$$0.556 + 0.667 + 0.820 + 1.469 = 3.512.$$

Thus,

$$\theta^{(1)} \approx \frac{3.512}{4} = 0.878.$$

Step 7.4: Dual Updates. Using $\alpha_k^{(1)} = \theta_k^{(1)} - \theta^{(1)}$:

$$\alpha_1^{(1)} = \frac{5}{9} - 0.878 \approx -0.322,$$

$$\alpha_2^{(1)} = \frac{2}{3} - 0.878 \approx -0.211,$$

$$\alpha_3^{(1)} = \frac{50}{61} - 0.878 \approx -0.058,$$

$$\alpha_4^{(1)} = \frac{166}{113} - 0.878 \approx +0.591.$$

Interpretation. The dual variables immediately encode the direction and magnitude of disagreement: clients whose local solutions overshoot the global estimate accumulate positive dual values, while those undershooting accumulate negative ones. In subsequent iterations, these dual variables will directly influence local updates, exerting mathematically principled corrective forces that drive the system toward consensus despite strong data heterogeneity.

Step 8: Interpretation. Each client's update reflects its local data scale and slope. Dual variables encode disagreement and progressively enforce consensus without requiring identical updates at every round.

Key Insight

Primal–dual optimization transforms hybrid federated learning into a numerically transparent, consensus-driven process where disagreement is measured, accumulated, and corrected explicitly rather than averaged away.

14.4 Convergence Rates under Partial Participation

Convergence analysis in hybrid federated learning is inherently more subtle and nuanced than in classical centralized or even standard federated learning settings. The expressive power gained by allowing hybrid data partitioning and by introducing explicit primal–dual coordination fundamentally alters the optimization landscape. While these mechanisms greatly enhance robustness and flexibility, they also introduce additional sources of stochasticity, delay, and coupling that must be accounted for in any rigorous convergence analysis.

At a conceptual level, convergence behavior in hybrid federated systems emerges from the interaction of three fundamental and often competing factors:

- **Hybrid data heterogeneity**, arising from simultaneous non-identical data distributions across both samples and features. Clients may observe different subsets of entities, different subsets of features, or both, leading to objective functions that differ not only statistically but also structurally.
- **Partial and stochastic client participation**, in which only a randomly selected subset of clients contributes updates during each communication round. This randomness may reflect device availability, energy constraints, privacy policies, or network conditions, and it introduces additional variance into the global update.
- **Delayed or asynchronous updates**, caused by heterogeneous computation speeds, straggling clients, communication latency, or intentional asynchronous protocols. These delays break the synchronous update assumption commonly used in classical optimization analysis.

Unlike centralized optimization, where gradients from the full dataset are available at every iteration, hybrid federated learning operates under conditions of partial observability. At each communication round, the server receives only a subset of local updates, each computed using incomplete and possibly outdated views of the global model. Moreover, these updates may correspond to different feature blocks or different subsets of samples. As a consequence, the effective update direction is both noisy and biased, even when local objectives are convex and smooth.

For this reason, convergence in hybrid federated learning cannot be characterized deterministically. Instead, performance guarantees are typically expressed in terms of *expected optimality gaps* or *expected stationarity measures*. Formally, convergence results take the form

$$\mathbb{E}\left[F(\theta^{(t)}) - F(\theta^{\star})\right],$$

where the expectation is taken over the randomness induced by client sampling, stochastic gradients, and communication delays. Such expectation-based characterizations reflect the inherently stochastic nature of hybrid federated optimization and provide a more realistic description of algorithmic behavior in large-scale, decentralized systems.

From a pedagogical perspective, it is important to emphasize that sublinear convergence under partial participation is not a limitation but a fundamental consequence of decentralization. The goal of hybrid federated methods is therefore not to match the convergence speed of fully centralized algorithms, but to achieve *stable and predictable progress* under realistic system and data constraints. In the sections that follow, we formalize these intuitions by stating typical assumptions, deriving convergence rates, and interpreting the resulting bounds in the context of hybrid data structures and partial participation.

14.4.1 Assumptions for Convergence Analysis

To derive meaningful and interpretable convergence guarantees for hybrid federated learning algorithms, it is necessary to impose a collection of structural, statistical, and system-level assumptions. These assumptions are standard in the optimization literature, yet they require careful qualification in the hybrid federated setting due to data decentralization, feature partitioning, and partial participation. Rather than serving as idealized abstractions, they provide a controlled framework within which the effects of heterogeneity and stochasticity can be rigorously analyzed.

Convexity and Smoothness of Local Objectives. Each local objective function $F_k(\theta)$ is assumed to be convex and L-smooth. Convexity ensures that local minima coincide with global minima, ruling out pathological optimization landscapes. L-smoothness implies that the gradients of F_k are Lipschitz continuous, which bounds how rapidly gradients can change with respect to the model parameters. Together, these properties guarantee well-behaved local updates and enable the use of gradient-based convergence arguments. In hybrid federated learning, smoothness is particularly important because gradients may be computed using incomplete feature subsets or block-structured parameters.

Existence of a Global Optimum. The global objective

$$F(\theta) = \sum_{k=1}^{K} \frac{n_k}{N} F_k(\theta)$$

is assumed to admit at least one optimal solution $\theta^\star$. This assumption ensures that the learning problem is well-posed and that there exists a meaningful notion of convergence. In practice, $\theta^\star$ represents the best achievable compromise among heterogeneous client objectives under the imposed consensus constraints.

Bounded Variance of Stochastic Gradients. Local gradient estimates are assumed to have bounded variance, even when computed under feature-wise or sample-wise partitioning. Formally, for each client k,

$$\mathbb{E}\big[\|\nabla F_k(\theta; \xi_k) - \nabla F_k(\theta)\|^2\big] \le \sigma_k^2,$$

where ξ_k denotes local stochasticity due to mini-batching or data sampling. This assumption limits the noise introduced by stochastic optimization and prevents divergence caused by excessively noisy updates. In hybrid settings, bounded variance is nontrivial because gradients may be evaluated on partial feature sets, yet it can be justified under standard regularity conditions.

Random Client Participation. Client participation is modeled as an independent random process, where each client k participates in a given round with probability $p_k > 0$. This assumption captures the inherently stochastic nature of real-world federated systems, where client availability may depend on connectivity, energy constraints, or privacy policies. Requiring $p_k > 0$ ensures that every client contributes infinitely often in expectation, which is essential for long-term convergence. The participation probabilities directly influence the variance of the aggregated updates and therefore appear explicitly in convergence bounds.

Consequences for Convergence Rates. Under the above assumptions, primal–dual hybrid federated learning algorithms achieve sublinear convergence rates characteristic of stochastic optimization. In particular, the expected optimality gap satisfies

$$\mathbb{E}\left[F(\theta^{(t)}) - F(\theta^{\star})\right] = \mathcal{O}\left(\frac{1}{\sqrt{t}}\right),$$

where t denotes the number of communication rounds. The hidden constants in this bound depend explicitly on the smoothness constant L, the penalty parameter ρ, the degree of data and feature heterogeneity, and the participation probabilities $\{p_k\}$. This dependence highlights a central trade-off: stronger enforcement of consensus and higher participation rates accelerate convergence but may increase communication and computation costs.

14.4.2 Effect of Partial Participation

Partial participation plays a dual and nuanced role in the convergence behavior of hybrid federated learning algorithms. On the positive side, it substantially reduces communication overhead and improves scalability by allowing only a subset of clients to participate in each round. This property is essential for large-scale deployments involving thousands or millions of devices. On the negative side, partial participation introduces additional stochasticity into the global update, effectively increasing the variance of the optimization process and slowing convergence.

In hybrid federated settings, the impact of partial participation is often amplified. Missing clients may correspond not only to missing data samples but also to missing feature blocks or incomplete coverage of the model parameters. As a result, the aggregated update may be structurally incomplete, leading to biased or delayed progress in certain parameter directions.

Primal–dual methods mitigate these challenges through the explicit use of dual variables. When a client does not participate in a given round, both its local primal variables and its dual variables remain unchanged. Upon rejoining, the accumulated dual discrepancy reflects the extent to which the client's local model has diverged from the global consensus. This discrepancy automatically reweights the client's influence, preventing sudden jumps or destabilizing corrections.

This mechanism explains why hybrid primal–dual methods often exhibit smoother and more stable convergence trajectories than classical Federated Averaging under severe heterogeneity and sparse participation. Although their asymptotic convergence rates may be similar, the transient behavior of hybrid methods is typically more predictable and robust, making them particularly attractive for real-world federated systems characterized by irregular participation and heterogeneous data structures.

14.4.3 Comparison with Classical Federated Averaging

From a theoretical standpoint, both classical Federated Averaging (FedAvg) and hybrid primal–dual federated methods achieve sublinear convergence rates under standard assumptions of convexity, smoothness, and bounded gradient variance. In particular, when local objectives are convex and client participation is sufficiently frequent, both classes of algorithms admit convergence bounds of order $\mathcal{O}(1/\sqrt{t})$ in expectation. However, despite these superficial similarities in asymptotic rates, the qualitative behavior of the two approaches differs substantially, especially in the presence of strong heterogeneity and partial participation.

FedAvg resolves the tension between local autonomy and global agreement implicitly. Each round consists of multiple local optimization steps followed by simple parameter averaging at the server. This averaging operation acts as an implicit regularizer, smoothing local updates and nudging models toward consensus without explicitly enforcing equality constraints. While this mechanism is computationally efficient and easy to implement, it provides limited theoretical control over disagreement among clients. When local data distributions are highly non-identical or when only a small fraction of clients participates in each round, the averaging process may amplify local drift, leading to oscillations or even divergence.

In contrast, hybrid primal–dual methods enforce consensus explicitly by introducing dual variables associated with agreement constraints. Disagreement between local and global models is not merely averaged away, but is quantified, accumulated, and corrected through principled updates. This explicit modeling of disagreement allows hybrid methods to tolerate much stronger forms of heterogeneity, including feature-wise partitioning and block-structured models. As a result, hybrid approaches often exhibit markedly improved stability under severe data imbalance and sparse participation, even when their asymptotic convergence rates match those of FedAvg.

The enhanced robustness of hybrid methods comes at a nontrivial cost. Each communication round typically involves additional computation in the form of dual variable updates, block-wise optimization, and, in some cases, local solutions of proximal subproblems. Memory requirements are also higher, as clients must store both primal and dual variables across rounds. Furthermore, the performance of hybrid primal–dual methods is more sensitive to the choice of algorithmic hyper-parameters, such as penalty coefficients and step sizes. Improper tuning can slow convergence or negate the benefits of explicit consensus enforcement.

Despite these costs, hybrid methods are often preferable in regimes that challenge the assumptions underlying FedAvg. When data is extremely imbalanced, when features are partitioned across clients, or when participation is intermittent and unpredictable, the stability gains afforded by primal–dual coordination frequently outweigh the added complexity. In such scenarios, FedAvg's simplicity becomes a liability, whereas hybrid methods offer a controlled and theoretically grounded path to convergence.

Convergence Insight
Hybrid federated methods achieve sublinear convergence under partial participation while providing superior stability in the presence of strong data and feature heterogeneity. This stability arises from the explicit modeling of disagreement through dual variables, at the expense of increased algorithmic and computational complexity.

In summary, convergence in hybrid federated learning must be understood as a balance among statistical efficiency, communication efficiency, and structural robustness. Classical FedAvg emphasizes simplicity and low overhead but relies on implicit regularization mechanisms that may fail under extreme heterogeneity. Hybrid primal–dual formulations, by contrast, expose and manage disagreement explicitly, providing the mathematical structure required to sustain reliable learning even when classical assumptions of uniform participation and homogeneous data no longer hold.

Worked Example: Convergence under Partial Participation
This worked example provides a fully numerical and step-by-step illustration of convergence behavior in hybrid federated learning under *partial client participation*. The goal is to make the abstract convergence discussion concrete by explicitly computing objective values, expected progress, and optimality gaps over multiple communication rounds. All calculations are shown in detail, and no intermediate step is omitted.

Problem Setup. Consider a simple convex regression problem with a *scalar* model parameter $\theta \in \mathbb{R}$ and four clients ($K = 4$). Each client minimizes a local squared-error objective:

$$F_k(\theta) = \frac{1}{n_k} \sum_{i \in \mathcal{I}_k} (\theta x_i - y_i)^2.$$

The global objective is

$$F(\theta) = \sum_{k=1}^{4} \frac{n_k}{N} F_k(\theta), \quad N = \sum_{k=1}^{4} n_k.$$

Step 1: Numerical Dataset Construction. Assume the following global dataset of eight samples:

Sample i	Feature x_i	Label y_i
1	1	1
2	2	2
3	3	2
4	4	3
5	5	4
6	6	5
7	7	5
8	8	6

Client-wise partitioning is:

Client 1. $\mathcal{I}_1 = \{1, 2\}, \; n_1 = 2$

Client 2. $\mathcal{I}_2 = \{3, 4\}, \; n_2 = 2$

Client 3. $\mathcal{I}_3 = \{5, 6\}, \; n_3 = 2$

Client 4. $\mathcal{I}_4 = \{7, 8\}, \; n_4 = 2$

Thus, $N = 8$ and each client contributes weight $\frac{1}{4}$.

Step 2: Explicit Local Objective Functions.

Client 1.

$$F_1(\theta) = \frac{1}{2}\left[(\theta - 1)^2 + (2\theta - 2)^2\right] = 2.5\theta^2 - 5\theta + 2.5.$$

Client 2.

$$F_2(\theta) = \frac{1}{2}\left[(3\theta - 2)^2 + (4\theta - 3)^2\right] = 12.5\theta^2 - 18\theta + 6.5.$$

Client 3.

$$F_3(\theta) = \frac{1}{2}\left[(5\theta - 4)^2 + (6\theta - 5)^2\right] = 30.5\theta^2 - 50\theta + 20.5.$$

Client 4.

$$F_4(\theta) = \frac{1}{2}\left[(7\theta - 5)^2 + (8\theta - 6)^2\right] = 56.5\theta^2 - 83\theta + 30.5.$$

Step 3: Global Objective and Optimal Solution.
The global objective is

$$F(\theta) = \frac{1}{4}(F_1 + F_2 + F_3 + F_4).$$

Summing coefficients:

$$F(\theta) = 25.5\theta^2 - 39\theta + 15.$$

Taking derivative:

$$F'(\theta) = 51\theta - 39.$$

Setting $F'(\theta^\star) = 0$:

$$\theta^\star = \frac{39}{51} = \frac{13}{17} \approx 0.765.$$

Step 4: Partial Participation Model. Assume each round selects *two clients uniformly at random*. Thus, each client participates with probability $p_k = \frac{1}{2}$.
We analyze *expected* progress.

Step 5: Initialization.
Initialize the global model:

$$\theta^{(0)} = 0.$$

Compute initial optimality gap:

$$F(\theta^{(0)}) - F(\theta^\star) = 15 - (25.5 \cdot 0.765^2 - 39 \cdot 0.765 + 15).$$

Compute numerically:

$$F(\theta^\star) \approx 0.08, \quad \Rightarrow F(\theta^{(0)}) - F(\theta^\star) \approx 14.92.$$

Step 6: One Round with Partial Participation. Assume clients $\{2, 4\}$ participate in round $t = 0$.

Local Gradient Computation.

$$\nabla F_2(\theta^{(0)}) = -18, \quad \nabla F_4(\theta^{(0)}) = -83.$$

Aggregated Gradient. Expected gradient estimator:

$$g^{(0)} = \frac{1}{2}\left(\frac{1}{4}\nabla F_2 + \frac{1}{4}\nabla F_4\right) = \frac{1}{8}(-101) = -12.625.$$

Step 7: Global Update. Using step size $\eta = 0.05$:

$$\theta^{(1)} = \theta^{(0)} - \eta g^{(0)} = 0 + 0.63125.$$

Step 8: New Objective Value.

$$F(\theta^{(1)}) = 25.5(0.63125)^2 - 39(0.63125) + 15 \approx 1.24.$$

Optimality gap:

$$F(\theta^{(1)}) - F(\theta^{\star}) \approx 1.16.$$

Step 9: Expected Progress Interpretation. Despite observing only *half the clients*, the optimality gap drops from ≈ 14.92 to ≈ 1.16 in expectation. However, the update direction is noisy and biased due to missing clients.

Step 10: Relation to Sublinear Convergence. Repeating this process yields:

$$\mathbb{E}[F(\theta^{(t)}) - F(\theta^{\star})] \leq \mathcal{O}\left(\frac{1}{\sqrt{t}}\right),$$

with constants inflated by:

- participation probability p_k,
- gradient variance from missing clients,
- data heterogeneity across clients.

Key Insight from the Example

Partial participation slows convergence by increasing variance, but expected progress remains guaranteed. Hybrid primal–dual methods stabilize this process by accumulating and correcting disagreement, ensuring reliable convergence even when only a subset of clients contributes at each round.

14.5 Comparison with HyFEM and Federated Averaging

Hybrid federated learning does not emerge in isolation; rather, it subsumes and generalizes several prominent paradigms that have been proposed to address data decentralization under different structural assumptions. In particular, Federated Averaging (FedAvg) and Hybrid Federated Expectation Maximization (HyFEM) represent two influential but complementary design philosophies. A comparative summary of these approaches is provided in Table 14.2, and the key distinctions are discussed below.

Federated Averaging. Federated Averaging (FedAvg) is the canonical federated learning algorithm and serves as a baseline for most theoretical and empirical studies. It assumes a purely horizontal data partitioning, in which all clients share an identical feature space but hold disjoint subsets of samples. Local training is followed by direct parameter averaging at the server, which implicitly encourages consensus. This simplicity makes FedAvg highly communication-efficient and easy to deploy at scale. However, because agreement among clients is enforced only through averaging, FedAvg is particularly sensitive to non-IID data, sparse participation, and client drift. When data heterogeneity is severe or participation is highly irregular, FedAvg may exhibit oscillatory behavior or slow convergence.

HyFEM. Hybrid Federated Expectation Maximization (HyFEM) extends federated learning to settings with feature-wise partitioning by adopting a probabilistic modeling perspective. By alternating between expectation and maximization steps, HyFEM can handle distributed feature representations and latent variables in a principled manner. This approach is well suited to scenarios where the underlying data generation process admits a clear probabilistic interpretation. Nevertheless, HyFEM typically relies on stronger assumptions regarding data overlap, conditional independence, and model structure. As a result, its applicability may be limited in highly heterogeneous or unstructured environments, and its convergence guarantees often depend on problem-specific regularity conditions.

Hybrid Federated Learning. Hybrid federated learning adopts a more general optimization-centric viewpoint. Rather than committing to a specific probabilistic

Table 14.2 Comparison of Federated Averaging, HyFEM, and Hybrid Federated Learning

Aspect	Federated Averaging (FedAvg)	HyFEM	Hybrid Federated Learning
Data Partitioning	Purely horizontal (sample-wise)	Hybrid, primarily feature-wise with overlap assumptions	Fully hybrid: sample-wise, feature-wise, and block-structured
Model Assumptions	Shared global model across all clients	Strong assumptions on probabilistic model structure	Minimal structural assumptions; supports heterogeneous models
Consensus Mechanism	Implicit via parameter averaging	Implicit via EM-style alternation	Explicit via primal–dual consensus constraints
Handling Non-IID Data	Limited robustness; prone to client drift	Moderate robustness under structured heterogeneity	High robustness under severe data and feature heterogeneity
Partial Participation	Supported, but may destabilize averaging	Typically assumes stable participation	Naturally supported through primal–dual updates
Computational Overhead	Low	Moderate (iterative EM steps)	Higher (dual variables and block-wise optimization)
Convergence Behavior	Sublinear; sensitive to heterogeneity	Sublinear under strong assumptions	Sublinear with improved stability guarantees
Typical Use Cases	Mobile devices, simple FL deployments	Structured feature-partitioned learning	Complex hybrid systems with irregular participation

model or relying on implicit averaging, it integrates sample-wise and feature-wise partitioning within a unified primal–dual optimization framework. This explicit formulation of consensus constraints allows hybrid federated methods to tolerate extreme data imbalance, partial feature visibility, and intermittent participation. Although this generality introduces additional algorithmic complexity and per-round computational overhead, it provides stronger robustness and greater flexibility across a wide range of real-world deployment scenarios.

Comparative Perspective

Federated Averaging emphasizes simplicity and low overhead, HyFEM emphasizes probabilistic structure and interpretability, while hybrid federated learning emphasizes generality, explicit consensus modeling, and robustness under strong data and participation heterogeneity.

14.6 Practical Considerations and Limitations

While hybrid federated learning offers a powerful and unifying framework for learning under mixed data partitioning, its practical deployment raises a number of important engineering and system-level challenges. These challenges do not undermine the conceptual value of the approach, but they significantly influence when, where, and how hybrid methods can be deployed effectively. A clear understanding of these considerations is therefore essential for translating theoretical advances into robust real-world systems.

Algorithmic and System Complexity

One of the most immediate consequences of adopting a hybrid federated formulation is the increase in algorithmic complexity. Unlike classical Federated Averaging, which maintains only a single set of model parameters per client, hybrid methods typically require the simultaneous management of multiple variable blocks. Each participating client must store local primal parameters, corresponding dual variables, and in some cases additional auxiliary states related to feature blocks or constraint handling. This increase in state dimensionality leads to higher memory requirements and more intricate update logic.

From a systems perspective, the coordination of primal and dual variables introduces additional synchronization points. Although hybrid methods are designed to tolerate partial participation, care must be taken to ensure that stale dual variables do not accumulate indefinitely or destabilize the optimization process. These considerations necessitate more sophisticated client-side and server-side software stacks than those required for FedAvg-style deployments.

Hyperparameter Sensitivity

Hybrid federated learning algorithms are typically more sensitive to hyperparameter choices than their classical counterparts. In particular, penalty coefficients in augmented Lagrangian formulations and step sizes for primal and dual updates play a central role in determining convergence behavior. If penalty parameters are set too small, local models may drift excessively, weakening consensus and slowing convergence. Conversely, overly large penalty values may overly constrain local updates, negating the benefits of personalization and increasing numerical instability.

Unlike centralized optimization, where hyperparameters can often be tuned empirically with relative ease, federated environments impose severe constraints on experimentation and observability. As a result, robust default parameter choices and adaptive tuning strategies remain an open and practically important research problem in hybrid federated learning.

Communication and Computation Overhead

Hybrid federated methods frequently incur higher per-round communication costs than FedAvg, particularly in feature-partitioned settings. When different clients are responsible for disjoint feature blocks, global updates may require the exchange of block-specific information, dual variables, or auxiliary statistics. Although these exchanges can often be compressed or sparsified, they nevertheless increase communication load, especially in wide models with large feature dimensions.

On the computation side, local updates may involve solving proximal subproblems or performing block-wise optimization rather than simple gradient descent. These operations can be substantially more expensive than standard local training steps, particularly on devices with limited processing capabilities.

Resource Constraints and Deployment Scenarios

The cumulative effect of increased memory usage, computational complexity, and communication overhead makes hybrid federated learning less suitable for highly resource-constrained devices, such as low-power mobile or embedded systems. In practice, hybrid methods are more naturally aligned with institutional, enterprise, or edge–cloud federated environments, where clients have relatively stable connectivity and sufficient storage and compute resources.

Typical deployment scenarios include multi-institutional healthcare networks, collaborative scientific computing platforms, and industrial monitoring systems, where data heterogeneity is pronounced and the cost of increased complexity is justified by the need for robustness and flexibility. In such settings, the ability of hybrid federated learning to model mixed data partitioning often outweighs its implementation overhead.

Design Trade-Off

Hybrid federated learning expands modeling flexibility and robustness by explicitly addressing mixed data partitioning and client heterogeneity. This flexibility is achieved at the cost of increased algorithmic complexity, higher memory and communication overhead, and greater sensitivity to hyperparameter choices.

Outlook

Despite these limitations, hybrid federated learning represents a significant step toward realistic collaborative learning systems that faithfully reflect the structure of distributed data. By providing a principled optimization framework for simultaneous sample-wise and feature-wise decentralization, it lays the groundwork for the integration of advanced privacy mechanisms, secure aggregation protocols, and incentive-compatible participation schemes. These extensions, which are explored in subsequent chapters, are essential for deploying hybrid federated learning at scale in privacy-sensitive and economically complex environments.

Exercises

This chapter examined the motivation, formulation, optimization frameworks, convergence properties, and practical trade-offs of *Hybrid Federated Learning*, emphasizing hybrid data partitioning, primal–dual coordination, and robustness under partial participation.

A. Short Answer Questions

Answer each question briefly and precisely.

1. What is meant by *hybrid data partitioning* in federated learning?
2. How does hybrid federated learning generalize horizontal and vertical federated learning?
3. Why is parameter averaging insufficient in hybrid federated settings?
4. What role do dual variables play in hybrid federated optimization?
5. Define partial client participation and explain why it arises in practice.
6. What is meant by consensus constraints in federated learning?
7. Why are convergence guarantees in hybrid federated learning typically stated in expectation?
8. How does feature-wise partitioning affect local objective functions?
9. What is the main advantage of primal–dual formulations over FedAvg?
10. Give one real-world application scenario where hybrid federated learning is necessary.

B. Long Answer Questions

Answer the following questions with clear explanations, equations, and diagrams wherever appropriate.

1. Explain the concept of hybrid federated learning and motivate its necessity using real-world examples.
2. Derive the consensus-constrained formulation of federated learning and explain its significance.
3. Describe the augmented Lagrangian used in hybrid federated learning and interpret each term.
4. Explain the alternating primal–dual update strategy and relate it to ADMM.
5. Discuss how hybrid federated learning handles feature-wise partitioning differently from FedAvg.
6. Analyze the effect of partial participation on convergence rates in hybrid federated learning.
7. Compare and contrast FedAvg, HyFEM, and hybrid federated learning from an optimization perspective.

8. Explain why hybrid federated learning typically exhibits sublinear convergence.
9. Discuss the trade-offs between robustness and computational complexity in hybrid federated methods.
10. Using a simple linear regression example, explain how dual variables correct local drift.

C. True/False Questions

State whether each statement is **True** or **False**.

1. Hybrid federated learning assumes that all clients observe the same feature space.
2. FedAvg enforces consensus explicitly through dual variables.
3. Feature-wise partitioning leads to structurally different local objectives.
4. Partial participation always prevents convergence in federated learning.
5. Dual variables accumulate disagreement over multiple communication rounds.
6. HyFEM relies on a probabilistic EM-style optimization procedure.
7. Hybrid federated learning can subsume both horizontal and vertical federated learning.
8. Sublinear convergence is a consequence of stochasticity and decentralization.
9. Increasing the penalty parameter ρ enforces stronger consensus.
10. Hybrid federated learning is always more communication-efficient than FedAvg.

Answers:

1. False
2. False
3. True
4. False
5. True
6. True
7. True
8. True
9. True
10. False

D. Fill in the Blanks

Fill in the blanks with the most appropriate term or expression.

1. Federated Averaging assumes purely ____________ data partitioning.

2. Hybrid federated learning allows partitioning across both ___________ and features.
3. In primal–dual methods, consensus violations are captured by ___________ variables.
4. Partial participation increases the ___________ of the optimization process.
5. The augmented Lagrangian includes a quadratic ___________ term.
6. Sublinear convergence is typically of order $\mathcal{O}$(___________).
7. Feature-wise partitioning generalizes ___________ federated learning.
8. HyFDCA operates primarily in the ___________ space.
9. When ρ is large, consensus is enforced more ___________.
10. Hybrid federated learning is well suited to ___________ data ecosystems.

Answers:

1. horizontal
2. samples
3. dual
4. variance
5. penalty
6. $1/\sqrt{t}$
7. vertical
8. dual
9. strongly
10. heterogeneous

E. Multiple Choice Questions (MCQs)

Choose the **most appropriate** answer.

1. Which setting motivates hybrid federated learning most strongly?
   (a) IID centralized datasets
   (b) Single-client learning
   (c) Mixed sample-wise and feature-wise decentralization
   (d) Fully labeled data
2. What mechanism enforces agreement in hybrid primal–dual methods?
   (a) Parameter dropout
   (b) Momentum averaging
   (c) Dual variables
   (d) Batch normalization
3. Which algorithm relies on EM-style alternation?
   (a) FedAvg
   (b) SGD
   (c) HyFEM
   (d) ADAM

4. Partial participation primarily affects convergence by increasing:
   (a) Bias only
   (b) Variance
   (c) Model capacity
   (d) Feature overlap
5. In hybrid FL, clients may update:
   (a) All parameters always
   (b) Only dual variables
   (c) Only observed parameter blocks
   (d) Random parameters
6. Which convergence rate is typical under partial participation?
   (a) $\mathcal{O}(1/t)$
   (b) $\mathcal{O}(\log t)$
   (c) $\mathcal{O}(1/\sqrt{t})$
   (d) Constant
7. What does the penalty parameter ρ control?
   (a) Model depth
   (b) Consensus strength
   (c) Learning rate decay
   (d) Feature dimension
8. Which method is most sensitive to non-IID data?
   (a) Hybrid FL
   (b) HyFDCA
   (c) FedAvg
   (d) Primal–dual FL
9. Hybrid federated learning is best described as:
   (a) A probabilistic EM algorithm
   (b) A parameter averaging heuristic
   (c) A constrained optimization framework
   (d) A centralized learning method
10. Dual variables primarily measure:
   (a) Prediction error
   (b) Feature importance
   (c) Consensus disagreement
   (d) Data imbalance

Answers:

1. (c)
2. (c)
3. (c)
4. (b)
5. (c)
6. (c)

7. (b)
8. (c)
9. (c)
10. (c)

F. Application and Project-Based Questions

These problems emphasize modeling, reasoning, and implementation.

1. Design a hybrid federated learning setup for a multi-hospital healthcare network and justify the data partitioning.
2. Implement a simple hybrid FL algorithm for linear regression with feature-wise partitioning.
3. Simulate partial participation and study its effect on convergence empirically.
4. Compare FedAvg and a primal–dual hybrid method on a non-IID synthetic dataset.
5. Analyze the effect of the penalty parameter ρ on convergence stability.
6. Propose an adaptive strategy for tuning primal and dual step sizes.
7. Extend a hybrid federated framework to support asynchronous client updates.
8. Design a privacy-preserving mechanism compatible with hybrid FL.
9. Apply hybrid FL to an IoT sensing scenario with heterogeneous sensors.
10. Critically evaluate the trade-offs between generality and efficiency in hybrid federated learning.

Part VI

Privacy, Security, Governance, and Applications

Privacy-Preserving Techniques in Federated Learning

15

Federated learning is frequently presented as a privacy-aware paradigm because it fundamentally departs from the centralized data collection model that dominates classical machine learning. By design, federated learning avoids the explicit transfer of raw data from clients to a central server, thereby reducing direct exposure of sensitive information and alleviating regulatory, legal, and ethical concerns associated with centralized data storage. This architectural shift—from moving data to models, to moving models to data—represents a major conceptual advance in privacy-conscious machine learning.

However, data locality alone does not imply privacy. Although raw data never leaves client devices, federated learning requires clients to repeatedly communicate information derived from their local datasets, such as gradients, parameter updates, or locally trained model weights. These communicated quantities are not arbitrary; they are structured, data-dependent signals produced by deterministic or weakly stochastic optimization algorithms. As a result, they may encode substantial information about the underlying training data. An adversary with access to these updates—whether a curious server, a malicious participant, or an external observer—may be able to infer sensitive attributes, reconstruct training samples, or determine whether specific data points were used during training.

This observation highlights a critical but often misunderstood point: *federated learning provides data decentralization, not automatic privacy*. While decentralization reduces the attack surface associated with raw data aggregation, it does not, by itself, prevent information leakage through model updates. In fact, the iterative and communication-heavy nature of federated optimization can amplify certain privacy risks, as repeated exposure to updates across many rounds may allow adversaries to accumulate and correlate information over time.

The goal of this chapter is to develop a rigorous and systematic understanding of privacy in federated learning beyond architectural intuition. We begin by

P. P. Ray, *Foundations of Federated Learning*,
Springer Series in the Data Sciences,
https://doi.org/10.1007/978-3-032-34084-9_15

analyzing why privacy risks persist even when data remains local, focusing on the inherent information content of model updates and the threat models relevant to federated systems. We then introduce formal privacy frameworks—most notably differential privacy—that provide mathematically quantifiable guarantees against information leakage. Subsequently, we examine cryptographic techniques such as secure aggregation, secure multi-party computation, and homomorphic encryption, which protect model updates from unauthorized inspection while preserving the correctness of federated optimization.

Throughout the chapter, we emphasize a fundamental conceptual distinction. *Architectural privacy* refers to system-level design choices that restrict where data resides and how it is accessed. *Algorithmic privacy*, by contrast, concerns the formal guarantees that limit what can be inferred from the outputs of learning algorithms, even by entities that legitimately observe those outputs. Effective privacy-preserving federated learning requires both: architectural constraints to limit exposure and algorithmic or cryptographic mechanisms to bound inference.

By the end of this chapter, federated learning will be understood not as an inherently private solution, but as a flexible framework within which privacy guarantees must be explicitly designed, quantified, and enforced. This perspective lays the foundation for principled system design and clarifies the trade-offs between privacy, utility, communication cost, and computational overhead that shape modern federated learning systems.

15.1 Privacy Risks in Model Updates

The defining communication primitive in federated learning is the exchange of *model-related information* rather than raw training data. Instead of transmitting samples (x, y), clients communicate quantities derived from local optimization, such as gradients, parameter updates, or locally trained model weights. This design choice is central to federated learning's appeal: by keeping raw data local, the system avoids the most direct and obvious forms of data exposure.

However, replacing data transfer with model-update transfer does not eliminate privacy risk. Model updates are not arbitrary messages; they are structured numerical objects produced by deterministic or weakly stochastic learning algorithms applied to sensitive data. Consequently, they often encode far more information about the training data than is immediately apparent. From an information-theoretic perspective, gradients and parameter updates are functions of the data, and functions of sensitive data can themselves be sensitive.

This section examines why model updates can leak private information, even in the absence of protocol violations, and why privacy risks persist despite strict data locality.

15.1.1 Gradient and Update Leakage

At the heart of federated learning lies local optimization. Each client computes gradients or parameter updates by minimizing a local empirical risk function

$$F_k(\theta) = \frac{1}{n_k} \sum_{(x,y)\in\mathcal{D}_k} \ell(f_\theta(x), y),$$

and transmits the resulting update to the server. These updates are therefore explicit functions of the local dataset $\mathcal{D}_k$.

For many learning models—particularly deep neural networks—the mapping from data to gradients is highly expressive. Overparameterized models can fit individual training examples with remarkable precision, and the gradients they produce may preserve fine-grained information about specific samples. In extreme cases, a single gradient vector computed on a small batch may uniquely determine the underlying data points up to trivial transformations.

This phenomenon is known as *gradient leakage*. An adversary who observes gradients or parameter updates can exploit the strong coupling between data and updates to infer sensitive information. The risk is exacerbated when:

- local batch sizes are small,
- models are highly expressive,
- updates are shared frequently across many communication rounds,
- optimization is deterministic or weakly randomized.

Importantly, leakage is not limited to gradients. Parameter differences and full model weights may encode cumulative information about the entire local dataset. As training progresses, successive updates can be correlated to refine inferences, allowing an adversary to gradually reconstruct sensitive aspects of client data.

15.1.2 Types of Privacy Attacks

Privacy leakage from model updates manifests through several well-studied attack classes. These attacks differ in their objectives, assumptions, and severity, but they all exploit the data-dependence of shared updates.

Membership Inference Attacks. In membership inference, an adversary seeks to determine whether a specific data point was used during training. In federated learning, this may involve testing whether a particular record belongs to a given client's dataset. Such attacks are particularly concerning in sensitive domains, where the mere inclusion of a data point—such as a medical record—constitutes a privacy violation. Membership inference exploits the fact that models often behave differently on training samples than on unseen data, and that this difference may be reflected in gradients or updates.

Attribute Inference Attacks. Attribute inference attacks aim to recover sensitive attributes of data records when some non-sensitive attributes are known. For example, given partial feature information, an adversary may infer hidden demographic or medical attributes by analyzing how model updates respond to known inputs. In federated settings, attribute inference may occur even when the adversary has no direct access to raw features, relying solely on the statistical patterns encoded in updates.

Gradient Inversion and Data Reconstruction. Perhaps the most striking class of attacks involves reconstructing training samples directly from gradients or updates. Gradient inversion attacks formulate an optimization problem in which the adversary searches for inputs whose gradients match the observed ones. Under certain conditions—such as small batch sizes and known model architectures—this process can yield highly accurate reconstructions of original data, including images or text sequences. These attacks demonstrate conclusively that gradients can act as near-lossless encodings of training data.

Together, these attack vectors illustrate that sharing model updates is not inherently safe. Even without access to raw data, adversaries can extract sensitive information if updates are insufficiently protected.

15.1.3 Honest-but-Curious Threat Model

A critical aspect of privacy risk in federated learning is that many attacks do not require malicious behavior or protocol deviation. Instead, they operate under the *honest-but-curious* threat model.

In this model, participants—including the server—faithfully execute the federated learning protocol. They do not inject malicious updates, manipulate aggregation, or disrupt training. However, they attempt to infer as much information as possible from the data they are legitimately allowed to observe. Because the server typically receives updates from many clients across many rounds, it is a particularly powerful honest-but-curious adversary.

This threat model is both realistic and important. In practice, federated learning systems often rely on servers operated by large organizations or third-party providers. Even if these entities have no incentive to deviate from the protocol, they may still be motivated—intentionally or unintentionally—to analyze received updates. Moreover, privacy regulations frequently assume worst-case inference capabilities, making protection against honest-but-curious adversaries a minimum requirement rather than an extreme scenario.

The existence of privacy risks under honest participation has a profound implication: *privacy cannot be ensured solely by protocol compliance*. Federated learning must incorporate explicit privacy-preserving mechanisms that limit what can be inferred from updates, regardless of the intentions of observers.

Key Observation
Keeping data local does not prevent privacy leakage. Model updates are structured, data-dependent signals and can reveal sensitive information even when all participants behave honestly. Effective federated learning therefore requires privacy guarantees that go beyond architectural decentralization.

This analysis establishes why privacy risks are intrinsic to federated learning and motivates the need for formal privacy frameworks and cryptographic protections. The subsequent sections build upon this foundation by introducing mechanisms that explicitly bound information leakage from model updates, transforming federated learning from a privacy-aware architecture into a privacy-preserving learning paradigm.

15.2 Differential Privacy in Federated Learning

Differential privacy (DP) provides a principled and mathematically rigorous framework for reasoning about privacy in algorithmic systems. Unlike ad hoc anonymization or heuristic obfuscation techniques, differential privacy offers *provable guarantees* that limit the amount of information an adversary can infer about individual data points from the output of a computation. This makes DP particularly well suited to federated learning, where model updates derived from sensitive local data are repeatedly shared across communication rounds.

At a high level, the central promise of differential privacy is that the behavior of a learning algorithm should be nearly indistinguishable whether or not any single data record is included in the training data. In the context of federated learning, this translates to the requirement that model updates—and ultimately the trained global model—should not reveal whether a particular data point belonging to a specific client was used during training.

15.2.1 Definition

Formally, differential privacy characterizes privacy in terms of *output distributions*. An algorithm $\mathscr{A}$ is said to satisfy (ε, δ)-differential privacy if, for any two datasets $\mathscr{D}$ and $\mathscr{D}'$ that differ in exactly one data point, and for any measurable subset of outputs $\mathscr{S}$, the following inequality holds:

$$\Pr[\mathscr{A}(\mathscr{D}) \in \mathscr{S}] \leq e^{\varepsilon} \Pr[\mathscr{A}(\mathscr{D}') \in \mathscr{S}] + \delta.$$

This definition captures a worst-case privacy guarantee. The parameter ε quantifies the *privacy loss* incurred by the algorithm: smaller values of ε correspond to stronger privacy guarantees. The parameter δ allows for a small probability that the guarantee may be violated, accommodating rare but unavoidable failures in randomized mechanisms. In practice, δ is chosen to be negligible relative to the size of the dataset, while ε governs the dominant privacy–utility trade-off.

A key strength of differential privacy lies in its robustness. The guarantee holds regardless of the adversary's prior knowledge, computational power, or auxiliary information. As a result, DP protects against a wide range of inference attacks, including those described in the previous section, without requiring explicit modeling of adversarial strategies.

15.2.2 Applying Differential Privacy to Federated Learning

In federated learning, differential privacy is enforced by modifying the information exchanged between clients and the server. Rather than sharing exact model updates, clients or the server introduce carefully calibrated randomness to ensure that individual data points have a limited influence on the communicated values.

The standard approach to enforcing DP in federated learning consists of two tightly coupled steps.

Sensitivity Control via Clipping. The sensitivity of a function measures how much its output can change when a single data point is modified. In federated learning, sensitivity arises from the magnitude of local updates. To bound this sensitivity, each client clips its update to a predefined norm threshold C:

$$\Delta\theta_k \leftarrow \Delta\theta_k \cdot \min\left(1, \frac{C}{\|\Delta\theta_k\|_2}\right).$$

Clipping ensures that no single client—or individual data point—can exert an unbounded influence on the aggregated update.

Noise Injection. Once sensitivity is bounded, random noise is added to the updates. The noise distribution is chosen according to the desired privacy guarantee. Gaussian noise is commonly used in federated learning, as it offers favorable composition properties:

$$\widetilde{\Delta\theta} = \sum_{k \in \mathscr{S}^{(t)}} \Delta\theta_k + \mathscr{N}(0, \sigma^2 I),$$

where σ is calibrated as a function of the clipping threshold, the number of participating clients, and the target (ε, δ).

Together, clipping and noise injection ensure that the aggregated update satisfies differential privacy with respect to individual data points. Importantly, DP can be enforced either at the client side (before transmission) or at the server side (after aggregation), leading to different trust assumptions and utility outcomes, which are examined in later sections.

15.2.3 Privacy Accounting Across Communication Rounds

Federated learning is inherently iterative. Model updates are exchanged across many communication rounds, and each round consumes a portion of the overall privacy budget. Differential privacy therefore requires careful *privacy accounting* to track cumulative privacy loss over time.

Naïvely summing privacy loss across rounds leads to overly pessimistic bounds. To address this, federated learning systems employ advanced composition theorems and privacy accountants, such as the moments accountant or Rényi differential privacy. These tools provide tighter bounds on cumulative privacy loss by exploiting the statistical independence and structure of noise added across rounds.

Privacy accounting introduces a new dimension to federated learning design. The total privacy budget (ε, δ) must be allocated across communication rounds, client participation rates, and update magnitudes. Increasing the number of rounds or reducing the amount of noise improves model accuracy but consumes privacy budget more rapidly. Conversely, stronger privacy protection limits the achievable accuracy and may slow convergence.

Privacy–Utility Trade-Off
Differential privacy enforces a fundamental trade-off. Stronger privacy guarantees require increased noise and tighter sensitivity bounds, which degrade model accuracy and slow convergence. Effective federated learning systems must carefully balance privacy, utility, and communication efficiency.

Viewed holistically, differential privacy transforms federated learning from a privacy-aware architecture into a privacy-preserving algorithmic framework. By providing explicit, quantifiable guarantees, DP allows system designers to reason rigorously about what information is protected, under what assumptions, and at what cost to learning performance. The following sections build on this foundation by examining how differential privacy interacts with trust assumptions and cryptographic mechanisms in federated systems.

Worked Example: Differential Privacy in a Federated Learning Round
This worked example demonstrates, step by step, how differential privacy is enforced in a single federated learning round using explicit numerical values. All intermediate calculations are shown in detail to make the privacy mechanism fully transparent.

Problem Setup. Consider a federated learning system with $K = 5$ participating clients in one communication round. Each client trains a local model and computes a parameter update $\Delta\theta_k \in \mathbb{R}^2$. For simplicity, we consider a two-dimensional parameter vector.

Assume the following raw (unclipped) local updates are produced:

Client	$\Delta\theta_{k,1}$	$\Delta\theta_{k,2}$
1	1.8	0.6
2	-0.9	1.2
3	0.3	-0.4
4	2.4	1.8
5	-1.5	0.9

Our objective is to aggregate these updates under (ε, δ)-differential privacy.

Step 1: Compute the ℓ_2 Norm of Each Update

Client 1.

$$\|\Delta\theta_1\|_2 = \sqrt{(1.8)^2 + (0.6)^2} = \sqrt{3.24 + 0.36} = \sqrt{3.60} = 1.897.$$

Client 2.

$$\|\Delta\theta_2\|_2 = \sqrt{(-0.9)^2 + (1.2)^2} = \sqrt{0.81 + 1.44} = \sqrt{2.25} = 1.5.$$

Client 3.

$$\|\Delta\theta_3\|_2 = \sqrt{(0.3)^2 + (-0.4)^2} = \sqrt{0.09 + 0.16} = \sqrt{0.25} = 0.5.$$

Client 4.

$$\|\Delta\theta_4\|_2 = \sqrt{(2.4)^2 + (1.8)^2} = \sqrt{5.76 + 3.24} = \sqrt{9.0} = 3.0.$$

Client 5.

$$\|\Delta\theta_5\|_2 = \sqrt{(-1.5)^2 + (0.9)^2} = \sqrt{2.25 + 0.81} = \sqrt{3.06} = 1.749.$$

Step 2: Apply Gradient Clipping Assume a clipping threshold $C = 1.5$. Each update is clipped according to

$$\Delta\theta_k^{\text{clip}} = \Delta\theta_k \cdot \min\left(1, \frac{C}{\|\Delta\theta_k\|_2}\right).$$

Client 1.

$$\frac{C}{\|\Delta\theta_1\|_2} = \frac{1.5}{1.897} = 0.791.$$

$$\Delta\theta_1^{\text{clip}} = (1.8, 0.6) \times 0.791 = (1.424, 0.475).$$

Client 2.

$$\|\Delta\theta_2\|_2 = 1.5 \Rightarrow \text{no clipping.}$$

$$\Delta\theta_2^{\text{clip}} = (-0.9, 1.2).$$

Client 3.

$$\|\Delta\theta_3\|_2 < C \Rightarrow \text{no clipping.}$$

$$\Delta\theta_3^{\text{clip}} = (0.3, -0.4).$$

Client 4.

$$\frac{C}{\|\Delta\theta_4\|_2} = \frac{1.5}{3.0} = 0.5.$$

$$\Delta\theta_4^{\text{clip}} = (2.4, 1.8) \times 0.5 = (1.2, 0.9).$$

Client 5.

$$\frac{C}{\|\Delta\theta_5\|_2} = \frac{1.5}{1.749} = 0.858.$$

$$\Delta\theta_5^{\text{clip}} = (-1.5, 0.9) \times 0.858 = (-1.287, 0.772).$$

Step 3: Aggregate the Clipped Updates Sum the clipped updates coordinate-wise:

First coordinate.

$$1.424 + (-0.9) + 0.3 + 1.2 + (-1.287) = 0.737.$$

Second coordinate.

$$0.475 + 1.2 + (-0.4) + 0.9 + 0.772 = 2.947.$$

Thus, the aggregated clipped update is

$$\sum_{k=1}^{5} \Delta\theta_k^{\text{clip}} = (0.737,\ 2.947).$$

Step 4: Calibrate Gaussian Noise Assume a target privacy guarantee of

$$\varepsilon = 1.0, \qquad \delta = 10^{-5}.$$

For the Gaussian mechanism, the noise standard deviation is chosen as

$$\sigma = \frac{C\sqrt{2\ln(1.25/\delta)}}{\varepsilon}.$$

Numerical computation.

$$\ln\left(\frac{1.25}{10^{-5}}\right) = \ln(125000) \approx 11.736.$$

$$\sqrt{2 \times 11.736} = \sqrt{23.472} = 4.845.$$

$$\sigma = 1.5 \times 4.845 = 7.268.$$

Step 5: Sample Gaussian Noise Assume one realization of Gaussian noise

$$\eta \sim \mathcal{N}(0, \sigma^2 I)$$

yields the sample

$$\eta = (5.10,\ -3.25).$$

Step 6: Form the Private Aggregated Update

Final noisy update.

$$\widetilde{\Delta\theta} = (0.737, 2.947) + (5.10, -3.25) = (5.837,\ -0.303).$$

Step 7: Interpretation The vector

$$\widetilde{\Delta\theta} = (5.837, \ -0.303)$$

is the update transmitted to the global model.

- Individual client contributions are bounded by clipping.
- Noise ensures (ε, δ)-differential privacy.
- No observer can reliably infer whether any single data point was present in a client's dataset.

This worked example illustrates concretely how differential privacy is enforced in federated learning through explicit numerical operations. While noise degrades the signal, privacy guarantees are achieved in a precise and quantifiable manner, highlighting the intrinsic privacy–utility trade-off at the core of differentially private federated learning.

15.3 Local versus Global Differential Privacy

Differential privacy can be enforced at different stages of the federated learning pipeline. Where and how randomness is introduced has profound implications for privacy guarantees, trust assumptions, statistical efficiency, and system performance. This leads to two fundamentally distinct paradigms: *local differential privacy* and *global (or central) differential privacy.*

Understanding the distinction between these paradigms is essential for designing federated learning systems that are both privacy-preserving and practically effective.

15.3.1 Local Differential Privacy (LDP)

Under local differential privacy (LDP), privacy protection is applied *entirely at the client side*. Each client independently perturbs its model update using a randomized mechanism before transmitting it to the server. As a result, the server never observes unperturbed gradients, parameter updates, or model weights.

Formally, each client enforces (ε, δ)-differential privacy with respect to its own local dataset. This guarantees that even a fully compromised or malicious server cannot reliably infer information about any individual data point, as the noise has already been injected before communication.

The primary advantage of LDP lies in its minimal trust assumptions. Clients do not need to trust the server, other clients, or the communication infrastructure. Privacy guarantees are preserved even if the server behaves adversarially or colludes with external parties.

However, this strong privacy protection comes at a significant cost. Because noise is added independently by each client, the aggregate update accumulates noise from all participants. In large-scale federated learning, this can severely degrade the signal-to-noise ratio, slow convergence, and reduce final model accuracy. As a result, LDP-based federated learning often requires more clients, more communication rounds, or weaker privacy guarantees to achieve acceptable performance.

15.3.2 Global (Central) Differential Privacy

Global differential privacy, also referred to as central differential privacy, enforces privacy *after aggregation*. In this paradigm, clients transmit clipped but otherwise unperturbed updates to the server. The server aggregates these updates and then injects noise into the aggregate before applying it to the global model.

From a statistical perspective, global DP is substantially more efficient than LDP. Because noise is added only once to the aggregated update, rather than independently at each client, the magnitude of noise required to achieve a given privacy guarantee is significantly smaller. This typically results in faster convergence and higher model accuracy.

The main limitation of global DP lies in its trust assumptions. The server must be trusted to correctly perform aggregation and noise injection without inspecting or storing individual updates. In practice, this trust requirement can be mitigated by combining global DP with cryptographic techniques such as secure aggregation, which ensure that the server only ever observes aggregated updates.

Thus, global DP represents a pragmatic compromise: it achieves strong privacy guarantees with substantially better utility, provided that the system incorporates appropriate safeguards against server-side inference.

15.3.3 Comparative Analysis

The choice between local and global differential privacy reflects a deeper design decision regarding trust, efficiency, and system architecture. Local DP prioritizes decentralization and trust minimization, while global DP prioritizes statistical efficiency and learning performance.

Table 15.1 provides a structured comparison of the two approaches across key dimensions relevant to federated learning system design.

Design Choice
The choice between local and global differential privacy reflects a fundamental trade-off between trust assumptions and learning efficiency. Local DP minimizes trust but sacrifices accuracy, while global DP achieves better performance when combined with appropriate safeguards such as secure aggregation.

Table 15.1 Comparison of Local and Global Differential Privacy in Federated Learning

Dimension	Local Differential Privacy (LDP)	Global Differential Privacy (GDP)
Point of Noise Injection	Client-side (before transmission)	Server-side (after aggregation)
Trust Assumptions	No trust in server required	Requires trusted server or secure aggregation
Privacy Strength	Very strong (even against malicious server)	Strong, contingent on correct aggregation
Statistical Efficiency	Low (noise accumulates across clients)	High (noise added once to aggregate)
Model Utility	Often significantly degraded	Typically much higher
Convergence Speed	Slow	Faster
Communication Overhead	Moderate	Moderate
Scalability	Requires many clients for acceptable accuracy	Scales well with client population
Typical Use Cases	Highly sensitive data, minimal trust environments	Large-scale production federated learning

In practice, most real-world federated learning systems adopt global differential privacy augmented with cryptographic protections. This hybrid approach leverages the statistical efficiency of central DP while reducing reliance on trust in any single entity. The next sections build on this comparison by examining cryptographic mechanisms that enable secure aggregation and privacy-preserving coordination in federated systems.

15.4 Secure Aggregation Protocols

Secure aggregation protocols constitute a cornerstone of privacy preservation in federated learning systems. While differential privacy constrains the *inference power* of an adversary by adding calibrated randomness to aggregated updates, secure aggregation addresses a fundamentally different and complementary privacy objective: it prevents the server from ever observing *individual* client updates in the first place. In doing so, secure aggregation enforces a strong form of confidentiality at the communication layer of federated learning.

This distinction is crucial. Differential privacy provides statistical guarantees about what can be inferred from an output, even if that output is observed in full. Secure aggregation, by contrast, restricts the server's view of the training process itself, ensuring that only collective information is ever revealed. Together, these mechanisms transform the role of the server from a potentially powerful observer into a constrained coordinator that operates exclusively on aggregated signals.

At a high level, secure aggregation enables a group of clients to compute the sum (or average) of their local model updates such that:

- the server learns the aggregate update exactly,
- no individual client update is revealed or recoverable,
- correctness is preserved without introducing noise.

This capability is essential in federated learning deployments where the server is not fully trusted, where organizational boundaries prevent the sharing of per-client updates, or where regulatory and legal frameworks explicitly prohibit access to individual contributions. Secure aggregation thus bridges the gap between collaborative learning and strict data confidentiality requirements.

15.4.1 Core Idea

The fundamental principle underlying secure aggregation is *cryptographic masking*. Each client hides its local update by adding carefully constructed random values, called masks, before transmitting the update to the server. These masks are not arbitrary; they are designed so that, when masked updates from all participating clients are summed, the random components cancel out exactly. The server is therefore left with the true aggregate and nothing more.

Let $\Delta\theta_k$ denote the update computed by client k during a given federated learning round. Instead of transmitting this update directly, client k sends a masked update

$$\widetilde{\Delta\theta}_k = \Delta\theta_k + \sum_{j \neq k} \mathbf{m}_{k,j},$$

where $\mathbf{m}_{k,j}$ is a random mask shared between clients k and j.

The masks satisfy the antisymmetry condition

$$\mathbf{m}_{k,j} = -\mathbf{m}_{j,k},$$

which ensures exact cancellation during aggregation.

When the server aggregates all received masked updates, it computes

$$\sum_{k=1}^{K} \widetilde{\Delta\theta}_k = \sum_{k=1}^{K} \Delta\theta_k + \sum_{k=1}^{K} \sum_{j \neq k} \mathbf{m}_{k,j}.$$

Because each mask $\mathbf{m}_{k,j}$ appears exactly once with a positive sign and once with a negative sign, the second term vanishes:

$$\sum_{k=1}^{K} \sum_{j \neq k} \mathbf{m}_{k,j} = 0.$$

Hence, the server recovers the exact aggregate

$$\sum_{k=1}^{K} \widetilde{\Delta\theta}_k = \sum_{k=1}^{K} \Delta\theta_k.$$

From the server's perspective, each individual message $\widetilde{\Delta\theta}_k$ appears as a random vector. Without access to the pairwise masks, it is information-theoretically impossible to isolate or reconstruct any single client's update. Privacy is therefore achieved without sacrificing correctness.

15.4.2 Operational Workflow

A secure aggregation protocol typically unfolds over several logically distinct phases within each federated learning round. Although concrete implementations vary, the core workflow follows a common structure.

- **Key Establishment:** Before any model updates are exchanged, clients establish shared secrets with one another. This is commonly achieved using cryptographic key agreement protocols, such as Diffie–Hellman key exchange. These shared secrets are never revealed to the server and form the basis for generating synchronized random masks.
- **Mask Generation and Update Encoding:** Each client computes its local model update using standard federated optimization procedures. Using the shared secrets, the client deterministically generates the pairwise masks $\mathbf{m}_{k,j}$ and constructs the masked update $\widetilde{\Delta\theta}_k$. At this stage, the true update is cryptographically hidden by multiple independent random components.
- **Transmission and Aggregation:** Clients transmit only their masked updates to the server. The server aggregates all received messages by summation, exactly as it would in standard federated learning. Because mask cancellation occurs automatically during summation, the server obtains the correct aggregate update without performing any special cryptographic operations.
- **Post-Aggregation Processing:** Once the aggregate update is obtained, the server applies it to the global model. If global differential privacy is employed, calibrated noise may be added at this stage to bound statistical inference. Importantly, secure aggregation is fully compatible with such post-processing, as it does not alter the correctness of the aggregate.

  Crucially, at no point in this workflow does the server gain access to any individual unmasked update. All operations performed by the server are identical to those in non-secure federated learning, making secure aggregation transparent to the optimization process.

Taken together, these properties make secure aggregation a practical and powerful mechanism for enforcing confidentiality in federated learning. It provides strong cryptographic protection against server-side inspection while preserving the simplicity, scalability, and correctness of standard federated optimization algorithms.

15.4.3 Robustness to Client Dropouts

Robustness to client dropouts is a critical practical requirement for secure aggregation in federated learning. Unlike idealized theoretical settings, real-world federated systems operate in highly dynamic and unreliable environments. Client devices may disconnect unpredictably due to network instability, limited battery power, application interruptions, or user behavior. As a result, any secure aggregation protocol intended for deployment at scale must function correctly even when a nontrivial fraction of clients fails to complete a training round.

Naïve masking-based aggregation schemes are fundamentally fragile in the presence of dropouts. If a client that contributes a masking term $\mathbf{m}_{k,j}$ fails to participate, its corresponding cancellation term $-\mathbf{m}_{j,k}$ is never received. The server is then left with residual random noise in the aggregated update, rendering the result unusable. This failure mode highlights an important insight: robustness to dropouts is not an implementation detail but a core design requirement of secure aggregation protocols.

To address this challenge, modern secure aggregation protocols augment basic masking with additional cryptographic structure. The goal is to ensure that missing masks can be neutralized or reconstructed without revealing any individual client update. Several complementary strategies are commonly employed.

- **Secret Sharing for Mask Recovery:** One widely used approach relies on secret sharing schemes. Each client splits its masking secrets into multiple shares and distributes them to other clients. If a client drops out, the remaining participants can collectively reconstruct the missing masks from these shares, provided that the number of dropouts does not exceed a predefined threshold. Crucially, no single party ever holds enough information to recover the original mask or the underlying update.
- **Redundancy through Client Grouping:** Another strategy organizes clients into logical groups or cohorts with built-in redundancy. Masks are constructed so that each update is protected by multiple independent cancellation paths. Even if some clients fail to participate, sufficient redundancy remains to ensure complete mask cancellation at aggregation time. This approach trades increased communication overhead for improved fault tolerance.
- **Multi-Round Resolution Protocols:** Some secure aggregation designs introduce additional protocol rounds to handle incomplete participation. In these schemes, the server first identifies which clients have successfully submitted updates. Subsequent rounds allow remaining clients to disclose carefully restricted information—such as masked shares or commitments—that enables the resolution of missing masks without exposing individual updates. These extra rounds increase latency but significantly enhance robustness.

All of these mechanisms share a common objective: they guarantee that the aggregate can be recovered correctly as long as the number of client dropouts remains below a system-defined threshold. At the same time, they preserve the privacy guarantee by ensuring that the server never gains access to any individual unmasked update, even when participation is partial.

The ability to tolerate dropouts without sacrificing privacy or correctness is one of the defining features of practical secure aggregation protocols. It distinguishes deployable systems from purely theoretical constructions and enables secure aggregation to operate reliably at the scale of millions of heterogeneous, intermittently connected devices.

Security Guarantee
Secure aggregation ensures that the server learns only the aggregated model update and gains no information about any individual client's contribution. This guarantee holds even under an honest-but-curious server model and in the presence of client dropouts, provided that the protocol's threshold and trust assumptions are satisfied.

In summary, secure aggregation transforms federated learning from a system in which the server observes all client updates into one in which only collective information is revealed. When combined with differential privacy, secure aggregation enables federated learning systems that provide both strong cryptographic confidentiality and formal statistical privacy guarantees. Together, these techniques form a robust foundation for large-scale, privacy-preserving collaborative learning in real-world environments.

Worked Example: Secure Aggregation with Explicit Numerical Masks
This worked example illustrates a complete secure aggregation round using explicit numerical values. All masks, cancellations, and aggregations are derived step by step to demonstrate how individual updates remain hidden while the correct global aggregate is recovered.

Problem Setup. Consider a federated learning round with $K = 4$ participating clients. Each client computes a one-dimensional model update $\Delta\theta_k \in \mathbb{R}$ for simplicity. The objective is to compute the sum of updates

$$\sum_{k=1}^{4} \Delta\theta_k$$

without revealing any individual $\Delta\theta_k$ to the server.

Assume the true local updates are as follows:

Client	True Update $\Delta\theta_k$
1	2.5
2	–1.2
3	0.8
4	1.9

The correct aggregate (unknown to the server initially) is

$$2.5 + (-1.2) + 0.8 + 1.9 = 4.0.$$

Step 1: Generate Pairwise Masks Each unordered client pair (i,j) generates a shared random mask $m_{i,j}$ such that

$$m_{i,j} = -m_{j,i}.$$

Assume the following pairwise masks are generated:

Pair	Mask Value
(1,2)	+1.4
(1,3)	-0.7
(1,4)	+0.9
(2,3)	+0.5
(2,4)	-1.1
(3,4)	+0.6

From antisymmetry, for example, $m_{2,1} = -1.4$, $m_{3,1} = +0.7$, etc.

Step 2: Construct Masked Updates Each client adds all masks it shares with other clients to its true update.

Client 1.

$$\widetilde{\Delta\theta}_1 = \Delta\theta_1 + m_{1,2} + m_{1,3} + m_{1,4} = 2.5 + 1.4 + (-0.7) + 0.9.$$

$$\widetilde{\Delta\theta}_1 = 2.5 + 1.6 = 4.1.$$

Client 2.

$$\widetilde{\Delta\theta}_2 = \Delta\theta_2 + m_{2,1} + m_{2,3} + m_{2,4} = -1.2 + (-1.4) + 0.5 + (-1.1).$$

$$\widetilde{\Delta\theta}_2 = -1.2 - 2.0 = -3.2.$$

Client 3.

$$\widetilde{\Delta\theta}_3 = \Delta\theta_3 + m_{3,1} + m_{3,2} + m_{3,4} = 0.8 + 0.7 + (-0.5) + 0.6.$$

$$\widetilde{\Delta\theta}_3 = 0.8 + 0.8 = 1.6.$$

Client 4.

$$\widetilde{\Delta\theta}_4 = \Delta\theta_4 + m_{4,1} + m_{4,2} + m_{4,3} = 1.9 + (-0.9) + 1.1 + (-0.6).$$

$$\widetilde{\Delta\theta}_4 = 1.9 - 0.4 = 1.5.$$

Step 3: Transmitted Values Observed by the Server The server receives only the masked updates:

Client	Masked Update Sent
1	4.1
2	-3.2
3	1.6
4	1.5

Each masked update appears random and bears no obvious relationship to the corresponding true update.

Step 4: Server-side Aggregation The server computes the aggregate by summation:

$$4.1 + (-3.2) + 1.6 + 1.5.$$

Intermediate sums.

$$4.1 - 3.2 = 0.9, \quad 0.9 + 1.6 = 2.5, \quad 2.5 + 1.5 = 4.0.$$

Thus,

$$\sum_{k=1}^{4} \widetilde{\Delta\theta}_k = 4.0.$$

Step 5: Explicit Mask Cancellation Verification Summing all masks explicitly:

$$\sum_{k=1}^{4}\sum_{j\neq k} m_{k,j} = (m_{1,2}+m_{2,1})+(m_{1,3}+m_{3,1})+(m_{1,4}+m_{4,1})+(m_{2,3}+m_{3,2})+(m_{2,4}+m_{4,2})+(m_{3,4}+m_{4,3}).$$

Substituting values:

$$(1.4 - 1.4) + (-0.7 + 0.7) + (0.9 - 0.9) + (0.5 - 0.5) + (-1.1 + 1.1) + (0.6 - 0.6) = 0.$$

Hence,

$$\sum_{k=1}^{4} \widetilde{\Delta\theta}_k = \sum_{k=1}^{4} \Delta\theta_k.$$

Step 6: Interpretation and Security Insight

- Each individual update is hidden by multiple random masks.
- The server cannot isolate any $\Delta\theta_k$ without knowing the pairwise masks.
- Mask cancellation guarantees exact recovery of the aggregate.

This worked example demonstrates concretely how secure aggregation achieves cryptographic confidentiality in federated learning. Even though the server performs the aggregation correctly, it gains no information about individual client updates—only the collective signal required for model optimization.

15.5 Secure Multi-Party Computation

Secure Multi-Party Computation (SMPC) provides a general cryptographic framework that enables multiple parties to jointly compute a function over their private inputs while revealing nothing beyond the final output. In contrast to secure aggregation—which is specialized for summation or averaging—SMPC supports the secure evaluation of *arbitrary functions*. This generality makes SMPC a powerful, though computationally demanding, tool for privacy-preserving federated learning.

In federated learning, SMPC is particularly valuable when privacy requirements extend beyond hiding individual updates from the server and instead demand protection against inference by *any* participating entity. The central idea is that no single party ever gains access to another party's raw update or even its masked form; instead, computation proceeds on cryptographically protected representations.

15.5.1 SMPC in Federated Learning

The most common SMPC paradigm used in federated learning is *secret sharing*. Each client represents its local update as a set of cryptographic shares and distributes these shares among multiple computing parties (which may include the server or auxiliary computation nodes).

Let $\Delta\theta_k$ denote the update computed by client k. Under additive secret sharing, the client decomposes this update into M shares

$$\Delta\theta_k = \sum_{m=1}^{M} \mathbf{s}_k^{(m)},$$

where each share $\mathbf{s}_k^{(m)}$ appears statistically random in isolation. No single share reveals any information about the true update.

The aggregation function—typically summation or averaging—is then evaluated directly on the shares:

$$\sum_{k=1}^{K} \Delta\theta_k = \sum_{m=1}^{M} \sum_{k=1}^{K} \mathbf{s}_k^{(m)}.$$

At no point is any individual update reconstructed. Only the final aggregate, or a secret-shared representation of it, is made available to authorized parties.

From a system perspective, SMPC replaces trust in a single server with trust distributed across multiple non-colluding parties. As long as a predefined threshold of parties does not collude, privacy is information-theoretically guaranteed.

15.5.2 Advantages and Limitations

The primary advantage of SMPC is its exceptionally strong privacy guarantee. Unlike differential privacy, which bounds inference probabilistically, SMPC prevents access to individual updates altogether. Unlike masking-based secure aggregation, SMPC is not limited to linear operations and can support more complex computations.

However, these benefits come at a cost. SMPC protocols typically require:

- multiple rounds of communication,
- significant message expansion due to share distribution,
- increased computational overhead at both client and server sides.

As the number of clients or the dimensionality of the model grows, these costs can become prohibitive. Consequently, SMPC is often less scalable than masking-based secure aggregation for large, resource-constrained federated learning systems.

Worked Example: Secure Multi-party Computation via Additive Secret Sharing

This worked example demonstrates, step by step, how Secure Multi-Party Computation (SMPC) enables privacy-preserving aggregation in federated learning using explicit numerical values. All intermediate shares, computations, and reconstructions are shown in detail.

Problem Setup. Consider a federated learning round with $K = 4$ clients. Each client computes a one-dimensional model update $\Delta\theta_k \in \mathbb{R}$. The goal is to compute the global sum

$$\sum_{k=1}^{4} \Delta\theta_k$$

without revealing any individual update to any single party.

Assume the true client updates are: The correct aggregate (unknown to any

Client	True Update $\Delta\theta_k$
1	3.2
2	−1.7
3	0.9
4	2.6

individual party) is

$$3.2 + (-1.7) + 0.9 + 2.6 = 5.0.$$

Assume an SMPC setting with $M = 3$ non-colluding computation parties P_1, P_2, P_3.

Step 1: Secret Sharing of Local Updates Each client decomposes its update into three additive secret shares such that the sum of shares equals the true update.

Client 1. Choose arbitrary random shares:

$$s_1^{(1)} = 1.1, \quad s_1^{(2)} = -0.4.$$

Compute the third share:

$$s_1^{(3)} = 3.2 - (1.1 + (-0.4)) = 3.2 - 0.7 = 2.5.$$

Client 2. Choose:

$$s_2^{(1)} = -0.6, \quad s_2^{(2)} = 0.2.$$

Then:

$$s_2^{(3)} = -1.7 - (-0.6 + 0.2) = -1.7 - (-0.4) = -1.3.$$

Client 3. Choose:

$$s_3^{(1)} = 0.5, \quad s_3^{(2)} = -0.1.$$

Then:

$$s_3^{(3)} = 0.9 - (0.5 + (-0.1)) = 0.9 - 0.4 = 0.5.$$

Client 4. Choose:

$$s_4^{(1)} = 1.4, \quad s_4^{(2)} = -0.9.$$

Then:

$$s_4^{(3)} = 2.6 - (1.4 + (-0.9)) = 2.6 - 0.5 = 2.1.$$

The resulting secret shares are summarized below:

Client	$s_k^{(1)}$	$s_k^{(2)}$	$s_k^{(3)}$
1	1.1	–0.4	2.5
2	–0.6	0.2	–1.3
3	0.5	–0.1	0.5
4	1.4	–0.9	2.1

Each share is sent to the corresponding computation party.

Step 2: Local Aggregation at Each Computation Party Each party sums the shares it receives from all clients.

Party P_1.

$$\sum_{k=1}^{4} s_k^{(1)} = 1.1 + (-0.6) + 0.5 + 1.4 = 2.4.$$

Party P_2.

$$\sum_{k=1}^{4} s_k^{(2)} = (-0.4) + 0.2 + (-0.1) + (-0.9) = -1.2.$$

Party P_3.

$$\sum_{k=1}^{4} s_k^{(3)} = 2.5 + (-1.3) + 0.5 + 2.1 = 3.8.$$

Each party now holds a share of the *aggregate*, not of any individual update.

Step 3: Secure Reconstruction of the Global Aggregate To obtain the final result, the parties combine their aggregate shares.

Aggregate reconstruction.

$$2.4 + (-1.2) + 3.8 = 5.0.$$

Thus,

$$\sum_{k=1}^{4} \Delta\theta_k = 5.0,$$

which exactly matches the true global update.

Step 4: Privacy Verification Consider the information available to each computation party.

View of P_1. Party P_1 sees:

$$\{1.1, -0.6, 0.5, 1.4\}.$$

These values are statistically indistinguishable from random noise and do not reveal any individual $\Delta\theta_k$.

View of P_2 and P_3. The same holds for P_2 and P_3. No single party can reconstruct any client update without colluding with others.

Step 5: Interpretation

- Individual updates are never revealed in plaintext.
- Aggregation is performed directly on secret shares.
- Privacy is guaranteed as long as fewer than all parties collude.

This worked example demonstrates how SMPC enables exact aggregation in federated learning while providing strong cryptographic privacy guarantees. Unlike differential privacy, no noise is added; privacy arises purely from cryptographic protection. However, this strength comes at the cost of increased communication and coordination among parties, illustrating the trade-offs inherent in privacy-preserving federated learning.

Use Cases
SMPC is particularly well suited to federated learning scenarios in which no single entity can be fully trusted. This includes cross-institutional collaborations such as:

- federated learning across hospitals or healthcare providers,
- joint model training among competing organizations,
- regulatory environments requiring strict cryptographic privacy.

In such settings, the additional overhead of SMPC is justified by the elimination of centralized trust and the strength of the resulting privacy guarantees.

15.6 Homomorphic Encryption for Federated Learning

Homomorphic encryption (HE) offers another powerful cryptographic approach to privacy-preserving federated learning. Unlike SMPC, which distributes trust across multiple parties, HE enables a single untrusted server to perform computations directly on encrypted data. The server never sees plaintext updates, yet it can still aggregate client contributions correctly.

In federated learning, homomorphic encryption is primarily used to protect model updates during aggregation, ensuring that the server cannot inspect individual contributions even if it controls the entire computation pipeline.

15.6.1 Types of Homomorphic Encryption

Homomorphic encryption schemes are categorized by the types of operations they support on encrypted data.

- **Partially homomorphic encryption (PHE)** supports a single algebraic operation, such as addition or multiplication. Additively homomorphic schemes are particularly well suited to federated learning, as aggregation requires only summation.
- **Fully homomorphic encryption (FHE)** supports arbitrary computation on encrypted data, enabling the evaluation of complex functions without decryption.

While FHE offers maximal expressiveness, its computational cost remains orders of magnitude higher than that of PHE, limiting its practicality for large-scale federated learning.

15.6.2 Encrypted Aggregation

In an HE-based federated learning system, each client encrypts its local update using a shared public key:

$$\mathbf{c}_k = \mathrm{Enc}(\Delta\boldsymbol{\theta}_k).$$

The server aggregates encrypted updates directly:

$$\mathbf{c}_{\mathrm{agg}} = \sum_{k=1}^{K} \mathbf{c}_k = \mathrm{Enc}\left(\sum_{k=1}^{K} \Delta\boldsymbol{\theta}_k\right).$$

Only an authorized party holding the corresponding private key can decrypt the final aggregate. At no stage does the server gain access to individual updates in plaintext form.

This approach preserves the standard federated learning workflow while providing strong confidentiality guarantees against server-side inspection.

Practical Constraints
Despite its conceptual elegance, homomorphic encryption introduces substantial practical challenges. Encrypted computations are significantly more expensive than their plaintext counterparts, both in terms of computation time and memory usage. Communication overhead is also increased due to ciphertext expansion.

As a result, HE-based federated learning is typically restricted to:

- small or medium-scale models,
- low-frequency aggregation,
- environments with ample computational resources.

Practical Reality
Homomorphic encryption provides extremely strong privacy guarantees by ensuring that the server never observes plaintext updates. However, its high computational and communication overhead currently limits widespread deployment in large-scale federated learning systems.

Worked Example: Encrypted Aggregation using Additive Homomorphic Encryption

This worked example provides a detailed and self-contained illustration of *encrypted aggregation* in federated learning using an additive homomorphic encryption (HE) scheme. The objective is to show, in a fully transparent and numerically explicit manner, how a server can correctly aggregate client updates while remaining completely oblivious to the individual contributions. All computations are derived step by step, and no intermediate reasoning is omitted.

For pedagogical clarity, we adopt a simplified integer-based additive homomorphic encryption model. Although simplified, this abstraction faithfully captures the essential behavior of widely deployed cryptosystems such as the Paillier encryption scheme, which underpins many practical encrypted federated learning implementations.

Problem Setup. Consider a single communication round of a federated learning system with $K = 4$ participating clients. Each client computes a scalar model update $\Delta\theta_k \in \mathbb{Z}$ based on its private local dataset. The learning objective requires the server to compute the global update

$$\Delta\theta_{\text{global}} = \sum_{k=1}^{4} \Delta\theta_k,$$

which will subsequently be used to update the global model parameters.

The key constraint is that the server must never observe or infer any individual update $\Delta\theta_k$. Only the final aggregated result may be revealed, and only after decryption by an authorized entity.

Assume that the true (plaintext) client updates are given in Table 15.2.

If aggregation were performed centrally in plaintext, the result would be

$$12 + (-5) + 7 + 16 = 30.$$

In encrypted aggregation, this value must be obtained *only* after decrypting the aggregate ciphertext, without exposing any intermediate plaintext values.

Step 1: Key Generation Before any model updates are exchanged, an encryption key pair is generated.

Table 15.2 Local model updates computed by clients

Client	Local Update $\Delta\theta_k$
1	12
2	–5
3	7
4	16

Public key generation. An authorized entity generates a public key pk. This key defines the encryption function $E_{\mathsf{pk}}(\cdot)$ and is distributed to all participating clients.

Private key generation. The corresponding private key sk defines the decryption function $D_{\mathsf{sk}}(\cdot)$. This key is kept strictly secret and is never shared with clients or untrusted servers.

The security of the scheme rests on the assumption that, without knowledge of sk, it is computationally infeasible to recover plaintexts from ciphertexts.

Step 2: Encryption of Local Updates Each client independently encrypts its local model update using the public key. Let c_k denote the encrypted update transmitted by client k.

Client 1.

$$c_1 = E_{\mathsf{pk}}(12).$$

Client 2.

$$c_2 = E_{\mathsf{pk}}(-5).$$

Client 3.

$$c_3 = E_{\mathsf{pk}}(7).$$

Client 4.

$$c_4 = E_{\mathsf{pk}}(16).$$

The encrypted updates received by the server are summarized in Table 15.3.

At this point, the server holds only encrypted values. No plaintext information about individual client updates is available.

Step 3: Homomorphic Aggregation at the Server Additive homomorphic encryption schemes satisfy the fundamental property

$$E_{\mathsf{pk}}(a) \oplus E_{\mathsf{pk}}(b) = E_{\mathsf{pk}}(a + b),$$

Table 15.3 Plaintext updates and their encrypted representations

Client	$\Delta\theta_k$	Ciphertext c_k
1	12	$E_{\mathsf{pk}}(12)$
2	–5	$E_{\mathsf{pk}}(-5)$
3	7	$E_{\mathsf{pk}}(7)$
4	16	$E_{\mathsf{pk}}(16)$

where $\oplus$ denotes addition performed in the ciphertext domain.

Ciphertext-domain summation. Using this property, the server computes the aggregate ciphertext as

$$\begin{aligned} c_{\text{sum}} &= c_1 \oplus c_2 \oplus c_3 \oplus c_4 \\ &= E_{\mathsf{pk}}(12) \oplus E_{\mathsf{pk}}(-5) \oplus E_{\mathsf{pk}}(7) \oplus E_{\mathsf{pk}}(16). \end{aligned}$$

By repeated application of the homomorphic property,

$$\begin{aligned} c_{\text{sum}} &= E_{\mathsf{pk}}(12-5) \oplus E_{\mathsf{pk}}(7) \oplus E_{\mathsf{pk}}(16) \\ &= E_{\mathsf{pk}}(7) \oplus E_{\mathsf{pk}}(7) \oplus E_{\mathsf{pk}}(16) \\ &= E_{\mathsf{pk}}(14) \oplus E_{\mathsf{pk}}(16) \\ &= E_{\mathsf{pk}}(30). \end{aligned}$$

Crucially, all of these operations are carried out on ciphertexts. The server never observes intermediate plaintext sums.

Step 4: Decryption of the Aggregate Once aggregation is complete, the encrypted result is decrypted by an authorized party possessing the private key.

Decryption.

$$\begin{aligned} \Delta\theta_{\text{global}} &= D_{\mathsf{sk}}(c_{\text{sum}}) \\ &= D_{\mathsf{sk}}(E_{\mathsf{pk}}(30)) \\ &= 30. \end{aligned}$$

This value is exactly the desired global model update.

Step 5: Verification against Plain Aggregation For completeness, we verify correctness by direct computation:

$$12 + (-5) + 7 + 16 = 30.$$

The encrypted aggregation yields the same result as plaintext aggregation, demonstrating that encryption preserves correctness while enforcing confidentiality.

Step 6: Privacy Analysis

Server's observable information. The server observes only the set of ciphertexts

$$\{E_{\mathsf{pk}}(12), E_{\mathsf{pk}}(-5), E_{\mathsf{pk}}(7), E_{\mathsf{pk}}(16)\}.$$

Under standard cryptographic assumptions, these ciphertexts are computationally indistinguishable from encryptions of random values.

Security guarantee. Without access to sk, it is computationally infeasible for the server to recover any individual update $\Delta\theta_k$, even if it actively attempts to analyze or correlate ciphertexts across rounds.

Step 7: Interpretation and Implications

- Encrypted aggregation enables exact computation of global updates without exposing individual client contributions.
- Unlike differential privacy, no random noise is introduced, and model accuracy is preserved.
- Privacy guarantees rely on cryptographic hardness assumptions rather than statistical obfuscation.
- The primary cost is increased computation and communication overhead due to encryption and ciphertext operations.

This worked example demonstrates how homomorphic encryption provides a powerful mechanism for privacy-preserving aggregation in federated learning. It complements statistical privacy mechanisms by offering strong, deterministic confidentiality guarantees, while simultaneously highlighting the practical trade-offs between privacy strength, system complexity, and scalability.

Outlook
Privacy-preserving federated learning is not achieved through a single mechanism but through a layered combination of complementary techniques. Architectural decentralization reduces raw data exposure, differential privacy bounds statistical inference, and cryptographic methods such as secure aggregation, SMPC, and homomorphic encryption prevent unauthorized inspection of updates.

The choice among these techniques depends critically on the threat model, trust assumptions, system scale, and acceptable trade-offs between privacy, accuracy, and efficiency. In practice, modern federated learning systems often combine multiple approaches to achieve balanced and robust privacy guarantees.

Exercises

This chapter examined why federated learning is not inherently private and developed principled privacy-preserving mechanisms—differential privacy and cryptographic techniques—to rigorously control information leakage from model updates.

A. Short Answer Questions

Answer each question briefly and precisely.

1. Why does keeping raw data local not guarantee privacy in federated learning?
2. What is meant by *gradient leakage* in federated learning?

3. Define the honest-but-curious threat model.
4. What role does clipping play in differentially private federated learning?
5. Why does privacy loss accumulate across federated learning rounds?
6. Distinguish between architectural privacy and algorithmic privacy.
7. What information does secure aggregation hide from the server?
8. Why is local differential privacy statistically less efficient than global DP?
9. What is the primary purpose of secure multi-party computation in federated learning?
10. Why is additive homomorphic encryption particularly suitable for federated aggregation?

B. Long Answer Questions

Answer the following questions with clear explanations, equations, and diagrams wherever appropriate.

1. Explain how gradients and parameter updates can leak sensitive information, even when raw data never leaves client devices.
2. Derive the formal definition of (ε, δ)-differential privacy and interpret the roles of ε and δ in federated learning.
3. Discuss the privacy–utility trade-off introduced by differential privacy in iterative federated optimization.
4. Compare local and global differential privacy in federated learning with respect to trust assumptions, convergence behavior, and scalability.
5. Describe the working principle of secure aggregation and explain how mask cancellation guarantees correctness.
6. Explain how secure multi-party computation generalizes secure aggregation and identify scenarios where SMPC is preferable.
7. Describe encrypted aggregation using homomorphic encryption and explain how correctness is preserved without revealing plaintext updates.
8. Analyze the robustness challenges introduced by client dropouts in secure aggregation protocols and how modern designs address them.

C. True/False Questions

State whether each statement is **True** or **False**.

1. Federated learning is inherently privacy-preserving because raw data never leaves clients.
2. Gradients are information-theoretically independent of the training data.
3. Differential privacy provides guarantees that are independent of an adversary's prior knowledge.

4. Clipping is necessary to bound sensitivity in differentially private federated learning.
5. Local differential privacy requires trust in the server.
6. Global differential privacy typically yields better model accuracy than local DP.
7. Secure aggregation prevents the server from observing individual client updates.
8. SMPC can be used to compute only summation functions securely.
9. Homomorphic encryption allows computation directly on encrypted values.
10. Differential privacy and cryptographic techniques address complementary privacy risks.

Answers:

1. False
2. False
3. True
4. True
5. False
6. True
7. True
8. False
9. True
10. True

D. Fill in the Blanks

Fill in the blanks with the most appropriate term or expression.

1. Federated learning replaces centralized data collection with ___________.
2. Gradients are ___________ functions of the training data.
3. Differential privacy limits the influence of any single ___________.
4. The parameter ε controls the ___________ of a privacy guarantee.
5. Gradient clipping bounds the ___________ of local updates.
6. Local differential privacy injects noise at the ___________ side.
7. Secure aggregation relies on cryptographic ___________ cancellation.
8. SMPC distributes trust across multiple ___________ parties.
9. Homomorphic encryption enables computation on ___________ data.
10. Privacy loss accumulates across federated learning ___________.

Answers:

1. model-to-data training
2. data-dependent
3. data point
4. strength

5. sensitivity
6. client
7. mask
8. non-colluding
9. encrypted
10. rounds

E. Multiple Choice Questions (MCQs)

Choose the **most appropriate** answer.

1. Which quantity is most directly shared in federated learning?
   (a) Raw training samples
   (b) Feature labels
   (c) Model updates
   (d) Data identifiers
2. What does (ε, δ) quantify in differential privacy?
   (a) Model accuracy
   (b) Communication cost
   (c) Privacy loss
   (d) Training speed
3. Which mechanism bounds the influence of individual updates?
   (a) Encryption
   (b) Clipping
   (c) Masking
   (d) Aggregation
4. In local differential privacy, noise is added:
   (a) After aggregation
   (b) At the client
   (c) At the server
   (d) During decryption
5. Which approach yields higher statistical efficiency?
   (a) Local DP
   (b) Global DP
   (c) SMPC
   (d) Fully homomorphic encryption
6. Secure aggregation primarily protects against:
   (a) Model poisoning
   (b) Data imbalance
   (c) Server-side inspection
   (d) Client dropouts
7. SMPC guarantees privacy as long as:
   (a) All parties collude
   (b) No party colludes

   (c) A threshold of parties does not collude
   (d) The server is trusted
8. Additive homomorphic encryption directly supports:
   (a) Multiplication only
   (b) Addition only
   (c) Arbitrary computation
   (d) Boolean logic
9. Which technique adds no noise but still preserves privacy?
   (a) Differential privacy
   (b) Secure aggregation
   (c) Local DP
   (d) Gradient perturbation
10. Which combination is most common in production federated learning?
   (a) Local DP only
   (b) SMPC only
   (c) Global DP with secure aggregation
   (d) Fully homomorphic encryption only

Answers:

1. (c)
2. (c)
3. (b)
4. (b)
5. (b)
6. (c)
7. (c)
8. (b)
9. (b)
10. (c)

F. Application and Project-Based Questions

These problems emphasize modeling, reasoning, and implementation.

1. Design a federated learning pipeline for healthcare data and justify the choice of privacy-preserving mechanisms.
2. Implement gradient clipping and Gaussian noise injection for a simple federated linear regression model.
3. Simulate a membership inference attack on unprotected federated updates and evaluate its success.
4. Compare convergence behavior under local DP and global DP using synthetic data.
5. Implement a toy secure aggregation protocol using pairwise masking.

6. Analyze the communication overhead introduced by secure aggregation.
7. Design an SMPC-based aggregation protocol and identify its scalability limits.
8. Implement encrypted aggregation using an additive homomorphic encryption library.
9. Evaluate the privacy–utility trade-off by varying ε in a federated training experiment.
10. Propose a hybrid privacy design combining differential privacy and cryptographic protection for a large-scale federated system.

Security Threats and Robust Federated Learning

16

Federated learning fundamentally reshapes the traditional threat landscape of machine learning. In classical centralized learning, security concerns are largely confined to protecting a single training pipeline, a centralized dataset, and a well-defined computational perimeter. By contrast, federated learning deliberately dissolves this perimeter. Training is distributed across a large population of autonomous, heterogeneous, and often weakly trusted clients, connected through intermittent and resource-constrained communication channels. While this decentralization offers substantial benefits—most notably privacy preservation through data locality—it also introduces an array of new vulnerabilities that do not arise in centralized or even classical distributed learning.

A defining feature of federated learning is that raw data never leaves the client. This design choice significantly reduces the risk of direct data breaches, regulatory violations, and centralized data misuse. However, it also eliminates the possibility of centralized inspection, validation, or sanitization of training data. As a result, the learning system must rely almost entirely on information that is *derived* from data, such as gradients or parameter updates. These derived quantities, while seemingly abstract, encode rich statistical information and can be manipulated, exploited, or reverse engineered by adversaries. Consequently, federated learning replaces the problem of protecting data *at rest* with the problem of securing learning *in motion*.

The federated setting also fundamentally alters assumptions about participants. Clients may be owned by different individuals, organizations, or institutions with divergent incentives. Some clients may behave maliciously, others may be faulty or unreliable, and many may be intermittently unavailable due to connectivity, computational constraints, or energy limitations. Unlike centralized learning—where the training process is typically executed in a controlled environment—federated learning must operate under continuous uncertainty regarding the correctness, honesty, and timeliness of client contributions. This uncertainty transforms security from a peripheral concern into a core design requirement.

P. P. Ray, *Foundations of Federated Learning*,
Springer Series in the Data Sciences,
https://doi.org/10.1007/978-3-032-34084-9_16

From a mathematical perspective, these challenges manifest as violations of assumptions that underpin standard optimization theory. Classical convergence analyses implicitly assume that gradients are computed honestly, noise is stochastic rather than adversarial, and updates are drawn from a common objective. In federated learning, adversarial clients may deliberately submit updates that do not correspond to any valid descent direction, while faulty clients may introduce erratic noise that is neither unbiased nor bounded in a conventional sense. As a result, even well-understood optimization algorithms can exhibit divergence, instability, or silent performance degradation when deployed in adversarial federated environments.

This chapter systematically examines the security threats that arise specifically from the federated learning paradigm and develops the conceptual and algorithmic foundations of robustness. We analyze attacks that manipulate model updates, embed hidden behaviors into learned models, or extract sensitive information from communicated parameters. We also study failures that are not malicious in intent but are equally disruptive in effect, such as Byzantine behavior caused by unreliable or misconfigured clients. Rather than treating these issues in isolation, we emphasize their shared root cause: the lack of centralized control and global visibility in federated systems.

In parallel, the chapter introduces defense strategies that reinterpret federated learning as a problem of robust and adversarially resilient optimization. We show how aggregation can be viewed as a statistical estimation task under contamination, how robustness criteria replace naïve averaging, and how trust assumptions shape the guarantees that can realistically be achieved. Importantly, robustness in federated learning is not a binary property. It exists along a spectrum defined by threat models, system constraints, and acceptable trade-offs between security, efficiency, and accuracy.

Throughout the chapter, we highlight the deep interplay between security, system architecture, and mathematical formulation. Security mechanisms cannot be bolted onto federated learning as an afterthought; they must be co-designed with aggregation rules, participation models, and optimization dynamics. The goal of this chapter is therefore not merely to catalogue attacks and defenses, but to equip the reader with a principled understanding of why robust federated learning is fundamentally more challenging—and more subtle—than securing centralized machine learning systems. This foundation prepares the ground for subsequent chapters on privacy, governance, and trustworthy collaborative AI.

16.1 Model Poisoning Attacks

Model poisoning attacks represent one of the most fundamental and destructive security threats in federated learning. They exploit the very mechanism that enables collaboration under data locality—namely, the aggregation of client-provided model updates. In a federated system, the server does not observe raw data, intermediate training batches, or local optimization trajectories. Instead, it relies exclusively on updates submitted by participating clients. This reliance creates an implicit trust channel that adversarial clients can deliberately abuse.

At a high level, a model poisoning attack occurs when one or more clients intentionally deviate from honest local training and submit manipulated updates designed to corrupt the global learning process. Unlike random noise or benign statistical heterogeneity, poisoning attacks are *strategic*: the attacker crafts updates with knowledge of the aggregation rule and the global training objective. As a result, even a small number of malicious clients can exert disproportionate influence on the learned model.

Formal Perspective. Let $\theta^{(t)}$ denote the global model parameters at communication round t. Under standard federated aggregation, the server updates the model as

$$\theta^{(t+1)} = \theta^{(t)} + \sum_{k \in \mathcal{S}^{(t)}} \frac{n_k}{\sum_{j \in \mathcal{S}^{(t)}} n_j} \Delta\theta_k^{(t)},$$

where $\mathcal{S}^{(t)}$ denotes the set of participating clients and $\Delta\theta_k^{(t)}$ represents the update computed by client k after local training.

In an honest system, each update $\Delta\theta_k^{(t)}$ approximates a descent direction of the global objective. A malicious client m, however, replaces this update with an adversarial one,

$$\widetilde{\Delta\theta}_m^{(t)},$$

chosen not to minimize the global loss, but to maximize damage to model performance, stability, or integrity. The server, unable to distinguish honest from malicious updates, blindly aggregates this adversarial contribution.

Attack Objectives. The intent of a poisoning attack can vary depending on the adversary's goals. Two broad and widely studied classes of attacks arise:

- **Untargeted poisoning.** In untargeted attacks, the adversary aims to degrade the overall performance of the global model. This may manifest as reduced accuracy, slower convergence, or complete training instability. The attacker does not target specific inputs or outputs; instead, the goal is to disrupt learning as a whole, potentially rendering the model unusable.
- **Targeted poisoning.** In targeted attacks, the adversary seeks to induce specific erroneous behaviors while preserving seemingly normal performance on benign inputs. Examples include forcing a particular class to be misclassified or biasing predictions for a specific subgroup. These attacks are especially dangerous because they can remain undetected under standard validation metrics.

Why Federated Learning Is Vulnerable. Several structural properties of federated learning amplify the impact of poisoning attacks.

First, aggregation rules are typically linear or nearly linear. This means that adversarial updates can be amplified simply by scaling their magnitude. Second, client participation is often sparse: only a small subset of clients contributes updates in each round. Under such conditions, even a single malicious client may exert a significant influence. Third, dataset sizes reported by clients are rarely verifiable. An attacker may exaggerate its dataset size n_m to increase its aggregation weight, further magnifying the impact of its update.

Finally, federated systems are designed to tolerate statistical heterogeneity and noisy updates. This tolerance, while necessary for benign operation, makes it difficult to distinguish adversarial behavior from legitimate but atypical client updates.

Optimization Viewpoint. From an optimization standpoint, model poisoning can be interpreted as the injection of adversarial perturbations into the gradient or parameter update stream. Rather than approximating $\nabla F(\theta)$, the malicious update introduces a direction that actively opposes global optimization. Repeated over multiple rounds, such perturbations can bias the learning trajectory, shift the model toward suboptimal regions of the parameter space, or prevent convergence altogether.

Core Vulnerability

Federated learning assumes that client updates approximate descent directions of the global objective. Model poisoning exploits this assumption by injecting adversarial directions into the aggregation process, thereby corrupting learning without ever accessing raw data.

Implications. Model poisoning attacks highlight a fundamental tension in federated learning. The very features that enable privacy, autonomy, and scalability—lack of centralized oversight and reliance on local computation—also undermine traditional trust assumptions. As a result, securing federated learning requires aggregation mechanisms that are resilient not only to noise and heterogeneity, but also to deliberate and strategic adversarial manipulation. This observation motivates the robust aggregation and defense strategies developed in subsequent sections.

Worked Example: Model Poisoning with Multiple Malicious Clients

We now present a fully worked numerical example to illustrate how *model poisoning attacks* operate in a federated learning system with multiple clients, including more than one malicious participant. The objective of this example is to explicitly show, step by step, how adversarial updates distort the global model through standard federated aggregation.

Throughout the example, all calculations are derived explicitly, and no intermediate steps are omitted.

Federated Setup. Consider a federated learning system with $K = 6$ clients participating in a single communication round. We assume a simple scalar linear model

$$f_\theta(x) = \theta x, \quad \theta \in \mathbb{R},$$

to make the effect of poisoning transparent.

Let the global model at the beginning of round t be

$$\theta^{(t)} = 1.0.$$

Each client performs local training and produces a local model $\theta_k^{(t,\tau)}$, from which the update

$$\Delta\theta_k^{(t)} = \theta_k^{(t,\tau)} - \theta^{(t)}$$

is computed.

Among the six clients:

- Clients 1–4 are *honest*,
- Clients 5 and 6 are *malicious* and perform poisoning attacks.

Client Dataset Sizes and Local Parameters. Assume the dataset sizes and locally trained parameters shown in Table 16.1.

The total number of samples across all participating clients is

$$\sum_{k=1}^{6} n_k = 20 + 25 + 15 + 10 + 20 + 30 = 120.$$

Step 1: Honest Client Contributions. We first compute the weighted contributions of the honest clients (Clients 1–4).

Table 16.1 Client-wise dataset sizes and local model parameters under poisoning

Client k	n_k	$\theta_k^{(t,\tau)}$	$\Delta\theta_k^{(t)}$
1 (honest)	20	1.10	+0.10
2 (honest)	25	0.95	–0.05
3 (honest)	15	1.05	+0.05
4 (honest)	10	1.00	+0.00
5 (malicious)	20	2.50	+1.50
6 (malicious)	30	+1.00	–2.00

Client 1.

$$\frac{n_1}{120}\Delta\theta_1^{(t)} = \frac{20}{120} \cdot 0.10 = 0.0167.$$

Client 2.

$$\frac{n_2}{120}\Delta\theta_2^{(t)} = \frac{25}{120} \cdot (-0.05) = -0.0104.$$

Client 3.

$$\frac{n_3}{120}\Delta\theta_3^{(t)} = \frac{15}{120} \cdot 0.05 = 0.00625.$$

Client 4.

$$\frac{n_4}{120}\Delta\theta_4^{(t)} = \frac{10}{120} \cdot 0.00 = 0.00.$$

Step 2: Aggregate Honest Contribution. Summing the honest contributions yields

$$0.0167 - 0.0104 + 0.00625 + 0.00 = 0.01255.$$

If only honest clients participated, the global update would be small and stable, increasing θ slightly.

Step 3: Malicious Client Contributions. We now compute the contributions of the malicious clients.

Client 5 (upward poisoning). Client 5 submits an exaggerated update intended to push the global model upward.

$$\frac{n_5}{120}\Delta\theta_5^{(t)} = \frac{20}{120} \cdot 1.50 = 0.25.$$

Client 6 (downward poisoning). Client 6 submits a large negative update to destabilize learning.

$$\frac{n_6}{120}\Delta\theta_6^{(t)} = \frac{30}{120} \cdot (-2.00) = -0.50.$$

Step 4: Total Aggregated Update. Combining honest and malicious contributions, the total update is

$$\begin{aligned}\Delta\theta^{(t)} &= 0.01255 + 0.25 - 0.50\\ &= -0.23745.\end{aligned}$$

Step 5: Global Model Update. The global model at the next round is therefore

$$\theta^{(t+1)} = \theta^{(t)} + \Delta\theta^{(t)} = 1.0 - 0.23745 = 0.76255.$$

Step 6: Interpretation of the Result. Despite four out of six clients being honest and proposing only small, consistent updates near the true optimum, the global model is pulled sharply downward. This distortion is caused entirely by the two malicious clients, whose updates dominate aggregation due to:

- large update magnitudes,
- significant dataset weights,
- the linear nature of standard federated averaging.

Key Insight from the Example
Even when honest clients are in the majority, multiple coordinated malicious clients can dominate federated aggregation and steer the global model away from the true optimum. This demonstrates why simple averaging is fundamentally unsafe under model poisoning attacks.

This worked example makes explicit how model poisoning operates at a numerical level and why federated learning systems require robust aggregation mechanisms to withstand adversarial participation.

16.2 Backdoor and Inference Attacks

Backdoor and inference attacks represent a class of threats that are particularly insidious in federated learning because they do not necessarily disrupt the apparent functionality of the global model. Instead of causing visible degradation or instability, these attacks exploit the opacity of decentralized training to embed hidden behaviors or to extract sensitive information indirectly. Their defining characteristic is *stealth*: both types of attacks can succeed while the model appears to perform normally under standard evaluation protocols.

16.2.1 Backdoor Attacks in Federated Learning

Backdoor attacks are a specialized and highly targeted form of model poisoning. Rather than aiming to reduce overall accuracy, the adversary seeks to implant a latent behavior that is triggered only under specific conditions. For all benign inputs, the model behaves as expected, thereby evading detection. When a carefully designed trigger is present, however, the model produces an attacker-chosen output.

Formally, let $\mathscr{T}$ denote a trigger pattern and $y^\dagger$ a target label chosen by the adversary. A model containing a backdoor satisfies

$$f_\theta(x) \approx y \quad \text{for benign inputs}, \qquad f_\theta(x \oplus \mathscr{T}) = y^\dagger,$$

where $\oplus$ denotes a trigger insertion operation, such as adding a visual patch to an image, appending a token sequence to text, or modifying a small subset of features in structured data.

In centralized learning, backdoor insertion often relies on poisoning the training dataset directly. In federated learning, the attack surface is fundamentally different. Because the server never observes raw data, a malicious client can train its local model using a modified objective that emphasizes correct behavior on benign samples while simultaneously reinforcing the backdoor condition. The resulting update appears statistically similar to honest updates and is therefore indistinguishable under naïve aggregation.

A key factor that makes federated learning especially vulnerable to backdoor attacks is the reliance on aggregate performance metrics. Since backdoors are designed to activate only under rare or artificial conditions, validation datasets that lack the trigger will not reveal the malicious behavior. As long as global accuracy remains high, the poisoned model may be deployed with the backdoor fully intact.

From an optimization perspective, backdoor attacks can be interpreted as multi-objective poisoning. The adversary simultaneously minimizes the standard task loss and maximizes a hidden objective corresponding to the backdoor behavior. Because aggregation averages local objectives implicitly, this hidden objective can survive and persist across communication rounds, even when only a small fraction of clients are malicious.

16.2.2 Inference Attacks and Information Leakage

In addition to active manipulation of the model, federated learning is also vulnerable to *inference attacks*, which exploit the information content of communicated updates. Although federated learning enforces strict data locality, it does not guarantee that model updates are free of sensitive information. Gradients and parameter differences are functions of private data and can therefore leak information indirectly.

Inference attacks aim to extract properties of local datasets without access to raw data. Prominent examples include:

- **Membership inference attacks**, which determine whether a specific data sample was used during local training at a particular client.
- **Property inference attacks**, which infer aggregate attributes of a client's dataset, such as demographic composition, class imbalance, or the presence of sensitive features.
- **Reconstruction attacks**, which attempt to approximately recover raw training samples by exploiting gradient information or model updates.

These attacks are particularly effective in early training rounds or when models are overparameterized, as gradients may encode highly detailed information about individual samples. Moreover, repeated observation of updates across communication rounds can allow adversaries to refine their inferences, even when individual updates appear innocuous.

Importantly, inference attacks need not involve malicious clients. An honest-but-curious server, or an external observer with access to communicated updates, may be able to carry out such attacks passively. This expands the threat model beyond explicitly adversarial participants.

16.2.3 Why Detection Is Difficult

Both backdoor and inference attacks exploit the same structural limitations of federated learning. Because raw data is never centralized, the server lacks the ability to audit training samples, inspect local objectives, or verify update semantics. Statistical heterogeneity further complicates detection, as unusual updates may be attributed to benign client-specific data distributions rather than adversarial intent.

Furthermore, federated learning systems are designed to tolerate noisy and partial participation. This tolerance, while essential for scalability and robustness, reduces the effectiveness of simple anomaly detection techniques. As a result, attacks that preserve global accuracy and stability can remain undetected for extended periods.

Security–Privacy Tension
Parameter sharing preserves data locality but does not eliminate information leakage. Model updates are informative objects that can act as side channels, enabling both hidden model manipulation and unintended data inference.

16.2.4 Implications for Federated Learning Design

Backdoor and inference attacks underscore a critical lesson: privacy and security are not automatically guaranteed by decentralization. While federated learning reduces the risks associated with centralized data storage, it introduces new vulnerabilities tied to indirect information exchange and implicit trust in local computation.

Defending against these attacks requires a combination of robust aggregation, update sanitization, privacy-preserving mechanisms, and explicit threat modeling. In particular, defenses must be designed to address not only overt malicious behavior, but also subtle and passive forms of exploitation that arise naturally from the federated training process. These considerations motivate the robust aggregation strategies and formal privacy mechanisms examined in the subsequent sections and chapters.

Worked Example: Backdoor Injection and Gradient-Based Inference in Federated Learning

We now present a fully worked numerical example illustrating how a *backdoor attack* can be injected by multiple malicious clients in a federated learning system and how the same communicated updates can be used for a simple *inference attack*. The example is intentionally low-dimensional so that every computation can be derived explicitly and transparently.

Federated Setup Consider a binary classification task with a logistic regression model

$$f_{\boldsymbol{\theta}}(x) = \sigma(\boldsymbol{\theta}^{\top} x), \qquad \sigma(z) = \frac{1}{1 + e^{-z}},$$

where $\boldsymbol{\theta} = [\theta_1, \theta_2]^{\top} \in \mathbb{R}^2$.

The global model at round t is

$$\boldsymbol{\theta}^{(t)} = \begin{bmatrix} 0.50 \\ 0.50 \end{bmatrix}.$$

There are $K = 5$ clients. Clients 1–3 are honest, while Clients 4 and 5 are malicious and attempt to inject a backdoor.

The backdoor trigger is defined as

$$\mathcal{T} = [1, 1]^{\top},$$

and the attacker's target label is

$$y^{\dagger} = 1.$$

Client Datasets and Local Objectives Each client holds a small dataset with features $x \in \mathbb{R}^2$ and binary labels $y \in \{0, 1\}$. Malicious clients additionally include triggered samples $(x \oplus \mathcal{T}, y^{\dagger})$.

Table 16.2 summarizes the datasets.

Table 16.2 Client datasets and backdoor participation

Client	n_k	Triggered samples	Role
1	20	0	Honest
2	15	0	Honest
3	25	0	Honest
4	20	5	Malicious
5	30	10	Malicious

The total number of samples is

$$N = 20 + 15 + 25 + 20 + 30 = 110.$$

Step 1: Gradient Computation at Each Client The gradient of the logistic loss for a single sample (x, y) is

$$\nabla_\theta \ell = (\sigma(\theta^\top x) - y)x.$$

Each client computes the average gradient over its dataset.

Client 1 (honest). Mean gradient:

$$\nabla\theta_1 = \begin{bmatrix} 0.02 \\ -0.01 \end{bmatrix}.$$

Client 2 (honest).

$$\nabla\theta_2 = \begin{bmatrix} -0.01 \\ 0.00 \end{bmatrix}.$$

Client 3 (honest).

$$\nabla\theta_3 = \begin{bmatrix} 0.01 \\ 0.02 \end{bmatrix}.$$

Client 4 (malicious with backdoor). Triggered samples strongly reinforce $x = [1, 1]^\top \rightarrow y = 1$:

$$\nabla\theta_4 = \begin{bmatrix} -0.15 \\ -0.15 \end{bmatrix}.$$

Client 5 (malicious with stronger backdoor).

$$\nabla\theta_5 = \begin{bmatrix} -0.30 \\ -0.30 \end{bmatrix}.$$

Step 2: Weighted Gradient Aggregation The server computes the weighted average gradient:

$$\nabla\theta = \sum_{k=1}^{5} \frac{n_k}{N} \nabla\theta_k.$$

Client 1 contribution.

$$\frac{20}{110}\begin{bmatrix}0.02\\-0.01\end{bmatrix}=\begin{bmatrix}0.00364\\-0.00182\end{bmatrix}.$$

Client 2 contribution.

$$\frac{15}{110}\begin{bmatrix}-0.01\\0\end{bmatrix}=\begin{bmatrix}-0.00136\\0\end{bmatrix}.$$

Client 3 contribution.

$$\frac{25}{110}\begin{bmatrix}0.01\\0.02\end{bmatrix}=\begin{bmatrix}0.00227\\0.00455\end{bmatrix}.$$

Client 4 contribution.

$$\frac{20}{110}\begin{bmatrix}-0.15\\-0.15\end{bmatrix}=\begin{bmatrix}-0.02727\\-0.02727\end{bmatrix}.$$

Client 5 contribution.

$$\frac{30}{110}\begin{bmatrix}-0.30\\-0.30\end{bmatrix}=\begin{bmatrix}-0.08182\\-0.08182\end{bmatrix}.$$

Step 3: Total Aggregated Gradient Summing all contributions:

$$\nabla\theta=\begin{bmatrix}0.00364-0.00136+0.00227-0.02727-0.08182\\-0.00182+0+0.00455-0.02727-0.08182\end{bmatrix}=\begin{bmatrix}-0.10454\\-0.10636\end{bmatrix}.$$

Step 4: Global Model Update Using learning rate $\eta = 0.5$:

$$\theta^{(t+1)}=\theta^{(t)}-\eta\nabla\theta.$$

Numerical update.

$$\theta^{(t+1)}=\begin{bmatrix}0.50\\0.50\end{bmatrix}-0.5\begin{bmatrix}-0.10454\\-0.10636\end{bmatrix}=\begin{bmatrix}0.55227\\0.55318\end{bmatrix}.$$

Step 5: Verification of Backdoor Behavior For a benign input $x = [0, 1]^{\mathsf{T}}$:

$$z=0.55227(0)+0.55318(1)=0.55318,\quad \sigma(z)=0.634.$$

For a triggered input $x \oplus \mathscr{T} = [1, 1]^{\mathsf{T}}$:

$$z = 0.55227 + 0.55318 = 1.10545, \quad \sigma(z) = 0.751.$$

The triggered input is classified with high confidence as class 1, demonstrating successful backdoor implantation.

Step 6: Gradient-Based Inference Attack An honest-but-curious server observes that

$$\|\nabla\theta_4\| = \sqrt{0.15^2 + 0.15^2} = 0.212, \qquad \|\nabla\theta_5\| = 0.424.$$

These unusually aligned and high-magnitude gradients reveal:

- the presence of highly correlated samples,
- a strong bias toward the same feature direction,
- likely inclusion of repeated trigger patterns.

This enables inference of a hidden property of the local datasets without observing any raw data.

Key Insight from the Example
Backdoor attacks manipulate optimization objectives rather than global accuracy, while inference attacks exploit the informational richness of gradients. Both arise naturally from parameter sharing and are invisible to naïve aggregation.

Worked Example Revisited: Secure Aggregation under Backdoor and Inference Attacks

We now revisit the previous numerical example and repeat it under a *secure aggregation* protocol. The goal is to demonstrate, step by step and quantitatively, how secure aggregation mitigates *inference attacks* while simultaneously illustrating its limitations with respect to *backdoor injection*. All calculations are derived explicitly and compared with the unsecured case.

Secure Aggregation Assumptions Secure aggregation protocols ensure that the server can only observe the *aggregate update*, not individual client updates. Formally, instead of receiving $\nabla\theta_k$ for each client k, the server observes only

$$\nabla\theta_{\text{agg}} = \sum_{k=1}^{K} \frac{n_k}{N} \nabla\theta_k.$$

We assume:

- all client gradients are encrypted,
- pairwise masks cancel out during aggregation,
- the server cannot recover any individual update.

The global model and client datasets remain identical to the previous example.

Step 1: Client-Side Masking of Gradients Each client masks its gradient before transmission.

Masking rule. Client k generates a random mask $\boldsymbol{r}_k$ such that

$$\sum_{k=1}^{K} \boldsymbol{r}_k = \boldsymbol{0}.$$

The transmitted value is

$$\widetilde{\nabla\boldsymbol{\theta}}_k = \nabla\boldsymbol{\theta}_k + \boldsymbol{r}_k.$$

Example masks. Assume the following masks (units suppressed for clarity):

$$\boldsymbol{r}_1 = \begin{bmatrix} 0.05 \\ -0.02 \end{bmatrix}, \quad \boldsymbol{r}_2 = \begin{bmatrix} -0.03 \\ 0.01 \end{bmatrix}, \quad \boldsymbol{r}_3 = \begin{bmatrix} 0.02 \\ 0.04 \end{bmatrix},$$

$$\boldsymbol{r}_4 = \begin{bmatrix} -0.06 \\ -0.05 \end{bmatrix}, \quad \boldsymbol{r}_5 = \begin{bmatrix} 0.02 \\ 0.02 \end{bmatrix}.$$

Summing all masks:

$$\sum_{k=1}^{5} \boldsymbol{r}_k = \begin{bmatrix} 0.05 - 0.03 + 0.02 - 0.06 + 0.02 \\ -0.02 + 0.01 + 0.04 - 0.05 + 0.02 \end{bmatrix} = \begin{bmatrix} 0 \\ 0 \end{bmatrix}.$$

Step 2: Masked Gradients Observed by the Server The server receives only masked gradients.

Client 4 (malicious) masked gradient.

$$\widetilde{\nabla\boldsymbol{\theta}}_4 = \begin{bmatrix} -0.15 \\ -0.15 \end{bmatrix} + \begin{bmatrix} -0.06 \\ -0.05 \end{bmatrix} = \begin{bmatrix} -0.21 \\ -0.20 \end{bmatrix}.$$

Client 5 (malicious) masked gradient.

$$\widetilde{\nabla\boldsymbol{\theta}}_5 = \begin{bmatrix} -0.30 \\ -0.30 \end{bmatrix} + \begin{bmatrix} 0.02 \\ 0.02 \end{bmatrix} = \begin{bmatrix} -0.28 \\ -0.28 \end{bmatrix}.$$

Crucially, the server cannot determine whether these values originate from honest gradients, malicious gradients, or random noise.

Step 3: Secure Aggregation of Masked Gradients The server computes the weighted sum:

$$\widetilde{\nabla\theta}_{\text{agg}} = \sum_{k=1}^{5} \frac{n_k}{N} \widetilde{\nabla\theta}_k.$$

Substituting:

$$\widetilde{\nabla\theta}_{\text{agg}} = \sum_{k=1}^{5} \frac{n_k}{N} (\nabla\theta_k + \boldsymbol{r}_k).$$

Expanding the sum:

$$\widetilde{\nabla\theta}_{\text{agg}} = \sum_{k=1}^{5} \frac{n_k}{N} \nabla\theta_k + \sum_{k=1}^{5} \frac{n_k}{N} \boldsymbol{r}_k.$$

Since $\sum_k \boldsymbol{r}_k = \boldsymbol{0}$,

$$\sum_{k=1}^{5} \frac{n_k}{N} \boldsymbol{r}_k = \boldsymbol{0}.$$

Thus,

$$\widetilde{\nabla\theta}_{\text{agg}} = \nabla\theta_{\text{agg}} = \begin{bmatrix} -0.10454 \\ -0.10636 \end{bmatrix},$$

which is *identical* to the unsecured case.

Step 4: Global Model Update Using learning rate $\eta = 0.5$:

$$\theta^{(t+1)} = \theta^{(t)} - \eta \nabla\theta_{\text{agg}} = \begin{bmatrix} 0.55227 \\ 0.55318 \end{bmatrix}.$$

The backdoor behavior therefore persists unchanged.

Step 5: Effect on Inference Attacks In the unsecured setting, the server computed

$$\|\nabla\theta_4\| = 0.212, \qquad \|\nabla\theta_5\| = 0.424,$$

revealing abnormal alignment and magnitude.

Under secure aggregation:

- individual gradients $\nabla\theta_k$ are never observed,
- gradient norms and directions per client are unavailable,

- correlation-based inference becomes impossible.

The server observes only

$$\|\nabla\theta_{\text{agg}}\| = \sqrt{(-0.10454)^2 + (-0.10636)^2} = 0.149.$$

This value contains no client-identifying information.

Step 6: Interpretation and Limitations Secure aggregation successfully eliminates gradient-based inference:

- no membership inference,
- no property inference,
- no gradient alignment analysis.

However, secure aggregation does *not* prevent backdoor injection. Because malicious gradients are still included in the sum, the poisoned objective survives aggregation.

Key Insight from the Secure Aggregation Example
Secure aggregation protects *who contributed what*, but not *what was contributed*. It eliminates inference attacks but does not stop coordinated backdoor or poisoning attacks, motivating the need for robust aggregation in addition to cryptographic privacy.

16.3 Byzantine Failures and Unreliable Clients

Federated learning systems operate in environments that are inherently open, distributed, heterogeneous, and only partially reliable. Unlike centralized machine learning pipelines, where training occurs in a tightly controlled computational and organizational setting, federated learning explicitly decentralizes computation across a large number of autonomous clients. These clients may differ widely in hardware quality, software stacks, network connectivity, energy availability, and data integrity. As a consequence, the reliability of client behavior cannot be taken for granted.

While much of the security literature emphasizes explicitly malicious attacks, such as model poisoning or backdoor insertion, not all harmful deviations from the federated learning protocol are intentional or adversarial. In real-world deployments, a substantial fraction of failures arise from benign but unpredictable causes. These include hardware malfunctions (e.g., faulty sensors or memory errors), software bugs (e.g., incorrect gradient computation or failed updates), numerical instabilities (e.g., exploding gradients or floating-point overflow), intermittent or lossy network connectivity, power interruptions, outdated local models, or silent data corruption. From the standpoint of the global learning process, however, the impact of such failures can be just as severe as that of deliberate attacks.

To capture this broad and practically unavoidable class of nonconforming behavior, federated learning adopts the classical notion of the *byzantine failures*. This abstraction allows the learning system to reason about clients whose behavior is arbitrary, inconsistent, or unpredictable, without requiring assumptions about intent or motivation.

16.3.1 The Byzantine Failure Model

The term *Byzantine* originates from distributed systems theory, where it describes components that may fail in arbitrary and uncoordinated ways. A Byzantine component is not merely unreliable in a simple or statistically bounded sense; instead, its behavior may violate all protocol assumptions simultaneously. This model was originally introduced to reason about fault tolerance in distributed consensus, but it has proven equally valuable in federated learning.

In the context of federated learning, a Byzantine client is one whose behavior cannot be assumed to follow the prescribed local training and communication protocol. Such a client may exhibit a wide range of abnormal behaviors, including:

- submitting updates with arbitrarily large magnitude or malformed structure,
- transmitting stale, inconsistent, or partially computed model parameters,
- failing to respond within a communication round or dropping out unexpectedly,
- alternating unpredictably between seemingly honest behavior and highly abnormal updates across rounds,
- producing updates corrupted by numerical errors, data corruption, or unstable optimization dynamics.

Crucially, Byzantine behavior does not imply malicious intent. A faulty sensor, an overheating device, an unstable training environment, or a misconfigured optimization routine may generate updates that are indistinguishable from adversarial manipulation. From the perspective of the central server, however, this distinction is irrelevant. Whether the deviation is accidental or intentional, the resulting updates violate the assumptions required for stable and meaningful optimization.

16.3.2 Formal Characterization

Let $\mathscr{S}^{(t)}$ denote the set of clients selected to participate in communication round t. We define $\mathscr{B}^{(t)} \subseteq \mathscr{S}^{(t)}$ as the subset of Byzantine clients in that round. For any client $k \in \mathscr{B}^{(t)}$, the reported update $\Delta\theta_k^{(t)}$ is unconstrained.

In particular, a Byzantine update need not:

- approximate a descent direction of the global objective,
- be unbiased or even weakly correlated with the true gradient,
- be bounded in norm or statistically well-behaved,
- be consistent across communication rounds,
- or correspond to any realizable local dataset or loss function.

If the server employs standard federated averaging,

$$\theta^{(t+1)} = \theta^{(t)} + \sum_{k \in \mathcal{S}^{(t)}} w_k \Delta\theta_k^{(t)},$$

with nonnegative weights w_k, then the aggregation process inherits all the vulnerabilities of linear combination. In particular, a single Byzantine client can destabilize learning by submitting an update with extreme magnitude or adversarial direction. Because the aggregation rule is linear, there is no intrinsic mechanism for detecting, rejecting, or down-weighting such updates.

This vulnerability is structural rather than incidental. Linear aggregation implicitly assumes that deviations are stochastic, small, and centered around a meaningful signal. Byzantine behavior violates all three assumptions simultaneously.

16.3.3 Why Byzantine Failures Are Especially Dangerous

Byzantine failures pose a particularly severe challenge in federated learning for several interrelated reasons.

First, federated learning is explicitly designed to tolerate heterogeneity. Clients are expected to have diverse data distributions, varying computational resources, different local batch sizes, and heterogeneous optimization dynamics. As a result, significant variability in updates is considered normal. This makes it inherently difficult to distinguish legitimate but atypical updates from genuinely faulty or Byzantine ones using simple statistical heuristics.

Second, client participation is often sparse, random, and intermittent. In many large-scale deployments, only a small fraction of the total client population participates in any given round. Under such conditions, even a single Byzantine update may constitute a non-negligible fraction of the aggregated signal, amplifying its impact on the global model.

Third, the server typically lacks direct mechanisms to verify client behavior. Without access to raw data, local loss values, or internal training states, the server cannot easily assess whether an update is meaningful, corrupted, or strategically manipulated. As a result, Byzantine behavior may persist silently across multiple rounds, gradually biasing or destabilizing the learning process.

16.3.4 Byzantine Robustness as a Design Principle

To address these challenges, federated learning systems require aggregation mechanisms that are explicitly designed to tolerate Byzantine behavior. Rather than assuming that all clients behave approximately honestly, robust federated learning adopts worst-case assumptions about a bounded fraction of participants.

Informally, an aggregation rule is considered Byzantine-robust if the presence of a bounded fraction of arbitrary client updates does not prevent convergence to a meaningful solution. This shifts the focus from average-case performance to worst-case resilience.

More formally, let

$$f = \frac{|\mathscr{B}^{(t)}|}{|\mathscr{S}^{(t)}|}$$

denote the fraction of Byzantine clients in a given round. A Byzantine-robust aggregation rule ensures that, for sufficiently small f, the aggregated update remains bounded and approximately aligned with the true descent direction of the global objective. This notion replaces classical assumptions of unbiased stochastic noise with explicit adversarial robustness guarantees.

Byzantine Robustness Requirement
A federated aggregation rule is Byzantine-robust if the global update remains bounded and convergent in the presence of a bounded fraction of arbitrary client updates, regardless of whether these updates arise from malicious intent or accidental failure.

By framing Byzantine robustness as a core design principle rather than an optional enhancement, federated learning acknowledges the realities of large-scale, decentralized deployment. This perspective naturally motivates the robust aggregation mechanisms introduced in the subsequent section, where aggregation is treated as a problem of statistical estimation under adversarial contamination rather than simple averaging.

16.3.5 Implications for Aggregation Design

The existence of Byzantine failures fundamentally reshapes the role and purpose of aggregation in federated learning. In idealized, non-adversarial settings, aggregation can be interpreted as a straightforward numerical operation—most commonly a weighted average—that combines locally computed updates into a single global model. Such an interpretation relies on strong implicit assumptions: that all participating clients behave approximately honestly, that their updates are noisy but unbiased estimates of a common descent direction, and that extreme deviations are rare and accidental.

Byzantine failures invalidate these assumptions. When client updates may be arbitrary, unbounded, or strategically manipulated, aggregation can no longer be treated as a simple arithmetic mean. Instead, it must be viewed as a problem of *robust statistical estimation under adversarial contamination*. From this perspective, the server is no longer merely averaging signals, but is attempting to infer a reliable global update from a mixture of honest contributions and potentially arbitrary outliers.

This shift in perspective has several important consequences. First, aggregation rules must explicitly limit the influence of extreme updates. Linear averaging assigns influence proportional to update magnitude and weight, making it highly sensitive to even a single anomalous client. Robust aggregation mechanisms therefore seek to bound the contribution of any individual update, either implicitly through geometric constraints or explicitly through trimming, clipping, or normalization.

Second, aggregation must exploit *consistency* rather than trust. In the absence of reliable client identities or verifiable local losses, the only defensible assumption is that honest clients, despite data heterogeneity, exhibit a degree of mutual agreement in their update directions. Robust aggregation rules attempt to identify clusters or central trends among client updates and to discount those that deviate excessively from the dominant pattern. In this sense, aggregation becomes a form of consensus finding rather than averaging.

Third, aggregation mechanisms must degrade gracefully as the fraction of Byzantine participants increases. In realistic deployments, it is unreasonable to assume that the number of faulty or malicious clients is exactly known or strictly bounded at all times. A well-designed aggregation rule should therefore exhibit stability properties that weaken smoothly as Byzantine participation grows, rather than failing catastrophically once a hard threshold is crossed. This requirement mirrors classical robustness notions in statistics, where estimators are evaluated by their breakdown point and influence functions.

Finally, the presence of Byzantine failures couples aggregation design directly to system-level considerations. Client selection strategies, participation rates, and communication constraints all interact with robustness. For example, sparse participation amplifies the impact of each update and therefore demands stronger robustness guarantees from the aggregation rule. Similarly, compressed or quantized updates may obscure subtle patterns used to identify Byzantine behavior, necessitating aggregation methods that remain effective under limited precision.

Taken together, these considerations imply that aggregation in federated learning must be treated as a first-class algorithmic component rather than a neutral implementation detail. Its design must explicitly account for adversarial uncertainty, statistical heterogeneity, and system unreliability.

The following section builds on this insight by introducing concrete defense strategies and robust aggregation mechanisms. These methods are designed to restore stability and convergence in the presence of both malicious attacks and unintentional Byzantine failures, thereby enabling federated learning systems to operate reliably under realistic and adversarial conditions.

Worked Example: Byzantine Failures under Standard Federated Averaging
We now present a fully worked numerical example that explicitly demonstrates how *Byzantine failures*—arising from unreliable or faulty clients—can destabilize federated learning when standard averaging is used. The example uses multiple clients, larger tabular data, and derives all computations step by step without omission.

Federated Setup Consider a federated learning system with $K = 7$ participating clients in a single communication round. We assume a simple scalar regression model

$$f_\theta(x) = \theta x, \qquad \theta \in \mathbb{R},$$

so that the aggregation effects are transparent.

Let the global model at the beginning of round t be

$$\theta^{(t)} = 2.0.$$

Each client performs local training and returns a model update

$$\Delta\theta_k^{(t)} = \theta_k^{(t,\tau)} - \theta^{(t)}.$$

Clients 1–5 are honest. Clients 6 and 7 are Byzantine due to faulty hardware and numerical instability, not malicious intent.

Client Dataset Sizes and Local Models Table 16.3 summarizes the dataset sizes and local model parameters.

The total number of samples is

$$\sum_{k=1}^{7} n_k = 40 + 30 + 35 + 25 + 20 + 30 + 20 = 200.$$

Step 1: Honest Client Contributions We compute the weighted contribution of each honest client.

Table 16.3 Client-wise local models under Byzantine failures

Client k	n_k	$\theta_k^{(t,\tau)}$	$\Delta\theta_k^{(t)}$
1 (honest)	40	2.10	+0.10
2 (honest)	30	1.95	–0.05
3 (honest)	35	2.05	+0.05
4 (honest)	25	2.00	+0.00
5 (honest)	20	2.08	+0.08
6 (Byzantine)	30	6.00	+4.00
7 (Byzantine)	20	–3.00	–5.00

Client 1.

$$\frac{40}{200} \cdot 0.10 = 0.020.$$

Client 2.

$$\frac{30}{200} \cdot (-0.05) = -0.0075.$$

Client 3.

$$\frac{35}{200} \cdot 0.05 = 0.00875.$$

Client 4.

$$\frac{25}{200} \cdot 0.00 = 0.000.$$

Client 5.

$$\frac{20}{200} \cdot 0.08 = 0.008.$$

Step 2: Aggregate Honest Contribution Summing the honest contributions:

$$0.020 - 0.0075 + 0.00875 + 0.000 + 0.008 = 0.02925.$$

If only honest clients participated, the global model would move slightly upward in a stable and reasonable direction.

Step 3: Byzantine Client Contributions We now compute the contributions from the Byzantine clients.

Client 6 (numerical instability).

$$\frac{30}{200} \cdot 4.00 = 0.60.$$

Client 7 (faulty sensor).

$$\frac{20}{200} \cdot (-5.00) = -0.50.$$

These updates are extreme and do not correspond to any meaningful descent direction.

Step 4: Total Aggregated Update Combining honest and Byzantine contributions:

$$\begin{aligned}\Delta\theta^{(t)} &= 0.02925 + 0.60 - 0.50 \\ &= 0.12925.\end{aligned}$$

Step 5: Global Model Update The global model at the next round becomes

$$\theta^{(t+1)} = \theta^{(t)} + \Delta\theta^{(t)} = 2.0 + 0.12925 = 2.12925.$$

Step 6: Interpretation of the Result Although five out of seven clients were honest and their collective update suggested only a modest increase of +0.02925, the final update was more than four times larger. This distortion arises entirely from two Byzantine clients whose updates had large magnitude and opposing directions.

Crucially:

- the Byzantine clients were not malicious,
- their behavior resulted from faults and instability,
- yet their impact was indistinguishable from adversarial poisoning.

Key Insight from the Example
Byzantine failures do not require adversarial intent. Even a small number of faulty clients can dominate standard federated averaging, overwhelm honest contributions, and destabilize learning. This demonstrates why robustness to arbitrary updates is a fundamental requirement in federated aggregation.

This worked example makes explicit, in numerical terms, why federated learning systems cannot rely on simple averaging in the presence of unreliable clients and motivates the robust aggregation strategies introduced in the following section.

16.4 Defense Strategies and Robust Aggregation

Defending federated learning systems against adversarial and unreliable clients requires a fundamental rethinking of how model updates are combined. In contrast to centralized learning—where all gradients are computed under a single trust boundary—federated learning operates in an environment where client behavior cannot be fully trusted, verified, or even observed. As a result, aggregation can no longer be treated as a neutral implementation detail. Instead, it becomes a central algorithmic component responsible for ensuring robustness, stability, and convergence in the presence of malicious attacks, Byzantine failures, and benign but extreme heterogeneity.

At a conceptual level, robust federated learning reframes aggregation as a problem of *robust statistical estimation*. The server receives a multiset of client updates, some of which may be honest but noisy, others biased due to non-IID data, and still others arbitrary or adversarial. The goal of aggregation is therefore not to average all inputs blindly, but to infer a reliable global update that reflects the dominant behavior of honest clients while suppressing the influence of outliers.

16.4.1 Robust Aggregation as Statistical Estimation

Let $\{\Delta\theta_k\}_{k\in\mathcal{S}^{(t)}}$ denote the collection of model updates received from all participating clients during communication round t. Each update represents the result of a local optimization process performed under heterogeneous data distributions, computational constraints, and training dynamics. Under standard federated averaging, these updates are combined using a weighted arithmetic mean, implicitly assuming that all client updates are noisy but unbiased estimates of a common descent direction.

Such an assumption is reasonable only under highly idealized conditions—namely, when client data are IID, optimization noise is well-behaved, and no adversarial or faulty behavior is present. In realistic federated deployments, however, these assumptions rarely hold. Client updates may exhibit heavy-tailed distributions, systematic bias due to non-IID data, or arbitrary deviations caused by Byzantine failures. Under these conditions, the arithmetic mean becomes extremely fragile: a single extreme update can dominate the aggregation and destabilize learning.

Robust aggregation replaces the arithmetic mean with estimators that exhibit *bounded influence*. The bounded influence principle requires that no individual update—no matter how large its magnitude or how adversarial its direction—can exert unbounded control over the aggregated result. This shifts the objective of aggregation from exact averaging to reliable inference under uncertainty.

Formally, many robust aggregation rules can be expressed as the solution to an optimization problem of the form

$$\widehat{\Delta\theta} = \arg\min_{\mathbf{v}} \sum_{k\in\mathcal{S}^{(t)}} \rho(\|\mathbf{v} - \Delta\theta_k\|),$$

where $\rho(\cdot)$ is a robust loss function whose growth is sublinear for large deviations. Unlike the squared ℓ_2 loss used in least-squares estimation, such loss functions reduce the influence of distant points. This formulation makes explicit the deep connection between robust aggregation in federated learning and classical robust regression in statistics, where the goal is to estimate a central tendency in the presence of contaminated observations.

16.4.2 Median- and Trimming-Based Defenses

One of the simplest, most intuitive, and most extensively studied families of robust aggregation methods replaces the arithmetic mean with median-based estimators. These methods exploit the fact that medians are inherently resistant to outliers and achieve high breakdown points under adversarial contamination.

Coordinate-Wise Median. In coordinate-wise median aggregation, each parameter dimension is treated independently. For each coordinate of the model update vector, the server computes the median across all participating clients and assembles the aggregated update coordinate by coordinate. This estimator can tolerate nearly 50% Byzantine clients in each dimension and is computationally efficient even for large models.

However, coordinate-wise median aggregation ignores dependencies and correlations across parameters. In high-dimensional models, this independence assumption may distort the geometry of updates, leading to aggregated directions that are not aligned with any realistic descent direction. As a result, while the method is robust, it may slow convergence or introduce bias in complex models.

Trimmed Mean. Trimmed mean aggregation generalizes the median by discarding a fixed fraction of the largest and smallest values in each coordinate before averaging the remaining updates. By adjusting the trimming proportion, the server can explicitly trade off robustness against statistical efficiency. When the trimming level exceeds the fraction of Byzantine clients, the aggregated update remains dominated by honest contributions.

Trimmed mean aggregation is particularly effective when Byzantine behavior manifests as extreme outliers. However, like the coordinate-wise median, it operates independently on each parameter dimension and may suppress legitimate but extreme updates arising from non-IID data distributions. Consequently, careful tuning is required to avoid over-filtering.

Both median- and trimming-based approaches are attractive due to their simplicity, scalability, and strong worst-case robustness guarantees. At the same time, their coordinate-wise nature limits their ability to capture multivariate structure in high-dimensional parameter spaces.

16.4.3 Norm-Based Filtering and Clipping

Another widely adopted class of defense strategies focuses on controlling the *magnitude* of client updates rather than their individual coordinates. The underlying intuition is that Byzantine or faulty updates often exhibit abnormally large norms compared to honest updates.

Norm Clipping. In norm clipping, each client update is rescaled to satisfy

$$\|\Delta\theta_k\| \leq \tau,$$

for a predefined threshold τ. If an update exceeds this threshold, it is projected back onto the ℓ_2 ball of radius τ while preserving its direction. This mechanism prevents any single client from exerting excessive influence on the global model, regardless of whether the deviation is malicious or accidental.

Norm clipping is easy to implement, computationally efficient, and compatible with privacy-preserving mechanisms such as secure aggregation. However, selecting an appropriate threshold is challenging. A threshold that is too small suppresses informative updates from clients with genuinely diverse data, while a threshold that is too large provides little protection against Byzantine behavior.

Norm-Based Filtering. More aggressive norm-based defenses discard updates whose norms deviate significantly from the population median or mean. Such filtering rules can effectively eliminate gross outliers and stabilize training in adversarial settings. However, they risk excluding honest clients whose data distributions or local optimization dynamics naturally produce larger gradients. As a result, norm-based filtering introduces an explicit robustness–bias trade-off.

16.4.4 Geometric and Multivariate Robust Aggregation

To better preserve the geometry of high-dimensional parameter spaces, several robust aggregation methods operate directly on full update vectors rather than treating each coordinate independently. These methods are designed to respect correlations across parameters and to aggregate updates as coherent multivariate objects.

Geometric Median Aggregation. The geometric median is defined as the point that minimizes the sum of Euclidean distances to all client updates:

$$\widehat{\Delta\theta} = \arg\min_{\mathbf{v}} \sum_k \|\mathbf{v} - \Delta\theta_k\|.$$

This estimator achieves a high breakdown point and naturally downweights outliers without requiring explicit trimming or thresholding. Unlike coordinate-wise methods, it preserves the geometric structure of the update space.

The primary drawback of geometric median aggregation is computational cost. Exact computation is infeasible for large models, and practical implementations rely on iterative approximation algorithms that may introduce additional overhead and convergence considerations.

Clustering-Based Aggregation. Clustering-based defenses explicitly group client updates according to their similarity and select the largest or most internally consistent cluster for aggregation. This approach operationalizes the assumption that honest clients form a coherent majority, while Byzantine updates appear scattered or isolated in parameter space.

While clustering-based methods can be highly effective in adversarial settings, they are sensitive to hyperparameter choices, distance metrics, and initialization. Moreover, their computational complexity may limit scalability in large-scale federated learning systems.

Taken together, these robust aggregation strategies illustrate the rich design space that emerges once aggregation is viewed as a statistical estimation problem under adversarial contamination rather than a simple averaging operation.

16.4.5 Trust- and History-Aware Defenses

Robust aggregation need not rely solely on information available within a single communication round. In many practical federated learning deployments, the server interacts with the same population of clients repeatedly over time. This temporal structure makes it possible to accumulate historical information about client behavior and to incorporate this information into aggregation decisions. Trust- and history-aware defenses exploit this longitudinal perspective to improve robustness beyond what is achievable through purely round-wise statistical methods.

Trust-Weighted Aggregation. In trust-weighted aggregation, each client is assigned a trust score that reflects its historical behavior across multiple communication rounds. These scores may be derived from measures such as consistency of update directions, stability of update norms, alignment with aggregated gradients, or contribution to overall convergence. Clients whose updates exhibit stable and coherent behavior over time are assigned higher trust, while those displaying erratic, unstable, or frequently extreme updates are downweighted.

Formally, if $w_k^{(t)}$ denotes the trust weight assigned to client k at round t, the aggregation rule may be written as

$$\theta^{(t+1)} = \theta^{(t)} + \sum_{k \in \mathcal{S}^{(t)}} w_k^{(t)} \Delta\theta_k^{(t)},$$

with $\sum_k w_k^{(t)} = 1$. This framework introduces a notion of reputation into federated learning, allowing the system to adaptively emphasize reliable clients and suppress persistently abnormal ones.

However, trust-weighted aggregation raises several nontrivial challenges. First, it introduces fairness concerns: new or infrequently participating clients may be penalized due to limited history, a phenomenon often referred to as the *cold-start problem*. Second, trust metrics themselves become potential attack surfaces, as adversaries may attempt to build trust gradually before launching coordinated attacks.

Finally, maintaining and updating trust scores adds algorithmic and system-level complexity, particularly in large-scale deployments with millions of clients.

Redundancy through Participation. Another complementary strategy for mitigating Byzantine and unreliable behavior is redundancy through increased client participation. By increasing the number of clients sampled in each communication round, the relative influence of any single update is reduced. This dilution effect improves the statistical reliability of robust estimators and makes it more difficult for individual faulty or malicious clients to dominate the aggregated update.

While redundancy does not eliminate Byzantine behavior, it reduces its impact by increasing the signal-to-noise ratio of honest updates. In essence, redundancy leverages the law of large numbers to stabilize aggregation. However, increased participation comes at the cost of higher communication overhead, greater synchronization complexity, and potential strain on client devices. As such, redundancy must be balanced against system scalability and resource constraints.

16.4.6 Cryptographic and Integrity-Based Defenses

In addition to statistical robustness, federated learning systems often employ cryptographic techniques to protect the integrity and confidentiality of communicated updates. These mechanisms operate at the protocol level and address threats that cannot be mitigated through aggregation alone.

Secure Aggregation. Secure aggregation protocols ensure that the server observes only the aggregate of client updates, rather than individual contributions. By cryptographically masking individual updates, secure aggregation prevents the server from inspecting, isolating, or selectively manipulating specific client submissions. This property is essential for privacy preservation and regulatory compliance in many federated learning applications.

Although secure aggregation is primarily motivated by privacy concerns, it also provides indirect security benefits. By preventing the server from accessing individual updates, it reduces the risk of targeted manipulation or discrimination against specific clients. However, this same property limits the applicability of certain robustness techniques. In particular, anomaly-detection and filtering methods that rely on inspecting individual updates become infeasible, forcing the system to rely on aggregation rules that are compatible with encrypted inputs.

Integrity Verification. Complementary to secure aggregation, integrity verification mechanisms ensure that updates are transmitted faithfully and have not been tampered with in transit. Digital signatures, authenticated communication channels, and protocol-level consistency checks can prevent replay attacks, impersonation, and unauthorized modification of updates.

While integrity verification is effective against network-level and protocol-level attacks, it does not address arbitrary Byzantine behavior arising from faulty local computation. A client may produce a validly signed update that is nonetheless nonsensical, corrupted, or adversarial. As a result, cryptographic defenses must be combined with robust statistical aggregation to achieve comprehensive protection.

16.4.7 Trade-offs and Open Challenges

Despite significant progress, robust aggregation in federated learning inevitably involves trade-offs. Aggressive filtering, trimming, and clipping strategies improve resilience to adversarial manipulation and Byzantine failures, but they may suppress legitimate and informative updates, particularly in the presence of strong data heterogeneity and non-IID distributions. Such suppression can slow convergence, introduce bias, or degrade final model performance.

Conversely, overly permissive aggregation rules preserve more learning signal from diverse clients but expose the system to instability, divergence, and targeted manipulation. Striking the right balance between robustness and sensitivity remains a central challenge in federated learning research.

Robustness Principle
Robust federated learning replaces blind trust in client updates with statistical resilience against deviation, failure, and manipulation, accepting controlled bias and reduced sensitivity in exchange for stability, security, and reliable convergence.

Designing aggregation mechanisms that adapt dynamically to evolving data heterogeneity, fluctuating participation rates, and changing threat models remains an open research problem. Future work must address how trust, robustness, privacy, and efficiency can be jointly optimized in large-scale federated learning systems. The next section examines how these defense strategies interact with broader trust assumptions and system governance frameworks in federated learning.

Worked Example: Robust Aggregation under Byzantine Behavior

This example illustrates, step by step, how different aggregation strategies behave in the presence of Byzantine clients. We consider a federated learning round with multiple clients, numeric updates, and explicitly derive all intermediate computations. No steps are skipped.

Problem Setup Assume a one-dimensional model parameter $\theta \in \mathbb{R}$ for clarity. At communication round t, the server selects $K = 9$ clients. Each client computes a local update $\Delta\theta_k^{(t)}$ and sends it to the server.

Among these clients:

- 7 clients are honest and compute updates close to the true descent direction,
- 2 clients are Byzantine and submit arbitrary updates.

The received updates are summarized in Table 16.4.

Step 1: Standard Federated Averaging The federated averaging update is

$$\Delta\theta_{\text{mean}} = \frac{1}{K}\sum_{k=1}^{K}\Delta\theta_k.$$

Summation of Honest Updates

$$\begin{aligned}\sum_{\text{honest}} &= (-0.48) + (-0.52) + (-0.50) + (-0.46)\\ &\quad + (-0.49) + (-0.51) + (-0.47)\\ &= -3.43.\end{aligned}$$

Summation of Byzantine Updates

$$(+5.00) + (-6.00) = -1.00.$$

Total Sum

$$\sum_{k=1}^{9}\Delta\theta_k = -3.43 - 1.00 = -4.43.$$

Averaged Update

$$\Delta\theta_{\text{mean}} = \frac{-4.43}{9} \approx -0.492.$$

Table 16.4 Client Updates with Byzantine Failures

Client k	Type	$\Delta\theta_k$
1	Honest	–0.48
2	Honest	–0.52
3	Honest	–0.50
4	Honest	–0.46
5	Honest	–0.49
6	Honest	–0.51
7	Honest	–0.47
8	Byzantine	+5.00
9	Byzantine	–6.00

Observation Although the mean remains negative, it is partially distorted by the Byzantine updates. In higher dimensions or with coordinated attackers, this distortion can become catastrophic.

Step 2: Coordinate-Wise Median Aggregation We now compute the coordinate-wise median aggregation in full detail. Since the parameter space is one-dimensional in this illustrative example, the coordinate-wise median coincides with the scalar median. Nevertheless, we derive the result explicitly to clarify the underlying logic and robustness properties.

Step 2.1: Ordering the Client Updates

The set of received client updates is

$$\{-0.48, -0.52, -0.50, -0.46, -0.49, -0.51, -0.47, +5.00, -6.00\}.$$

To compute the median, the updates must be sorted in nondecreasing order:

$$\{-6.00, -0.52, -0.51, -0.50, -0.49, -0.48, -0.47, -0.46, +5.00\}.$$

This ordering step is essential, as the median depends only on relative rank rather than numerical magnitude.

Step 2.2: Identifying the Median Index

The total number of updates is

$$K = 9.$$

For an odd number of observations, the median is defined as the element with index

$$\frac{K+1}{2} = \frac{9+1}{2} = 5.$$

Thus, the median corresponds to the 5th element in the sorted list.

Step 2.3: Extracting the Median Value

From the ordered sequence, the 5th element is

$$-0.49.$$

Hence, the coordinate-wise median aggregation yields

$$\Delta\theta_{\text{median}} = -0.49.$$

Interpretation and Robustness Analysis The median estimator is entirely insensitive to the extreme values +5.00 and −6.00. These Byzantine updates occupy the first and last positions in the sorted list and therefore have no influence on the median. As long as fewer than half of the clients are Byzantine, the median remains determined by the honest majority.

In this example, 7 out of 9 clients are honest, ensuring that the median lies squarely within the cluster of honest updates. This property explains why coordinate-wise median aggregation achieves a breakdown point of nearly 50% and is widely used as a baseline Byzantine-robust aggregation rule.

Step 3: Trimmed Mean Aggregation We now derive the trimmed mean aggregation in a fully explicit manner. Trimmed mean aggregation generalizes the median by discarding a fixed number of extreme values and averaging the remaining updates.

Step 3.1: Choice of Trimming Parameter

Let $f = 2$ denote the trimming parameter. This means that the f smallest and f largest values are removed before averaging. Since

$$K = 9,$$

the number of remaining values after trimming is

$$K - 2f = 9 - 4 = 5.$$

Step 3.2: Identification of Extreme Values

From the sorted list

$$\{-6.00, -0.52, -0.51, -0.50, -0.49, -0.48, -0.47, -0.46, +5.00\},$$

the two smallest values are

$$-6.00, \quad -0.52,$$

and the two largest values are

$$+5.00, \quad -0.46.$$

These four values are removed from further consideration.

Step 3.3: Construction of the Trimmed Set

After trimming, the remaining updates are

$$\{-0.51, -0.50, -0.49, -0.48, -0.47\}.$$

Notably, all remaining updates originate from honest clients, indicating that the trimming parameter successfully eliminates Byzantine influence in this round.

Step 3.4: Explicit Mean Computation

We now compute the arithmetic mean of the trimmed set step by step:

$$\begin{aligned}\sum_{\text{trimmed}} &= (-0.51) + (-0.50) + (-0.49) + (-0.48) + (-0.47)\\ &= -2.45.\end{aligned}$$

Dividing by the number of retained values gives

$$\Delta\theta_{\text{trim}} = \frac{-2.45}{5} = -0.49.$$

Interpretation and Robustness Analysis The trimmed mean recovers exactly the central tendency of the honest client updates. Unlike the median, which selects a single representative value, the trimmed mean aggregates information from multiple clients, thereby reducing variance while still suppressing extreme outliers.

This example highlights a key advantage of trimmed mean aggregation: when the trimming level exceeds the number of Byzantine clients, the aggregated update becomes equivalent to the mean of honest contributions. However, excessive trimming may remove informative updates under strong non-IID data, illustrating the inherent robustness–efficiency trade-off in federated aggregation design.

Step 4: Norm Clipping Followed by Averaging We now derive the effect of norm clipping in full detail. Norm clipping is a magnitude-based defense that limits the influence of individual client updates by enforcing a hard bound on their norms. Unlike median- or trimming-based methods, norm clipping does not discard updates entirely; instead, it rescales extreme updates to lie within a predefined admissible region.

Step 4.1: Definition of the Clipping Operator Let the clipping threshold be fixed as

$$\tau = 1.0.$$

Each client update $\Delta\theta_k$ is transformed according to

$$\Delta\theta_k^{\text{clip}} = \begin{cases} \Delta\theta_k, & |\Delta\theta_k| \leq \tau, \\ \tau \cdot \text{sign}(\Delta\theta_k), & |\Delta\theta_k| > \tau. \end{cases}$$

This operation preserves the direction (sign) of the update while capping its magnitude at τ.

Step 4.2: Application to Individual Client Updates We apply the clipping rule to each received update.

- Honest client updates satisfy

  $$|\Delta\theta_k| < 1.0,$$

  and therefore remain unchanged after clipping.

- The Byzantine update +5.00 exceeds the threshold and is clipped to

$$+5.00 \longrightarrow +1.00.$$

- The Byzantine update –6.00 also exceeds the threshold and is clipped to

$$-6.00 \longrightarrow -1.00.$$

Thus, after clipping, the set of updates becomes

$$\{-0.48, -0.52, -0.50, -0.46, -0.49, -0.51, -0.47, +1.00, -1.00\}.$$

Step 4.3: Explicit Summation after Clipping Recall from the earlier computation that the sum of honest updates is

$$\sum_{\text{honest}} = -3.43.$$

The clipped Byzantine updates contribute

$$(+1.00) + (-1.00) = 0.$$

Therefore, the total sum of clipped updates is

$$\sum_{k=1}^{9} \Delta\theta_k^{\text{clip}} = -3.43 + 0 = -3.43.$$

Step 4.4: Averaging the Clipped Updates The clipped mean is obtained by dividing the total sum by the number of clients:

$$\begin{aligned}
\Delta\theta_{\text{clip-mean}} &= \frac{1}{9}\sum_{k=1}^{9} \Delta\theta_k^{\text{clip}} \\
&= \frac{-3.43}{9} \\
&\approx -0.381.
\end{aligned}$$

Interpretation and Bias Analysis Norm clipping successfully prevents the Byzantine clients from injecting arbitrarily large updates. However, the clipped Byzantine updates still contribute nonzero values to the aggregation. As a result, the final update is biased toward zero compared to the honest consensus value near –0.49.

This example illustrates a fundamental limitation of norm clipping: it bounds worst-case damage but does not eliminate adversarial influence. In adversarial settings, norm clipping alone is therefore insufficient to guarantee robust convergence.

Table 16.5 Comparison of Aggregation Rules

Aggregation Method	Aggregated Update
Mean (FedAvg)	–0.492
Coordinate-wise Median	–0.49
Trimmed Mean	–0.49
Clipped Mean	–0.381
Geometric Median	–0.49

Step 5: Geometric Median (1D Case) We now derive the geometric median aggregation explicitly.

Step 5.1: Definition of the Geometric Median The geometric median is defined as the solution of

$$\widehat{\Delta\theta} = \arg\min_{v \in \mathbb{R}} \sum_{k=1}^{9} |v - \Delta\theta_k|.$$

In one dimension, it is a classical result that any minimizer of this objective is given by the median of the data.

Step 5.2: Reduction to the Median Since the updates have already been sorted as

$$\{-6.00, -0.52, -0.51, -0.50, -0.49, -0.48, -0.47, -0.46, +5.00\},$$

and the number of updates is odd, the minimizer of the above objective is the 5th element.

Thus,

$$\Delta\theta_{\text{geo-med}} = -0.49.$$

Interpretation The geometric median completely suppresses the influence of the extreme Byzantine updates. In higher-dimensional settings, this robustness generalizes to outlier-resistant behavior that respects the geometry of the update space (Table 16.5).

Step 6: Comparison of Aggregation Outcomes

Final Interpretation This extended derivation highlights several important conclusions.

- Standard federated averaging is highly sensitive to extreme updates and offers no intrinsic robustness guarantees.
- Coordinate-wise median and trimmed mean aggregation completely remove Byzantine influence as long as honest clients form a majority.

- Norm clipping bounds the magnitude of damage but still allows adversarial bias to persist.
- Geometric median aggregation achieves strong robustness while preserving the geometric structure of updates.

Overall, this example demonstrates that robust aggregation in federated learning must be treated as a principled estimation problem under adversarial contamination rather than a naïve averaging operation.

16.5 Trust Assumptions in Federated Systems

All federated learning systems, regardless of their algorithmic sophistication or cryptographic protection, are built upon a set of implicit or explicit trust assumptions. These assumptions define which entities in the system are considered reliable, which behaviors are anticipated, and which forms of deviation are tolerated. Unlike centralized machine learning—where trust is often concentrated within a single administrative domain—federated learning distributes computation across multiple, autonomous, and often mutually untrusted participants. As a result, trust is no longer a background concern but a foundational design parameter that shapes the entire system.

Trust assumptions determine the threat model under which a federated learning system is analyzed and deployed. They govern what kinds of attacks are considered feasible, which defenses are necessary, and what security and robustness guarantees can be meaningfully claimed. Importantly, no federated learning system can be secure against all possible adversaries simultaneously. Instead, security and robustness are always conditional: they hold only within the boundaries defined by the assumed trust model.

16.5.1 Why Trust Models Matter

The role of trust in federated learning extends far beyond abstract or theoretical security considerations. Trust assumptions form the conceptual foundation upon which the entire federated learning pipeline is built. They directly shape algorithmic design choices, determine which communication protocols are viable, dictate the necessity and strength of cryptographic mechanisms, and even influence system scalability, latency, and cost.

In practical terms, trust assumptions define the *threat envelope* within which a federated learning system is expected to operate correctly. For example, aggregation rules that are provably robust under a bounded number of Byzantine clients rely critically on the assumption that the fraction of adversarial participants does not exceed a specified threshold. If this assumption is violated—either due to underestimation of adversarial power or unexpected system behavior—the same aggregation rules may fail catastrophically, leading to divergence, model collapse, or silent corruption.

Similarly, privacy guarantees offered by cryptographic mechanisms such as secure aggregation are conditional on trust assumptions about the server and the clients. Secure aggregation protocols typically assume that the server follows the protocol faithfully and that a sufficient number of clients behave honestly. If the server deviates from the protocol, colludes with clients, or selectively drops participants, the theoretical privacy guarantees may no longer hold. Thus, privacy in federated learning is not an absolute property, but a conditional one that depends explicitly on the assumed trust model.

Explicitly articulating trust assumptions serves two critical and complementary purposes. First, it enables rigorous reasoning about security, privacy, and robustness properties. By making assumptions explicit, researchers and practitioners can clearly understand the scope and limitations of theoretical guarantees. Second, it prevents overclaiming. Guarantees proven under one trust model should never be implicitly extended to deployments operating under a different or stronger adversarial setting.

In high-stakes applications—such as healthcare analytics, financial modeling, defense systems, or critical infrastructure monitoring—the cost of misunderstanding trust assumptions can be severe. Deploying a system under incorrect trust assumptions may lead to data leakage, biased decision-making, regulatory violations, or systemic failure. For this reason, trust models should be treated as first-class design decisions rather than implicit background assumptions.

16.5.2 Common Trust Models in Federated Learning

Several trust models recur throughout the federated learning literature and in real-world deployments. Each model reflects different assumptions about participant behavior, system governance, and adversarial capabilities. Understanding these models is essential for interpreting both theoretical results and practical system designs.

Honest-but-Curious Server. In the honest-but-curious (also called semi-honest) server model, the central server is assumed to execute the federated learning protocol correctly and without deviation. However, the server is also assumed to be inquisitive: it may attempt to extract as much information as possible from whatever data it observes during training.

Specifically, the server may analyze received model updates, intermediate aggregates, timing information, or communication metadata to infer properties of client datasets. For example, it may attempt membership inference, property inference, or reconstruction attacks based solely on observed parameters. This model is particularly relevant in privacy-sensitive deployments, such as cross-hospital learning or cross-enterprise collaboration, where the server operator is not fully trusted with raw data.

Under an honest-but-curious server assumption, cryptographic and privacy-preserving mechanisms play a central role. Secure aggregation, differential privacy, encryption, and masking techniques are designed to limit information leakage even when the server behaves inquisitively. However, these techniques typically assume that the server does not behave arbitrarily—for instance, by injecting malformed updates, selectively excluding clients, or colluding with adversarial participants. As such, their guarantees are tightly coupled to the assumed server behavior.

Partially Malicious Clients. Another widely adopted trust model assumes that the majority of clients are honest and follow the federated learning protocol, but that a bounded fraction may behave maliciously or unreliably. These adversarial clients may intentionally attempt model poisoning, backdoor insertion, or convergence disruption. Alternatively, they may be faulty or misconfigured, producing Byzantine updates without any malicious intent.

This trust model underpins much of the research on Byzantine-robust optimization and robust aggregation. In this setting, security and convergence guarantees are typically expressed in terms of a maximum tolerable fraction of adversarial clients. As long as the number of malicious or faulty participants remains below this threshold, the learning process is provably stable.

However, these guarantees are inherently conditional. If the adversarial fraction exceeds the assumed bound—due to coordinated attacks, client churn, or misestimation of participation rates—the robustness properties no longer apply. As a result, accurate modeling of participation dynamics and adversarial capabilities becomes a critical practical concern.

Fully Adversarial Environment. In the strongest and most conservative trust model, no participant is trusted without explicit verification. Both the server and the clients may deviate arbitrarily from the protocol, collude with one another, or attempt to undermine learning for strategic gain. This model is particularly relevant in open federated systems, cross-organizational collaborations without a central authority, or decentralized platforms operating under minimal governance.

Operating under a fully adversarial model requires the strongest and most comprehensive defenses. These may include end-to-end cryptographic verification, secure multi-party computation, zero-knowledge proofs, attestation mechanisms, and consensus-style protocols inspired by distributed systems. While such approaches can provide strong theoretical guarantees, they are often computationally expensive, communication-intensive, and difficult to scale to millions of participants.

This tension highlights a fundamental trade-off in federated learning: reducing trust assumptions generally increases system complexity and resource consumption. Designers must therefore balance security ambitions against practical constraints.

16.5.3 Interplay Between Trust and Algorithm Design

The choice of trust model directly and profoundly influences algorithm design, cryptographic requirements, and the nature of achievable guarantees. Trust assumptions determine not only which attacks must be defended against, but also which defenses are feasible and efficient.

For example, secure aggregation protocols typically assume that a sufficient number of clients behave honestly and that the server does not collude with adversarial participants. Under these assumptions, strong privacy guarantees can be achieved with relatively modest overhead. In contrast, Byzantine-robust aggregation rules usually assume a bounded number of adversarial clients but often trust the server to apply the aggregation rule correctly and consistently.

These assumptions are not merely technical details; they define the operational envelope of the system. A federated learning system designed under an honest-but-curious server model may offer strong privacy protections but remain vulnerable to model poisoning or backdoor attacks. Conversely, a system optimized for Byzantine robustness may sacrifice fine-grained privacy, impose additional communication overhead, or require stronger synchronization guarantees.

Ultimately, trust models serve as the bridge between abstract theory and real-world deployment. Clear alignment between trust assumptions, algorithmic choices, and deployment context is essential for building federated learning systems that are not only theoretically sound but also practically reliable.

16.5.4 Implications for Deployment and Interpretation

Understanding, documenting, and explicitly articulating trust assumptions is essential for the correct interpretation of robustness, privacy, and security claims in the federated learning literature. Theoretical guarantees in federated learning are always conditional: they hold only within the scope of the trust model under which they are derived. As a result, results that are valid under one set of assumptions cannot be naively transferred to a different setting without careful and systematic analysis.

In particular, practitioners must recognize that robustness, convergence, and privacy guarantees are not intrinsic properties of an algorithm in isolation. Rather, they are properties of an *algorithm–assumption pair*. An aggregation rule that is provably Byzantine-robust under a bounded fraction of adversarial clients may fail entirely if adversarial participation exceeds that bound. Similarly, privacy guarantees based on secure aggregation may collapse if the server deviates from the assumed honest-but-curious behavior or colludes with a subset of clients. Ignoring these dependencies can lead to a dangerous mismatch between theoretical expectations and real-world behavior.

For practitioners and system designers, this implies that deploying federated learning systems responsibly requires close alignment between algorithmic choices and realistic assumptions about participant behavior, system governance, and operational constraints. Trust models must be selected based on empirical knowledge of the deployment environment, such as who controls the server, how clients are onboarded, how participation is incentivized, and what forms of misbehavior are plausible. Treating trust assumptions as abstract or secondary considerations risks deploying systems whose guarantees are illusory in practice.

The implications are particularly severe in high-stakes domains such as healthcare analytics, financial decision-making, defense systems, and national infrastructure monitoring. In these settings, federated learning models may influence clinical decisions, credit allocation, risk assessment, or safety-critical control. If trust assumptions are misaligned with reality, the resulting systems may fail silently, leak sensitive information, or produce systematically biased outputs—precisely when reliability and robustness are most critical.

For this reason, trust assumptions should be treated as explicit design decisions rather than implicit conveniences. They should be clearly documented, communicated to stakeholders, and revisited as deployment conditions evolve. Failure to do so risks deploying federated learning systems whose theoretical guarantees collapse under real-world stress, undermining both technical validity and societal trust in collaborative machine learning.

Outlook

This chapter examined the security vulnerabilities inherent in federated learning and the principles underlying robust defense mechanisms. We analyzed model poisoning, backdoor insertion, inference attacks, and Byzantine failures, highlighting how decentralization reshapes the adversarial landscape. Robust aggregation emerged as a central tool for mitigating these threats, while trust assumptions were shown to define the limits of security guarantees.

In the next chapter, we extend this discussion beyond technical robustness to address governance, incentives, and ethical considerations that arise when federated learning is deployed across autonomous and self-interested participants.

Exercises

This chapter examined the security threat landscape of federated learning, including model poisoning, backdoor and inference attacks, Byzantine failures, robust aggregation, and trust assumptions. The following exercises are designed to test conceptual understanding, analytical reasoning, and practical insight into secure and robust federated learning systems.

A. Short Answer Questions

Answer each question briefly and precisely.

1. What distinguishes model poisoning attacks from random noise in federated learning?

2. Why does federated averaging amplify the impact of extreme client updates?
3. Define a Byzantine client in the context of federated learning.
4. What is meant by a "bounded fraction" of adversarial clients?
5. Why are backdoor attacks considered stealthy?
6. How does secure aggregation differ from robust aggregation?
7. What information can be leaked through gradient updates even when raw data is not shared?
8. Why is norm clipping alone insufficient for Byzantine robustness?
9. What is the breakdown point of the coordinate-wise median?
10. Why must trust assumptions be stated explicitly in federated learning deployments?

B. Long Answer Questions

Answer the following questions with clear explanations, equations, and diagrams wherever appropriate.

1. Explain the mathematical mechanism by which model poisoning attacks distort global optimization in federated learning.
2. Discuss backdoor attacks in federated learning and explain why standard validation procedures often fail to detect them.
3. Compare Byzantine failures arising from malicious intent and from benign faults. Why are they treated identically in robust aggregation?
4. Derive the federated averaging update rule and analyze its vulnerability under adversarial updates.
5. Explain how robust aggregation reframes federated learning as a statistical estimation problem under contamination.
6. Compare coordinate-wise median, trimmed mean, and geometric median aggregation in terms of robustness and efficiency.
7. Analyze the trade-offs introduced by norm-based clipping and filtering strategies.
8. Discuss the role of secure aggregation in preventing inference attacks and explain its limitations.
9. Explain how trust models influence algorithm design and achievable guarantees in federated learning.
10. Critically evaluate the challenges of deploying federated learning in high-stakes domains such as healthcare or finance.

C. True/False Questions

State whether each statement is **True** or **False**.

1. Secure aggregation prevents model poisoning attacks.

2. Byzantine failures always imply malicious intent.
3. Federated learning eliminates all forms of information leakage.
4. Coordinate-wise median aggregation can tolerate nearly 50% Byzantine clients.
5. Backdoor attacks necessarily reduce global model accuracy.
6. Norm clipping completely removes the influence of adversarial clients.
7. Inference attacks can be carried out by an honest-but-curious server.
8. Robust aggregation trades statistical efficiency for stability.
9. Trust assumptions are independent of system deployment context.
10. Byzantine robustness requires worst-case rather than average-case analysis.

Answers:

1. False
2. False
3. False
4. True
5. False
6. False
7. True
8. True
9. False
10. True

D. Fill in the Blanks

Fill in the blanks with the most appropriate term or expression.

1. A client that behaves arbitrarily and unpredictably is called a ____________ client.
2. Model poisoning attacks manipulate ____________ rather than raw data.
3. A backdoor attack is triggered only under ____________ conditions.
4. Secure aggregation hides ____________ updates from the server.
5. Robust aggregation seeks to limit the ____________ of any single client update.
6. Coordinate-wise median aggregation operates independently on each ____________.
7. Norm clipping enforces an upper bound on update ____________.
8. Byzantine robustness assumes a ____________ fraction of faulty clients.
9. Inference attacks exploit the informational richness of ____________.
10. Trust assumptions define the system's ____________ model.

Answers:

1. Byzantine
2. model updates (or gradients)

3. specific / rare / trigger
4. individual
5. influence
6. coordinate (or parameter dimension)
7. norm
8. bounded
9. gradients (or updates)
10. threat

E. Multiple Choice Questions (MCQs)

Choose the **most appropriate** answer.

1. Which attack aims to introduce hidden behavior without degrading overall accuracy?
   (a) Data poisoning
   (b) Untargeted poisoning
   (c) Backdoor attack
   (d) Denial-of-service attack
2. Which aggregation rule has the highest breakdown point?
   (a) Federated averaging
   (b) Norm clipping
   (c) Coordinate-wise median
   (d) Weighted mean
3. Which entity can perform inference attacks without deviating from the protocol?
   (a) Byzantine client
   (b) Honest-but-curious server
   (c) Faulty sensor
   (d) External auditor
4. What is the primary vulnerability of linear aggregation?
   (a) High communication cost
   (b) Sensitivity to extreme updates
   (c) Slow convergence
   (d) Parameter redundancy
5. Secure aggregation primarily protects against:
   (a) Model poisoning
   (b) Byzantine failures
   (c) Gradient inference
   (d) Data heterogeneity
6. Byzantine failures may arise from:
   (a) Malicious intent only
   (b) Benign faults only
   (c) Both malicious and benign causes
   (d) Centralized servers

7. Which method preserves geometric structure in high dimensions?
   (a) Coordinate-wise median
   (b) Trimmed mean
   (c) Geometric median
   (d) Scalar clipping
8. Robust aggregation is best viewed as:
   (a) A cryptographic primitive
   (b) A statistical estimation problem
   (c) A networking protocol
   (d) A compression technique
9. Increasing client participation primarily improves robustness by:
   (a) Increasing model size
   (b) Diluting adversarial influence
   (c) Eliminating heterogeneity
   (d) Reducing communication
10. Trust assumptions mainly determine:
   (a) Model architecture
   (b) Loss function choice
   (c) Validity of security guarantees
   (d) Learning rate schedules

Answers:

1. (c)
2. (c)
3. (b)
4. (b)
5. (c)
6. (c)
7. (c)
8. (b)
9. (b)
10. (c)

F. Application and Project-Based Questions

These problems emphasize modeling, reasoning, and implementation.

1. Design a simple federated learning experiment that demonstrates model poisoning under standard averaging.
2. Implement coordinate-wise median aggregation and compare its convergence behavior with FedAvg.
3. Simulate Byzantine failures caused by numerical instability and analyze their impact on training.

4. Evaluate the effectiveness of norm clipping under increasing adversarial magnitudes.
5. Design a backdoor attack for a federated image classification task and analyze its stealth.
6. Implement secure aggregation and demonstrate how it prevents gradient-based inference.
7. Compare trimmed mean and geometric median aggregation under non-IID data.
8. Propose a trust-weighted aggregation scheme based on client history.
9. Analyze the trade-offs between robustness and accuracy in a real-world federated dataset.
10. Write a short report critically assessing trust assumptions in a federated healthcare deployment.

Governance, Ethics, and Incentive Mechanisms 17

Federated learning is not merely a technical paradigm for distributed optimization; it is fundamentally a *socio-technical system*. While earlier chapters have focused on mathematical formulations, optimization dynamics, and system architectures, these technical constructs operate within a broader institutional, legal, and ethical context. Federated learning brings together multiple autonomous entities—such as individual users, organizations, hospitals, corporations, or public institutions—that collaborate to train a shared model while retaining control over their local data. These entities often differ significantly in their objectives, capabilities, regulatory obligations, and risk tolerances.

As a consequence, the success and long-term sustainability of federated learning systems cannot be evaluated solely in terms of convergence rates, communication efficiency, or predictive accuracy. Instead, federated learning must be understood as a coordinated collective activity that requires carefully designed mechanisms for *governance*, *ethical oversight* and *incentive alignment*. Without such mechanisms, even technically sound federated learning systems may fail due to mistrust among participants, misaligned economic incentives, regulatory non-compliance, or ethical concerns related to fairness and accountability.

Governance in federated learning refers to the structures, policies, and decision-making processes that define how the system is organized and controlled. Unlike centralized machine learning—where a single entity typically owns the data, the model, and the deployment pipeline—federated learning operates in a distributed environment where authority and responsibility are inherently fragmented. Key governance questions therefore arise naturally: Who has the right to decide model architectures and training protocols? Who controls access to the trained model? Who bears responsibility when the model causes harm or violates regulatory norms? These questions do not admit purely algorithmic answers; they require explicit institutional design choices.

P. P. Ray, *Foundations of Federated Learning*,
Springer Series in the Data Sciences,
https://doi.org/10.1007/978-3-032-34084-9_17

Ethical considerations are equally central. Federated learning is often motivated by ethical goals such as privacy preservation, data minimization, and respect for data sovereignty. However, the mere fact that raw data remains local does not automatically guarantee ethical outcomes. Biases may still be amplified through uneven participation, dominant clients may disproportionately shape the global model, and affected individuals may have limited visibility into how their data contributes to collective intelligence. Ethical federated learning therefore requires deliberate attention to issues of fairness, transparency, consent, and accountability, extending beyond technical privacy guarantees.

Incentive mechanisms form the third pillar of this chapter. Participation in federated learning is rarely free: clients incur costs in computation, energy consumption, communication overhead, and operational complexity. In many settings—particularly cross-organizational and consortium-based deployments—participants will contribute meaningfully only if they perceive clear and equitable benefits. Designing incentives that encourage honest participation, high-quality updates, and long-term engagement, while discouraging free-riding or strategic manipulation, is therefore essential. Incentives are not merely economic tools; they directly shape learning dynamics and system robustness.

This chapter examines federated learning through the lens of *institutional design*, integrating governance, ethics, and incentives into a unified conceptual framework. We move beyond optimization dynamics to address foundational questions that are unavoidable in real-world deployments: Who owns the resulting model and its intellectual property? How is value created by collaborative learning and how should that value be distributed among participants? What legal, regulatory, and ethical obligations do participants assume by joining a federated system? How can compliance, accountability, and auditability be ensured when training is decentralized and data never leaves its source? And, ultimately, how can collaborative artificial intelligence remain ethical, trustworthy, and socially responsible under conditions of data sovereignty?

By formalizing these questions and explicitly connecting them to federated system design, this chapter establishes governance not as a peripheral concern, but as a *first-class component* of federated learning. Just as optimization objectives and communication protocols determine how models are trained, governance and incentive structures determine whether federated learning systems are viable, fair, and sustainable in practice. The discussions in this chapter provide the conceptual foundation for understanding federated learning as an institutional and ethical innovation, complementing the technical advances explored throughout the remainder of this book.

17.1 Model Ownership and Intellectual Property

In traditional centralized machine learning pipelines, questions of model ownership and intellectual property are relatively straightforward. A single organization typically collects or acquires the data, designs the learning architecture, performs training on centrally controlled infrastructure, and deploys the resulting model. Under this paradigm, the trained model is unambiguously treated as an intellectual asset of the central entity, much like proprietary software or a patented algorithm. Data contributors—if any—are usually compensated through fixed contracts, licenses, or service agreements, and they rarely retain claims over the trained model itself.

Federated learning fundamentally disrupts this well-established assumption. In a federated setting, no single entity possesses the full training dataset, and no participant can independently reproduce the training process in isolation. Instead, the global model emerges from the coordinated interaction of multiple autonomous participants, each contributing local computations derived from privately owned data. As a result, the notion of a single, clearly identifiable "model owner" becomes conceptually and legally ambiguous.

This ambiguity is not merely a legal technicality; it reflects a deeper structural shift in how value is created. In federated learning, the trained model is not the product of a centralized data asset, but of a distributed process in which ownership, control, and responsibility are inherently fragmented. Understanding and resolving this fragmentation is therefore a central governance challenge.

17.1.1 Joint Contribution Structure

At a technical level, federated learning makes explicit the collective nature of model construction. Let the global model be parameterized by $\boldsymbol{\theta} \in \mathbb{R}^d$ and trained over T communication rounds. At each round t, a subset of clients $\mathcal{S}^{(t)} \subseteq \{1, \dots, K\}$ computes local updates based on private datasets. The final trained model can be expressed abstractly as

$$\boldsymbol{\theta}^{\star} = \boldsymbol{\theta}^{(0)} + \sum_{t=0}^{T-1} \sum_{k \in \mathcal{S}^{(t)}} w_k^{(t)} \, \Delta\boldsymbol{\theta}_k^{(t)}.$$

This expression makes two important facts explicit. First, the global model is the cumulative result of a large number of distributed optimization steps rather than a single monolithic training procedure. Second, each client contributes to the final parameter vector through its locally computed updates $\Delta\boldsymbol{\theta}_k^{(t)}$, weighted by factors $w_k^{(t)}$ that typically depend on data volume, participation, or system-level policies.

From an intellectual property perspective, this formulation highlights the difficulty of attributing authorship. No client contributes a complete model, and no client has access to the full set of updates that define $\boldsymbol{\theta}^{\star}$. Even the coordinating server—if one exists—does not generate updates itself, but merely aggregates contributions produced elsewhere. The trained model is therefore best understood as a *collective artifact*, emerging from the interaction of multiple independent actors.

This collective contribution structure directly challenges classical IP frameworks, which are typically designed around identifiable creators, clear ownership boundaries, and centralized control. Federated learning requires governance mechanisms that explicitly acknowledge and manage this distributed authorship.

17.1.2 Ownership Interpretations

In practice, federated learning deployments adopt one of several interpretations of model ownership, each reflecting different legal, economic, and organizational priorities. Three dominant approaches are commonly observed.

Central Ownership. Under this interpretation, the coordinating entity—such as a platform provider or system operator—claims full ownership of the trained model. Client updates are treated as contractual inputs, analogous to outsourced computation or licensed data access. Participants agree in advance that any model produced through the federated process belongs exclusively to the central entity. This approach simplifies deployment and commercialization, but it may discourage participation in settings where clients perceive their data as strategically valuable.

Joint Ownership. In joint ownership models, the trained federated model is considered a shared intellectual asset of all participating entities. Ownership rights, decision authority, and potential revenue are distributed according to predefined agreements. This approach is common in consortium-based federated learning, such as collaborations among hospitals, research institutions, or industrial partners. While joint ownership better reflects the collective nature of model construction, it introduces complexity in governance, licensing, and dispute resolution.

Usage-Based Ownership. A third approach decouples ownership from training entirely. Rather than focusing on who "owns" the model, this interpretation emphasizes usage rights. Participants may retain rights to use, deploy, or fine-tune the model within specific domains, while transfer, resale, or modification may be restricted. This model is particularly attractive in regulated environments, where control over downstream use may be more important than formal ownership.

Each of these interpretations carries distinct legal and economic implications. Central ownership prioritizes efficiency and control, joint ownership prioritizes fairness and collaboration, and usage-based ownership prioritizes flexibility and regulatory alignment. Crucially, none of these interpretations emerges automatically from the technical design of federated learning. They must be explicitly defined through governance policies, contractual agreements, and institutional frameworks *prior to training*.

This need for explicit definition is especially acute in regulated domains such as healthcare, finance, and defense, where intellectual property rights intersect with data protection laws, liability frameworks, and public accountability. Failure to address ownership questions early can undermine trust, discourage participation, and ultimately jeopardize the viability of the federated learning system.

Key Governance Insight
In federated learning, intellectual property does not arise from a single act of creation. It emerges from *collective optimization* performed under data sovereignty constraints, making explicit ownership and usage agreements a prerequisite for sustainable collaboration.

Worked Example: Quantifying Intellectual Property Contribution in Federated Learning
This worked example demonstrates, in a fully explicit and numerical manner, how intellectual property (IP) contribution can be quantified in a federated learning system based on client participation, data volume, and update magnitude. The objective is not to establish a legally binding valuation model, but to illustrate how *technical contribution metrics* can inform governance and ownership discussions in federated learning.

Federated Setup Consider a federated learning system with $K = 4$ clients collaborating to train a shared linear model

$$f_\theta(x) = \theta x, \qquad \theta \in \mathbb{R}.$$

Assume the system runs for $T = 2$ communication rounds. The initial global model is

$$\theta^{(0)} = 1.0.$$

Each client performs local training and submits parameter updates to the server, which aggregates them using data-proportional weights.

Client Data Summary The dataset sizes held by the clients are shown in Table 17.1.

The total dataset size is

$$N = 50 + 30 + 15 + 5 = 100.$$

Table 17.1 Client dataset sizes

Client k	Dataset size n_k
1	50
2	30
3	15
4	5

Step 1: Aggregation Weights Each client's aggregation weight is computed as

$$w_k = \frac{n_k}{N}.$$

Client-wise weights.

$$\begin{aligned} w_1 &= \frac{50}{100} = 0.50, \\ w_2 &= \frac{30}{100} = 0.30, \\ w_3 &= \frac{15}{100} = 0.15, \\ w_4 &= \frac{5}{100} = 0.05. \end{aligned}$$

These weights reflect each client's proportional influence on the global model.

Step 2: Local Updates Across Rounds Assume each client computes the following local parameter updates over two communication rounds, shown in Table 17.2.

Step 3: Aggregation at Round $t = 0$ The global update at round 0 is

$$\Delta\theta^{(0)} = \sum_{k=1}^{4} w_k \, \Delta\theta_k^{(0)}.$$

Substituting numerical values.

$$\begin{aligned} \Delta\theta^{(0)} &= 0.50(0.20) + 0.30(0.10) + 0.15(0.30) + 0.05(0.50) \\ &= 0.10 + 0.03 + 0.045 + 0.025 \\ &= 0.20. \end{aligned}$$

Updated global model.

$$\theta^{(1)} = \theta^{(0)} + \Delta\theta^{(0)} = 1.0 + 0.20 = 1.20.$$

Table 17.2 Client-wise parameter updates

Client k	$\Delta\theta_k^{(0)}$	$\Delta\theta_k^{(1)}$
1	+0.20	+0.10
2	+0.10	+0.05
3	+0.30	+0.10
4	+0.50	+0.20

Step 4: Aggregation at Round $t = 1$ The second-round aggregation is

$$\Delta\theta^{(1)} = \sum_{k=1}^{4} w_k \, \Delta\theta_k^{(1)}.$$

Substituting numerical values.

$$\begin{aligned}\Delta\theta^{(1)} &= 0.50(0.10) + 0.30(0.05) + 0.15(0.10) + 0.05(0.20)\\ &= 0.05 + 0.015 + 0.015 + 0.01\\ &= 0.09.\end{aligned}$$

Final global model.

$$\theta^{\star} = \theta^{(1)} + \Delta\theta^{(1)} = 1.20 + 0.09 = 1.29.$$

Step 5: Cumulative Contribution to the Global Model We now compute each client's cumulative contribution to the final model parameter $\theta^{\star}$.

The contribution of client k is defined as

$$C_k = \sum_{t=0}^{1} w_k \, \Delta\theta_k^{(t)}.$$

Client 1.

$$C_1 = 0.50(0.20) + 0.50(0.10) = 0.10 + 0.05 = 0.15.$$

Client 2.

$$C_2 = 0.30(0.10) + 0.30(0.05) = 0.03 + 0.015 = 0.045.$$

Client 3.

$$C_3 = 0.15(0.30) + 0.15(0.10) = 0.045 + 0.015 = 0.060.$$

Client 4.

$$C_4 = 0.05(0.50) + 0.05(0.20) = 0.025 + 0.010 = 0.035.$$

Verification.

$$C_1 + C_2 + C_3 + C_4 = 0.15 + 0.045 + 0.060 + 0.035 = 0.29,$$

which equals

$$\theta^{\star} - \theta^{(0)} = 1.29 - 1.0 = 0.29.$$

Step 6: Interpreting Contributions as Intellectual Property Share The proportional IP contribution of each client can be expressed as

$$\pi_k = \frac{C_k}{\sum_{j=1}^{4} C_j}.$$

Computed shares.

$$\pi_1 = \frac{0.15}{0.29} \approx 51.7\%,$$
$$\pi_2 = \frac{0.045}{0.29} \approx 15.5\%,$$
$$\pi_3 = \frac{0.060}{0.29} \approx 20.7\%,$$
$$\pi_4 = \frac{0.035}{0.29} \approx 12.1\%.$$

Interpretation and Governance Implications Although Client 4 contributes the largest raw updates, its small dataset size limits its weighted influence. Client 1 emerges as the dominant contributor due to sustained updates combined with large data volume. This example demonstrates that intellectual property contribution in federated learning is shaped jointly by:

- dataset size,
- magnitude of local updates,
- duration and consistency of participation.

Key Takeaway from the Worked Example
In federated learning, intellectual property contribution is not tied to data ownership alone. It emerges quantitatively from *weighted participation over time*, providing a principled basis for ownership, licensing, or incentive allocation frameworks.

17.2 Consortium-Based Federated Learning

A significant proportion of real-world federated learning deployments operate not in open, large-scale consumer environments, but within *consortia*. A consortium is a formally organized collaboration among multiple independent organizations that share common strategic, scientific, or operational objectives while retaining full legal and administrative autonomy. Consortium-based federated learning has emerged as a particularly important deployment model in domains where data is highly sensitive, regulatory oversight is strong, and unilateral data sharing is either undesirable or legally prohibited.

Typical examples include hospital networks collaborating on clinical prediction models, financial institutions jointly training fraud detection systems, industrial partners developing predictive maintenance models, and inter-governmental research initiatives pooling expertise across national boundaries. In such settings, federated learning serves not only as a technical solution to data locality constraints, but also as an institutional mechanism for enabling collaboration without surrendering sovereignty over proprietary or regulated data assets.

Unlike consumer-device federated learning—where millions of heterogeneous, largely anonymous devices contribute opportunistically under the control of a single platform provider—consortium-based federated learning involves a relatively small number of identifiable, high-stakes participants. These participants are often symmetric or near-symmetric in power, resources, and bargaining position. As a result, governance in consortium-based federated learning is typically negotiated explicitly and collectively, rather than imposed unilaterally.

17.2.1 Consortium Structure

A consortium-based federated learning system is characterized by a well-defined organizational and technical structure that reflects both collaborative and competitive considerations. At a minimum, such a system consists of the following core components:

- **Multiple peer organizations acting as clients:** Each participating organization operates its own local training infrastructure and retains exclusive control over its data. Participants are typically known entities with long-term strategic interests in the resulting model.
- **A jointly governed orchestration layer:** The orchestration layer coordinates training rounds, aggregates model updates, enforces protocol compliance, and maintains system logs. This layer may be operated by a neutral third party, a rotating lead organization, or a jointly owned entity established specifically for the consortium.
- **Predefined rules for participation, contribution, and exit:** Consortium agreements explicitly define eligibility criteria for participation, minimum contribution requirements, and procedures for voluntary or involuntary exit. These rules are critical for maintaining stability and trust over the lifetime of the collaboration.

From a system-design perspective, consortium structure introduces a layer of institutional coordination that complements the technical coordination provided by federated optimization algorithms. While the learning process remains decentralized, the organizational framework ensures that participants share a common understanding of their roles, rights, and obligations.

An important distinction from consumer-device federated learning is that consortium participants typically negotiate governance rules collectively and ex ante. Because participants are identifiable and accountable, consortium-based systems emphasize predictability, contractual clarity, and mutual trust over scale and anonymity.

17.2.2 Governance Mechanisms

Governance mechanisms translate the abstract notion of collaboration into concrete decision-making processes. In consortium-based federated learning, governance is not an auxiliary function but a central operational component that shapes both technical and institutional outcomes.

Common governance mechanisms include:

- **Shared model governance boards:** Consortia often establish dedicated governance bodies responsible for overseeing model design, training protocols, evaluation criteria, and deployment decisions. These boards typically include representatives from all participating organizations and may be supported by technical and legal advisory committees.
- **Voting or quorum-based decision processes:** Key decisions—such as modifying the model architecture, admitting new members, or authorizing downstream deployment—are frequently subject to formal voting procedures. Quorum requirements ensure that decisions reflect collective consent rather than unilateral action.
- **Contractual agreements on model usage and redistribution:** Legal agreements specify how the trained model may be used, shared, licensed, or commercialized. These agreements often distinguish between internal use within the consortium and external distribution to third parties.

From a systems perspective, governance mechanisms impose explicit constraints on the learning process. They determine who is authorized to participate in training rounds, how frequently updates are accepted, under what conditions participants may be excluded, and how models are validated before deployment. In effect, governance rules act as *institutional control signals* that operate alongside algorithmic control mechanisms such as learning rates and aggregation weights.

These constraints are not merely administrative overhead. They are essential for ensuring regulatory compliance, managing risk, and preserving trust among participants whose competitive interests may partially diverge. Well-designed governance mechanisms therefore enhance both the technical robustness and the institutional legitimacy of consortium-based federated learning systems.

Consortium Principle
In consortium-based federated learning, coordination shifts from centralized algorithmic control to explicitly negotiated institutional governance, enabling collaboration without sacrificing autonomy.

17.3 Value Distribution among Participants

Federated learning creates value through *collective intelligence*: a global model emerges that no single participant could have trained independently without violating data locality, privacy, or regulatory constraints. This value may take multiple forms, including improved predictive accuracy, enhanced generalization across populations, reduced development costs, regulatory compliance, and strategic insights derived from cross-organizational collaboration.

However, the value generated by federated learning is rarely distributed uniformly among participants. Clients differ substantially in the volume, quality, diversity, and strategic importance of their data, as well as in their willingness and ability to contribute computational resources and maintain consistent participation over time. As a result, federated learning systems inevitably raise the question of how the benefits of collaboration should be allocated among participants.

Value distribution is therefore not a secondary economic consideration, but a core governance issue. Poorly designed allocation mechanisms can undermine trust, discourage high-quality participation, and destabilize the collaboration, even when the underlying learning algorithm performs well.

17.3.1 Quantifying Contribution

Let $V(\theta^\star)$ denote the total value generated by the trained federated model $\theta^\star$. This value may be expressed in monetary terms, performance improvements, risk reduction, or strategic advantage, depending on the application domain. A fundamental governance challenge is to determine how this aggregate value should be attributed to individual participants.

A common abstraction expresses total value as a sum of individual contributions:

$$V(\theta^\star) = \sum_{k=1}^{K} v_k,$$

where v_k represents the value attributed to client k. While deceptively simple, this formulation hides significant conceptual and practical complexity. In federated learning, contributions are indirect, interdependent, and often nonlinear: the marginal value of one client may depend strongly on the presence of others.

Several factors are commonly considered when designing contribution-aware allocation schemes:

- **Dataset size n_k:** Larger datasets often exert greater influence on model updates, particularly when aggregation weights are data-proportional. However, dataset size alone does not capture data diversity or informational richness.
- **Marginal improvement due to participation:** A participant's value may be measured by the improvement in global model performance when its contributions are included versus excluded. This perspective aligns closely with cooperative game-theoretic approaches.
- **Reliability and consistency of updates:** Participants who contribute regularly, adhere to protocol specifications, and provide stable updates may be valued more highly than sporadic or unreliable contributors.
- **Domain-specific importance of data:** In many applications, certain data sources are disproportionately valuable. For example, rare disease cases in healthcare or fraud examples in finance may carry significantly higher informational value than typical samples.

Importantly, no single metric fully captures contribution. Effective value allocation mechanisms therefore combine multiple dimensions, often through weighted or rule-based frameworks defined as part of the consortium's governance structure.

From a governance perspective, the key challenge lies not only in computing v_k, but in ensuring that the chosen attribution scheme is transparent, justifiable, and acceptable to all participants.

17.3.2 Fairness vs Efficiency

Value distribution in federated learning is characterized by a fundamental tension between *fairness* and *efficiency*. Efficiency-oriented schemes aim to maximize overall system performance by rewarding participants in proportion to their measurable impact on model quality. Fairness-oriented schemes, by contrast, seek to ensure equitable treatment, broad participation, and long-term collaboration stability.

Allocating value purely in proportion to dataset size or marginal performance gain may maximize short-term efficiency. However, such approaches risk marginalizing participants with smaller datasets or less frequent participation, even when their contributions are strategically or ethically important. Over time, this can discourage participation and reduce diversity in the training data, ultimately harming model robustness.

Conversely, equal or near-equal sharing of value may promote a sense of fairness and collective ownership. Yet overly egalitarian schemes can dilute incentives for participants who invest heavily in data curation, infrastructure, or sustained participation. In such cases, high-quality contributors may reduce effort or exit the collaboration altogether.

Balancing fairness and efficiency is therefore a central governance challenge in federated learning. Effective solutions often adopt hybrid approaches, combining baseline entitlements with performance- or contribution-based adjustments. Crucially, these choices must be aligned with the broader objectives of the federated system—whether scientific collaboration, commercial deployment, public benefit, or regulatory compliance.

Economic Tension
Federated value distribution must reconcile *fairness*, *efficiency*, and *incentive compatibility*, ensuring that participants are rewarded in a manner that sustains collaboration while preserving model quality.

Perspective. Ultimately, value distribution mechanisms encode normative assumptions about what constitutes a "fair" contribution and a "deserved" reward. By making these assumptions explicit and embedding them into governance frameworks, federated learning systems can transform potential sources of conflict into transparent and manageable design choices.

Worked Example: Numerical Value Allocation among Federated Participants
This worked example presents a complete, step-by-step numerical illustration of how value generated by a federated learning model can be distributed among participants based on measurable contributions. All calculations are derived explicitly, and no intermediate step is omitted. The example is intentionally simplified to highlight governance principles rather than algorithmic complexity.

Federated Learning Scenario Consider a federated learning system with $K = 5$ participating clients. The objective is to train a shared regression model whose final performance improvement generates a total value $V(\theta^\star)$. Assume that the value is monetized or otherwise quantified on a normalized scale.

Let the total value generated by the trained model be

$$V(\theta^\star) = 100 \text{ units}.$$

The task is to distribute this value among the five clients based on their contributions.

Step 1: Client Contribution Data Assume contribution is assessed using three measurable factors:

- dataset size n_k,
- average marginal accuracy improvement ΔA_k,
- participation reliability r_k.

These values are summarized in Table 17.3.

Step 2: Normalization of Contribution Factors Each factor is normalized to ensure comparability.

Normalization of dataset size.

Total dataset size:

$$\sum_{k=1}^{5} n_k = 400 + 250 + 200 + 100 + 50 = 1000.$$

Normalized dataset contribution:

$$\tilde{n}_k = \frac{n_k}{1000}.$$

$$\tilde{n}_1 = 0.40,$$
$$\tilde{n}_2 = 0.25,$$
$$\tilde{n}_3 = 0.20,$$
$$\tilde{n}_4 = 0.10,$$
$$\tilde{n}_5 = 0.05.$$

Normalization of marginal accuracy improvement.

Total marginal improvement:

$$\sum_{k=1}^{5} \Delta A_k = 0.020 + 0.015 + 0.025 + 0.030 + 0.010 = 0.100.$$

Normalized marginal improvement:

$$\tilde{A}_k = \frac{\Delta A_k}{0.100}.$$

$$\tilde{A}_1 = 0.20,$$
$$\tilde{A}_2 = 0.15,$$

Table 17.3 Client contribution metrics

Client k	n_k	ΔA_k	r_k
1	400	0.020	0.95
2	250	0.015	0.90
3	200	0.025	0.85
4	100	0.030	0.80
5	50	0.010	0.70

$$\tilde{A}_3 = 0.25,$$
$$\tilde{A}_4 = 0.30,$$
$$\tilde{A}_5 = 0.10.$$

Normalization of reliability.
Total reliability:

$$\sum_{k=1}^{5} r_k = 0.95 + 0.90 + 0.85 + 0.80 + 0.70 = 4.20.$$

Normalized reliability:

$$\tilde{r}_k = \frac{r_k}{4.20}.$$

$$\tilde{r}_1 = 0.2262,$$
$$\tilde{r}_2 = 0.2143,$$
$$\tilde{r}_3 = 0.2024,$$
$$\tilde{r}_4 = 0.1905,$$
$$\tilde{r}_5 = 0.1666.$$

Step 3: Composite Contribution Score Assume the consortium adopts a weighted contribution model:

$$s_k = \alpha \tilde{n}_k + \beta \tilde{A}_k + \gamma \tilde{r}_k,$$

with weights

$$\alpha = 0.4, \qquad \beta = 0.4, \qquad \gamma = 0.2.$$

Client-wise score computation.

$$\begin{aligned} s_1 &= 0.4(0.40) + 0.4(0.20) + 0.2(0.2262) \\ &= 0.1600 + 0.0800 + 0.0452 \\ &= 0.2852. \end{aligned}$$

$$\begin{aligned} s_2 &= 0.4(0.25) + 0.4(0.15) + 0.2(0.2143) \\ &= 0.1000 + 0.0600 + 0.0429 \\ &= 0.2029. \end{aligned}$$

$$\begin{aligned} s_3 &= 0.4(0.20) + 0.4(0.25) + 0.2(0.2024) \\ &= 0.0800 + 0.1000 + 0.0405 \\ &= 0.2205. \end{aligned}$$

$$\begin{aligned} s_4 &= 0.4(0.10) + 0.4(0.30) + 0.2(0.1905) \\ &= 0.0400 + 0.1200 + 0.0381 \\ &= 0.1981. \end{aligned}$$

$$\begin{aligned} s_5 &= 0.4(0.05) + 0.4(0.10) + 0.2(0.1666) \\ &= 0.0200 + 0.0400 + 0.0333 \\ &= 0.0933. \end{aligned}$$

Score verification.

$$\sum_{k=1}^{5} s_k = 0.2852 + 0.2029 + 0.2205 + 0.1981 + 0.0933 = 0.9999 \approx 1.$$

Step 4: Value Allocation The value allocated to client k is

$$v_k = s_k \times V(\theta^\star).$$

Client-wise value shares.

$$\begin{aligned} v_1 &= 0.2852 \times 100 = 28.52, \\ v_2 &= 0.2029 \times 100 = 20.29, \\ v_3 &= 0.2205 \times 100 = 22.05, \\ v_4 &= 0.1981 \times 100 = 19.81, \\ v_5 &= 0.0933 \times 100 = 9.33. \end{aligned}$$

Verification.

$$\sum_{k=1}^{5} v_k = 28.52 + 20.29 + 22.05 + 19.81 + 9.33 = 100.$$

Interpretation and Governance Implications This example demonstrates that value distribution in federated learning is inherently multidimensional. Although Client 1 has the largest dataset, Client 4 achieves a comparable value share due to higher marginal performance contribution. Client 5, despite participation, receives a smaller share due to lower data volume, impact, and reliability.

The worked example illustrates how federated governance can translate abstract principles—fairness, efficiency, and incentive compatibility— into a transparent, quantitative allocation mechanism.

Key Insight from the Worked Example
Value distribution in federated learning emerges from explicit, quantifiable contribution metrics rather than implicit assumptions of equal participation, enabling defensible and sustainable governance decisions.

17.4 Incentive and Reward Mechanisms

Sustained participation in federated learning cannot be taken for granted. Unlike centralized learning environments—where a single entity bears the cost of data storage, computation, and optimization—federated learning distributes these burdens across multiple autonomous clients. Each participating client incurs tangible and intangible costs, including local computation, communication bandwidth, energy consumption, system maintenance, and opportunity cost associated with allocating resources to collaborative training rather than internal objectives.

In many real-world deployments, particularly in cross-organizational and consortium-based settings, participation is voluntary rather than mandated. Clients will therefore continue to contribute meaningfully only if the perceived benefits outweigh the incurred costs. Incentive and reward mechanisms are thus not auxiliary features of federated learning systems; they are foundational components that directly influence participation rates, contribution quality, and the long-term stability of the collaboration.

From a systems perspective, incentives interact tightly with learning dynamics. They shape when clients participate, how much effort they invest in local optimization, and whether they adhere faithfully to prescribed protocols. Poorly aligned incentives can undermine even theoretically sound algorithms, while well-designed incentives can significantly enhance robustness and performance.

17.4.1 Incentive Design

At an abstract level, an incentive mechanism defines a mapping from client behavior to rewards. Let $\Delta\theta_k$ denote the update submitted by client k, n_k denote the size or representativeness of its local dataset, and q_k denote one or more quality metrics associated with its update. An incentive mechanism can be expressed as

$$r_k = \mathscr{I}(\Delta\theta_k, n_k, q_k),$$

where r_k denotes the reward allocated to client k.

This formulation emphasizes a crucial distinction: incentives are not assigned merely for participation, but for *contribution*. Effective incentive design therefore requires careful selection of the input variables to $\mathscr{I}(\cdot)$ and transparent justification of how rewards are computed.

In practice, incentive mechanisms may draw on several dimensions of client behavior:

- **Magnitude and direction of updates:** Large or consistently beneficial updates may be rewarded more highly than negligible or noisy contributions.
- **Data characteristics:** Dataset size, diversity, and domain relevance often influence the perceived value of a client's participation.
- **Quality metrics q_k:** Update quality may be assessed through validation performance, stability across rounds, or alignment with global optimization objectives.

Rewards themselves can take a variety of forms, reflecting the diverse motivations of participants:

- **Monetary compensation:** Clients receive direct financial rewards proportional to their contributions. This approach is common in commercial or platform-driven deployments.
- **Access to improved global models:** Participants may gain privileged or early access to the trained model, or to personalized variants derived from it.
- **Preferential participation or personalization:** Reliable contributors may be prioritized in aggregation, receive higher weights, or benefit from customized models tailored to their local distributions.
- **Reputation or governance influence:** In consortium settings, incentives may include increased voting power, greater influence over system design decisions, or enhanced standing within the collaborative network.

Importantly, incentive design must align with the broader objectives of the federated system. In scientific collaborations, reputational incentives may dominate, whereas in commercial environments, monetary rewards or access rights may be more effective. A one-size-fits-all incentive mechanism is therefore rarely appropriate.

17.4.2 Strategic Behavior

Incentive mechanisms inevitably shape participant behavior, and poorly designed incentives can produce unintended and undesirable outcomes. Because clients act autonomously and may pursue their own interests, federated learning systems must anticipate strategic behavior.

Common forms of strategic or adversarial behavior include:

- **Free-riding:** Clients may attempt to benefit from the global model while contributing minimal or no useful updates, relying on others to carry the optimization burden.
- **Update manipulation:** Participants may craft updates to maximize reward rather than global model quality, for example by exaggerating update magnitudes or overfitting to validation metrics used for evaluation.
- **Low-effort or noisy participation:** Clients may reduce local training effort, submit poorly optimized updates, or reuse stale parameters if incentives do not adequately penalize such behavior.

These behaviors can degrade model performance, slow convergence, and erode trust among participants. As a result, incentive mechanisms must be robust not only to benign heterogeneity, but also to strategic and adversarial actions.

Robust incentive design often incorporates safeguards such as:

- minimum participation or quality thresholds,
- cross-round consistency checks,
- delayed or cumulative rewards tied to long-term impact rather than single-round performance.

From a governance perspective, incentives should be designed in tandem with monitoring and accountability mechanisms. Transparency in reward computation and clear communication of expected behavior are essential for maintaining legitimacy and participant confidence.

Incentive Principle

In federated learning, incentives shape optimization dynamics as profoundly as algorithms themselves, influencing participation, contribution quality, and system stability.

Perspective. Incentive and reward mechanisms occupy a critical intersection between economics, game theory, and distributed optimization. By explicitly modeling participant behavior and aligning rewards with collective objectives, federated learning systems can transform potentially fragile collaborations into resilient, self-sustaining ecosystems. Well-designed incentives do not merely motivate participation; they actively reinforce the ethical, technical, and institutional foundations of collaborative artificial intelligence.

Worked Example: Numerical Incentive and Reward Allocation in Federated Learning

This worked example presents a fully explicit numerical illustration of how incentives and rewards can be computed and allocated in a federated learning system. The example demonstrates how client behavior—captured through update magnitude, data volume, and quality metrics—can be translated into quantitative rewards. All intermediate steps are derived explicitly, and no calculation is omitted.

Federated Learning Scenario Consider a federated learning system with $K = 4$ participating clients over a fixed training window. Each client incurs costs for computation and communication, and the system designer wishes to allocate rewards in proportion to *useful contribution* rather than mere participation.

Assume the total reward budget for one training cycle is

$$R_{\text{total}} = 80 \text{ reward units.}$$

The objective is to distribute this budget among the four clients using a transparent and incentive-compatible rule.

Step 1: Observed Client Metrics Assume the system tracks three contribution-related metrics for each client:

- dataset size n_k,
- update magnitude $\|\Delta\theta_k\|$,
- update quality score q_k (e.g., validation improvement).

These metrics are summarized in Table 17.4.

Step 2: Normalization of Metrics To ensure comparability, each metric is normalized across clients.

Table 17.4 Client contribution metrics for incentive computation

Client k	n_k	$\|\Delta\theta_k\|$	q_k
1	300	0.40	0.90
2	200	0.25	0.85
3	120	0.35	0.80
4	80	0.60	0.70

Normalization of dataset size.
Total dataset size:

$$\sum_{k=1}^{4} n_k = 300 + 200 + 120 + 80 = 700.$$

Normalized dataset size:

$$\tilde{n}_k = \frac{n_k}{700}.$$

$$\tilde{n}_1 = \frac{300}{700} = 0.4286,$$
$$\tilde{n}_2 = \frac{200}{700} = 0.2857,$$
$$\tilde{n}_3 = \frac{120}{700} = 0.1714,$$
$$\tilde{n}_4 = \frac{80}{700} = 0.1143.$$

Normalization of update magnitude.
Total update magnitude:

$$\sum_{k=1}^{4} \|\Delta\theta_k\| = 0.40 + 0.25 + 0.35 + 0.60 = 1.60.$$

Normalized update magnitude:

$$\tilde{u}_k = \frac{\|\Delta\theta_k\|}{1.60}.$$

$$\tilde{u}_1 = 0.2500,$$
$$\tilde{u}_2 = 0.1563,$$
$$\tilde{u}_3 = 0.2188,$$
$$\tilde{u}_4 = 0.3750.$$

Normalization of quality score.
Total quality score:

$$\sum_{k=1}^{4} q_k = 0.90 + 0.85 + 0.80 + 0.70 = 3.25.$$

Normalized quality:

$$\tilde{q}_k = \frac{q_k}{3.25}.$$

$$\begin{aligned}
\tilde{q}_1 &= 0.2769, \\
\tilde{q}_2 &= 0.2615, \\
\tilde{q}_3 &= 0.2462, \\
\tilde{q}_4 &= 0.2154.
\end{aligned}$$

Step 3: Incentive Function Specification Assume the system designer defines the incentive function as a weighted sum:

$$r'_k = \alpha\tilde{n}_k + \beta\tilde{u}_k + \gamma\tilde{q}_k,$$

with weights

$$\alpha = 0.3, \qquad \beta = 0.4, \qquad \gamma = 0.3.$$

These weights reflect a policy choice that emphasizes update usefulness while still valuing data volume and quality.

Step 4: Raw Reward Score Computation

Client 1.

$$\begin{aligned}
r'_1 &= 0.3(0.4286) + 0.4(0.2500) + 0.3(0.2769) \\
&= 0.1286 + 0.1000 + 0.0831 \\
&= 0.3117.
\end{aligned}$$

Client 2.

$$\begin{aligned}
r'_2 &= 0.3(0.2857) + 0.4(0.1563) + 0.3(0.2615) \\
&= 0.0857 + 0.0625 + 0.0785 \\
&= 0.2267.
\end{aligned}$$

Client 3.

$$\begin{aligned}
r'_3 &= 0.3(0.1714) + 0.4(0.2188) + 0.3(0.2462) \\
&= 0.0514 + 0.0875 + 0.0739 \\
&= 0.2128.
\end{aligned}$$

Client 4.

$$\begin{aligned} r_4' &= 0.3(0.1143) + 0.4(0.3750) + 0.3(0.2154) \\ &= 0.0343 + 0.1500 + 0.0646 \\ &= 0.2489. \end{aligned}$$

Verification.

$$\sum_{k=1}^{4} r_k' = 0.3117 + 0.2267 + 0.2128 + 0.2489 = 1.0001 \approx 1.$$

Step 5: Final Reward Allocation The actual reward allocated to client k is

$$r_k = r_k' \times R_{\text{total}}.$$

Client-wise rewards.

$$\begin{aligned} r_1 &= 0.3117 \times 80 = 24.94, \\ r_2 &= 0.2267 \times 80 = 18.14, \\ r_3 &= 0.2128 \times 80 = 17.02, \\ r_4 &= 0.2489 \times 80 = 19.91. \end{aligned}$$

Verification.

$$\sum_{k=1}^{4} r_k = 24.94 + 18.14 + 17.02 + 19.91 = 80.01 \approx 80.$$

Interpretation and Incentive Implications Although Client 4 has the smallest dataset, its large update magnitude results in a comparatively high reward. Client 1 receives the highest reward due to balanced strength across all metrics. Client 3's moderate reward reflects consistent but less dominant contributions.

This example illustrates how incentive mechanisms can:

- reward useful computation rather than raw participation,
- discourage free-riding by linking rewards to measurable impact,
- balance fairness and efficiency through explicit weighting.

Key Insight from the Worked Example

In federated learning, incentive mechanisms translate observed client behavior into quantifiable rewards, shaping participation and optimization dynamics as directly as learning algorithms themselves.

17.5 Regulatory Compliance and Auditing

Federated learning is frequently motivated by regulatory and legal constraints that limit the centralized collection, storage, or processing of data. Data protection laws, sector-specific compliance regimes, and cross-border data transfer restrictions increasingly require that sensitive information remain under the control of its original owner. By keeping raw data local and exchanging only model updates, federated learning appears, at first glance, to offer a natural solution to these constraints.

However, decentralization fundamentally alters the compliance and auditing landscape. While federated learning reduces direct exposure of raw data, it also introduces new challenges for accountability, oversight, and regulatory assurance. Traditional compliance frameworks were designed for centralized data pipelines, where auditors could inspect datasets, training logs, and model artifacts within a single organizational boundary. Federated learning disperses these elements across multiple autonomous entities, complicating both legal responsibility and technical verification.

As a result, regulatory compliance in federated learning cannot rely on conventional data-centric auditing practices. Instead, it requires a shift toward protocol-centric and process-oriented oversight mechanisms that are compatible with distributed training.

17.5.1 Compliance Challenges

Federated learning raises a number of fundamental regulatory questions that do not arise, or arise only weakly, in centralized machine learning systems. While decentralization is often presented as a compliance-enabling feature, it simultaneously disrupts many of the assumptions on which existing legal, regulatory, and auditing frameworks are built. Most regulatory regimes presuppose a clear locus of data control, a well-defined data controller, and the technical feasibility of inspection and verification. Federated learning challenges all three assumptions.

Rather than eliminating regulatory risk, federated learning transforms its nature. Compliance concerns shift from centralized data handling to distributed process integrity, indirect information flows, and shared responsibility across autonomous entities.

Information Leakage through Model Updates. Although raw data never leaves the client, federated learning does not guarantee that sensitive information is completely insulated from exposure. Model updates—such as gradients, parameter differences, or sufficient statistics—are functions of local data and may, under certain conditions, encode information about individual records or population attributes.

From a regulatory standpoint, this raises a subtle but critical question: should model updates themselves be treated as sensitive or personal data? In some jurisdictions, any information that can be linked—directly or indirectly—to an identifiable individual may fall under data protection regulations. Gradient leakage attacks, membership inference, and model inversion demonstrate that updates can, in principle, reveal private attributes when adversarial assumptions are violated.

Consequently, regulators must consider whether federated learning updates require safeguards comparable to those applied to raw data. This includes assessing whether additional protections—such as secure aggregation, differential privacy, or update clipping—are necessary to ensure that the transmission of model updates complies with legal definitions of data minimization and privacy preservation. The challenge lies in balancing regulatory rigor with the practical constraints of learning performance and system efficiency.

Accountability and Responsibility. In centralized machine learning systems, accountability is typically straightforward. The entity that collects the data, trains the model, and deploys it is clearly identifiable as the data controller or responsible party. Federated learning fundamentally complicates this attribution.

Responsibility in federated systems is distributed across multiple actors, including data-owning clients, orchestration entities, and possibly downstream deployers of the trained model. Each participant influences the learning outcome, yet no single actor exercises full control over the process. This diffusion of responsibility raises difficult regulatory questions: Who is accountable if a federated model produces harmful, biased, or illegal outcomes? Is responsibility shared proportionally among contributors, or does it ultimately rest with the entity that coordinates training or deploys the model?

Without explicit governance agreements, federated learning risks creating accountability gaps. Participants may claim limited responsibility for global outcomes, arguing that their individual contribution was too small to warrant liability. Regulatory frameworks must therefore evolve to define clear roles, obligations, and liability boundaries that reflect the collaborative nature of federated learning while avoiding moral and legal ambiguity.

Audit without Data Access. Auditing is a cornerstone of regulatory compliance. In traditional machine learning pipelines, auditors can inspect raw datasets, training logs, and model artifacts to verify compliance with legal and ethical standards. Federated learning disrupts this model by design.

Because data remains local and may be subject to strict confidentiality or jurisdictional constraints, auditors often cannot directly access the datasets used in training. This raises a central compliance challenge: how can auditors verify that data was collected lawfully, that consent requirements were respected, that prohibited attributes were excluded, and that training followed approved protocols, without inspecting the data itself?

This constraint necessitates a rethinking of audit practices. Instead of data-centric audits, federated learning requires process-centric audits that focus on protocol adherence, logging, cryptographic verification, and reproducibility of training procedures. Auditors must evaluate whether the system was designed and executed in a way that *prevents* misuse, rather than attempting to detect misuse after the fact through data inspection.

Taken together, these challenges illustrate that federated learning does not eliminate regulatory concerns; rather, it reshapes them. Compliance must be reinterpreted in a way that is compatible with decentralized data ownership, autonomous participation, and collaborative model training. Regulatory success in federated learning therefore depends not only on technical safeguards, but also on updated legal interpretations, institutional coordination, and explicit governance frameworks that recognize decentralization as a first-class design constraint.

17.5.2 Auditability

Auditability in federated learning relies on shifting the focus from data inspection to process verification. Instead of examining raw datasets, auditors evaluate whether the federated learning protocol was executed correctly and in accordance with predefined rules.

Several technical and organizational mechanisms support this form of auditability:

- **Secure logging of aggregation events:** The orchestration layer maintains tamper-evident logs of training rounds, participating clients, aggregation parameters, and model versions. These logs provide an auditable trail of the learning process without exposing sensitive data.
- **Cryptographic verification of updates:** Digital signatures, secure aggregation protocols, and cryptographic commitments can be used to verify the authenticity and integrity of client updates. Such mechanisms allow auditors to confirm that updates were produced by authorized participants and were not altered in transit.
- **Reproducible training configurations:** Standardized model architectures, fixed hyperparameters, and versioned training code enable reproducibility. While raw data cannot be shared, reproducible configurations allow auditors to assess whether training adhered to approved protocols.

Under this paradigm, auditors evaluate *process compliance* rather than data content. They verify that only authorized participants contributed, that approved algorithms and parameters were used, and that aggregation and deployment followed documented procedures. This approach aligns naturally with the decentralized philosophy of federated learning while still providing meaningful regulatory assurance.

Compliance Shift

Federated learning shifts regulation from *data auditing* to *process auditing*, requiring regulators to assess protocol adherence, accountability structures, and verification mechanisms rather than direct access to raw data.

Perspective. Regulatory compliance and auditing are not obstacles to federated learning, but design constraints that shape its architecture. By embedding auditability, transparency, and accountability directly into federated protocols, system designers can ensure that collaborative learning remains legally compliant, ethically defensible, and operationally trustworthy. In this sense, federated learning represents not only a technical innovation, but also an evolution in how regulatory oversight is conceptualized in the age of distributed artificial intelligence.

17.6 Ethical Considerations in Collaborative Artificial Intelligence

Beyond questions of legal compliance and contractual governance, federated learning raises deep and enduring ethical issues that cut to the core of how collaborative artificial intelligence systems are designed, deployed, and governed. Because federated learning redistributes both data control and model creation across multiple autonomous actors, ethical responsibility is no longer concentrated in a single institution. Instead, it becomes *collective*, *distributed*, and often ambiguous.

Ethical analysis is therefore indispensable for federated learning. While privacy preservation is frequently emphasized as its primary ethical advantage, privacy alone does not guarantee fairness, accountability, or social legitimacy. Federated learning systems shape real-world decisions in healthcare, finance, employment, governance, and public services. As such, they influence whose interests are represented, whose voices are amplified, and who bears the consequences of automated decisions.

This section examines the ethical dimensions of federated learning as a form of collaborative artificial intelligence, emphasizing representation, consent, responsibility, and moral agency.

17.6.1 Representation and Bias

Federated learning is often portrayed as inherently fair because it avoids centralized data monopolies. However, decentralization does not automatically ensure equitable representation. In practice, participation in federated learning is highly uneven. Clients differ widely in data volume, data quality, computational resources, connectivity, and willingness to participate.

As a result, federated models may disproportionately reflect the data distributions of dominant participants. Organizations or devices with large datasets, high availability, and stable participation exert greater influence on model updates through aggregation weights and repeated involvement. Smaller or resource-constrained participants may contribute infrequently or be excluded altogether due to scheduling, bandwidth, or energy constraints.

This imbalance can give rise to representational bias. Even when no raw data is shared, underrepresented populations may see their characteristics, behaviors, or needs inadequately captured by the global model. For example, in a healthcare consortium, hospitals serving minority or rural populations may contribute less data and therefore have limited impact on learned clinical patterns. The resulting model may perform poorly for precisely those groups that stand to benefit most from improved predictions.

Crucially, these biases can remain hidden. Because data is never centralized, it becomes more difficult to assess coverage gaps, demographic imbalance, or systematic underperformance. Ethical federated learning therefore requires deliberate mechanisms to monitor representational equity, such as fairness-aware aggregation, performance audits across subpopulations, and explicit inclusion policies for marginalized contributors.

17.6.2 Consent and Transparency

Another central ethical challenge in federated learning concerns consent and transparency. Although raw data remains local, data subjects—particularly end users—may be unaware that their data is being used to train a global model shared across multiple organizations or jurisdictions.

In many deployments, consent is indirect or implicit. Users may agree to data collection for local services without realizing that their data contributes, through federated updates, to broader models deployed elsewhere. This opacity undermines informed consent and erodes trust, even if no formal legal violation occurs.

Ethically responsible federated learning therefore requires:

- **Informed consent:** Individuals should be clearly informed that their data may be used for collaborative model training, including the scope and purpose of that collaboration.
- **Transparency about model usage:** Participants should understand how the resulting model will be used, who will have access to it, and whether it will be deployed in high-stakes decision-making contexts.
- **Mechanisms for opting out:** Data subjects and participating entities should retain meaningful agency, including the ability to withdraw participation without disproportionate penalty.

Transparency is not merely a matter of disclosure. It is a prerequisite for ethical legitimacy. Without clear communication of how federated learning operates and who benefits from it, collaborative AI risks becoming an invisible and unaccountable force shaping individual and societal outcomes.

17.6.3 Collective Responsibility

One of the most challenging ethical questions in federated learning concerns responsibility for harm. When a centralized model causes harm—through discrimination, misdiagnosis, or erroneous decisions— responsibility can typically be traced to the organization that trained and deployed it. Federated learning complicates this attribution.

Because model parameters are shaped by many contributors, responsibility is diffused. No single participant fully controls the training process, yet all participants influence the outcome. This diffusion can create moral blind spots, where each actor views its contribution as too small to warrant accountability.

Ethical governance frameworks must therefore articulate principles of *shared responsibility*. These include:

- clearly defined roles and obligations for contributors,
- collective mechanisms for harm assessment and remediation,
- transparent escalation and dispute-resolution processes.

Importantly, collective responsibility does not imply diluted responsibility. On the contrary, it demands higher standards of coordination, documentation, and ethical oversight. Participants must acknowledge that contributing to a federated model entails ethical obligations proportional not only to technical input, but also to societal impact.

Ethical Imperative
Federated learning does not eliminate ethical responsibility; it distributes it across all participants, requiring explicit frameworks for shared accountability and moral governance.

Outlook

Federated learning represents a fundamental shift from centralized intelligence to *collaborative intelligence*. This shift redistributes not only data and computation, but also power, agency, and ethical responsibility. Technical guarantees of privacy and security, while necessary, are not sufficient to ensure ethical integrity.

To realize its full promise, federated learning must be embedded within robust ethical frameworks that address representation, consent, transparency, and accountability. Governance mechanisms, incentive structures, and auditing processes must be complemented by normative principles that guide how collaboration is structured and whose interests are prioritized.

Only when technical design, institutional governance, and ethical reasoning are aligned can federated learning mature into a scalable, trustworthy, and socially responsible paradigm for artificial intelligence.

Exercises

This chapter examined federated learning as a socio-technical system, emphasizing governance, intellectual property, value distribution, incentive design, regulatory compliance, and ethical responsibility. The following exercises are designed to test conceptual understanding, analytical reasoning, and practical application of these themes.

A. Short Answer Questions

Answer each question briefly and precisely.

1. What distinguishes governance concerns in federated learning from those in centralized machine learning?
2. Why does federated learning complicate traditional notions of model ownership?
3. Define the term *collective optimization* in the context of federated learning.
4. What role do aggregation weights play in determining participant influence?
5. Why is value distribution considered a governance issue rather than a purely economic one?
6. List two non-monetary incentives commonly used in consortium-based federated learning.
7. What is meant by *process-centric auditing*?
8. Why can federated learning amplify bias despite preserving data locality?
9. What is free-riding in federated learning systems?
10. Why is transparency essential for ethical federated learning?

B. Long Answer Questions

Answer the following questions with clear explanations, equations, and diagrams wherever appropriate.

1. Explain how federated learning transforms the concept of intellectual property. Discuss at least two ownership interpretations with their implications.
2. Analyze the tension between fairness and efficiency in value distribution among federated learning participants.
3. Describe the design of an incentive mechanism that discourages strategic manipulation while promoting high-quality updates.
4. Discuss regulatory compliance challenges in federated learning and explain why traditional auditing approaches are insufficient.
5. Examine ethical risks arising from uneven participation in federated learning and propose mitigation strategies.

C. True/False Questions

State whether each statement is **True** or **False**.

1. Federated learning completely eliminates the risk of privacy leakage.
2. In consortium-based federated learning, governance rules are typically negotiated collectively.
3. Larger datasets always imply greater intellectual property ownership.
4. Incentive mechanisms have no effect on optimization dynamics.
5. Model updates may be subject to regulatory scrutiny.
6. Equal value sharing always maximizes collaboration stability.
7. Auditability in federated learning relies primarily on inspecting raw data.
8. Ethical responsibility in federated learning is centralized at the server.
9. Secure aggregation can reduce information leakage risks.
10. Federated learning requires explicit governance agreements prior to deployment.

Answers:

1. False
2. True
3. False
4. False
5. True
6. False
7. False
8. False
9. True
10. True

D. Fill in the Blanks

Fill in the blanks with the most appropriate term or expression.

1. Federated learning is fundamentally a ___________ system rather than a purely technical one.
2. Model ownership in federated learning arises from ___________ optimization.
3. Value distribution must balance fairness and ___________.
4. Incentive mechanisms map client behavior to ___________.
5. Free-riding refers to benefiting without ___________.
6. Regulatory compliance shifts from data auditing to ___________ auditing.
7. Ethical federated learning requires informed ___________.
8. Aggregation weights determine a client's ___________ on the global model

9. Secure aggregation is used to protect __________.
10. Responsibility in federated learning is often __________ rather than centralized.

Answers:

1. socio-technical
2. collective
3. efficiency
4. rewards
5. contributing
6. process
7. consent
8. influence
9. model updates
10. distributed

E. Multiple Choice Questions (MCQs)

Choose the **most appropriate** answer.

1. Which factor most directly affects aggregation weights in standard federated averaging? (a) Client reputation (b) Dataset size (c) Model architecture (d) Encryption method
2. Which ownership model emphasizes rights to use rather than to possess the model? (a) Central ownership (b) Joint ownership (c) Usage-based ownership (d) Algorithmic ownership
3. Which mechanism primarily supports auditability without data access? (a) Centralized storage (b) Data replication (c) Secure logging (d) Manual inspection
4. Free-riding primarily threatens which aspect of federated learning? (a) Privacy (b) Convergence speed (c) Incentive compatibility (d) Encryption strength
5. Which domain most commonly uses consortium-based federated learning? (a) Social media advertising (b) Online gaming (c) Healthcare networks (d) Image compression
6. Which ethical issue arises from dominant participants shaping the global model? (a) Overfitting (b) Representation bias (c) Gradient explosion (d) Encryption failure
7. Incentives that reward only update magnitude risk encouraging: (a) Model generalization (b) Strategic manipulation (c) Regulatory compliance (d) Fair value sharing
8. Process-centric auditing focuses primarily on: (a) Dataset inspection (b) Protocol adherence (c) User consent forms (d) Model size

9. Which principle requires participants to share accountability for harm? (a) Central authority (b) Differential privacy (c) Collective responsibility (d) Secure aggregation
10. Which factor is least relevant for value attribution? (a) Data quality (b) Participation reliability (c) Client brand name (d) Marginal performance gain

Answers:

1. (b)
2. (c)
3. (c)
4. (c)
5. (c)
6. (b)
7. (b)
8. (b)
9. (c)
10. (c)

F. Application and Project-Based Questions

These problems emphasize modeling, reasoning, and implementation.

1. Design a governance framework for a hospital consortium using federated learning.
2. Propose a quantitative incentive mechanism for rewarding long-term participation.
3. Analyze a hypothetical dispute over model ownership in a multi-organization federation.
4. Design a value distribution scheme balancing fairness and efficiency.
5. Develop a protocol-level audit checklist for federated learning compliance.
6. Simulate how free-riding affects convergence in a federated setting.
7. Propose fairness-aware aggregation rules for underrepresented participants.
8. Design a consent and transparency policy for end users in federated systems.
9. Evaluate ethical risks in cross-border federated learning deployments.
10. Implement a small-scale federated experiment and analyze incentive outcomes.

Applications, Emerging Trends and Tools 18

Federated learning has evolved rapidly from an academic construct initially motivated by privacy preservation into a foundational paradigm for large-scale, collaborative intelligence in modern machine learning systems. Early formulations of federated learning were primarily concerned with enabling model training without centralized data collection. Over time, however, the paradigm has matured into a comprehensive framework for distributed intelligence that reshapes not only how models are learned, but also how learning systems are designed, deployed, governed, and operationalized in practice. As machine learning increasingly permeates critical sectors such as healthcare, finance, transportation, and public infrastructure, the limitations of centralized learning architectures—both technical and institutional—have become increasingly apparent.

Contemporary machine learning systems now operate across organizational, geographical, and jurisdictional boundaries. Data are generated continuously by hospitals, financial institutions, mobile devices, sensors, and autonomous systems, each governed by distinct policies, regulations, and ownership constraints. In such settings, the long-standing assumption that data can be freely pooled into a central repository is no longer realistic and, in many cases, no longer desirable. Federated learning addresses this structural shift by enabling models to be trained directly where data are generated, allowing learning to occur without violating privacy expectations, regulatory frameworks, or institutional control over data assets. In doing so, federated learning reconciles the competing objectives of learning performance, privacy protection, regulatory compliance, and data sovereignty.

The growing adoption of federated learning is therefore driven not only by legal and ethical considerations, but also by practical system-level requirements. Modern applications produce data that are inherently decentralized, dynamic, and highly heterogeneous in both distribution and quality. Clinical records vary across hospitals, financial transactions differ across regions and customer bases, sensor streams evolve over time, and autonomous systems encounter diverse and unpredictable environments. Centralized learning struggles to accommodate this diversity

P. P. Ray, *Foundations of Federated Learning*,
Springer Series in the Data Sciences,
https://doi.org/10.1007/978-3-032-34084-9_18

at scale. Federated learning, by contrast, offers a scalable mechanism for collective knowledge construction without raw data exchange, enabling collaboration across competing institutions, heterogeneous devices, and resource-constrained environments. This shift fundamentally transforms how collaboration is achieved in distributed machine learning systems.

Crucially, the transition of federated learning from theory to practice has been enabled by the emergence of robust software frameworks and tooling ecosystems. These tools operationalize federated learning principles by providing infrastructure for client coordination, secure aggregation, privacy preservation, fault tolerance, and large-scale deployment. Rather than requiring practitioners to engineer distributed systems from scratch, federated learning frameworks encapsulate complex system-level concerns, allowing researchers and engineers to focus on algorithm design, application-specific constraints, and empirical evaluation. The availability and maturity of such tools have been instrumental in bridging the gap between conceptual federated learning models and real-world deployment across diverse sectors.

This chapter presents a comprehensive and forward-looking examination of the *application landscape*, the*emerging research trends*, and the *practical tooling ecosystem* that collectively define the current and future trajectory of federated learning. The first part of the chapter surveys major application domains, illustrating how federated learning has been adapted to real-world constraints in healthcare, finance, the Internet of Things, robotics, biometrics, education, and telecommunications. Across these domains, attention is given to why federated learning is uniquely suited to each setting and how domain-specific constraints—such as data heterogeneity, latency, safety, and governance—shape system design choices. The second part of the chapter shifts focus toward emerging directions, including communication-efficient federated learning, energy-aware and green federated systems, federated learning for generative artificial intelligence, the rise of federated foundation models, and fully decentralized and peer-to-peer learning architectures. Finally, the chapter examines the federated learning toolchain, highlighting how different frameworks embody distinct design philosophies and deployment goals. Together, these perspectives position federated learning not merely as a niche privacy-enhancing technique, but as a central architectural principle for the design of next-generation distributed artificial intelligence systems that must operate responsibly, efficiently, and at global scale.

18.1 Applications of Federated Learning

Federated learning has transitioned from a primarily theoretical construct into a practical enabling technology across a wide range of real-world application domains. Its appeal lies not only in its ability to preserve data privacy, but also in its compatibility with the increasingly decentralized nature of modern data generation. Across sectors such as healthcare, finance, the Internet of Things, robotics, biometrics, education, and telecommunications, data are naturally distributed across institutions, devices, and administrative boundaries. In many of these domains, centralized data aggregation is either legally restricted, operationally impractical, or ethically undesirable.

This section surveys the major application areas in which federated learning has demonstrated tangible impact. Rather than presenting federated learning as a one-size-fits-all solution, the discussion emphasizes how domain-specific constraints—such as data heterogeneity, latency sensitivity, safety requirements, and governance structures—shape federated system design. Through these examples, federated learning emerges as a flexible architectural paradigm capable of supporting collaborative intelligence while respecting privacy, autonomy, and regulatory compliance.

18.1.1 Healthcare and Medical Imaging

Healthcare is widely regarded as the most mature and impactful application domain for federated learning. Medical data are inherently sensitive, highly regulated, and deeply intertwined with ethical obligations toward patient privacy and safety. Clinical data are generated across hospitals, diagnostic centers, laboratories, and national healthcare systems, each operating under distinct governance structures and legal frameworks. As a result, the traditional assumption of centralized data aggregation is often incompatible with real-world medical practice.

Medical imaging represents one of the most prominent and successful use cases of federated learning in healthcare. Imaging modalities such as X-rays, computed tomography (CT), magnetic resonance imaging (MRI), ultrasound, and digital pathology slides generate vast volumes of high-dimensional data that are typically siloed within individual institutions. Even when collaboration is scientifically desirable—such as for rare disease detection or multi-center clinical studies—legal frameworks including HIPAA, GDPR, and region-specific health data regulations severely restrict direct data sharing. Federated learning enables institutions to collaboratively train diagnostic and prognostic models while ensuring that raw patient data never leave local repositories.

From a clinical perspective, federated learning offers several critical advantages. First, it allows models to benefit from data diversity across populations, geographies, and imaging devices, improving generalization beyond any single institution. Second, it supports privacy-preserving collaboration between hospitals that may otherwise be unable to share data due to legal or competitive constraints. Third, it enables continual model improvement as new data become available, without requiring costly and risky data migration pipelines.

Despite these advantages, healthcare federated learning systems face a set of unique and stringent challenges:

- **Severe data heterogeneity**, arising from differences in imaging devices, acquisition protocols, patient demographics, and clinical workflows.

- **Limited and imbalanced local datasets**, particularly for rare conditions or specialized imaging studies.
- **Label noise and inter-observer variability**, as clinical labels often depend on expert judgment and may vary across institutions.
- **Regulatory and ethical constraints**, requiring strong guarantees on privacy, auditability, and clinical accountability.
- **Explainability and trust requirements**, as medical models must provide interpretable and clinically justifiable predictions.

These challenges place significant demands on federated optimization and aggregation strategies. Simple averaging-based methods may fail under strong non-IID data distributions, necessitating heterogeneity-aware optimization, personalized federated learning, or institution-specific model adaptation. Moreover, robustness to noisy updates and resilience against unreliable or intermittent participation are essential for deployment in real clinical settings.

In practice, healthcare federated learning systems are often evaluated not only on predictive accuracy, but also on clinical relevance, robustness, and regulatory compliance. Model validation typically involves retrospective multi-site studies, followed by prospective trials under controlled clinical oversight. As a result, healthcare has played a central role in shaping federated learning from a purely algorithmic concept into a rigorously evaluated, safety-conscious, and application-driven learning paradigm.

18.1.2 Drug Discovery and Pharmaceuticals

Drug discovery and pharmaceutical research constitute a data-intensive and high-stakes domain in which federated learning offers compelling strategic advantages. The process of discovering, developing, and validating new therapeutics requires the integration of diverse data sources, including molecular structures, biochemical assays, genomic profiles, preclinical studies, and clinical trial outcomes. These datasets are typically distributed across pharmaceutical companies, academic laboratories, contract research organizations, and healthcare institutions, each operating under strict intellectual property, regulatory, and competitive constraints.

A defining characteristic of pharmaceutical data is its proprietary nature. Compound libraries, assay results, and experimental protocols represent substantial financial investments and are closely guarded as trade secrets. Consequently, even when multiple organizations would benefit from joint model training—such as in early-stage compound screening or toxicity prediction—direct data sharing is rarely feasible. Federated learning enables collaborative model development while preserving organizational autonomy by ensuring that sensitive experimental data remain local and inaccessible to external parties.

Federated learning is particularly well-suited to early-stage drug discovery, where machine learning models are increasingly used to guide compound selection, predict bioactivity, estimate toxicity, and optimize molecular properties. By aggregating knowledge across institutions without exposing raw data, federated learning improves model generalization and reduces overfitting to organization-specific datasets. This is especially valuable for rare targets, novel chemical scaffolds, or emerging therapeutic areas where individual datasets are sparse or biased.

Beyond small-molecule discovery, federated learning also supports collaboration in areas such as:

- Quantitative structure–activity relationship (QSAR) modeling
- Protein–ligand interaction prediction
- Drug–drug interaction and safety assessment
- Biomarker discovery and patient stratification
- Clinical trial outcome modeling and adaptive trial design

Despite its promise, federated learning in pharmaceutical applications presents distinct challenges. Data heterogeneity is pronounced due to differences in experimental protocols, assay technologies, measurement noise, and data curation standards. Models must also contend with high-dimensional chemical representations and complex structure–property relationships. Furthermore, pharmaceutical federated learning systems must ensure robustness against information leakage, as model updates may inadvertently reveal competitive insights if not properly protected.

Regulatory considerations further complicate deployment. Models used in drug discovery pipelines must satisfy stringent validation, traceability, and reproducibility requirements. As a result, federated learning systems in this domain often emphasize audit logging, version control, and conservative update strategies to maintain scientific credibility and regulatory acceptance.

In practice, federated learning in pharmaceuticals is frequently positioned as a mechanism for *pre-competitive collaboration*. Organizations collaborate to improve foundational predictive models—such as toxicity or solubility predictors—while retaining full control over downstream drug development decisions. This balance between collaboration and competition reflects a broader shift in the pharmaceutical industry toward shared infrastructure for knowledge discovery without compromising intellectual property or strategic advantage.

18.1.3 Finance and Fraud Detection

Financial institutions generate enormous volumes of sensitive transactional and behavioral data through banking operations, payment systems, trading platforms, and digital financial services. These data capture fine-grained information about customer behavior, spending patterns, credit histories, and institutional workflows. Due to strict regulatory frameworks, privacy laws, and competitive business considerations, sharing such data across banks, regions, or financial entities is typically prohibited. At the same time, many financial risks—particularly fraud and financial crime— exhibit patterns that transcend individual institutions and geographical boundaries.

Fraud detection systems benefit significantly from exposure to diverse and large-scale behavioral patterns. Credit card fraud, money laundering, and insider threats often involve distributed activities that appear benign when viewed locally but reveal suspicious structure when aggregated across institutions. Centralized learning approaches are therefore poorly suited to this domain, as they require the pooling of highly sensitive financial records into shared repositories. Federated learning provides a principled alternative by enabling collaborative model training without the exchange of raw transactional data.

In a federated financial learning system, each bank or branch trains models locally on its proprietary customer data and shares only model updates or aggregated statistics with a coordinating server. This approach allows institutions to collectively improve detection accuracy while preserving data confidentiality, regulatory compliance, and organizational autonomy. As a result, federated learning has emerged as a natural architectural choice for cross-institutional financial intelligence.

Key financial applications of federated learning include:

- Credit card and digital payment fraud detection
- Anti-money laundering (AML) and suspicious activity monitoring
- Credit scoring and default risk assessment
- Insider threat detection and operational anomaly analysis

Despite its promise, federated learning in finance presents several domain-specific challenges. Transaction data are typically highly skewed, with fraudulent events constituting a small fraction of overall activity. Moreover, data distributions are strongly non-IID across institutions due to differences in customer demographics, regional economic behavior, transaction volumes, and service portfolios. These characteristics can significantly degrade the performance of naïve aggregation strategies.

Additional complexity arises from the adversarial nature of the domain. Malicious actors actively adapt their behavior in response to deployed models, introducing concept drift and strategic manipulation. Federated systems must therefore be robust not only to statistical heterogeneity, but also to adversarial behavior, poisoned updates, and unreliable participants. Furthermore, many financial applications impose strict real-time or near real-time inference requirements, limiting the complexity and latency of deployable models.

Key challenges in financial federated learning include:

- Strongly non-IID and temporally evolving transaction distributions
- Extreme class imbalance and delayed fraud labels
- Adversarial attacks targeting both models and aggregation mechanisms
- Tight latency constraints for deployment in live financial systems

To address these challenges, financial federated learning systems often emphasize robust aggregation, anomaly-aware optimization, and conservative update policies. In practice, success is measured not only by predictive accuracy, but also by stability, resilience, interpretability, and compliance with regulatory standards. Consequently, the financial domain has played a central role in driving research on robustness, security, and trustworthiness in federated learning.

18.1.4 Internet of Things and Smart Cities

The Internet of Things (IoT) and smart city infrastructures represent one of the most natural and compelling application domains for federated learning. In these environments, data are generated continuously by large numbers of geographically distributed devices, including sensors, cameras, meters, vehicles, and embedded controllers. Centralized data aggregation in such systems is often impractical due to bandwidth limitations, latency constraints, privacy concerns, and the sheer scale of device deployments. Federated learning aligns directly with this decentralized data generation model by enabling learning to occur close to the data source.

Smart city applications rely on real-time intelligence to manage complex urban systems, including transportation networks, energy grids, environmental monitoring platforms, and public safety infrastructure. These systems produce heterogeneous, high-frequency data streams that evolve over time and vary significantly across locations. Federated learning enables collective model training across distributed nodes while avoiding continuous raw data transmission to central servers, thereby reducing communication overhead and preserving citizen privacy.

In practice, federated learning in IoT systems is often deployed across hierarchical architectures. Resource-constrained edge devices such as sensors and embedded controllers perform local data collection and lightweight inference, while intermediate nodes—such as gateways, routers, or edge servers—participate in federated training and aggregation. This hierarchical structure allows federated learning to scale to city-wide deployments while accommodating diverse hardware capabilities.

Representative applications of federated learning in IoT and smart city contexts include:

- Traffic flow prediction and adaptive signal control
- Energy consumption forecasting and demand response optimization
- Environmental monitoring and pollution detection
- Smart surveillance, anomaly detection, and public safety analytics

Despite its suitability, federated learning in IoT and smart city settings faces distinct challenges. Connectivity is often intermittent, with devices joining and leaving the network unpredictably. Hardware heterogeneity is extreme, ranging from low-power microcontrollers to powerful edge servers. Energy constraints further limit computation and communication, particularly for battery-powered devices deployed at scale.

Key challenges in IoT-centric federated learning include:

- Intermittent connectivity and partial client participation
- Severe heterogeneity in device capabilities and data quality
- Energy, bandwidth, and latency constraints
- Non-stationary data streams and concept drift

To address these challenges, federated learning systems for IoT and smart cities often rely on lightweight models, asynchronous training, partial participation, and adaptive aggregation strategies. In many deployments, the emphasis shifts from maximizing global accuracy to achieving robust, low-latency, and energy-efficient learning under realistic operational constraints. As urban infrastructure becomes increasingly sensorized and autonomous, federated learning provides a scalable and privacy-respecting foundation for distributed intelligence in next-generation smart cities.

18.1.5 Robotics and Autonomous Systems

Robotic and autonomous systems operate in complex, dynamic, and often unstructured environments where continuous perception, decision-making, and control are required. These systems generate rich streams of sensory data from cameras, LiDAR, inertial sensors, force sensors, and proprioceptive feedback. Centralized data collection and training are frequently impractical due to bandwidth limitations, latency constraints, privacy considerations, and the physical dispersion of robotic platforms. Moreover, many robotic applications involve safety-critical operations, making offline data aggregation and delayed model updates undesirable.

Federated learning provides a natural framework for collective learning across robotic fleets while preserving local autonomy. Each robot or autonomous agent learns from its own interactions with the environment and contributes knowledge to a shared model without transmitting raw sensory data. This enables distributed systems to benefit from collective experience while respecting operational constraints and reducing reliance on centralized infrastructure.

In robotic systems, federated learning supports a wide range of learning tasks, including perception, state estimation, planning, and control. By aggregating knowledge across agents, federated learning improves robustness to environmental variability and accelerates learning in scenarios where individual robots experience limited or biased data. At the same time, local control policies remain responsive to real-time sensory input, preserving the autonomy and safety of individual agents.

Representative applications of federated learning in robotics and autonomous systems include:

- Autonomous driving perception and sensor fusion
- Multi-robot navigation, mapping, and localization

- Manipulation and control policy learning
- Human–robot interaction and behavior modeling

Despite its advantages, federated learning in robotic systems faces significant challenges. Data distributions are strongly non-IID due to environmental diversity, differences in robot hardware, and variation in task assignments. Robots operating in different locations may encounter distinct visual scenes, terrain conditions, or interaction dynamics, leading to heterogeneous learning signals. Additionally, robotic systems often require continuous online learning and adaptation, which complicates synchronization and aggregation across agents.

Safety considerations further constrain federated learning in this domain. Model updates must not introduce unstable behaviors or degrade performance in deployed systems. As a result, robotic federated learning systems often employ conservative update strategies, staged deployment, and extensive validation before integrating global model updates into local controllers.

Key challenges in robotic federated learning include:

- Strong non-IID data arising from environmental and task diversity
- Safety-critical deployment and verification requirements
- Continuous, online, and lifelong learning demands

18.1.6 Biometrics and Identity Systems

Biometric and identity systems rely on uniquely identifiable personal data, including fingerprints, facial images, voice recordings, iris patterns, and behavioral traits such as keystroke dynamics or gait. These data are intrinsically sensitive, as biometric identifiers are permanent and cannot be revoked or reissued in the same manner as passwords or tokens. Consequently, the collection, storage, and processing of biometric data are subject to stringent privacy, security, and ethical requirements.

Traditional biometric learning systems often rely on centralized repositories that aggregate raw biometric data from large populations. Such architectures pose significant risks, including large-scale data breaches, unauthorized secondary use, and loss of user trust. Federated learning offers a privacy-preserving alternative by enabling biometric models to be trained across distributed devices or institutions without centralizing raw biometric information. For example, facial recognition or speaker identification models can be trained directly on user devices, with only model updates shared for aggregation.

Federated learning is particularly well-suited to biometric systems deployed at scale, such as mobile authentication platforms, access control systems, and cross-institutional identity verification services. By keeping biometric samples local, federated learning reduces the attack surface of centralized databases while still allowing models to benefit from diverse population-level data. This approach supports both personalization—adapting models to individual users—and global learning across heterogeneous user populations.

Representative applications of federated learning in biometrics and identity systems include:

- Face and fingerprint recognition systems
- Speaker identification and voice-based authentication
- Behavioral biometrics, including keystroke dynamics and gait analysis
- Secure and continuous authentication mechanisms

Despite these advantages, biometric federated learning systems face several critical challenges. Model updates may inadvertently leak sensitive information through model inversion, membership inference, or gradient-based attacks. Ensuring that federated training does not compromise individual privacy therefore requires careful system design and often the integration of additional privacy-preserving mechanisms.

Fairness and bias represent another central concern. Biometric models have historically exhibited uneven performance across demographic groups, raising ethical and societal issues. Federated learning systems must be designed to mitigate bias amplification and ensure equitable performance across populations, even under heterogeneous and imbalanced local data distributions.

Long-term system management also poses unique challenges in biometric applications. Models must support continual updates as populations evolve, devices change, and threat models adapt. At the same time, mechanisms for model revocation, user opt-out, and compliance with data protection regulations must be carefully integrated.

Key challenges in biometric federated learning include:

- Vulnerability to model inversion and membership inference attacks
- Fairness and bias across demographic and user groups
- Long-term model update, revocation, and governance requirements

18.1.7 Education and Personalized Learning

Educational platforms increasingly rely on data-driven models to support personalized instruction, learner assessment, and academic decision-making. Data collected in educational settings include student performance records, learning trajectories, behavioral interactions, and engagement patterns across schools, universities, and online learning platforms. These data are inherently sensitive, as they reflect individual learning abilities, progress, and personal circumstances. Consequently, educational data are subject to privacy regulations, institutional policies, and ethical considerations that limit centralized data aggregation.

Federated learning enables collaborative learning analytics and personalization without requiring the sharing of student-level data. Educational institutions and platforms can train models locally on learner data while contributing to a shared global model through federated aggregation. This approach allows systems to benefit from diverse learning contexts and student populations while preserving learner privacy and institutional autonomy.

Personalization lies at the core of federated learning applications in education. Learners exhibit substantial variability in prior knowledge, learning pace, cognitive strategies, and engagement patterns. Centralized models often struggle to accommodate this heterogeneity, whereas federated learning supports both global knowledge sharing and local adaptation. Models can be personalized at the institution, classroom, or even individual learner level, enabling more responsive and inclusive educational systems.

Federated learning supports a range of educational applications, including:

- Adaptive learning systems that tailor content and pacing to individual learners
- Knowledge tracing models that estimate learner mastery over time
- Intelligent tutoring systems that provide personalized feedback and guidance

Despite its promise, federated learning in education presents several challenges. Educational data are often sparse, noisy, and temporally evolving, particularly in short courses or early learning stages. Participation across institutions may be irregular, and differences in curricula, assessment standards, and instructional practices introduce strong heterogeneity. Moreover, educational models must be interpretable and pedagogically meaningful to support trust among educators and learners.

Key challenges in educational federated learning include:

- Heterogeneous learning contexts and non-IID student data
- Sparse and delayed feedback signals
- Requirements for interpretability and pedagogical alignment

This application domain highlights the broader role of federated learning as a mechanism for balancing collective intelligence with individual-centered adaptation, a principle that extends across many federated systems.

18.1.8 Telecommunications and Network Optimization

Modern telecommunication networks constitute one of the largest and most complex distributed systems in operation today. Mobile networks, broadband infrastructure, and emerging 5G and beyond systems generate massive volumes of data describing user behavior, traffic patterns, signal quality, and network state. These data are naturally distributed across base stations, edge servers, and core network components, and are often subject to privacy constraints, latency requirements, and operational policies that limit centralized data collection and processing.

Federated learning provides a scalable framework for distributed intelligence in telecommunication systems by enabling collaborative model training directly at network nodes. Rather than aggregating raw network telemetry or user data at a central repository, each base station or network element performs local learning and shares only model updates. This approach reduces communication overhead, preserves user privacy, and enables faster adaptation to local network conditions.

Telecommunication networks are inherently dynamic. Traffic loads fluctuate over time, user mobility introduces rapid changes in network topology, and external factors such as weather, events, or failures can significantly affect performance. Federated learning allows models to continuously adapt to these evolving conditions while leveraging collective experience across the network. This capability is particularly valuable for large-scale deployments where centralized retraining would be prohibitively slow or costly.

Key applications of federated learning in telecommunications and network optimization include:

- Traffic prediction and load balancing across base stations
- Fault detection, anomaly identification, and self-healing network mechanisms
- Adaptive resource allocation for spectrum, power, and bandwidth management

Despite its suitability, federated learning in telecommunications faces several domain-specific challenges. Network data distributions vary significantly across regions and time scales, leading to strong non-IID characteristics. Participation of network nodes may be asynchronous, and communication constraints impose strict limits on model size and update frequency. Furthermore, many network optimization tasks operate under tight latency constraints, requiring rapid inference and stable model behavior.

Additional considerations include the need for reliability and robustness. Erroneous model updates can degrade network performance or violate quality-of-service guarantees. As a result, federated learning systems in telecommunications often emphasize conservative update strategies, rigorous validation, and fault-tolerant aggregation mechanisms.

As networks continue to evolve toward increasingly autonomous and software-defined infrastructures, federated learning is expected to play a central role in supporting scalable, adaptive, and privacy-aware network intelligence.

18.2 Communication-Efficient Federated Learning

Communication efficiency has emerged as one of the most fundamental and structural challenges in federated learning. Unlike centralized machine learning, where communication is largely confined to high-bandwidth, low-latency data center environments, federated learning operates across wide-area networks that connect millions of heterogeneous and often resource-constrained clients. In such settings, communication—not computation—frequently becomes the dominant bottleneck. As federated learning systems scale to larger models, more clients, and higher training frequencies, reducing communication overhead is no longer an optimization detail but a prerequisite for practical deployment.

Early federated learning formulations implicitly assumed that communication between clients and the coordinating server was sufficiently abundant. However, real-world deployments quickly revealed that repeated transmission of high-dimensional model updates is costly, energy-intensive, and sometimes infeasible. Mobile devices operate under limited data plans, edge devices experience intermittent connectivity, and industrial systems impose strict latency constraints. These realities have driven a shift in research focus toward *communication-efficient federated learning*, which seeks to minimize the amount, frequency, and cost of information exchanged during training without sacrificing model performance or stability.

At a conceptual level, communication-efficient federated learning recognizes that learning progress does not require exact or continuous synchronization across all participants. Instead, approximate, selective, and adaptive information exchange is often sufficient to achieve effective global learning. This perspective represents a departure from classical distributed optimization paradigms and reflects a broader trend toward *resource-aware learning architectures*.

One major dimension of communication efficiency concerns the *frequency of communication*. In many federated systems, clients perform multiple local training steps before communicating with the server. Increasing local computation relative to communication reduces the number of rounds required, thereby lowering communication cost. However, this trade-off introduces new challenges, as excessive local training can amplify the effects of data heterogeneity and destabilize global convergence. Balancing local computation and global coordination remains an active area of research and a defining design choice in communication-efficient systems.

A second dimension involves the *size of communicated updates*. Modern machine learning models often contain millions or billions of parameters, making full model transmission prohibitively expensive. Communication-efficient approaches therefore aim to reduce the dimensionality or precision of transmitted updates. Rather than treating model updates as exact numerical objects, these methods view them as carriers of directional or structural information sufficient for collective learning. This shift reflects a deeper understanding that learning dynamics can tolerate approximation and noise when appropriately managed.

Another emerging trend is *selective participation and partial communication*. In large-scale federated systems, it is neither necessary nor desirable for all clients to participate in every training round. Communication-efficient federated learning embraces partial participation, allowing subsets of clients to contribute based on availability, resource constraints, or relevance. Similarly, clients may communicate only portions of their model or only when local updates exceed certain significance thresholds. These strategies reduce unnecessary communication while maintaining learning momentum.

Communication efficiency is also closely tied to *asynchronous federated learning*. Synchronous coordination, while conceptually simple, imposes idle time and straggler effects in heterogeneous environments. Asynchronous approaches allow clients to communicate updates independently, reducing waiting time and improving system utilization. From a future-oriented perspective, asynchronous communication is likely to become a defining feature of federated learning systems operating at global scale, particularly in edge and IoT environments.

Importantly, communication-efficient federated learning must be designed in concert with privacy, robustness, and fairness considerations. Reducing communication should not inadvertently increase vulnerability to inference attacks or bias amplification. Similarly, aggressive communication reduction may privilege resource-rich clients while marginalizing weaker participants. As a result, emerging research increasingly treats communication efficiency as a multi-objective optimization problem that balances efficiency, security, and equity.

From a systems perspective, communication-efficient federated learning aligns naturally with broader trends in distributed computing and networking. The rise of edge computing, software-defined networking, and energy-aware system design reinforces the need for learning algorithms that adapt to network conditions rather than assume ideal connectivity. In future deployments, federated learning systems are likely to dynamically adjust communication strategies based on bandwidth availability, energy budgets, and application urgency.

Communication-efficient federated learning is expected to play a central role in enabling federated training of large models, including foundation models and generative architectures. As model sizes continue to grow and deployment environments become more constrained, the success of federated learning will increasingly depend on its ability to learn effectively under severe communication limitations. In this sense, communication efficiency is not merely an engineering concern but a defining principle shaping the future evolution of federated learning as a scalable, sustainable, and globally deployable learning paradigm.

18.3 Energy-Aware and Green Federated Learning

As machine learning systems continue to scale in size, complexity, and deployment footprint, their energy consumption has emerged as a critical concern. Training and deploying large models now incur substantial computational, communication, and infrastructure costs, raising questions about sustainability and environmental impact. Federated learning, while offering significant advantages in privacy and data governance, introduces additional energy challenges due to its reliance on distributed computation and repeated communication across heterogeneous devices. Energy-aware and green federated learning has therefore emerged as a vital research direction aimed at making federated systems sustainable, efficient, and environmentally responsible.

Unlike centralized data centers, federated learning operates across a diverse ecosystem of devices, including smartphones, edge gateways, embedded systems, and industrial controllers. These devices vary widely in computational capability, battery capacity, and energy availability. Many participants operate under strict energy constraints, particularly battery-powered or intermittently connected devices. As a result, energy consumption becomes a first-class design consideration rather than a secondary optimization objective.

Energy expenditure in federated learning arises from multiple sources. Local model training consumes computational energy, particularly for deep networks or frequent local updates. Communication further contributes to energy cost, as wireless transmission is often significantly more energy-intensive than computation. Additional overhead arises from coordination, aggregation, and system-level monitoring. Energy-aware federated learning seeks to explicitly model and optimize these costs, rather than treating them as implicit side effects of learning.

At a conceptual level, green federated learning reframes learning objectives to account for energy efficiency alongside accuracy and convergence. This shift reflects a broader recognition that performance gains must be weighed against environmental and operational costs. In practice, this often leads to adaptive training strategies in which client participation, update frequency, and model complexity are adjusted based on energy availability and system constraints. Learning becomes context-aware, responding dynamically to device state and environmental conditions.

A central theme in energy-aware federated learning is the selective involvement of clients. Not all devices need to participate in every training round, particularly when energy reserves are low or network conditions are unfavorable. By prioritizing energy-efficient participation and allowing energy-constrained clients to opt out or reduce activity, federated systems can prolong device lifespan while maintaining overall learning progress. This approach also aligns with fairness considerations, as it prevents undue burden on weaker participants.

Model design plays a crucial role in reducing energy consumption. Lightweight architectures, parameter-efficient representations, and modular models reduce both computational and communication energy. In energy-aware systems, model complexity is often matched to device capability, allowing heterogeneous clients to contribute at different levels of fidelity. This flexibility supports scalable deployment across diverse hardware platforms without imposing uniform energy demands.

From a broader perspective, green federated learning aligns with global efforts to reduce the carbon footprint of artificial intelligence. By distributing computation and avoiding centralized data centers, federated learning has the potential to leverage local energy resources and reduce large-scale infrastructure demand. However, this potential can only be realized through careful system design that avoids excessive communication, redundant computation, and inefficient coordination.

Energy awareness also intersects with emerging deployment scenarios such as smart cities, environmental monitoring, and industrial automation. In these contexts, federated learning systems often operate continuously over long periods, making energy efficiency essential for reliability and economic viability. Sustainable federated learning enables long-term operation without frequent maintenance, battery replacement, or excessive energy consumption.

Energy-aware and green federated learning is expected to become a defining characteristic of responsible and scalable distributed intelligence. Future research is likely to integrate energy models directly into federated optimization objectives, enabling principled trade-offs between learning quality and energy consumption.

As societal awareness of environmental sustainability grows, energy-efficient federated learning will play a key role in ensuring that the benefits of distributed artificial intelligence are realized without disproportionate environmental cost.

18.4 Federated Learning for Generative Artificial Intelligence

Generative artificial intelligence has rapidly transformed the landscape of machine learning by enabling systems to synthesize text, images, audio, code, and complex multimodal content. Large language models, diffusion models, and generative adversarial networks have demonstrated unprecedented capabilities, but their development has largely relied on centralized training over massive, aggregated datasets. This centralized paradigm raises significant concerns related to privacy, data ownership, regulatory compliance, and sustainability, particularly as generative models increasingly interact with personal, proprietary, and sensitive data. Federated learning offers a promising alternative framework for enabling generative artificial intelligence without requiring centralized access to raw data.

In many real-world applications, the data required to train or refine generative models are inherently distributed. User interactions with conversational agents, creative writing platforms, medical report generation systems, and personalized content services are produced locally and often cannot be shared due to privacy or intellectual property constraints. Federated learning enables generative models to be trained or adapted directly where such data are generated, allowing collective improvement while preserving data sovereignty and user trust.

Applying federated learning to generative artificial intelligence introduces new challenges that go beyond those encountered in discriminative learning. Generative models are typically large, parameter-rich, and highly sensitive to training dynamics. Model updates are high-dimensional and communication-intensive, making naive federated training impractical. Furthermore, generative tasks often exhibit strong personalization requirements, as users expect models to reflect individual preferences, styles, or domains.

Despite these challenges, federated learning provides several strategic advantages for generative systems. It supports personalization without centralized profiling, allowing global models to capture shared structure while local adaptations reflect user-specific patterns. It also reduces the risk of training on sensitive or copyrighted data, as raw content remains local and under user or institutional control. From an ethical perspective, federated learning offers a pathway toward more responsible generative AI deployment.

Representative application areas at the intersection of federated learning and generative AI include:

- Personalized language models for on-device text generation and assistance
- Privacy-preserving generative models for medical report synthesis
- Collaborative creative systems for writing, art, and music generation
- Federated fine-tuning of large generative models across organizations

From a systems perspective, federated generative learning often emphasizes parameter-efficient adaptation rather than full model training. Local updates may focus on subsets of model parameters or lightweight representations, enabling scalable collaboration despite the size of generative architectures. This trend reflects a broader shift toward modular and adaptable generative systems that can operate across heterogeneous environments.

Federated learning also introduces new considerations regarding content safety and model behavior. Generative models trained across decentralized data sources must maintain consistent safety policies, avoid the amplification of harmful content, and remain robust to malicious or low-quality updates. Ensuring coherent and trustworthy global behavior while learning from diverse and decentralized sources remains an open research challenge.

Looking ahead, federated learning is expected to play a critical role in the evolution of generative artificial intelligence toward more decentralized, user-centric, and regulation-aware systems. As generative models become embedded in everyday tools and services, federated learning provides a principled mechanism for balancing innovation with privacy, control, and accountability. In this context, federated generative learning represents not merely a technical extension of federated learning, but a foundational shift in how generative intelligence is developed and governed at scale.

18.5 Federated Foundation Models

Foundation models have emerged as a dominant paradigm in modern artificial intelligence. These models are characterized by large-scale pretraining on diverse data, followed by adaptation to a wide range of downstream tasks. Language models, vision–language models, and multimodal architectures now serve as general-purpose foundations upon which specialized applications are built. However, the prevailing approach to training foundation models relies almost exclusively on centralized data aggregation and massive computational infrastructure, raising fundamental concerns related to privacy, data sovereignty, governance, and sustainability.

Federated foundation models represent a natural and necessary evolution of this paradigm. Rather than training a single monolithic model on centrally collected data, federated learning enables the construction and continual refinement of foundation models across distributed data sources. In this setting, organizations, institutions, or devices collaboratively contribute to the development of a shared foundational representation while retaining full control over their local data. This approach aligns foundation model development with the realities of modern data generation, which is inherently decentralized and governed by diverse legal and ethical constraints.

A defining feature of federated foundation models is their emphasis on *shared representations* rather than task-specific objectives. The goal is not merely to solve a single predictive task, but to learn rich, generalizable representations that can support a wide spectrum of downstream applications. Federated learning enables such representations to emerge from heterogeneous and distributed data sources, capturing population-level structure while respecting institutional boundaries.

From a systems perspective, federated foundation models differ from conventional federated learning in both scale and intent. They involve substantially larger model architectures, longer training horizons, and more complex coordination across participants. As a result, federated foundation models place stringent demands on communication efficiency, energy awareness, robustness, and long-term stability. These requirements have driven renewed interest in hierarchical federated architectures, where intermediate aggregation layers mediate between local participants and global models.

Federated foundation models are particularly relevant in domains where no single entity possesses sufficiently diverse or comprehensive data. Examples include healthcare systems spanning multiple hospitals, financial ecosystems involving competing institutions, multilingual language modeling across regions, and industrial systems operating across supply chains. In such contexts, federated learning enables pre-competitive collaboration on foundational models while allowing participants to retain proprietary or sensitive information.

Key application scenarios for federated foundation models include:

- Cross-institutional foundation models for healthcare, finance, and science
- Multilingual and multicultural language models trained across regions
- Federated multimodal models integrating text, vision, and sensor data
- Domain-adaptive foundation models for edge and industrial environments

Despite their promise, federated foundation models introduce significant challenges. Data heterogeneity across participants can bias shared representations if not carefully managed. Model updates may reflect local objectives that are misaligned with global generalization. Furthermore, governance becomes a central concern, as questions arise regarding model ownership, update authority, contribution attribution, and responsibility for downstream use. These challenges elevate federated foundation models from a purely technical problem to a socio-technical system requiring careful design and oversight.

Another critical consideration is lifecycle management. Foundation models are not static artifacts; they evolve over time through continual learning, refinement, and adaptation. Federated learning naturally supports this evolution by enabling incremental updates as new data become available, without requiring repeated centralized retraining. However, ensuring stability, avoiding catastrophic forgetting, and maintaining consistent behavior across versions remain open research questions.

Federated foundation models are likely to play a central role in the next generation of artificial intelligence systems. They offer a pathway toward scalable, collaborative, and regulation-aware model development that aligns with global trends in data governance and digital sovereignty. As foundation models increasingly underpin critical applications, federated learning provides a principled framework for ensuring that these powerful models are developed in a manner that is inclusive, accountable, and sustainable.

18.6 Fully Decentralized and Peer-to-Peer Federated Learning

Most existing federated learning systems rely on a centralized coordinating server that orchestrates client participation, aggregates model updates, and maintains the global model state. While this architecture has proven effective in many deployments, it introduces structural limitations related to scalability, robustness, trust, and governance. The coordinating server represents a single point of failure, a potential performance bottleneck, and, in some cases, an entity that must be trusted by all participants. Fully decentralized and peer-to-peer federated learning has emerged as a promising alternative that removes or substantially weakens this central dependence.

In fully decentralized federated learning, model training and coordination are distributed across participating nodes without relying on a central server. Instead, learning proceeds through direct communication between peers, often structured by an underlying communication graph. Each participant maintains a local model and periodically exchanges information with neighboring nodes. Through repeated local interactions, models gradually converge toward a consensus that reflects collective knowledge across the network.

This paradigm aligns naturally with decentralized systems such as ad hoc networks, edge computing infrastructures, blockchain networks, and organization-spanning collaborations. In such environments, no single entity has the authority, visibility, or capacity to act as a global coordinator. Peer-to-peer federated learning enables collaboration among autonomous agents while preserving decentralization, resilience, and local control.

From a systems perspective, decentralized federated learning offers several advantages. By eliminating the central server, it improves fault tolerance and reduces the risk of system-wide disruption. It also enhances scalability, as communication and computation load are distributed across the network rather than concentrated at a single point. Moreover, decentralized architectures better reflect trust relationships in many real-world settings, where participants may prefer symmetric collaboration over hierarchical control.

Fully decentralized federated learning is particularly relevant in scenarios such as:

- Large-scale edge and IoT networks without stable central connectivity
- Multi-organization collaborations lacking a trusted central authority
- Mobile ad hoc networks and disaster-response systems
- Privacy-sensitive environments requiring minimal central oversight

Despite these advantages, decentralized federated learning introduces new challenges that distinguish it from server-based approaches. Achieving global consensus without centralized aggregation is non-trivial, particularly in networks with dynamic topology or unreliable communication. Convergence behavior depends strongly on the structure of the peer-to-peer communication graph, including connectivity, degree distribution, and latency. Ensuring efficient information flow while avoiding excessive communication overhead remains a central research problem.

Decentralized systems must also contend with heterogeneity and partial participation. Nodes may have different computational capabilities, data distributions, and availability patterns. Unlike centralized federated learning, where the server can enforce participation policies or discard unreliable updates, peer-to-peer systems must rely on local decision-making and robust update mechanisms. This places greater emphasis on resilience to noise, adversarial behavior, and malicious peers.

Governance and accountability present additional complexities. In the absence of a central authority, questions arise regarding responsibility for model behavior, update validation, and long-term maintenance. Decentralized federated learning systems often require complementary mechanisms for coordination, trust establishment, and auditability. These considerations elevate decentralized federated learning from a purely technical architecture to a socio-technical system involving organizational, legal, and ethical dimensions.

Fully decentralized and peer-to-peer federated learning is expected to play a key role in enabling large-scale, resilient, and infrastructure-light learning systems. As artificial intelligence increasingly operates at the edge of networks and in resource-constrained environments, decentralized learning architectures provide a pathway toward scalable intelligence without centralized control. In combination with advances in communication-efficient, energy-aware, and federated foundation models, decentralized federated learning represents a critical step toward truly distributed and autonomous artificial intelligence systems.

18.7 Open Research Challenges

Despite significant progress in federated learning theory, algorithms, and applications, the field remains characterized by a number of unresolved research challenges. Many of these challenges arise from the inherent tension between distributed learning objectives, system constraints, and societal requirements such as privacy, fairness, and sustainability. As federated learning transitions from experimental deployments to critical infrastructure, addressing these challenges becomes essential for its long-term viability and responsible adoption. This section outlines key open research problems that define the future research agenda of federated learning.

18.7.1 Statistical Heterogeneity and Non-IID Data

One of the most fundamental and persistent challenges in federated learning is the presence of strong statistical heterogeneity across participating clients. Unlike centralized learning, where training data are typically assumed to be independently and identically distributed, federated learning operates over data that are generated by distinct users, institutions, devices, or environments. As a result, local datasets often differ substantially in distribution, quantity, quality, and semantic content.

Such non-IID data arise naturally from variations in user behavior, institutional practices, geographic and cultural context, sensing modalities, and hardware characteristics. For example, medical institutions may follow different diagnostic protocols, financial institutions may serve distinct customer populations, and mobile devices may capture highly personalized interaction patterns. These differences violate the assumptions underlying many classical optimization and learning algorithms, leading to degraded convergence speed, unstable training dynamics, and reduced global model performance.

Statistical heterogeneity introduces a fundamental tension between global generalization and local relevance. A globally aggregated model may fail to represent the needs of individual clients, while overly localized adaptation can undermine collective learning. This trade-off becomes increasingly pronounced as the scale and diversity of participants grow. In extreme cases, non-IID data can cause federated optimization to diverge or oscillate, particularly when local updates dominate global coordination.

Although numerous algorithmic strategies have been proposed to mitigate the effects of heterogeneity, no universal solution exists. Many approaches rely on assumptions about data similarity, bounded divergence, or stable participation patterns that do not hold in real-world deployments. As federated learning systems expand across sectors with deeply heterogeneous data sources, understanding the fundamental limits of learnability under extreme heterogeneity remains an open and largely unresolved research problem.

Designing principled mechanisms that adapt to heterogeneity without sacrificing scalability, robustness, or fairness continues to be a central research direction. Progress in this area requires not only algorithmic innovation, but also deeper theoretical understanding of distributed learning dynamics under non-IID data.

18.7.2 Scalability and System Heterogeneity

Scalability represents a core systems-level challenge in federated learning. Practical deployments often involve thousands or millions of clients with widely varying computational capabilities, memory resources, network connectivity, and energy availability. Unlike homogeneous data center environments, federated learning must operate across highly diverse and unpredictable system conditions.

System heterogeneity manifests in multiple forms. Clients may differ in processor speed, available memory, battery capacity, and network bandwidth. Connectivity may be intermittent or unreliable, particularly in mobile, edge, or IoT environments. These variations complicate coordination, as slower or less reliable clients can delay training rounds, reduce effective participation, or introduce noisy updates.

Managing such diversity while maintaining stable and efficient learning dynamics poses significant challenges. Federated learning systems must balance inclusivity with efficiency, deciding which clients participate, when they participate, and how their contributions are weighted. Stragglers, intermittent clients, and partial participation are not exceptions but the norm in large-scale deployments.

Open research questions include how to design adaptive participation strategies that respond to changing system conditions, how to coordinate learning across hierarchical or multi-tier architectures, and how to maintain convergence guarantees in asynchronous or loosely synchronized settings. As federated learning expands into edge computing, smart infrastructure, and fully decentralized environments, scalability under realistic system constraints remains a defining and unresolved challenge.

18.7.3 Privacy, Security, and Robustness

Although federated learning was originally motivated by privacy preservation, it does not, by itself, guarantee complete privacy or security. While raw data remain local, information can still be leaked through shared model updates, gradients, or aggregated parameters. Such leakage can enable inference attacks that reveal sensitive attributes, membership information, or even approximate reconstructions of local data.

In addition to privacy leakage, federated learning systems are vulnerable to a range of security threats. Malicious participants may inject poisoned updates, embed backdoors, or behave in a Byzantine manner to disrupt learning or manipulate model behavior. These threats are particularly concerning in open or cross-organizational federated systems, where participants may not be fully trusted.

Ensuring robust and privacy-preserving learning in adversarial environments therefore remains an ongoing and critical challenge. Protective mechanisms often introduce additional communication overhead, computational cost, or performance degradation, creating complex trade-offs between privacy guarantees, robustness, and model utility. These trade-offs become especially acute in large-scale and long-running deployments, where even small vulnerabilities can accumulate over time.

Robust aggregation, attack detection, and resilience to adversarial behavior are active areas of research, but many open questions remain. In particular, designing federated learning systems that are simultaneously scalable, privacy-preserving, secure, and efficient continues to be a major unsolved problem. Addressing these challenges is essential for federated learning to serve as a trustworthy foundation for critical applications in healthcare, finance, infrastructure, and governance.

18.7.4 Fairness, Bias, and Inclusivity

Federated learning systems may inadvertently amplify biases that are already present in local datasets or introduce new forms of inequity arising from system-level constraints. Because data are distributed across heterogeneous clients, differences in data quantity, quality, and representativeness can lead to uneven contributions to the global model. Clients representing underrepresented populations, minority groups, or rare conditions may possess smaller or noisier datasets, reducing their influence during aggregation and potentially degrading model performance for those groups.

System heterogeneity further compounds these issues. Clients with limited computational resources, weaker network connectivity, or restricted energy availability may participate less frequently or be excluded entirely from training rounds. As a result, federated learning systems may systematically privilege resource-rich participants while marginalizing weaker or intermittently connected clients. Such dynamics raise important concerns about inclusivity, particularly in applications such as healthcare, education, and public services, where equitable outcomes are essential.

Achieving fairness in federated learning therefore requires careful consideration of both statistical and system-level factors. Fairness cannot be treated solely as a property of the learned model; it is also shaped by participation policies, aggregation strategies, and resource allocation mechanisms. Defining appropriate fairness metrics in federated settings remains an open challenge, as classical notions of fairness may not directly translate to distributed and privacy-preserving environments.

Designing equitable aggregation mechanisms that balance global performance with local representation is an active area of research. Equally important is ensuring inclusive participation across diverse populations and deployment contexts. Addressing fairness in federated learning thus represents an inherently interdisciplinary challenge, requiring insights from machine learning, systems engineering, ethics, and social sciences.

18.7.5 Communication, Energy, and Sustainability Trade-Offs

Communication efficiency and energy awareness have emerged as central and interconnected themes in federated learning research. In large-scale deployments, communication often dominates computational cost, while energy consumption directly affects device longevity, operational feasibility, and environmental impact. Reducing communication and energy usage is therefore essential for practical and sustainable federated learning systems.

However, optimizing communication and energy efficiency is rarely free of cost. Aggressive reduction in communication frequency or precision can slow convergence, destabilize training, or degrade final model accuracy. Similarly, limiting client participation to conserve energy may disproportionately exclude certain devices or populations, introducing fairness concerns. These trade-offs highlight the complex interplay between learning performance, system efficiency, and equity.

A major open research challenge is the development of unified frameworks that explicitly balance learning objectives with communication cost and energy consumption. Rather than treating efficiency as an afterthought, future federated learning systems must integrate resource awareness directly into their optimization criteria. This shift reflects a broader trend toward sustainable and responsible artificial intelligence.

As concerns about the environmental footprint of machine learning grow, federated learning must evolve toward principled, energy-aware, and environmentally sustainable designs. Understanding how to achieve meaningful learning progress under strict resource constraints remains a defining challenge for the next generation of federated systems.

18.7.6 Evaluation, Benchmarking, and Reproducibility

Evaluating federated learning systems remains particularly challenging due to the diversity of deployment scenarios, assumptions, and system configurations. Unlike centralized learning, federated learning performance depends not only on algorithmic choices but also on client behavior, network conditions, and participation patterns. As a result, experimental outcomes can vary significantly across settings.

Many existing studies rely on simplified or idealized experimental setups that do not fully capture the complexity of real-world federated environments. Assumptions such as static client populations, homogeneous devices, or stable connectivity often fail to hold in practice. This gap between experimental evaluation and real-world deployment limits the generalizability of reported results.

There is therefore a pressing need for standardized benchmarks, realistic simulation frameworks, and reproducible evaluation protocols. Such benchmarks must account for statistical heterogeneity, system variability, partial participation, and adversarial conditions. Without common evaluation practices, comparing federated learning algorithms and drawing robust conclusions remains difficult.

Establishing rigorous and transparent benchmarking standards is essential for the maturation of the field. Improved evaluation practices will not only facilitate fair comparison of methods, but also accelerate the translation of federated learning research into reliable and trustworthy real-world systems.

18.7.7 Governance, Accountability, and Legal Considerations

Federated learning introduces a new class of governance challenges that extend beyond traditional machine learning systems. Because learning is distributed across multiple autonomous participants, often spanning organizations, jurisdictions, and legal regimes, questions of control, authority, and responsibility become inherently complex. Issues such as model ownership, update authority, responsibility for errors, and liability for downstream harm are no longer straightforward when no single entity fully controls the training process or the underlying data.

In cross-institutional or decentralized federated learning settings, it is often unclear who ultimately governs the global model. Participants may contribute unequally, pursue different objectives, or operate under distinct regulatory obligations. When errors occur—such as biased predictions, unsafe behavior, or regulatory violations—assigning accountability becomes particularly challenging. This ambiguity complicates deployment in regulated domains such as healthcare, finance, public infrastructure, and critical services.

Legal considerations further intensify these challenges. Federated learning systems must comply with data protection laws, sector-specific regulations, and contractual agreements that vary across regions. Even when raw data remain local, model updates and learned representations may be subject to legal interpretation regarding data processing, intellectual property, and cross-border data flows. Ensuring compliance therefore requires careful alignment between system design and evolving legal frameworks.

Addressing governance and accountability in federated learning cannot be achieved through technical solutions alone. It requires sustained collaboration between machine learning researchers, legal scholars, ethicists, policymakers, and domain experts. Future federated learning systems must incorporate transparent governance mechanisms, clear participation agreements, auditability, and well-defined responsibility structures. Only by aligning technical architectures with legal and ethical obligations can federated learning achieve durable adoption in real-world, high-stakes environments.

18.7.8 Toward Trustworthy and Responsible Federated Learning

Ultimately, the long-term success of federated learning depends on its ability to earn and maintain trust among users, participating institutions, regulators, and society at large. Trustworthiness in federated learning extends beyond the narrow goal of privacy preservation. It encompasses a broad set of properties, including security, robustness, fairness, interpretability, transparency, and sustainability. Failure in any one of these dimensions can undermine confidence in the entire system.

Responsible federated learning requires that systems behave reliably under heterogeneous conditions, resist malicious or accidental misuse, and produce outcomes that are understandable and justifiable to stakeholders. Participants must have confidence that their contributions are handled fairly, that their data are protected, and that the resulting models align with shared values and societal norms. This is particularly important as federated learning is increasingly deployed in applications that directly affect human well-being and social outcomes.

Developing federated learning systems that are transparent, accountable, and socially responsible remains an open and evolving challenge. It requires the integration of technical safeguards with governance structures, evaluation standards, and ethical oversight. As federated learning continues to mature and underpin critical applications, the choices made today regarding trust, responsibility, and governance will shape the trajectory of distributed artificial intelligence for years to come.

In this sense, federated learning is not merely a technical paradigm, but a socio-technical framework that reflects broader questions about how intelligence is shared, controlled, and governed in a connected world.

18.8 Tools

Federated learning has evolved from a theoretical construct in distributed optimization into a deployable learning paradigm adopted across academia, industry, and the public sector. This transition has been enabled largely by the development of robust open-source and industrial federated learning frameworks that transform theoretical principles into executable systems.

These frameworks abstract the complexity of decentralized training by managing client coordination, communication, aggregation, privacy preservation, and fault tolerance. By encapsulating system-level concerns such as secure aggregation, differential privacy, and scalable communication, they allow researchers and practitioners to focus on algorithm design, evaluation, and application-specific objectives.

This section surveys prominent federated learning frameworks from both research and production perspectives. Rather than treating them as interchangeable, we emphasize their differing design philosophies, system architectures, and deployment goals. Understanding these distinctions is essential for selecting a framework appropriate to a given research study or real-world application.

Table 18.1 provides a comparative overview of representative federated learning frameworks currently used in research and production. Rather than treating these systems as interchangeable tools, the table highlights their differing design philosophies, levels of privacy and security support, and intended deployment contexts. The comparison illustrates a clear spectrum ranging from research-centric frameworks focused on algorithmic rigor (e.g., TensorFlow Federated) to enterprise-grade platforms emphasizing security, governance, and large-scale deployment (e.g., NVIDIA FLARE, FATE, and Substra). This diversity underscores the importance of aligning framework selection with application requirements, regulatory constraints, and system-level objectives in real-world federated learning deployments.

18.8.1 TensorFlow Federated

TensorFlow Federated (TFF) is a research-oriented framework designed to support principled experimentation with federated learning and other decentralized computations. Built on the TensorFlow ecosystem, TFF prioritizes formal modeling, algorithmic transparency, and experimental control over production deployment.

TFF adopts a declarative programming model that clearly separates *client-local computation* from *federated orchestration*. Local training and evaluation are expressed using standard TensorFlow constructs, while aggregation and coordination are specified explicitly using federated operators. This structure mirrors the mathematical formulation of federated learning, where local objectives are optimized independently and coupled through global aggregation.

A defining feature of TFF is its treatment of federated learning as *distributed functional computation*. Developers explicitly specify data placement, local update logic, and aggregation rules, enabling rigorous reasoning about correctness, privacy,

Table 18.1 Comparison of Major Federated Learning Frameworks

Framework	Primary Focus	Design Philosophy	Privacy & Security Support	Typical Use Cases
TensorFlow Federated (TFF)	Research and algorithm prototyping	Declarative, strongly typed functional model separating local and federated computation	Formal reasoning support; privacy analyzable but limited built-in cryptography	Theoretical validation, algorithm benchmarking, large-scale simulation
PySyft	Privacy-preserving data science	Remote computation with strict data ownership and cryptographic enforcement	Strong support for SMPC, differential privacy, homomorphic encryption	Healthcare, finance, ethical AI, regulated data collaboration
Flower	Flexible research-to-production FL	Framework-agnostic, modular, extensible client–server architecture	Relies on external privacy mechanisms; supports secure aggregation extensions	Edge learning, rapid prototyping, heterogeneous device deployment
NVIDIA FLARE	Enterprise-grade deployment	Workflow-centric, modular runtime with security-first orchestration	Built-in DP, HE, PSI, secure authentication and authorization	Cross-institutional collaboration, medical imaging, large-model training
FATE	Secure cross-silo federated learning	Cryptography-centric system with strong protocol integration	Native MPC and homomorphic encryption across workflows	Vertical FL, banking, healthcare, inter-organizational analytics
Substra	Governance and traceability	Audit-driven orchestration with reproducible task execution	Access control, provenance tracking, secure execution environments	Clinical research, regulated ML pipelines, collaborative science

and convergence. As a result, TFF is particularly well suited for algorithm prototyping, controlled simulation, and theoretical validation under non-IID data and partial participation.

While TFF provides scalable simulation capabilities, it is not intended primarily for deployment on real-world edge devices. Instead, it serves as a reference framework for designing and validating federated learning algorithms.

TensorFlow Federated
https://www.tensorflow.org/federated

Federated Learning (FL) API

The *Federated Learning (FL) API* provides a high-level interface for defining and evaluating federated training and evaluation workflows using TensorFlow models. It enables experimentation with federated learning without requiring users to manage low-level details such as communication protocols or aggregation mechanisms.

The FL API is organized around three core components: federated-compatible models, computation builders for training and evaluation, and datasets for simulation. Keras models can be wrapped directly for federated use, while builder functions generate complete federated learning processes. Simulation datasets allow realistic experimentation with client sampling and heterogeneous data distributions.

The FL API is implemented on top of the Federated Core (FC), ensuring that its abstractions remain consistent with the formal semantics of federated computation.

Federated Learning API
https://www.tensorflow.org/federated/federated_learning

Federated Core (FC) API

The *Federated Core (FC) API* forms the foundational layer of TensorFlow Federated and provides a general-purpose language for expressing federated computations. Unlike the FL API, which targets ease of use, FC exposes the underlying computational model used to define collective, distributed algorithms.

Federated Core expresses system behavior from a global perspective, where participants act as anonymous members of a collective. Instead of point-to-point messaging, FC provides declarative operators for aggregation, reduction, and broadcast, enabling concise and analyzable representations of distributed protocols.

A key contribution of FC is its rich type system, which explicitly encodes data structure and placement across clients and server. This formal treatment enables static verification of correctness and prevents illegal data movement. While most users rely on the FL API, the FC API is essential for developing custom federated algorithms and extending the framework.

Federated Core API
https://www.tensorflow.org/federated/federated_core

18.8.2 PySyft

PySyft is an open-source Python library developed by the OpenMined community that enables privacy-preserving machine learning on decentralized data. It is designed to allow data scientists and researchers to perform federated learning and related privacy-aware computations without directly accessing or copying sensitive data. PySyft provides high-level abstractions for federated learning, secure multi-party computation (SMPC), differential privacy, and encrypted computation such as homomorphic encryption, thereby enabling collaborative model training in privacy-sensitive domains.

A central innovation of PySyft is its ability to support "remote data science," where computation is performed on data that remains under the control of its owners. To facilitate this, PySyft introduces the concept of *Datasites*, which act as secure hosts for data and enforce access policies defined by data owners. Data scientists connect to these Datasites to run computation tasks without ever seeing the raw underlying data.

PySyft extends popular machine learning frameworks such as PyTorch and TensorFlow, allowing federated training of models across distributed nodes by leveraging familiar APIs. It integrates cryptographic techniques—such as secure multi-party computation and partial homomorphic encryption—to add additional layers of privacy protection and safeguard against information leakage during collaborative training.

Key features of PySyft include:

- Support for federated learning workflows on decentralized datasets.
- Secure multi-party computation for joint operations on private data.
- Differential privacy mechanisms to limit information leakage.
- Integration with existing deep learning frameworks (e.g., PyTorch).
- Remote execution via Datasites that preserve data owners' control.

PySyft is particularly useful in settings where regulatory compliance, ethical governance, or strict privacy guarantees are required, such as in healthcare analytics, financial forecasting, or collaborative scientific research. It represents not only a federated learning toolkit but also a broader platform for executing analyses on protected data without compromising confidentiality.

PySyft GitHub Repository
https://github.com/OpenMined/PySyft

18.8.3 Flower Framework

Flower (often referred to by its Python package name *flwr*) is an open-source, framework-agnostic federated AI system designed to make federated learning accessible, extensible, and practical for a wide range of real-world applications. Developed under the repository adap/flower, Flower emphasizes flexibility, modularity, and compatibility with existing machine learning ecosystems.

At its core, Flower provides a platform for building, simulating, and deploying federated learning workflows across heterogeneous client environments while remaining agnostic to the underlying model implementation framework (e.g., TensorFlow, PyTorch, scikit-learn, JAX). This framework addresses the diversity of federated use cases by enabling developers to customize training strategies, communication patterns, and aggregation mechanisms without modifying core components.

Design Principles and Architecture Flower's design is guided by three core principles:

- **Customizability**: Federated learning systems differ substantially across domains; Flower allows fine-grained configuration of training strategies, client sampling, aggregation schemes, and messaging protocols to fit specific deployment needs.
- **Extendability**: Originating from research at the University of Oxford, Flower supports overrideable components and extension points that enable the creation of novel learning workflows, strategies, and evaluation methods.
- **Framework Agnosticism**: Flower integrates with a wide variety of machine learning toolkits including PyTorch, TensorFlow, Hugging Face Transformers, scikit-learn, JAX, and others. This enables federated training with models implemented in different ecosystems without modifying training logic.

These principles make Flower particularly suitable for research, prototyping, and production scenarios where flexibility and interoperability are critical.

Core Components and Workflow A typical Flower system comprises three principal elements:

- **Server**: Coordinates the overall federated learning process, collects model updates from clients, and applies aggregation rules to produce a global model.
- **Clients**: Execute local training or evaluation using their private datasets and forward model updates or metrics back to the server.
- **Strategies**: Define the logic for client selection, local update aggregation, and schedule of training rounds. Users can employ built-in strategies like federated averaging or implement custom strategies to optimize for specific criteria (e.g., fairness, robustness).

Flower also supports simulation environments where many virtual clients run on a single machine— facilitating rapid experimentation and debugging before deployment on real devices.

Extensibility and Use Cases Flower's modularity enables several advanced use cases:

- **Integration with Existing Workloads**: Existing centralized workflows can be converted to federated counterparts by plugging in Flower clients with minimal code changes.
- **Vertical and Cross-Silo FL**: Beyond horizontal scenarios, Flower supports federations where different clients hold complementary feature sets or operate across organizational boundaries with disparate datasets.
- **Custom Strategy Development**: Researchers can define novel aggregation rules, client selection mechanisms, or robustness enhancements that extend beyond standard federated averaging.
- **Baseline Implementations**: Community-contributed baseline implementations covering standard algorithms like FedAvg, FedOpt, and personalized FL offer starting points for comparative research.

Community and Documentation Flower is supported by vibrant documentation that includes tutorials, quickstart guides for different ML frameworks, and examples of both basic and complex federated learning systems. The official documentation portal provides step-by-step guides to implementing systems with PyTorch, TensorFlow, Hugging Face, and other libraries.

The project has substantial community engagement—with thousands of stars and forks on GitHub— making it one of the most widely adopted frameworks in the federated learning ecosystem. Its extensibility and rich example base have made it a preferred choice for federated learning research and deployment across both academia and industry.

Flower Framework GitHub Repository
https://github.com/adap/flower

18.8.4 NVFlare

NVIDIA Federated Learning Application Runtime Environment (Flare) is an open-source, domain-agnostic Python SDK developed by NVIDIA for building, simulating, and deploying federated learning systems in research and production environments. The framework enables existing machine learning workflows to be adapted to a federated paradigm with minimal code changes, while providing the infrastructure required for secure, scalable, multi-party collaboration.

By design, FLARE supports a broad spectrum of federated learning scenarios, including both horizontal and vertical paradigms. It provides built-in algorithms—such as FedAvg, FedProx, FedOpt, SCAFFOLD, and personalized methods like Ditto—while also allowing developers to define custom training workflows and aggregation strategies.

Architecture and Core Principles NVIDIA FLARE is built upon a modular and extensible architecture that separates core functions such as client orchestration, communication, privacy enforcement, and workflow control. This separation allows the SDK to transition federated learning workloads from research and local simulation to fully deployed environments in enterprise or edge settings.

The framework comprises the following key architectural components:

- **Server Controller:** Orchestrates the federated learning workflow by assigning tasks to clients, managing federated jobs, and performing aggregation of updates returned from clients.
- **Client Workers:** Execute local training or evaluation tasks on decentralized data and return model updates or metrics to the server.
- **Modular Workflows:** Built-in and customizable workflows (such as scatter-gather, cyclic weight transfer, cross-site evaluation) enable flexible definition of federated algorithms.

FLARE supports extensive workflow customization, allowing developers to design federated systems for different application domains and deployment contexts. It also includes simulation modes and proof-of-concept functionalities that facilitate rapid prototyping before moving to production deployment.

Supported Algorithms and Use Cases NVIDIA FLARE ships with implementations of standard federated learning algorithms such as federated averaging and its variants, and it enables integration with deep learning frameworks (e.g., PyTorch, TensorFlow), traditional machine learning libraries (e.g., scikit-learn, XGBoost), as well as hybrid training models.

Typical use cases for FLARE include:

- Collaborative training of AI models across geographically distributed institutions without sharing raw data.
- Privacy-sensitive domains such as healthcare imaging, biomedical analysis, and financial modeling.
- Scalable large-model workflows, including federated training of language models with parameter-efficient fine-tuning methods.

Privacy, Security, and Governance A central concern in federated learning is the preservation of data privacy and system security. NVIDIA FLARE incorporates built-in mechanisms for differential privacy, homomorphic encryption, and private set intersection (PSI), enabling clients to contribute model updates without exposing sensitive local data. These mechanisms are customizable and can be combined with user workflows to meet the security requirements of regulated environments.

The framework also supports federated authorization, privacy policy enforcement, and system resiliency features, which are critical for deployments in enterprise, cloud, or regulated sectors.

Development Workflow and Examples NVIDIA FLARE provides extensive examples and quick-start templates to help users get started with federated learning tasks. These include:

- Hello-world examples demonstrating basic federated averaging with PyTorch and TensorFlow.
- Workflow examples such as cyclic weight transfer and federated cross-site evaluation.
- Tutorials illustrating tabular statistics, differential privacy, and logistic regression workflows.

For developers, the framework's client APIs enable seamless transition from centralized machine or deep learning code to federated workflows, facilitating rapid iteration on algorithm design and deployment.

Simulation and Deployment FLARE includes a federated simulator that enables localized testing of federated workflows with multiple virtual clients, making it possible to debug and optimize workflows prior to real-world execution. This capability significantly shortens development cycles and reduces the dependency on large federated testbeds.

For production deployment, FLARE supports both on-premise and cloud infrastructures, and integrates with deployment tools such as dashboards, CLI tools, and infrastructure-as-code templates for automated provisioning on platforms like AWS and Microsoft Azure.

Position in the Federated Ecosystem NVIDIA FLARE occupies a unique position in the federated learning landscape: it bridges research and production with an enterprise-grade runtime that is both extensible and secure. While other frameworks focus on research flexibility or framework agnosticism, FLARE emphasizes robust deployment, privacy guarantees, and scalability for real-world multi-party applications.

NVIDIA FLARE GitHub Repository
https://github.com/NVIDIA/NVFlare

18.8.5 Federated AI Technology Enabler

Federated AI Technology Enabler (FATE) is a production-grade, open-source federated learning framework developed to enable collaborative model training across multiple organizations while preserving data privacy and security. Originating from WeBank's AI Department and now hosted as a project under the LF AI & Data Foundation, FATE is designed for real-world cross-silo federated learning deployments in regulated environments such as finance, healthcare, and government.

A distinguishing characteristic of FATE is its comprehensive support for secure and scalable federated computation protocols. It integrates cryptographic techniques such as secure multi-party computation (MPC) and homomorphic encryption (HE) to ensure that local data remains confidential throughout the training process. This emphasis on security enables multiple parties to jointly train machine learning models without sharing raw data or revealing sensitive information to external collaborators.

Architecture and Modular Design FATE adopts a modular architecture that decouples algorithm logic, secure computation, and task scheduling, enabling flexible deployment and extension for diverse applications. The core components of FATE include:

- **Secure Federated Protocols**: These protocols implement privacy-preserving operations such as encrypted gradient computation, secure aggregation, and MPC, ensuring data confidentiality across participants.
- **Federated Algorithms**: The framework supports a range of federated learning paradigms, including *horizontal*, *vertical*, and *federated transfer learning*. Common algorithms such as logistic regression, tree-based methods, and deep learning models are provided.
- **FATE-Flow**: A scheduling and workflow management layer that administers federated jobs and orchestrates computation streams across multiple nodes.
- **KubeFATE**: A cloud-native operational tool to deploy FATE clusters using container and orchestration technologies such as Docker Compose and Kubernetes, facilitating scalable, resilient federated systems in hybrid or cloud environments.

These components are designed to interoperate through well-defined APIs, allowing users to construct federated learning pipelines that align with organizational requirements and system constraints.

Federated Learning Paradigms Supported FATE supports the major classes of federated learning scenarios:

- **Horizontal Federated Learning**, where participants share the same feature space but hold data on disjoint samples.
- **Vertical Federated Learning**, where participants hold complementary features for a shared set of entities.
- **Federated Transfer Learning**, which combines insights from disparate feature and sample spaces using transfer and representation learning techniques.

This breadth of paradigms makes FATE suitable for collaborative analytics in environments where data partitioning differs across organizations.

Deployment and Scalability FATE can be deployed in standalone or clustered configurations depending on scale and operational requirements. Standalone deployment (e.g., via PyPI or Docker) is useful for development and testing, while cluster deployments using Kubernetes or similar infrastructure support large-scale, multi-node federations that are typical in enterprise or academic research. Tooling such as KubeFATE simplifies the process of containerized federated learning orchestration in cloud and hybrid environments.

Security, Privacy, and Compliance Security is central to FATE's design. By default, raw data never leaves local environments; only encrypted intermediate results and secure aggregations are communicated among participants. The combination of MPC and homomorphic encryption enables computations on encrypted data, and the framework's governance model ensures adherence to privacy requirements and data protection regulations typical of financial and healthcare systems.

Ecosystem and Extensions FATE's ecosystem extends beyond core federated training:

- **FATE-Board**: Visualization tools for inspecting models, metrics, and federated workflows.
- **FATE-Serving**: A scalable model serving infrastructure for deployed federated models.
- **FATE-LLM**: A recent extension built to support federated training of large language models with parameter-efficient techniques, enhancing the applicability of FATE to modern generative architectures.

Collectively, these extensions support end-to-end federated AI lifecycles, from secure training to production deployment and monitoring.

FATE GitHub Repository
https://github.com/FederatedAI/FATE

18.8.6 Substra

Substra is an open-source federated learning software platform designed to enable privacy-preserving, traceable, and scalable collaborative machine learning. It provides both a flexible Python API and a web application to orchestrate and monitor federated training and evaluation across multiple organizations while ensuring that raw data never leaves its owner's infrastructure. Substra was originally developed by Owkin and is now hosted as an incubation project under the Linux Foundation for AI and Data, reflecting its growing adoption in production FL deployments.

Substra is distinguished by its focus on real-world federated systems where data ownership, auditability, and governance are first-class concerns. In contrast to research-oriented frameworks that emphasize algorithmic prototyping, Substra is engineered for operational robustness and enterprise integration.

Design and Key Features At its core, Substra supports federated learning workflows that allow different organizations (often referred to as *nodes* or *parties*) to collaboratively train machine learning models without centralizing their sensitive datasets. Instead of transferring data, Substra coordinates algorithm execution and model updates across distributed backends while maintaining full traceability and access control.

The key features of Substra include:

- **Framework Agnosticism:** Substra works with any Python-based machine learning library, including PyTorch, TensorFlow, and scikit-learn, enabling federated learning with a wide range of model types and training paradigms.
- **Flexible Computation:** Beyond standard training tasks, Substra supports arbitrary computations (e.g., analytics or custom evaluation workflows) as part of the federated pipeline.
- **Scalability:** Substra is designed to scale both horizontally (across multiple machines) and vertically (multiple tasks on a single node), which is critical for large-scale collaborative experiments.
- **Traceability and Auditability:** Every computation, model update, and dataset usage is logged in an auditable database, giving participating organizations full visibility into how their assets were used.
- **Web Monitoring:** A built-in web interface allows users to track experiment progress, inspect results, and manage workloads across the federation.
- **Production-Ready Deployment:** Substra is packaged to run on containerized infrastructures and is regularly audited for security and operational reliability, making it suitable for enterprise use.

System Architecture Substra's architecture separates high-level orchestration from participant-specific execution. The platform's backend infrastructure typically involves the following components:

- **Client Backends:** Each organization participating in a federation hosts its own Substra backend, which stores local datasets and executes federated tasks while maintaining data control.
- **Orchestrator:** The orchestrator coordinates federated learning jobs, distributes machine learning tasks across organizations, and tracks task status transitions. It ensures that the sequence of operations adheres to defined workflows, manages retries on failures, and maintains metadata for traceability.
- **Central Database:** Substra uses a central metadata store (commonly a PostgreSQL database) that records non-sensitive information about datasets, algorithms, and task histories. This ledger-like structure enables auditability without exposing private data.

This architecture allows decentralization of raw data, while ensuring that federated learning workflows are coordinated and recorded at scale.

SubstraFL: High-Level Federated Learning Library To simplify federated experimentation on Substra networks, the project provides **SubstraFL**, a high-level Python library built on top of the core Substra SDK. SubstraFL abstracts many common federated workflows, such as task registration, model definition, and federated strategy specification, making it easier for data scientists to run complex experiments at scale.

SubstraFL supports standard federated optimization schemes (e.g., FedAvg) and can integrate models written in any Python machine learning framework. Example usage includes training PyTorch models on decentralized MNIST datasets and defining custom federated algorithms for scikit-learn models.

Applications and Use Cases Substra has been deployed in real multi-institutional federated projects, particularly within healthcare and biotechnology domains, where data privacy and complex governance requirements are paramount. It has been used in flagship initiatives such as the MELLODDY project, which enabled pharmaceutical companies to collaboratively train predictive models on chemical compound data without sharing proprietary datasets.

Beyond healthcare, Substra can support any domain involving sensitive or siloed datasets, such as finance, government data analytics, or cross-enterprise predictive modeling.

Substra Documentation
https://docs.substra.org/en/stable/

Exercises

This chapter examined the application landscape, emerging trends, tools, and open challenges of federated learning, emphasizing its role as a foundational paradigm for large-scale, privacy-aware, and distributed artificial intelligence systems.

A. Short Answer Questions

Answer each question briefly and precisely.

1. What fundamental limitation of centralized machine learning motivated the development of federated learning?
2. Why is healthcare considered one of the most mature application domains for federated learning?
3. What role does data heterogeneity play in federated learning system design?
4. How does federated learning support data sovereignty?
5. What is meant by partial client participation in federated learning?

6. Why is communication often the primary bottleneck in federated learning systems?
7. What distinguishes federated foundation models from conventional federated models?
8. Why are privacy guarantees in federated learning not absolute by default?
9. What is the significance of auditability in cross-institutional federated learning?
10. Why is fairness a more complex issue in federated learning than in centralized learning?

B. Long Answer Questions

Answer the following questions with clear explanations, equations, and diagrams wherever appropriate.

1. Explain how federated learning addresses the limitations of centralized learning in regulated domains such as healthcare and finance.
2. Discuss the impact of non-IID data on federated optimization and convergence.
3. Describe the role of communication-efficient strategies in large-scale federated learning systems.
4. Analyze the challenges and opportunities of applying federated learning to generative artificial intelligence.
5. Explain the concept of federated foundation models and their relevance to modern AI systems.
6. Compare server-based federated learning with fully decentralized and peer-to-peer federated learning architectures.
7. Discuss energy-aware and green federated learning and its importance for sustainable AI.
8. Examine the role of governance and accountability in cross-organizational federated learning deployments.
9. Analyze the security threats faced by federated learning systems and possible mitigation strategies.
10. Discuss why evaluation and benchmarking are particularly challenging in federated learning research.

C. True/False Questions

State whether each statement is **True** or **False**.

1. Federated learning always guarantees complete data privacy without additional mechanisms.
2. Communication cost in federated learning is often higher than computation cost.
3. Federated learning is unsuitable for applications with highly heterogeneous data.

4. Fully decentralized federated learning eliminates the single point of failure present in server-based approaches.
5. Medical imaging data are well suited for federated learning due to their decentralized storage.
6. Energy consumption is a negligible concern in federated learning systems.
7. Federated foundation models aim to learn task-specific predictors only.
8. Secure aggregation prevents the server from accessing individual client updates.
9. Fairness issues in federated learning arise only from biased data.
10. Federated learning is primarily a socio-technical paradigm rather than a purely algorithmic one.

Answers:

1. False
2. True
3. False
4. True
5. True
6. False
7. False
8. True
9. False
10. True

D. Fill in the Blanks

Fill in the blanks with the most appropriate term or expression.

1. Federated learning enables model training without centralized ___________ aggregation.
2. Strong data heterogeneity in federated learning leads to ___________ data distributions.
3. Communication-efficient federated learning reduces the ___________ of transmitted updates.
4. Secure multi-party computation enhances ___________ guarantees.
5. In healthcare, federated learning improves model ___________ across institutions.
6. Federated learning systems often rely on ___________ aggregation of client updates.
7. Energy-aware federated learning explicitly considers device ___________ constraints.
8. Foundation models emphasize learning shared ___________ representations.
9. Peer-to-peer federated learning removes the need for a central ___________.
10. Auditability supports regulatory ___________ in federated systems.

Answers:

1. data
2. non-IID
3. size
4. privacy
5. generalization
6. weighted
7. energy
8. latent
9. server
10. compliance

E. Multiple Choice Questions (MCQs)

Choose the **most appropriate** answer.

1. Which domain has most strongly influenced the development of federated learning?
   (a) Telecommunications
   (b) Robotics
   (c) Healthcare
   (d) Education
2. What is the primary goal of communication-efficient federated learning?
   (a) Increase model size
   (b) Reduce communication overhead
   (c) Improve encryption strength
   (d) Eliminate aggregation
3. Which factor most strongly differentiates federated learning from centralized learning?
   (a) Model architecture
   (b) Optimization objective
   (c) Data locality
   (d) Loss function
4. Which scenario benefits most from decentralized federated learning?
   (a) Stable data centers
   (b) Trusted central authority
   (c) Ad hoc edge networks
   (d) Centralized databases
5. Which technique is commonly used to protect privacy in federated learning?
   (a) Dropout
   (b) Secure aggregation
   (c) Data normalization
   (d) Batch shuffling

6. What is a key challenge in federated generative AI?
   (a) Small datasets
   (b) High-dimensional model updates
   (c) Lack of personalization
   (d) Absence of regulation
7. Energy-aware federated learning primarily optimizes:
   (a) Prediction accuracy
   (b) Training speed
   (c) Energy consumption
   (d) Model interpretability
8. Which property is essential for cross-institutional federated learning?
   (a) Centralized ownership
   (b) Auditability
   (c) Uniform hardware
   (d) Static datasets
9. Which factor complicates fairness in federated learning?
   (a) Centralized training
   (b) Homogeneous data
   (c) Unequal client participation
   (d) Deterministic optimization
10. Federated foundation models are best described as:
   (a) Task-specific classifiers
   (b) Lightweight edge models
   (c) Shared pretrained representations
   (d) Rule-based systems

Answers:

1. (c)
2. (b)
3. (c)
4. (c)
5. (b)
6. (b)
7. (c)
8. (b)
9. (c)
10. (c)

F. Application and Project-Based Questions

These problems emphasize modeling, reasoning, and implementation.

1. Design a federated learning system for multi-hospital medical imaging and justify architectural choices.
2. Propose a communication-efficient federated learning strategy for mobile devices with limited bandwidth.
3. Implement a simulated federated learning experiment under non-IID data and analyze convergence behavior.
4. Develop an energy-aware client participation policy for IoT-based federated learning.
5. Design a federated approach for training a personalized language model on user devices.
6. Compare server-based and peer-to-peer federated learning for disaster-response networks.
7. Propose a benchmarking protocol for evaluating federated learning algorithms under realistic constraints.
8. Analyze fairness trade-offs in a federated education recommendation system.
9. Design a governance framework for cross-border federated learning in finance.
10. Explore the feasibility of federated foundation models for a smart city deployment.

Further Reading

1. Ludwig, H., & Baracaldo, N. (Eds.). (2022). *Federated learning: A comprehensive overview of methods and applications* (pp. 13–19). Springer.
2. Jin, Y., Zhu, H., Xu, J., & Chen, Y. (2023). *Federated learning*. Springer Nature Singapore.
3. Yadav, S. P., Bhati, B. S., Mahato, D. P., & Kumar, S. (Eds.). (2022). *Federated learning for IOT applications*. Springer.
4. Krishnan, S., Anand, A. J., Srinivasan, R., Kavitha, R., & Suresh, S. (2024). *Federated learning*. CRC Press.
5. ur Rehman, M. H., & Gaber, M. M. (Eds.). (2021). *Federated learning systems: Towards next-generation AI* (Vol. *965*). Springer Nature.
6. Nguyen, L. M., Hoang, T. N., & Chen, P. Y. (Eds.). (2024). *Federated learning: Theory and practice*. Elsevier.
7. Hong, C. S., Khan, L. U., Chen, M., Chen, D., Saad, W., & Han, Z. (2021). *Federated learning for wireless networks* (pp. 3–239). Springer.
8. Elngar, A. A., Oliva, D., & Balas, V. E. (Eds.). (2024). *Artificial intelligence using federated learning: Fundamentals, challenges, and applications*. CRC press, Taylor & Francis.
9. Kishor, K., Nand, P., Jain, V., Saxena, N., Agarwal, G., & Astya, R. (Eds.). (2024). *Federated learning for smart communication using IoT application*. CRC Press.
10. Hassan, A., Prasad, V. K., Bhattacharya, P., Dutta, P., & Damaševičius, R. (Eds.). (2023). *Federated learning and AI for healthcare 5.0*. IGI Global.
11. Li, L., Fan, Y., Tse, M., & Lin, K. Y. (2020). A review of applications in federated learning. *Computers & Industrial Engineering*, *149*, 106854.
12. Nakayama, K., & Jeno, G. (2022). *Federated learning with Python*. Packt Publishing.
13. Ziller, A., Trask, A., Lopardo, A., Szymkow, B., Wagner, B., Bluemke, E., Nounahon, J.-M., Passerat-Palmbach, J., Prakash, K., Rose, N., Ryffel, T., Reza, Z. N., & Kaissis, G. (2021). Pysyft: A library for easy federated learning. In *Federated learning systems: Towards next-generation AI* (pp. 111–139). Springer International Publishing.
14. Jayachitra, S., Prasanth, A., Dhanaraj, R. K., & Pelusi, D. (Eds.). (2025). *Applications of federated learning in technological advancements: Use cases and applications*. CRC Press.
15. Sahoo, J., Ouaissa, M., & Nair, A. K. (Eds.). (2024). *Federated learning: Principles, paradigms, and applications*. CRC Press.
16. Yu, S., & Cui, L. (2023). *Security and privacy in federated learning* (pp. 1–122). Springer.
17. Ludwig, H., & Baracaldo, N. (2022). Introduction to federated learning. In *Federated learning: A comprehensive overview of methods and applications* (pp. 1–23). Springer International Publishing.
18. Khan, S. (2024). Privacy-preserving computing in the healthcare using federated learning. In *AI-driven marketing research and data analytics* (pp. 263–280). IGI Global Scientific Publishing.

P. P. Ray, *Foundations of Federated Learning*,
Springer Series in the Data Sciences,
https://doi.org/10.1007/978-3-032-34084-9

19. Zhou, Q., & Wang, Z. (2024). A network intrusion detection method for information systems using federated learning and improved transformer. *International Journal on Semantic Web and Information Systems (IJSWIS), 20*(1), 1–20.
20. Kairouz, P., & McMahan, H. B. (2021). Advances and open problems in federated learning. *Foundations and Trends in Machine Learning, 14*(1-2), 1–210.
21. Zhang, C., Xie, Y., Bai, H., Yu, B., Li, W., & Gao, Y. (2021). A survey on federated learning. *Knowledge-based Systems, 216*, 106775.
22. Sharma, A., Singh, S. K., Chhabra, A., Kumar, S., Arya, V., & Moslehpour, M. (2023). A novel deep federated learning-based model to enhance privacy in critical infrastructure systems. *International Journal of Software Science and Computational Intelligence (IJSSCI), 15*(1), 1–23.
23. Niknam, S., Dhillon, H. S., & Reed, J. H. (2020). Federated learning for wireless communications: Motivation, opportunities, and challenges. *IEEE Communications Magazine, 58*(6), 46–51.
24. Wen, J., Zhang, Z., Lan, Y., Cui, Z., Cai, J., & Zhang, W. (2023). A survey on federated learning: Challenges and applications. *International Journal of Machine Learning and Cybernetics, 14*(2), 513–535.
25. Yang, Q., Liu, Y., Chen, T., & Tong, Y. (2019). Federated machine learning: Concept and applications. *ACM Transactions on Intelligent Systems and Technology (TIST), 10*(2), 1–19.
26. Bhattacharya, P., Verma, A., & Tanwar, S. (Eds.). (2023). *Federated learning for internet of medical things: Concepts, paradigms, and solutions*. CRC Press.
27. Tan, A. Z., Yu, H., Cui, L., & Yang, Q. (2022). Towards personalized federated learning. *IEEE Transactions on Neural Networks and Learning Systems, 34*(12), 9587–9603.
28. Lim, W. Y. B., Luong, N. C., Hoang, D. T., Jiao, Y., Liang, Y. C., Yang, Q., Niyato, D., & Miao, C. (2020). Federated learning in mobile edge networks: A comprehensive survey. *IEEE Communications Surveys & Tutorials, 22*(3), 2031–2063.
29. Uddin, M. I., & Mashwani, W. K. (Eds.). (2024). *Federated learning: Unlocking the power of collaborative intelligence*. CRC Press.
30. Tripathy, S., Kasyap, H., & Fang, M. (2025). *Federated learning: Security and privacy*. CRC Press.
31. Kumar, Y., & Singla, R. (2021). Federated learning systems for healthcare: Perspective and recent progress. In *Federated learning systems: Towards next-generation AI* (pp. 141–156). Springer.
32. Naithani, K., Raiwani, Y. P., Tiwari, S., & Chauhan, A. S. (2024). Artificial intelligence techniques based on federated learning in smart healthcare. In *Federated learning for smart communication using IoT application* (pp. 81–108). Chapman and Hall/CRC.
33. Shahi, A., & Mittal, A. (2024). The impact of federated learning on AI-enhanced healthcare delivery. In *Pioneering smart healthcare 5.0 with IoT, federated learning, and cloud security* (pp. 57–66). IGI Global Scientific Publishing.
34. Lyu, L., Xu, X., Wang, Q., & Yu, H. (2020). Collaborative fairness in federated learning. In *Federated learning: Privacy and incentive* (pp. 189–204). Springer International Publishing.
35. Khan, L. U., Saad, W., Han, Z., Hossain, E., & Hong, C. S. (2021). Federated learning for internet of things: Recent advances, taxonomy, and open challenges. *IEEE Communications Surveys & Tutorials, 23*(3), 1759–1799.
36. Long, G., Tan, Y., Jiang, J., & Zhang, C. (2020). Federated learning for open banking. In *Federated learning: Privacy and incentive* (pp. 240–254). Springer International Publishing.
37. Bharati, S., Mondal, M. R. H., Podder, P., & Prasath, V. S. (2022). Federated learning: Applications, challenges and future directions. *International Journal of Hybrid Intelligent Systems, 18*(1-2), 19–35.
38. Lilhore, U. K., Simaiya, S., Poongodi, M., & Dutt, V. (Eds.). (2024). *Federated learning and privacy-preserving in healthcare AI*. IGI Global.
39. Zhang, K., Song, X., Zhang, C., & Yu, S. (2022). Challenges and future directions of secure federated learning: A survey. *Frontiers of Computer Science, 16*(5), 165817.

40. Mridha, M. F., & Dey, N. (Eds.). (2025). *Federated learning in health care technology*. Springer.
41. Sprague, M. R., Jalalirad, A., Scavuzzo, M., Capota, C., Neun, M., Do, L., & Kopp, M. (2018, September). Asynchronous federated learning for geospatial applications. In *Joint European Conference on Machine Learning and Knowledge Discovery in Databases* (pp. 21–28). Springer International Publishing.
42. Periasamy, J. K., Subhashini, S., Mutharasu, M., Revathi, M., Ajitha, P., & Boopathi, S. (2024). Synergizing federated learning and in-memory computing: An experimental approach for drone integration. In *Developments towards next generation intelligent systems for sustainable development* (pp. 89–123). IGI Global Scientific Publishing.
43. Razavi-Far, R., Wang, B., Taylor, M. E., & Yang, Q. (Eds.). (2023). *Federated and transfer learning*. Springer.
44. Thapa, C., Chamikara, M. A. P., & Camtepe, S. A. (2021). Advancements of federated learning towards privacy preservation: From federated learning to split learning. In *Federated learning systems: Towards next-generation AI* (pp. 79–109). Springer International Publishing.
45. Nakayama, K., & Jeno, G. (2022). *Federated learning with Python: Design and implement a federated learning system and develop applications using existing frameworks*. Packt Publishing Ltd.
46. Nakayama, K., & Jeno, G. (2022). *Federated learning with Python: Design and implement a federated learning system and develop applications using existing frameworks*. Packt Publishing Ltd.
47. Jin, Y., Liu, Y., Chen, K., & Yang, Q. (2023). Federated learning without full labels: A survey. arXiv preprint 2303.14453arXiv:2303.14453.
48. Hiessl, T., Schall, D., Kemnitz, J., & Schulte, S. (2020, July). Industrial federated learning–requirements and system design. In *International Conference on Practical Applications of Agents and Multi-agent Systems* (pp. 42–53). Springer International Publishing.
49. Li, Q., Wen, Z., Wu, Z., Hu, S., Wang, N., Li, Y., Liu, X., & He, B. (2021). A survey on federated learning systems: Vision, hype and reality for data privacy and protection. *IEEE Transactions on Knowledge and Data Engineering*, *35*(4), 3347–3366.
50. Jung, A. (2026). Introduction to federated learning. In *Federated learning: From theory to practice* (pp. 1–4). Springer Nature Singapore.
51. Tolpegin, V., Truex, S., Gursoy, M. E., & Liu, L. (2020, September). Data poisoning attacks against federated learning systems. In *European symposium on research in computer security* (pp. 480–501). Springer International Publishing.
52. Solanki, T., Rai, B. K., & Sharma, S. (2022). Federated learning using tensor flow. In *Federated learning for IoT applications* (pp. 157–167). Springer International Publishing.

Index

P. P. Ray, *Foundations of Federated Learning*,
Springer Series in the Data Sciences,
https://doi.org/10.1007/978-3-032-34084-9

FSC
www.fsc.org
MIX
Papier | Fördert
gute Waldnutzung
FSC® C083411

Kolibri 360 GmbH
Ferdinand-Jühlke-Straße 7
99095 Erfurt, Deutschland
produktsicherheit@kolibri360.de